NUCLEAR SPECTROSCOPY OF ASTROPHYSICAL SOURCES

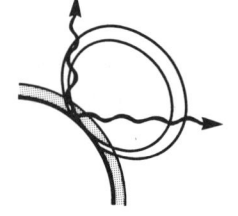 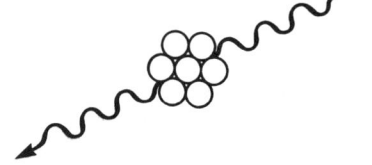 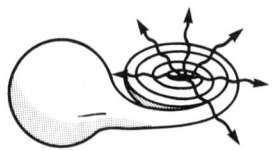

AIP CONFERENCE PROCEEDINGS 170

RITA G. LERNER
SERIES EDITOR

NUCLEAR SPECTROSCOPY OF ASTROPHYSICAL SOURCES
WASHINGTON, D.C. 1987

EDITORS:

NEIL GEHRELS
NASA/GODDARD SPACE
FLIGHT CENTER

GERALD H. SHARE
NAVAL RESEARCH
LABORATORY

AMERICAN INSTITUTE OF PHYSICS   NEW YORK 1988

Authorization to photocopy items for internal or personal use, beyond the free copying permitted under the 1978 US Copyright Law (see statement below), is granted by the American Institute of Physics for users registered with the Copyright Clearance Center (CCC) Transactional Reporting Service, provided that the base fee of $3.00 per copy is paid directly to CCC, 27 Congress St., Salem, MA 01970. For those organizations that have been granted a photocopy license by CCC, a separate system of payment has been arranged. The fee code for users of the Transactional Reporting Service is: 0094-243X/87 $3.00.

Copyright 1988 American Institute of Physics.

Individual readers of this volume and non-profit libraries, acting for them, are permitted to make fair use of the material in it, such as copying an article for use in teaching or research. Permission is granted to quote from this volume in scientific work with the customary acknowledgment of the source. To reprint a figure, table or other excerpt requires the consent of one of the original authors and notification to AIP. Republication or systematic or multiple reproduction of any material in this volume is permitted only under license from AIP. Address inquiries to Series Editor, AIP Conference Proceedings, AIP, 335 E. 45th St., New York, NY 10017.

L.C. Catalog Card No. 88-71625
ISBN 0-88318-370-6
DOE CONF-871245

Printed in the United States of America.

CONTENTS

Preface xi

HISTORICAL PERSPECTIVES

Gamma-Ray Spectroscopy: An Historical Perspective 1
  L.E. Peterson
Gamma-Ray Astronomy: An Historical Perspective 17
  R.E. Lingenfelter
Gamma-Ray Astronomy in the Nuclear Transition Region
(The Promise and the Problems) 24
  E.L. Chupp

GAMMA RADIATION FROM EXPLOSIVE NUCLEOSYNTHESIS: OBSERVATIONS

X-Ray Observations of SN 1987A with Ginga 34
  Y. Tawara
Observations of Supernova 1987A with Instrumentation on the
Soviet Space Station MIR 37
  C. Reppin, J. Englhauser. C. Hanson, W. Pietsch, J. Trümper, W. Voges,
  E. Kendziorra, M. Bezler, R. Staubert, R. Sunyaev, A. Kaniovskiy,
  V. Efremov, M. Grebenev, A. Kuznetsov, A. Melioranskiy, D. Stepanov,
  and I. Chulkov
NASA's Program for Observing Supernova 1987A 43
  A.N. Bunner
SMM Gamma-Ray Observations of SN 1987A 51
  S.M. Matz, G.H. Share, and E.L. Chupp
Imaging Observations of SN 1987A at Gamma-Ray Energies 60
  W.R. Cook, D.M. Palmer, T.A. Prince, S.M. Schindler, C.H. Starr,
  and E.C. Stone
High Resolution Observations of Gamma-Ray Line Emission
from SN 1987A 66
  W.G. Sandie, G.H. Nakano, L.F. Chase, Jr., G.J. Fishman, C.A. Meegan,
  R.B. Wilson, W.S. Paciesas, and G.P. Lasché
Hard X-ray Observations of SN 1987A 73
  R.B. Wilson, G.J. Fishman, C.A. Meegan, W.S. Paciesas, and G.N. Pendleton
50–500 MeV Gamma-Ray Emission in the Early Phase of SN 1987A 80
  R.K. Sood, J.A. Thomas, L. Waldron, R.K. Manchanda, P. Ubertini,
  A. Bazzano, C.D. La Padula, G.K. Rochester, T.J. Sumner, G. Frye,
  T. Jenkins, R. Koga, and P. Albats
Summary of Gamma-Ray Observations of SN 1987A 87
  N. Gehrels, M. Leventhal, and C.J. MacCallum
Comments on the Reported Detection of ^{56}Co Gamma Ray Lines
from SN 1987A 96
  M. Leventhal

Gamma Radiation from Explosive Nucleosynthesis: Theory

Gamma Producing Radioactivities from Supernovae 98
    S.E. Woosley and P.A. Pinto

Gamma-Ray Lines from Supernovae 110
    K.W. Chan and R.E. Lingenfelter

Particle Acceleration and Gamma-Ray Production in SN 1987A 116
    A.K. Harding, T.K. Gaisser, and T. Stanev

Iron Story 123
    M. Cassé

Gamma-Ray Line Diagnostics of Novae 130
    M.D. Leising

Nucleosynthesis of Gamma-Ray Emitters in Novae 136
    M.T. Wolff and M.D. Leising

Measurements of Astrophysical Reaction Rates for Radioactive Samples 143
    P.E. Koehler, H.A. O'Brien, and C.D. Bowman

Galactic Plane and Center

Gamma-Ray Spectroscopy of Radionuclei from Galactic Nucleosynthesis 149
    W.A. Mahoney

Observations of Diffuse Galactic Gamma Radiation 159
    G.H. Share

Theoretical Estimates of Diffuse Galactic Gamma-Ray Emissions 167
    J.C. Higdon

Constraints on Gamma-Ray Line and Continuum Emission from the Galactic Center Region at MeV Energies 175
    R. Diehl, P. v.Ballmoos, and V. Schönfelder

Spatial Distribution of Interstellar ^{26}Al Gamma-Rays: Preliminary Results Using SMM 181
    W.R. Purcell, M.P. Ulmer, G.H. Share, R.L. Kinzer, and E.L. Chupp

Origin of a Possible Diffuse Galactic Emission at 511 keV 187
    M. Signore and G. Vedrenne

Gamma-Ray Line Emissions as Tracers of Tenuous Phases of the Interstellar Medium 194
    J.C. Higdon

A Possible Galactic Center Positron Annihilation Medium: Neutral Hydrogen 196
    B.L. Brown and M. Leventhal

Solar System Gamma-Radiation

Gamma-Ray and Neutron Spectroscopy of Planetary Surfaces and Atmospheres 203
    R.C. Reedy

Observations of Nuclear Emissions in Solar Flares 211
    D.J. Forrest

Models of Gamma-Ray Production in Solar Flares 217
    R. Ramaty, J.A. Miller, X.-M. Hua, and R.E. Lingenfelter
Solar Abundances Using Gamma-Ray Line Spectroscopy of a
Solar Flare 228
    R.J. Murphy
The Neutron Capture Line and its Compton Tail as a Flare Diagnostic 234
    W.T. Vestrand
Nuclear Gamma-Ray Line Ratios as Spectral Diagnostics for Protons
Accelerated in Solar Flares 240
    C.J. Crannell and F.L. Lang
Solar Flare Gamma-Ray Line Shapes 246
    C.W. Werntz and F.L. Lang
Relative Timing of Solar Prompt Gamma-Ray Line and X-ray Emission
Produced by Flare Accelerated Ions and Electrons 252
    E. Hulot, N. Vilmer, and G. Trottet

GAMMA-RAY BURSTS

Gamma-Ray Burst Spectroscopy 258
    K. Hurley
Theories of Gamma-Ray Burst Spectra 265
    D.Q. Lamb
Spectral Diagnostics of Neutron Star Activity 285
    G.J. Hueter
Compatibility Between Gamma-Ray Burst Spectroscopy and
Source Models 293
    I.G. Mitrofanov and C. Barat
Spectral Modelling of the Soft Gamma-Ray Repeater 302
    E.P. Liang
Gamma-Ray Bursts from Neutron Star Detonation 307
    F.C. Michel
The High-Resolution Spectroscopy of Gamma-Ray Transients 312
    T.L. Cline

DISCRETE SOURCES

Gamma-Ray Emission of Cygnus X–1 315
    J.C. Ling
Gamma-Rays from Cygnus X–1: Modeling and Nonthermal
Pair Production 326
    C.D. Dermer and E.P. Liang
Properties of Hydrogen/Helium Accretion Plasmas 332
    N. Guessoum and C.D. Dermer
Nuclear Breakup Reactions and Gamma-Ray Lines in
High-Temperature Plasmas 338
    N. Guessoum
Compton Reflection of Gamma-Rays by Cold Electrons 345
    T.R. White, A.P. Lightman, and A.A. Zdziarski

Model of Gamma-Ray Emission from Large-Scale Jets 350
    M.W. Anyakoha, P.N. Okeke, and S.E. Okoye
High Energy Gamma-Rays from the Vela Pulsar: Long-Term Variability and
Energy Distribution 356
    I.A. Grenier, W. Hermsen, and J. Clear
Are the Gamma-Rays from the Vela Pulsar Polarized? 362
    P.A. Caraveo, G.F. Bignami, I. Mitrofanov, and G. Vacanti

FUTURE AND PROPOSED SATELLITE EXPERIMENTS

Capabilities of GRO for Observations of Supernovae and Novae 368
    J.D. Kurfess
The Gamma-Ray Burst Capabilities of BATSE and the
Gamma-Ray Observatory 378
    G.J. Fishman
On the Capabilities of GRO to Produce Surveys of the Galactic Plane
in the Light of Gamma-Ray Lines 389
    V. Schönfelder
GASOL: A New Experiment for the Study of Gamma-Ray Emission from
Solar Flares 395
    C. Barat, F. Cotin, M. Niel, R. Talon, G. Vedrenne, M. Pick, G. Trottet,
    N. Vilmer, K. Hurley, A. Kuznetzov, R. Sunyaev, and O. Terekhov
Spectroscopic Capabilities of the Solar-A Satellite 401
    M. Yoshimori
The High Energy Transient Experiment - HETE - a Multi-Wavelength
Survey Mission for the 1990's 407
    G. Ricker, J. Doty, S. Rappaport, K. Hurley, E. Fenimore, D. Roussel-Dupre,
    M. Niel, G. Vedrenne, D. Lamb, and S. Woosley
The Nuclear Astrophysics Explorer, A Proposed Explorer Mission
for the 1990's 417
    J.L. Matteson, B.J. Teegarden, and W.A. Mahoney
About the Ability of GRASP to Measure Diffuse Gamma-Ray
Line Sources 427
    P. Durouchoux, G. Bignami, A. Dean, N. Lund, B. McBreen, V. Schonfelder,
    B. Swanenburg, G. Vedrenne, and C. Winkler
ANGAS: A New Spaceborne High Resolution Gamma-Ray Spectrometer 432
    G.H. Nakano, J.R. Kilner, M.J. Murphy, M.H. Vartanian, and G.P. Lasché

NEW BALLOON-BORNE EXPERIMENTS

The Gamma-Ray Imaging Spectrometer (GRIS) Instrument and Plans for
Observing SN 1987A 439
    J. Tueller, S. Barthelmy, N. Gehrels, B.J. Teegarden, M. Leventhal,
    and C.J. MacCallum
Characteristics of EXITE and Plans for SN 1987A 444
    C.E. Covault, J. Braga, and J.E. Grindlay

Performance of LAPEX and its Spectroscopic Capabilities in the
20–300 keV Energy Band to Observe SN 1987A 451
    F. Frontera, A. Basili, D. Dal Fiume, T. Franceschini, G. Landini, E. Morelli,
    M. Pamini, J.M. Poulsen, S. Silvestri, E. Costa, D. Cardini, A. Emanuele,
    and A. Rubini

A High-Resolution Gamma-Ray, Hard X-ray, and Neutron Spectrometer for
Solar Flare Observations 456
    R.P. Lin

Prometheus I: Rice University's New Gamma-Ray Telescope 465
    R.C. Haymes, J.E. Fitch, and B. Sen

Development of a Liquid Xenon Time Projection Chamber for
Gamma-Ray Astronomy 472
    E. Aprile and M. Suzuki

INSTRUMENTS AND TECHNIQUES

The Gas Scintillation Drift Chamber as a Hard X-ray Detector 478
    A. Parsons, B. Sadoulet, S. Weiss, D. Smith, K. Hurley, and R.P. Lin

Pulse Shape Discrimination for Background Reduction in
Germanium Detectors 484
    D.M. Smith, M. Shapshak, R. Campbell, J.H. Primbsch, R.P. Lin, P.N. Luke,
    N.W. Madden, and R.H. Pehl

Performance of a Five-Segment Coaxial N-Type Germanium Detector 490
    L.S. Varnell, J.C. Ling, W.A. Mahoney, R.H. Pehl, C.P. Cork, D.A. Landis,
    P.N. Luke, N.W. Madden, and D.F. Malone

Bolometers as High-Resolution Gamma Spectrometers 498
    G. Simpson

Model Independent Spectral Deconvolution 505
    B.E. Schaefer

When and Why Background Subtraction Must be Done Before Data
Accumulation in Gamma-Ray Spectroscopy 511
    W.A. Wheaton, A.S. Jacobson, J.C. Ling, and W.A. Mahoney

PREFACE

A workshop on Nuclear Spectroscopy of Astrophysical Sources was held in Washington, D.C. in December 1987. This workshop brought together over 120 observers and theorists actively engaged in studying the sites of nuclear gamma-ray production in astrophysical sources and developing instrumentation for observing these photons. There are a great diversity of such sites including planets, the sun, supernovae, novae, close binary systems, the Galaxy, active galaxies, and gamma-ray bursters. New results were presented in all of these areas.

The exciting discovery of gamma-radiation from the radioactive decay of ^{56}Co produced in the LMC supernova highlighted the first day of the meeting. This discovery culminated an over 20-year effort to prove that supernovae are the birthplace of the heavy elements that are dispersed in the Universe. These Proceedings detail the original observations and theories, and summarize the extensive program developed by NASA and other organizations to observe this extraordinary event.

Another focus of the meeting was the observed Galactic 0.511 and 1.809 MeV line radiations. This topic provoked a lively debate concerning their spatial distributions and evidence for variability.

Spectral observations and new theoretical studies of cosmic gamma-ray bursts, Cygnus X-1, and the Vela pulsar are providing new insights into the physics of high-energy plasmas and intense magnetic fields. Such fields and plasmas are expected in the vicinity of neutron stars, black holes, and active galactic nuclei.

One of the purposes of the workshop was to tie together these studies of high-energy emissions from cosmic objects with sources closer to home, such as the Sun and planets. The same physical processes produce these emissions and the same experimental techniques are used to observe them. As with the cosmic sources, nuclear spectroscopy of the sun and planets is revealing elemental abundances which are not observable by other means.

Peppered throughout this meeting were reflections on the early days of gamma-ray spectroscopy. These personal reflections place the current discoveries in perspective and reveal the prescience of the pioneers in this field.

It was clear from the discussions that our instrumentation and theoretical tools are on the threshold of transforming this field into an important astronomical discipline. The next generation of satellite and balloon-borne observations will be launched within the next decade with sensitivities and capabilities an order of magnitude better than existing systems. Judging from the progress made since the last gamma-ray spectroscopy workshop in 1978, the future is indeed bright.

We wish to acknowledge Reuven Ramaty as the motivating force behind this workshop. Financial support was provided by NASA Headquarters, NASA's Solar Maximum Mission Project, and the Office of Naval Research. Many people contributed to making this workshop a successful educational and stimulating experience. We would especially like to thank Charlotte Pascoe for coordinating the conference and for her enthusiasm. The quality of these Proceedings is significantly improved by the tireless editing of Lois Williams. Evelyn Schronce, Dorothy Harbour, Jo Lyon, and Danny Messina provided technical support, for which we are grateful.

The scientific content of the workshop and review of the manuscripts in this volume were the responsibility of the Scientific Organizing Committee, which consisted of Michel Casse (CEN, Saclay), Edward Chupp (Univ. of New Hampshire), Neil Gehrels (NASA/GSFC, co-chair), Neil Johnson (NRL), Marvin Leventhal (AT&T Bell Labs.), Richard Lingenfelter (Univ. Calif., San Diego), William Mahoney (JPL), James Matteson (Univ. Calif., San Diego), Reuven Ramaty (NASA/GSFC), Gerald H. Share (NRL, co-chair), and Stan Woosley (Univ. Calif., Santa Cruz).

Neil Gehrels and Gerald H. Share
June 1988

Gamma-Ray Spectroscopy: An Historical Perspective

Laurence E. Peterson*
National Aeronautics and Space Administration
Washington, D. C. 20546

ABSTRACT

The possibility of MeV-range gamma-rays from extra-terrestrial sources had been speculated on by cosmic-ray physicists since the late 1940's. The first definitive detection occurred with balloon-borne cosmic-ray instrumentation during a class 2 solar flare in March 1958, apparently associated with the acceleration of a non-thermal particle population. Following this detection, physicists were motivated to develop instrumentation specific for observation of astronomical gamma-ray sources. Gamma-ray lines were also first observed during the flares of August 1972, apparently associated with accelerated particles undergoing nuclear interactions in the solar atmosphere. The development of low background, high resolution Ge counters has permitted construction of gamma-ray telescopes with unprecedented resolution and sensitivity. Even modest versions of these devices have measured discrete gamma-ray lines from sources as diverse as cosmic gamma-ray bursts, the galactic center and the galactic plane. Many other predictions are within the range of modern detectors.

INTRODUCTION

The development of gamma-ray astronomy and the spectroscopy of cosmic gamma-ray sources is associated with progress in the understanding of cosmic-rays, nucleosynthesis, compact objects, and other phenomena associated with high-energy astrophysics. This paper is a review of the evolution of gamma-ray spectroscopy from a personal perspective. Gamma-ray astronomy has emerged as an important activity of modern astrophysics; the scientific possibilities have exceeded the imagination of those who pioneered the field a generation ago. In parallel, the instruments of today use techniques which were even unknown in the early times. Now we are able to measure the intensities, spectra, and even the widths of gamma-ray lines associated with specific high-energy and nucleosynthetic phenomena.

THE BEGINNINGS

Cosmic-ray and nuclear physicists in the late forties and early fifties speculated on the possibilities of cosmic gamma-ray

*On leave from the University of California, San Diego to Jet Propulsion Laboratory, Pasadena, California.

astronomy. They must have realized that, in addition to a variety of continuum processes, there was the possibility that gamma-ray lines would be produced by nuclear reactions in astrophysical situations. Feenberg and Primakoff[1] first calculated the energy losses of cosmic-rays interacting with starlight in interstellar and intergalactic space, but apparently failed to estimate the resulting flux of photons at the earth. However, even if the fluxes had been estimated to be relatively intense, the possibility of measuring them must have seemed remote to the cosmic-ray investigator using balloon-borne experiments. The instrumentation available to detect primary gamma-rays was, at best, rudimentary.

The observational breakthrough came unexpectedly from serendipitous observations during the International Geophysical Year (IGY) in 1957 and 1958. I was a graduate student working with John Winckler at the University of Minnesota at the time, and he had developed a simple instrumentation set, based on a Neher ionization chamber and a small Geiger counter. The objective was to measure time variations of cosmic-rays on a series of high altitude (~100,000 ft.) balloon flights at various latitudes. We soon found these instruments could detect ionizing radiations other than cosmic-rays at high altitudes, and in fact X-rays from visible aurora were discovered during the first flight of the IGY from Minneapolis[2].

The initial observation of non-terrestrial gamma-rays occurred during a balloon flight of the IGY apparatus over Cuba on 20 March 1958[3]. At 1304.5 UT there was an 18 second increase in the rate of the ionization chamber and Geiger counter coincident with a Class II chromospheric flare on the sun and an associated microwave burst. The original data are shown in Figure 1. From the counter rates and

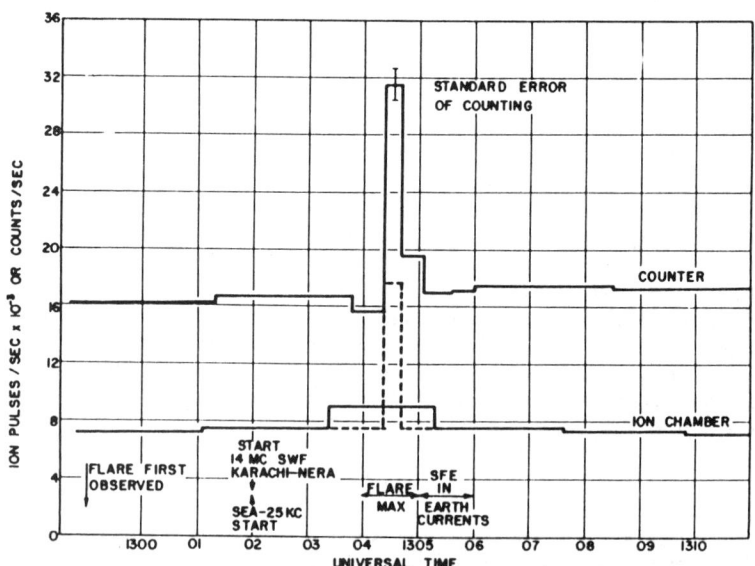

Figure 1 - The ionizing radiation and microwave data from the solar flare of 20 March 1958. A crude estimate of the nature of the spectrum was obtained from the ion chamber/G-M counter ratio.

their ratio one has essentially two parameters, and if one assumes (or knows a priori!) the emission process, one can obtain two coefficients of a parametized spectrum, such as a power law intensity and its slope, or a gamma-ray line energy and its flux.

We interpreted this burst as due to electrons accelerated during the flare, impinging on and stopping in the solar atmosphere, thereby emitting bremsstrahlung. In this model, the radio emission is due to gyro-synchrotion emission of the accelerated electrons in the high magnetic field associated with the sunspot region where the flare occurred.

It was not long before similar bursts were reported by other workers [4,5] and alternate interpretations were developed. These were based on the idea of a thermal plasma heated to multi-million degree temperatures in the flare region[5]. In fact, even with the rudimentary data available at the time (or because of it!) the interpretation of the observations were sufficiently controversial so that the "thermal" and "non-thermal" camps quickly formed, and even these had the "thick-target" versus "thin-target" splinters. The interest in and controversy about the new discovery indicated the phenomena had a rather profound significance.

Meanwhile, theorists working on cosmic-ray and other astrophysical phenomena were thinking seriously of the prospets for gamma-ray astronomy. Many of us experimentalists received our inspiration and motivation from the seminal paper of Morrison[6] on the subject. As if by design, this paper appeared in the Nuovo Cimento the same week as the initial solar gamma-ray burst occurred. A portion of the initial page of this paper is reproduced in Figure 2. Morrison discussed the various channels of astronomical information, i.e. radio, optical, cosmic-ray and now gamma-ray, and distinguished between continuum and line processes. He in fact suggested the active sun as a "first choice" for a source; others being the Crab Nebula and extragalactic radio sources. He even suggested a general background of cosmological origin! One notes however, that neither he, nor anyone else at the time wrote of the possibilities for X-ray astronomy.

Morrison also discussed the observational requirements. He inferred limits on fluxes over the range between 0.1 - 15 MeV to be about 1 MeV/cm^2-sec. He correctly predicted the background problem due to cosmic-rays (the splash albedo) and even suggested scintillation counters formed in a configuration which takes advantage of Compton scattering would be useful. Certainly the Compton telescope, like the rest of gamma-ray astronomy, has evolved far beyond Phil Morrison's dreams.

THE EARLY DEVELOPMENTS

Soon after Morrison's paper and the observations of solar gamma-ray bursts, several cosmic-ray groups became serious about exploratory work in cosmic gamma-ray astronomy. By the dawn of the sixties decade, Bill Kraushaar and George Clark at MIT were building an instrument for Explorer 11 to search for ~100 MeV gamma-rays, likely due to cosmic-ray collisions with the interstellar medium[7].

On Gamma-Ray Astronomy.

P. MORRISON

Department of Physics, Cornell University - Ithaca, N. Y.

(ricevuto il 22 Dicembre 1957)

Summary. — Photons in the visible range form the basis of astronomy. They move in straight lines, which preserves source information, but they arise only very indirectly from nuclear or high-energy processes. Cosmic-ray particles, on the other hand, arise directly from high-energy processes in astronomical objects of various classes, but carry no information about source direction. Radio emissions are still more complex in origin. But γ-rays arise rather directly in nuclear or high-energy processes, and yet travel in straight lines. Processes which might give rise to continuous and discrete γ-ray spectra in astronomical objects are described, and possible source directions and intensities are estimated. Present limits were set by observations with little energy or angular discrimination; γ-ray studies made at balloon altitudes, with feasible discrimination, promise valuable information not otherwise attainable.

Figure 2 - A reproduction of the leading page from the seminal paper on gamma-ray Astronomy. This paper appeared the same week as the first observation of solar gamma-rays.

One of their students, Tom Cline, was exploring the possibilities in the lower energy region[8]. John Simpson at Chicago had one of his students, Frank Jones, doing some work with balloon-borne scintillation counters[9]. Investigators in other parts of the world were doubtless doing similar preliminary investigations. At Minnesota I had considerable encouragement from John Winckler and Ed Ney to follow through on our discoveries, and I initiated a series of studies of the atmospheric background produced by cosmic rays. Simultaneously, Winckler and I had proposed successfully to do an exploratory 0.1 - 3.0 MeV experiment on the First Orbiting Solar Observatory (OSO-1)[10].

Although it was clear that a shielded or collimated scintillation counter configuration was the technique of choice for gamma-ray astronomy in the 1 MeV range, many difficulties were soon encountered. First, the fluxes from cosmic sources were likely lower than the original optimistic predictions. Second, the background was much higher than anticipated. The atmospheric counting rates in a simple 5 cm dia scintillation counter in the 50 keV to 10 MeV range were not even understood. Shielding and collimation proved a more difficult problem than was first anticipated.

I, among others, set about a series of experiments to determine the nature of the background in scintillation counters at balloon altitudes and to try various collimation schemes. Figure 3 shows a simple apparatus, consisting of a 5 cm dia x 5.5 cm long "phoswich" style Na(Tl) counter. This had lead and cadmium cylinders which could be dropped away, therefore no longer shielding the counter. In retrospect, two important results came from these and similar experiments.

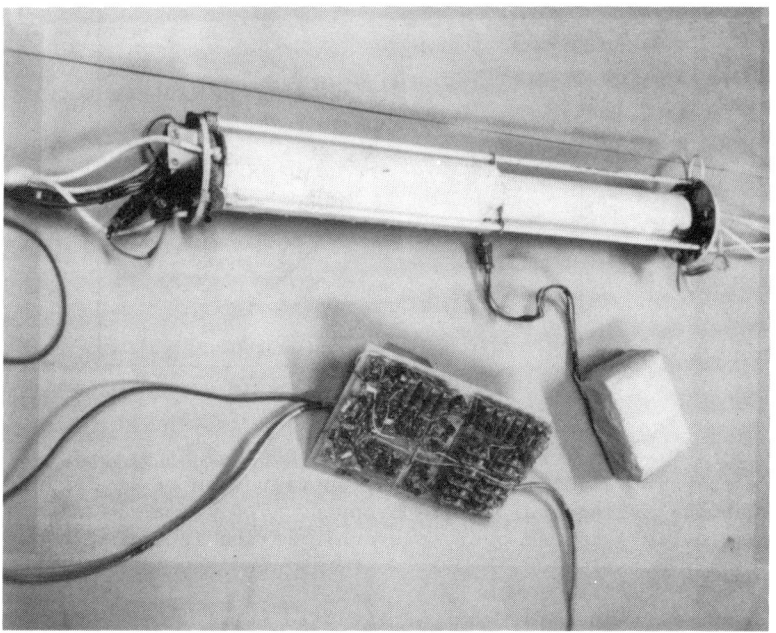

Figure 3 - A simple balloon apparatus used to study low-energy atmospheric gamma-rays and determine the properties of detectors and shielding configurations in the radiation environment of near space. The cylindrical cylinders of Cd and Pb, which surrounded a 5.5 x 5.0 cm diameter NaI "phoswich" scintillation counter, could be dropped during the flight.

First, the intensity and spectrum of low-energy gamma-rays in the atmosphere in equilibrium with the cosmic-ray beam was determined unambiguously, and the 511 keV line due to positrons annihilating in the atmosphere was resolved[11]. Figure 4 shows the latter result, which was the first detection of a gamma-ray line at a location essentially removed from the earth's surface.

Second, the effects of shielding, or more precisely local passive matter, around a counter were understood. No effect was observed when the Cd shield was dropped (thermal neutrons were not producing a large background in the NaI scintillation counter) and the rate above about 100 keV actually decreased when the Pb shield was dropped. Lead, and therefore high-Z materials, produced more background than they removed.

During this time I was building the instrument for OSO-1, and it became apparent that some of the concepts on which this experiment

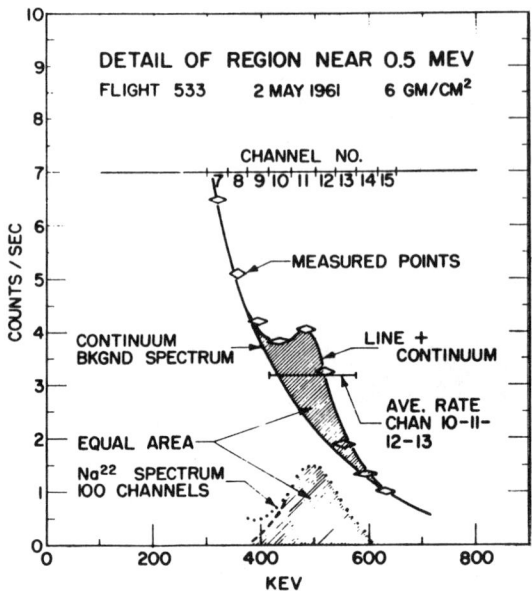

Figure 4 - The spectral region near 0.5 MeV of atmospheric gamma-rays as determined by the apparatus shown in Figure 3, with the cylindrical shield dropped. The gamma-ray line due to cosmic-ray produced positrons annihilating in the atmosphere is clearly established.

was based were invalid, and the expected results were unlikely to be obtained. Nevertheless, the experiment was launched in March 1962 and did measure hard x-ray bursts from the sun, put upper limits on various sources, and determined many background effects[12]. These included radioactivity induced in NaI by the trapped protons of the South Atlantic Anomaly (SAA)[13].

The real breakthrough on an instrumental concept, sorely needed at this time, was due to Ken Frost at GSFC. He conceived an instrument for the OSO-2 which featured the entire collimating shield made of CsI(Tl), connected in electrical anti-coincidence with the central detector[14]. Ken and I later collaborated in a balloon experiment to determine the properties of this device and make some preliminary observations. The apparatus and instrumentation, typical of the time, is shown in Figure 5[15]. For the first time, physicists trying to measure cosmic fluxes in the MeV range had a detector system whose properties in the radiation environment of space could, in principle, be predicted.

In 1962, I was making the transition from Minnesota to the University of California at San Diego, hoping to start a gamma-ray astronomy group. Jim Arnold, also at UCSD, had discovered an "interplanetary flux" of gamma-rays on Ranger 3[16]. Ken Frost was continuing his work on solar gamma-rays, and others, including Ed Chupp at New Hampshire and Bob Haymes at Rice, were developing various configurations of anti-coincidence shielded scintillation counters for gamma-ray astronomy.

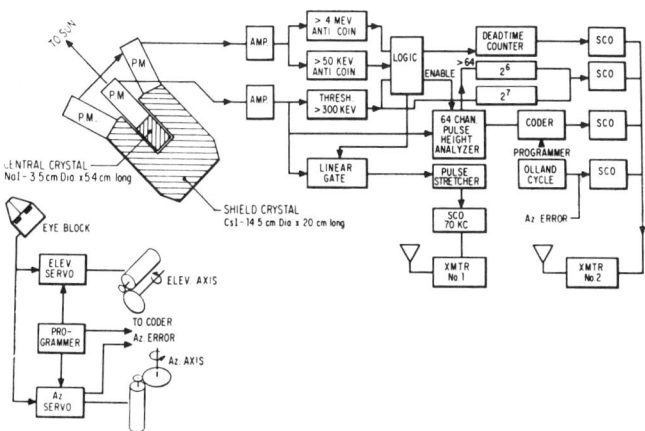

Figure 5 - The diagram of an early experiment to search for quiet-time solar gamma-rays conducted jointly by the University of Minnesota and the Goddard Space Flight Center. A similar detector was flown on the OSO-2 spacecraft. This was the first application of the active anti-coincidence shield to low-energy gamma-ray astronomy.

Meanwhile, all of us were searching for cosmic gamma-ray lines at the limit of resolution and sensitivity available from the scintillation counter configurations which could be constructed at the time. Several theorists were also writing on the potential of the emerging field of high energy astrophysics[17,18,19] and on the interpretation of the few results available at the time on the spectral continuum[20].

EMERGENCE OF SPECTROSCOPY

By the mid 1960's, it became apparent that the spectroscopy of gamma-rays from cosmic sources had real potential. First, following Morrison, Lingenfelter and Ramaty did serious calculations on the possibilities from solar flares[21]. They based their calculations on the idea that protons accelerated in the solar flare region would undergo nuclear interactions in the solar atmosphere and produce lines through inelastic scattering, radioactive decay, neutron capture and positron annihilation. Such protons were known to occur because of direct measurements made in interplanetary space.

Second, there was considerable speculation associated with supernovae explosions. The discussions centered around a particular isotope, Cf^{254}, which undergoes spontaneous fission with a half-life of 55 days, to explain the light curve of Type I supernova. Although the idea was first suggested by Burbidge, Burbidge, Christy, Fowler and Hoyle[22], Clayton and Craddock[23] developed the idea, and made calculations of the various heavy elements likely synthesized in equilibrium with the Cf^{254} and the decaying products which could produce gamma-rays. The expected fluxes at the earth due to heavy element decay, as revised by Jacobson[24], are shown in Figure 6 for an epoch 900 years after the explosion. Clayton and Craddock even put requirements on the spectral resolution and aperture properties of an instrument to measure the predicted fluxes.

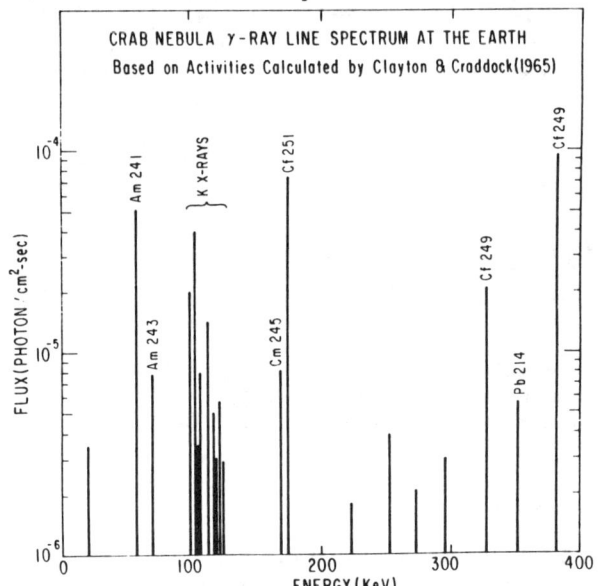

Figure 6 - The line gamma-ray spectrum from the Crab Nebula predicted by Clayton and Craddock. This spectrum is due to the decaying products of heavy elements produced in equilibrium with enough fissioning ^{254}Cf to account for the light source of type I Supernovae.

Over a decade would pass before detector systems having these specifications were to be flown on balloons; furthermore, the Cf^{254} hypothesis for supernova light curves was being replaced by the synthesis of Ni^{56} and its subsequent radioactive decay following a complete stellar collapse. Nevertheless, all these ideas proved a motivating factor for the instrumentalists, as did the remarkable discoveries in X-ray astronomy occurring almost weekly during this time.

On the observational (or rather experimental) side there was a singular advance in technology which provided the technique needed for cosmic gamma-ray spectroscopy, namely, the development of the Ge(Li) detector. At UCSD I engaged one of my students, Allen (Bud) Jacobson, to use one of these detectors in an active anti-coincidence shield to observe the Crab Nebula, and hopefully prove (or disprove) the Cf^{254} hypothesis as worked out by Clayton and Craddock. Detectors available at the time were small by present standards (~5-30 cm^3), they had to be kept cooled from birth (because the Li would "undrift"), and the shield and aperture configurations were rudimentary. The detector system used by Jacobson is shown in Figure 7[24].

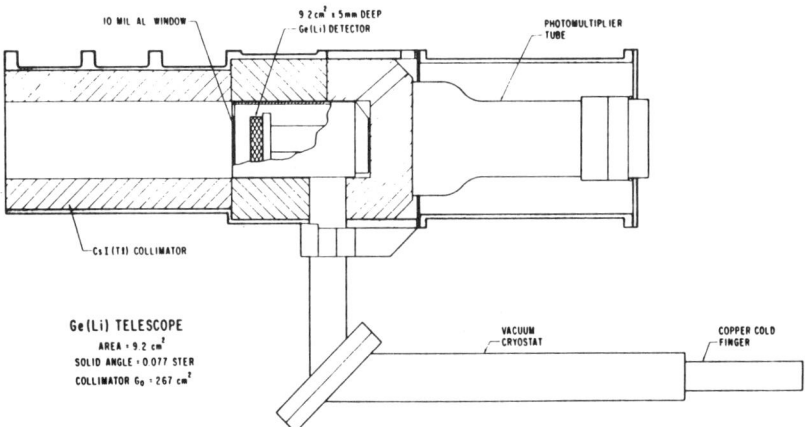

Figure 7 - The detector system used by Jacobson at UCSD to search for the gamma-rays predicted from the Crab Nebula. This was one of the early applications of cooled Germainium detectors to the problem of cosmic gamma-ray spectroscopy.

Another difficulty associated with the experiments of that era was the pointing control. Although servo principles and technology were well known in engineering circles, physicists preferred an intuitive approach, and the pointing techniques, while clever, often lacked sophistication. The gondola used by Jacobson was a predecessor of more advanced configurations, and consisted of an alt-azimuth pointing control, using the earth's magnetic and gravity fields as references. Part of the gondola mass, coupled through the suspension to the balloon, became the reaction inertia. The gondola configuration, shown in Figure 8, could point the detector and its LN_2 Dewar, which had a collective mass of many tens of kiligrams, to about 1° on the celestial sphere.

Other workers, notably Jim Overbeck and his student Al Womack at MIT[25] had also constructed a similar device, and were flying it contemporaneously. These experiments only resulted in upper limits of a few times 10^{-3} ph/cm^2-sec on the expected gamma-ray lines from the Crab Nebula, and from other sources. They did however point the way to the required detector characteristics, and one had the feeling that only a few factors in sensitivity would result in a real detection. Soon after, Haymes and co-workers, using NaI detectors with anticoincidence shields had indications of the 0.51 MeV annihilation radiation from the galactic center, as well as other lines from other sources[26].

THE SUCCESSFUL SEVENTIES

The first non-controversial detection of a cosmic gamma-ray line was made by Ed Chupp and his colleagues at the University of New Hampshire, who had placed a 7.5 x 7.5 cm NaI counter with a 2 pi anti-coincidence shield, on the OSO-7 to search for solar gamma-rays. This was launched 29 September 1971 and the detection came during the 2B solar flare of 4 August 1972. This large flare, along with a 3B

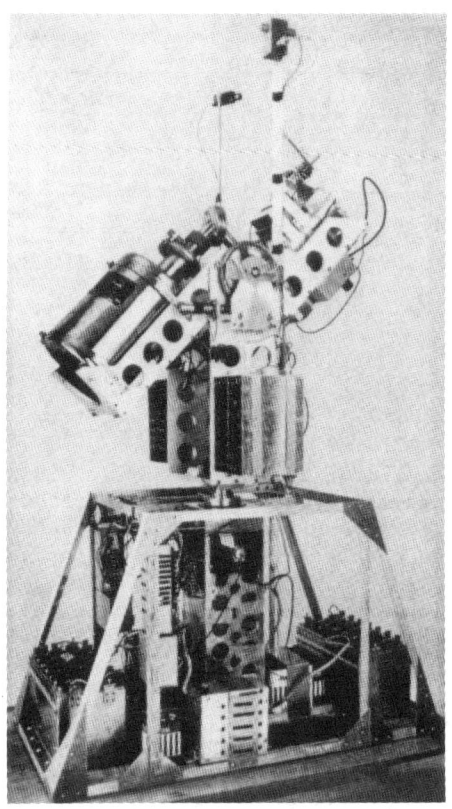

Figure 8 - The detector of figure 7 mounted in an alt-azimuth pointing control. The LN_2 Dewar could maintain the Ge(Li) cool for some 30 hours. Only upper limits on Crab gamma-rays were obtained from this, as well as other contemporaneous balloon investigations.

flare on 7 August, which also produced gamma-rays, was associated with the passage of an extremely active sunspot region across the solar disc. Chupp's instrument on the OSO-7 recorded a spectrum near the peak of the flare[27] which showed lines at 0.51 MeV due to positron annihilation, at 2.2 MeV due to neutron capture, and at 4.4 and 6.13 MeV due to neutron scattering on carbon and oxygen in the solar atmosphere.

This discovery was a remarkable confirmation of the predictions of Lingenfelter and Ramaty, and cosmic gamma-ray spectroscopy was born as an observational science. Other discoveries followed before the decade was out.

Shortly after this, Marvin Leventhal at Bell Labs and his colleagues at Sandia had developed a large volume (130 cm^3) Ge(Li) balloon borne detector to search nucleosynthetic lines, and in particular to study the 0.5 MeV line from the galactic center. This line and its associated continuum spectrum are a diagnostic for positronium formation in the interstellar medium. Their initial result[28], obtained in a balloon flight 11-12 November 1977 from Alice Springs, Australia is shown in Figure 9, and confirmed the earlier results of Haymes [26].

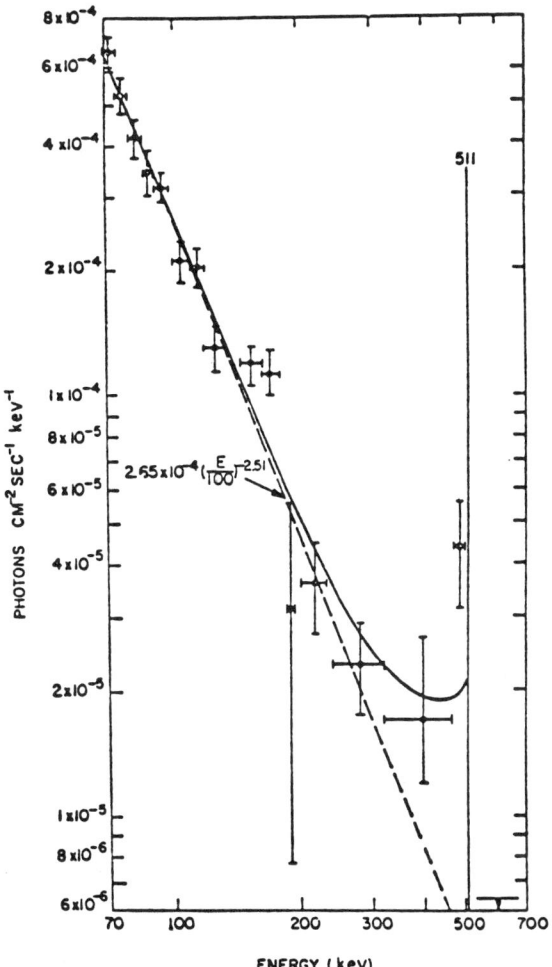

Figure 9 - The 0.511 MeV gamma-ray line observed from the galactic center by Leventhal and his colleagues using a 130 cm^3 shielded Ge detector. The solid line indicates a possible model based on positronium formation followed by an annihilation.

The next discovery of a discrete spectral feature was associated with an unexpected phenomena, quantized cyclotron emission due to relavistic electrons in the ~10^{12} gauss field of a neutron star. These were first observed by Trumper and his colleagues at the Max-Planck Institute, Garching. They studied the hard x-ray spectrum from the compact binary-ray emitting object Her-1 using large area

NaI counters on a series of balloon flights[29] and observed a feature near 35 KeV.

Similar features were seen in the spectra of cosmic gamma-ray bursts by Mazets and his associates with NaI detectors on Venera 11 and 12[30]. In fact, interpretation of the line feature as cyclotron emission is taken as almost conclusive proof that the gamma-ray burst phenomena has its origin near the surface of a neutron star. Figure 10 shows a spectrum from a burst, which, in addition to the

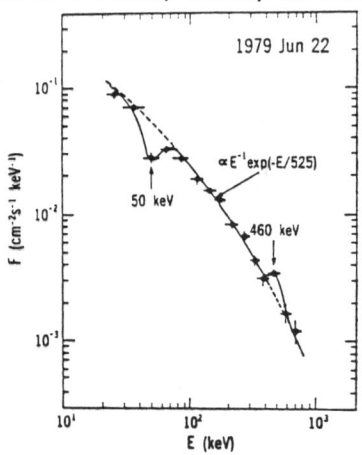

Figure 10 - A result obtained by Mazets on the Venera Spacecraft from a gamma-ray burst. The spectrum shows both a feature near 50 keV, possibly due to gyrosynchrotron emission in a 10^{12} gauss magnetic field, and near 420 keV. The latter may be due to red-shifted 0.511 MeV annihilation gamma-rays.

cyclotron feature, has a line-like feature near 420 keV. This is believed due to gravitationally red-shifted 0.511 MeV annihilation radiation produced at the surface of the neutron star[30]. Teegarden and Cline found weak evidence for other red-shifted nuclear gamma-ray lines in the 29 November 1978 burst, from the first cooled Ge detector flown in space on ISEE-3[31].

Jacobson, who went to JPL about 1969, had proposed a cooled germanium detector to be flown on one of NASA's High Energy Astronomical Observatories (HEAO). The instrument consisted of an array of four 90 cm^3 Ge detectors, surrounded by a CsI(Na) anticoincidence shield[32]. The detector, mounted in its cryostat as shown in Figure 11, was cooled by a solid-cryogen refrigerator. This, together with an instrument designed by Nakano for the satellite P-78-1 to perform a sky survey[33], were the first high sensitivity, extensively shielded Ge gamma-ray spectrometers flown in space. The HEAO-3 instrument, which operated for nearly a year from launch in Sept 1979, showed that the 0.511 MeV gamma-ray from the galactic center was variable on a six-month time scale[34]. This has profound implications for the source region[35]. Jacobson's instrument also discovered the predicted line from the galactic plane at 1.809 MeV, due to decay of ^{26}Al with a half-life of 7.5 x 10^5

Figure 11 - The cluster of four intrinsic Ge detectors of 80 cm^3 each flown on the HEAO-3 in September 1979. This instrument measured narrow lines from the galactic center region, solar flares, and the galactic plane, as well as many continuum sources.

years[36]; however, the flux was observed at about a factor of five greater than predicted[35]. ^{26}Al is likely produced in nucleosynthetic processes due to supernovae and novae, and dispersed through the interstellar medium following the explosive events.

PROLOGUE TO THE FUTURE

The observational discoveries during the early part of this decade and the associated theoretical developments have advanced spectroscopy of cosmic gamma-ray phenomena from a series of speculative and exploratory concepts to a branch of astrophysics standing on its own merit. Entire conferences and review papers have been devoted to the subject[35,37].

The observational advances have, of course, continued. The Solar Maximum Mission (SMM), which was launched on 14 February 1980, carried a follow-on to the New Hampshire solar gamma-ray experiment on OSO-7. This instrument has observed gamma-ray lines from dozens of solar flares, and provided new results on the acceleration, dynamics, and transport of energetic particles in the solar atmosphere. The UCSD/MIT Hard X-ray and Gamma-ray instrument on the First High Energy Astronomical Observatory (HEAO-1) and the series of Konus instruments on Venera spacecraft have observed cyclotron line features from several x-ray emitting binary systems involving neutron stars, and from dozens of cosmic gamma-ray bursts. The annihilation line from the galactic center and galactic plane has been measured by the SMM, which has also confirmed the ^{26}Al emission from the galactic plane. Increasingly complex instruments have been proposed and are being constructed as advances in the detector art proceed.

The recent supernova in the Large Magellanic Cloud has been the event which certified the future of gamma-ray line spectroscopy. Here the imploding core of a collapsing star produces a shell of ^{56}Ni, according to theories developed over the last decade[38]. This theory of stellar collapse was strikingly confirmed with the detection of neutrinos from the event. The ^{56}Ni decayed to ^{56}Co with a six-day half-life, and this is now decaying to ^{56}Fe with a 77-day

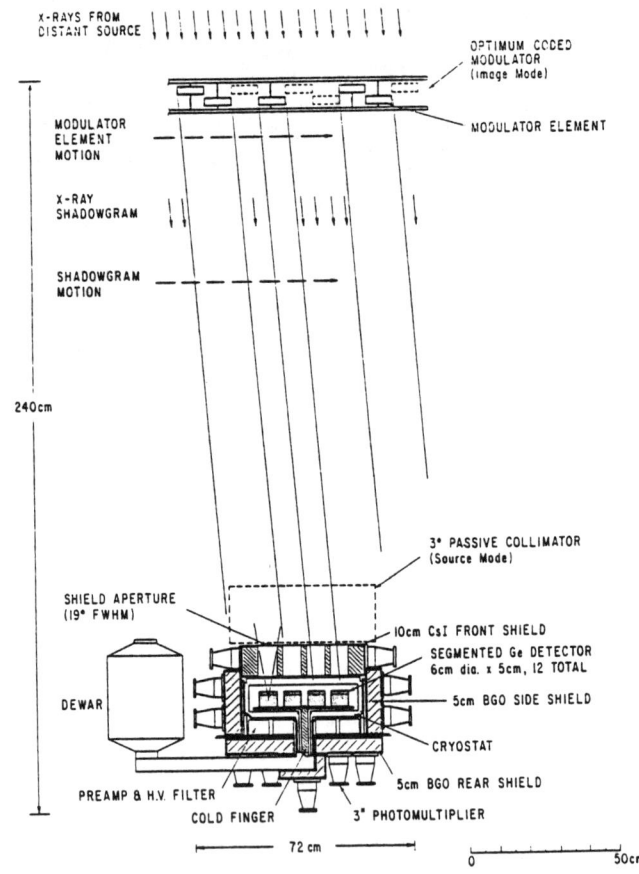

HIGH RESOLUTION SPECTROMETER

Figure 12 - A modern balloon system used in gamma-ray astronomy incorporates a cluster of Ge detectors in a collimating anti-coincidence shield with a coded-mask aperture to determine source locations to 0.1° over a 12° field-of-view.

half-life. Even as I write these words, the overlying shell is slowly expanding and thinning to reveal the decaying nuclei. The SMM, as well as balloon-borne Ge detectors, have already made positive detections of lines due to the decaying ^{56}Co. These will be reported in the proceedings of this conference. Clearly SN1987a occurred at a unique moment in history. But a few years earlier, the sensitive technology needed to study the direct and immediate consequences of explosive nucleosynthesis from a rare nearby supernova would not have been available.

For the present, and well into the near future, observations will be carried out with sophisticated arrays of Ge detectors in anticoincidence shields. These may be supplemented with coded mask imaging systems to obtain angular resolution in the few arc minute regime. Figure 12 shows a schematic of a modern balloon gondola being constructed at UCSD to observe SN 1987a, among other sources[39]. In the more distant future, a similar device may be placed on an orbiting spacecraft. Only in this way can the combination of low background, high sensitivity, and long observing times required to provide a qualitative advance in cosmic gamma-ray spectroscopy be obtained.

ACKNOWLEDGEMENTS

The author wishes to acknowledge the essential contributions of his many colleagues and students to the x-ray and gamma-ray astronomy program at the University of California, San Diego. I apologize to those whose important work during the unfolding of the saga of gamma-ray spectroscopy was inadvertently omitted in this review. This work was supported by the NASA Headquarters core grants NsG-318 and NAGW-449. I acknowledge support through the Visiting Senior Scientist program at the Jet Propulsion Laboratory while on leave from UCSD to NASA.

REFERENCES

1. E. Feenberg and H. Primakoff, Phys. Rev. 73, 449 (1948).
2. Winckler, J. R. and L. E. Peterson, Phys. Rev. 108, 903 (1957).
3. L. E. Peterson and J. R. Winckler, J. Geophys. Res. 64, 697 (1959).
4. J. I. Vette and F. G. Casal, Phys. Rev. Letters, 6, 334 (1961).
5. T. A. Chubb, H. Friedman and R. W. Kreplin, J. Geophys. Res. 65, 1831 (1960).6.P. Morrison, Il Nuovo Cimento VII, 858 (1958).
7. G. W. Clark, G. P. Garmire, and W. L. Kraushaar, Astrophys. J. (Lett), 153, L203 (1968).
8. T. L. Cline, Phys. Rev. Letters 7, 109 (1961).
9. F. C. Jones, J. Geophys. Res. 66, 2029 (1961).
10. L. E. Peterson and R. L. Howard, IRE Trans. Nuc. Sci., NS-8, 21(1961).
11. L. E. Peterson, J. Geophys. Res. 68, 979 (1963).
12. L. E. Peterson, UCSD Preprint UCSD SP-68-1, (1967).
13. L. E. Peterson, J. Geophy. Res. 70, 1762 (1965).
14. K. J. Frost and E. D. Rothe, IRE Trans. Nuc. Sci. NS-9, 381

(1962).
15. K. J. Frost, E. D. Rothe and L. E. Peterson, J. Geophy. Res. $\underline{71}$, 4079 (1966).
16. A. E. Metzger, E. C. Anderson, M. A. Van Dilla, and J. R. Arnold, Nature $\underline{204}$, 766 (1964).
17. R. J. Gould and G. R. Burbidge, Annales d'Astrophysics $\underline{28}$, 171 (1965).
18. P. Morrison, Handbuck der Physik $\underline{46/1}$, (Springer-Verlag, 1961), p.1.
19. V. L. Ginsberg and S. I. Syrovatski, The Origin of Cosmic Rays (Pergamon Press, London, 1964) p.   .
20. J. E. Felton and P. Morrison, Phys. Rev. Letters $\underline{10}$, 453 (1963).
21. R. E. Lingenfelter and R. Ramaty, High Energy Nuclear Radiations in Astrophysics, ed. B. S. P. Shen (W. A. Benjamin, N. Y., 1967) p. 99.
22. G. R. Burbidge, F. Hoyle, E. M. Burbidge, R. F. Christy, and W. A. Fowler, Phys. Rev. $\underline{103}$, 1145 (1956).
23. D. E. Clayton and W. L. Craddock, Ap. J. $\underline{142}$, 189 (1965).
24. A. S. Jacobson, PhD. thesis, University of California, San Diego, 1968, $\underline{\text{UCSD-SP 68-2}}$.
25. E. A. Womack and J. W. Overbeck, J. Geophy. Res. $\underline{75}$, (1971).
26. R. C. Haymes et al., Astrophy. J. $\underline{201}$, 592 (1975).
27. E. A. Chupp, D. J. Forrest, P. R. Higbie, A. N. Suri, and C. Tsai, Nature $\underline{241}$, 333 (1973).
28. M. Leventhal, C. J. MacCallum, and C. J. Stang, Astrophy. J. $\underline{225}$, L11 (1978).
29. J. Trumper et al., Astrophy. J. $\underline{219}$, L105 (1978).
30. E. P. Mazets et al., Astrophy. Space Sci. $\underline{80}$, 1 (1981).
31. B. J. Teegarden and T. L. Cline, Astrophy. J. $\underline{236}$, L67 (1980).
32. W. A. Mahoney, J. C. Ling, A. S. Jacobson, and R. M. Tapphorn, Nuc. Inst. and Methods. $\underline{178}$, 363 (1980).
33. W. L. Imhof and G. H. Nakano, Astrophy. J. $\underline{214}$, 38 (1977).
34. G. R. Riegler et al., Astrophy. J. $\underline{248}$, L13 (1981).
35. R. Ramaty and R. E. Lingenfelter, Ann. Rev. Nuc. Part. Sci. $\underline{32}$, (Annual Reviews Inc., Palo Alto, 1982).
36. W. A. Mahoney, J. C. Ling, W. A. Wheaton, and A. S. Jacobson, Astrophy. J. $\underline{286}$, 278 (1984).
37. International Gamma-Ray Symposium, Goddard Space Flight Center, 2-4 June 1976.
38. D. D. Clayton, S. A. Colgate, and G. J. Fishman, Astrophy. J. $\underline{155}$, 75 (1969).
39. J. L. Matteson et al., 19th Intl. Cosmic Ray Conf., 11-23 August 1985, La Jolla Calif., $\underline{\text{OG-3}}$, 326.

GAMMA-RAY ASTRONOMY: A HISTORICAL PERSPECTIVE

R. E. Lingenfelter
Center for Astrophysics and Space Sciences
University of California, San Diego, CA 92093

INTRODUCTION

Recently I have been involved in one of those periodic efforts at trying to predict future directions in gamma-ray astronomy, at least to the extent that theoretical expectations might define and motivate experimental programs in new directions. Thus I would like to take this opportunity to explore just how successful such theoretical expectations and predictions have been in determining fruitful new directions of research in gamma-ray astronomy in the past. In doing so, I will briefly review the course of gamma-ray astronomy over the past thirty years, examining what the theoretical expectations were; to what extent they were realized; how well they anticipated new directions of research; and alternatively, how often were new directions unexpected. I will also discuss what we can learn from it all; and what we might expect in the future. Obviously, there's some subjectivity in deciding what is a new direction, or whether a prediction is realized — especially when it may seem to be the right result for the wrong reason – but that's unavoidable.

THE BEGINNINGS OF GAMMA-RAY ASTRONOMY

Gamma-ray astronomy was born thirty years ago at 13:04:30 UT on March 20, 1958 in a balloon about 100,000 feet over Cuba with the detection by Peterson and Winckler[1] of $\sim$ 0.2 to 0.5 MeV gamma rays from a class 2 solar flare. But gamma-ray astronomy had in fact been conceived half a century earlier, following the twin discoveries in 1900 of penetrating "gamma" radiation by Villard[2] and atmospheric ionization by Wilson[3]. Wilson suggested that extraterrestrial gamma rays might be responsible for the ionization and twelve years later in balloon flights Hess[4] demonstrated the extraterrestrial and extrasolar origin of the ionizing rays, which soon came to be known as "cosmic rays". The cosmic rays were thus generally thought to be gamma rays until 1927, when Clay[5] discovered the geomagnetic latitude dependence of the cosmic ray flux on a voyage from Java to Genoa. This clearly established that the cosmic rays were charged particles, not gamma rays. By 1952 balloon observations had shown that gamma rays made up $<$ 1% of the cosmic ray flux. Then, in 1958 extraterrestrial gamma rays were at last discovered.

So, what were the theoretical expectations at the birth of gamma-ray astronomy in 1958? Morrison[6] summarized several of them in an *Il Nuovo Cimento* article dated just four days before the solar flare gamma-ray discovery. His foremost prediction was in fact that of solar flare gamma-ray continuum emission from electron bremsstrahlung, based on the suggestion by Boischot[7] that relativistic flare electrons were responsible for Type II radio bursts. Morrison predicted a flux of 0.1 to a few photons cm^{-2} s^{-1} in the energy range 10 to 100 MeV from a class 2 flare. Peterson and Winckler[1] measured a flux of $\sim$ 30 photons cm^{-2} s^{-1} at lower energies from just such a flare 4 days later.

© 1988 American Institute of Physics

Even though the theoretical expectation preceded the observation, it did not, however, motivate the observation, which had been planned for other reasons.

In addition, Morrison predicted solar flare gamma-ray line emission at 2.2 MeV from neutron capture on hydrogen, undoubtedly based on the earlier suggestion by Biermann et al.[8] that neutrons were also produced in solar flares. This suggestion did indeed stimulate an observational quest, supported by much more theoretical work[9-11] predicting many additional lines, that led to their ultimate discovery from the August 4, 1972 flare by Chupp et al.[12] with detectors on the OSO-7 satellite.

Morrison also predicted detectable nuclear gamma-ray lines from the decay of ^{226}Ra in the Crab Nebula as "fallout" from ^{254}Cf. This was obviously based on the suggestion by Burbidge et al.[13] that the exponential light curves of supernovae were powered by the spontaneous fission of ^{254}Cf synthesized in the explosion. The latter idea, however, was subsequently discarded after it was suggested, first by Pankey[14] in 1961 and later by Colgate and McKee[15], that ^{56}Ni→^{56}Co→^{56}Fe decay powered the light curves. This suggestion led Clayton, Colgate and Fishman[16] to predict the supernova gamma-ray lines from ^{56}Co decay that have just been discovered[17-19].

Morrison further predicted detectable gamma-ray line emission from positron annihilation and from π° decay from the extragalactic radio sources M87 and Cygnus A, assuming that such sources might be powered by matter-antimatter annihilation, as had been suggested by Burbidge and Hoyle[20]. Gamma-ray observations on the OSO-3 satellite by Kraushaar et al.[21] refuted this suggestion in 1972. Lastly, Morrison pointed out that upper limits on the possible cosmic flux of gamma rays from matter-antimatter annihilation, determined from rocket-borne detectors[22], set a critical limit on the rate of matter-antimatter creation in steady state cosmology much lower than that expected[20].

There were several other important theoretical expectations that Morrison did not discuss. In 1952 Hayakawa[23] had predicted diffuse galactic gamma-ray emission at > 10 MeV from the decay of π° produced by cosmic-ray nucleon interactions with the interstellar gas. That same year Hutchinson[24] predicted emission at the same energies from cosmic-ray electron bremsstrahlung. Diffuse galactic gamma rays, apparently resulting from both of these processes, were discovered by Clark et al.[25] with detectors on the OSO-3 satellite in 1967.

Hayakawa[26] in 1958 also predicted gamma rays up to 100 GeV from π° decay in the Crab Nebula, based on the suggestion by Burbidge[27] that the synchrotron radiating electrons, responsible for the optical emission, were secondaries resulting from pion production by relativistic protons. This idea was later discarded after the discovery of pulsars in 1968.

Finally, in 1958 Arnold[28] predicted detectable gamma-ray line emission from the moon resulting from natural radioactivity — ^{40}K, Th and U — and soon after predicted[29] additional lines resulting from cosmic ray induced nuclear excitation of lunar surface material. Arnold and his colleagues[30] subsequently observed such emission with detectors on Apollo-15 in 1971.

From this we see that out of the nine potential sources of gamma rays expected theoretically by 1958; five were confirmed by subsequent observations; two were refuted by gamma-ray observations; and two were discarded for other reasons. Thus the theoretical predictions were in fact correct in roughly half of the cases — a very respectable success rate — which greatly increased our understanding of high energy astrophysical processes and led to further predictions. Moreover, these theoretical expectations stimulated the development on new instrumentation, which, as we shall see, also led to the opening of other unexpected new directions of research.

To examine the role of theory further, its of interest to try to briefly list all of the different gamma-ray sources and processes discovered over the past thirty years with an eye toward determining which, in fact, were predicted and which were quite unexpected — owing their discovery to the development of more sensitive instrumentation.

PREDICTED GAMMA-RAY SOURCES

PREDICTED	SOURCES & PROCESSES	DISCOVERED
1952	Diffuse Galactic π^0 Gamma Rays *Hayakawa*[23]	1968 *Clark et al.*[25]
1952	Galactic Cosmic-Ray Bremsstrahlung *Hutchinson*[24]	1968 *Clark et al.*[25]
1958	Solar Flare Electron Bremsstrahlung *Morrison*[6]	1958 *Peterson & Winckler*[1]
1958	Solar Flare Neutron Capture Line *Morrison*[6]	1973 *Chupp et al.*[12]
1958	Lunar ^{40}K, Th & U Lines *Arnold*[28]	1973 *Metzger et al.*[30]
1961	Lunar Nuclear Excitation Lines *Arnold*[29]	1974 *Metzger et al.*[30]
1964	AGN Compton-Synchrotron Emission *Ginzburg et al.*[31]	1975 *Grindlay et al.*[32]
1964	Solar Flare Nuclear Excitation Lines *Chupp*[9]	1973 *Chupp et al.*[12]
1965	Solar Flare π^0 Gamma Rays *Dolan & Fazio*[10]	1985 *Forrest et al.*[33]
1969	Supernova ^{56}Co Lines *Clayton et al.*[16]	1987 *Matz et al.*[17]
1971	Pulsar Curvature Radiation *Sturrock*[34]	1971 *Kurfess*[35]/*Grindlay*[36]
1977	Galactic Nucleosynthetic ^{26}Al Line *Arnett*[37]/*Ramaty & Lingenfelter*[38]	1982 *Mahoney et al.*[39]

In this table and those that follow, the references are only for the first suggestion of a source, or process, although in most instances much more detailed theoretical predictions were subsequently made by the same or other authors. Similarly, the referenced discoveries are only for the first significant observation of a source, or process, although most of the sources have subsequently been observed in much more detail with greater sensitivity. The dates listed are the dates of publication, not those of the observations.

UNPREDICTED GAMMA-RAY SOURCES

DISCOVERED	SOURCES & PROCESSES	DISCOVERER
1962	Cosmic Background Gamma Rays	Arnold et al.[40]
1968	Crab Nebula Gamma Rays	Haymes et al.[41]
1972	Diffuse Galactic 511 keV Line	Johnson et al.[42]
1973	Gamma Ray Bursts	Klebesadel et al.[43]
1973	Very High Energy Accretion Sources	Vladimirsky et al.[44]
1975	High Energy Unidentified Sources	Kniffen et al.[45]
1977	Accreting Neutron Star Cyclotron Line	Trumper et al.[46]
1977	Pulsar Transient 400 keV Line	Leventhal et al.[47]
1978	Galactic Center 511 keV Line	Leventhal et al.[48]
1978	Gamma-Ray Line Transient	Jacobson et al.[49]
1979	Pulsar Transient Cyclotron Line	Ling et al.[50]
1979	Gamma-Ray Burst 400 keV Line	Mazets et al.[51]
1981	Gamma-Ray Burst Absorption Line	Mazets et al.[52]
1983	Accreting Black Hole 511 keV Line	Nolan & Matteson[53]
1984	Soft Repeating Gamma-Ray Bursters	Golenetskii et al.[54]
1987	High Energy Pulsar Glitch Transient	Grenier et al.[55]

We see from these two tallies of the gamma-ray sources and processes discovered over the past thirty years, that roughly half were expected theoretically and half were unexpected — 12 were predicted and 16 were unpredicted. Of those sources that were predicted, in two-thirds of the cases the time lag between their prediction and their discovery was more than 10 years — in fact the median was 14 years. This lag generally reflects the time required to develop both the instruments of sufficient sensitivity and the balloon or satellite observing programs in which the discoveries were made.

Moreover, there still remain many other astrophysical gamma-ray sources and processes, most of which were predicted at least 10 years ago, but have not yet been discovered.

PREDICTED GAMMA-RAY SOURCES YET TO BE DISCOVERED?

PREDICTED	SOURCES & PROCESSES	PREDICTOR
1961	Very High Energy $\gamma\gamma$ Absorption	*Nikishov*[56]
1964	Interstellar Gas Nuclear Lines	*Hayakawa et al.*[57]
1969	Galactic Supernova ^{44}Ti Lines	*Clayton et al.*[16]
1969	Diffuse Extragalactic ^{56}Co Lines	*Clayton & Silk*[58]
1970	Martian Nuclear Lines	*Metzger & Arnold*[59]
1971	Diffuse Galactic ^{60}Co Lines	*Clayton*[60]
1974	Galactic Nova ^{22}Na Lines	*Clayton & Hoyle*[61]
1974	Accreting Black Hole π° Gamma Rays	*Dahlbacka et al.*[62]
1974	Extragalactic Supernova ^{57}Co Lines	*Clayton*[63]
1976	Primordial Black Hole Gamma Rays	*Page & Hawking*[64]
1977	Interstellar Grain Nuclear Lines	*Lingenfelter & Ramaty*[65]
1977	Accreting Black Hole Nuclear Lines	*Lingenfelter & Higdon*[66]
1978	Pulsar 1.022 MeV Annihilation Line	*Kompaneets*[67]
1983	Pulsar Quarkonium Lines	*Kanbach & Schlickeiser*[68]

AND THERE ARE UNDOUBTEDLY AT LEAST AS MANY NOT YET IMAGINED!

CONCLUSIONS

• Thus we see that about half of the gamma-ray sources and processes that were theoretically predicted thirty or more ago have subsequently been discovered — increasing our understanding and opening new directions for further research.

• We also see that roughly half of the new gamma-ray sources and processes that have been discovered within the last thirty years were theoretically predicted and those predictions nearly all stimulated the development of more sensitive instrumentation that led to the confirming discoveries.

• The other half of the gamma-ray discoveries during the last thirty years were wholly unpredicted — resulting from the development of more sensitive instrumentation. These have also greatly expanded our knowledge of high energy phenomena and furthered our understanding.

- Moreover, the record of the past thirty years suggests that such unexpected discoveries may not just be something to hope for, but may in fact be counted upon.
- We also see that nearly all of the predicted, but as yet undiscovered, sources that remain require high resolution spectroscopic observations.
- Finally, since most of the gamma-ray sources known — solar flares, gamma-ray bursts, supernovae, active galactic nuclei, etc. — are either transient or highly variable, simultaneous observations are necessary for the confirmation of observations.

Acknowledgements. I thank Jim Arnold, Geoff Burbidge, Carl Fichtel, Josh Grindlay, Duane Gruber, Alice Harding, Hugh Hudson, Larry Peterson, Reuven Ramaty, Rick Rothschild, Bob Reedy and Stan Woosley for valuable comments and NASA for financial support of this work under grant NSG-7541.

REFERENCES

1. L. E. Peterson and J. R. Winckler, *Phys. Rev. Letters* **1**, 205 (1958).
2. P. Villard, *Compt. Rend. Acad. Sci. Paris* **130**, 1010 (1900).
3. C. T. R. Wilson, *Proc. Cambridge Phil. Soc.* **11**, 32 (1900).
4. V. F. Hess, *Physik. Z.* **13**, 1084 (1912).
5. J. Clay, *Proc. Nederlandsche Akad. v. Wet.* **30**, 1115 (1927).
6. P. Morrison, *Nuovo Cimento* **7**, 858 (1958).
7. A. Boischot, *Compt. Rend. Acad. Sci. Paris* **244**, 1326 (1957).
8. L. Biermann, O. Haxel and A. Schluter, *Z. Naturforsch.* **6a**, 47 (1951).
9. E. L. Chupp, *AAS-NASA Symp. on the Physics of Solar Flares* (NASA, Washington, 1964) p. 445.
10. J. F. Dolan and G. G. Fazio, *Rev. Geophys.* **3**, 319 (1965).
11. R. E. Lingenfelter and R. Ramaty, *High-Energy Nuclear Reactions in Astrophysics*, ed. B. S. P. Shen (Benjamin, Inc., New York, 1967) p. 99.
12. E. L. Chupp et al., *Nature* **241**, 333 (1973).
13. G. R. Burbidge et al., *Phys. Rev.* **103**, 1145 (1956).
14. T. Pankey, Ph. D. dissertation, Howard Univ. (1961).
15. S. A. Colgate and C. McKee, *Astrophys. J.* **157**, 623 (1969).
16. D. D. Clayton, S. A. Colgate and G. J. Fishman, *Astrophys. J.* **155**, 75 (1969).
17. S. M. Matz et al., this volume (1987).
18. W. Sandie et al., this volume (1987).
19. W. R. Cook, D. M. Palmer and T. A. Prince, this volume (1987).
20. G. R. Burbidge and F. Hoyle, *Nuovo Cimento* **4**, 558 (1956).
21. W. L. Kraushaar et al., *Astrophys. J.* **177**, 341 (1972).
22. G. W. Perlow and C. W. Kissinger, *Phys. Rev.* **84**, 572 (1952).
23. S. Hayakawa, *Prog. Theor. Phys.* **8**, 571 (1952).
24. G. W. Hutchinson, *Phil. Mag.* **43**, 847 (1952).
25. G. W. Clark, G. P. Garmire and W. L. Kraushaar, *Astrophys. J.* **153**, L203 (1968).
26. S. Hayakawa, *Prog. Theor. Phys.* **19**, 219 (1958).
27. G. R. Burbidge, *Astrophys. J.* **127**, 48 (1958).
28. J. R. Arnold, *Proc. Lunar & Planet. Explor. Colloq.* **1:3**, 28 (1958).
29. J. R. Arnold, The Selenographer's Lexicographer (unpublished) (1961).
30. A. E. Metzger et al., *Science* **179**, 800 (1973).

31. V. L. Ginzburg, L. Ozernoi and S. I. Syrovatski, *Doklady* **154**, 557 (1964).
32. J. E. Grindlay *et al.*, *Astrophys. J.* **197**, L9 (1975).
33. D. Forrest *et al.*, *19th Internat. Cosmic Ray Conf.* **4**, 146 (1985).
34. P. A. Sturrock, *Astrophys. J.* **164**, 529 (1971).
35. J. D. Kurfess, *Astrophys. J.* **168**, L39 (1971).
36. J. E. Grindlay, *Nature, Phys. Sci.* **234**, 153 (1971).
37. W. D. Arnett, *Ann. N.Y. Acad. Sci.* **302**, 90 (1977).
38. R. Ramaty and R. E. Lingenfelter, *Astrophys. J.* **213**, L5 (1977).
39. W. A. Mahoney *et al.*, *Astrophys. J.* **262**, 742 (1982).
40. J. R. Arnold *et al.*, *J. Geophys. Res.* **67**, 4876 (1962).
41. R. C. Haymes *et al.*, *Astrophys. J.* **151**, L9 (1968).
42. W. N. Johnson, F. R. Harnden and R. C. Haymes, *Astrophys. J.* **172**, L1 (1972).
43. R. W. Klebesadel *et al.*, *Astrophys. J.* **182**, L85 (1973).
44. B. M. Vladimirsky *et al.*, *13th Internat. Cosmic Ray Conf.* **1**, 456 (1973).
45. D. A. Kniffen *et al.*, *14th Internat. Cosmic Ray Conf.* **1**, 100 (1975).
46. J. Trumper *et al.*, *Ann. N.Y. Acad. Sci.* **302**, 538 (1977).
47. M. Leventhal, C. J. MacCallum and A. C. Watts, *Astrophys. J.* **216**, 491 (1977).
48. M. Leventhal, C. J. MacCallum and P. D. Stang, *Astrophys. J.* **225**, L11 (1978).
49. A. S. Jacobson *et al.*, *Gamma-Ray Spectroscopy in Astrophysics* (NASA, Greenbelt, 1978) p. 228.
50. J. C. Ling *et al.*, *Astrophys. J.* **231**, 896 (1979).
51. E. P. Mazets *et al.*, *Nature* **282**, 587 (1979).
52. E. P. Mazets *et al.*, *Nature* **290**, 378 (1981).
53. P. L. Nolan and J. L. Matteson, *Astrophys. J.* **265**, 389 (1983).
54. S. V. Golenetskii, V. N. Ilyinskii and E. P. Mazets, *Nature* **307**, 41 (1984).
55. I. A. Grenier *et al.*, *20th Internat. Cosmic Ray Conf.* **1**, 77 (1987).
56. A. I. Nikishov, *Zh. Eksp. Teor. Fiz.* **41**, 549 (1961).
57. S. Hayakawa, *Prog. Theor. Phys., Supp.* **30**, 153 (1964).
58. D. D. Clayton and J. Silk, *Astrophys. J.* **158**, L43 (1969).
59. A. E. Metzger and J. R. Arnold, *Appl. Opt.* **9**, 1289 (1970).
60. D. D. Clayton, *Nature* **234**, 291 (1971).
61. D. D. Clayton and F. Hoyle, *Astrophys. J.* **187**, L101 (1974).
62. G. H. Dahlbacka, G. F. Chapline and T. A. Weaver, *Nature* **250**, 36 (1974).
63. D. D. Clayton, *Astrophys. J.* **188**, 155 (1974).
64. D. N. Page and S. W. Hawking, *Astrophys. J.* **206**, 1 (1976).
65. R. E. Lingenfelter and R. Ramaty, *Astrophys. J.* **211**, L19 (1977).
66. R. E. Lingenfelter and J. C. Higdon, *Astrophys. J.* **215**, L53 (1977).
67. D. A. Kompaneets, *Pisma Astron. Zh.* **4**, 304 (1978).
68. G. Kanbach and R. Schlickeiser, *Positron-Electron Pairs in Astrophysics* (Am. Inst. Phys., New York, 1983) p. 204.

GAMMA-RAY ASTRONOMY IN THE NUCLEAR TRANSITION REGION (THE PROMISE AND THE PROBLEMS)

E. L. Chupp
University of New Hampshire, Durham, NH 03824

INTRODUCTION

Primary γ rays have been searched for since the earliest studies of cosmic rays from balloons. The well known, inspiring predictions by Morrison[1] stimulated enthusiastic groups of fledgling experimentalists, who to this day are striving to develop the promise of the field. The reason for this undaunted interest is the fact that the γ-ray lines and continuum γ rays carry unique information about many astrophysical phenomena. The γ-ray observations give insight into the acceleration of particles in solar flares, galactic nucleosynthesis and the dynamics of an evolving supernova. Further, we can hope to understand the enigmatic γ-ray bursts and other, yet unknown phenomena. The present list of accomplishments is meager for a field that offers so much promise, yet inspite of the difficulties inherent to the field the promise is even greater today. This is so, because only a few observations have brought forth many insights, as I will discuss. Also many unresolved questions remain concerning the interpretation of the exciting observations of the cosmic ^{26}Al and the positron-electron annihilation γ lines and as well the γ-ray lines from solar flares; this is a clear measure of the vitality of the field.

In the following we will review, briefly, some of the conflicts existing, at this time, on the observational scene and suggest approaches for future development of the field.

REFLECTIONS

Let me say at the outset, that it is a great pleasure to be the first speaker on the last day's session of this important workshop. Since you have heard, and will hear the details of experiments and γ-ray detection techniques from the other speakers, I would like to offer some views that may be of value for the future development of the field and some thoughts that our younger colleagues at the early stage of their careers may benefit from. First, let me tell you how I happened to end up in this field, since the path I followed was not predictable at the outset and was sometimes due to happenstance and not premeditation.

As a graduate student, at Berkeley, I studied cosmic-ray time-variations using a neutron monitor, which I built with very little outside assistance in our laboratory. The detector was sitting in a trailer in Oakland, California and I, watching it register day-to-day events, was hoping for "*the big one*," a ground level solar cosmic-ray event. It eventually came, but it occurred two years after I got my PhD in 1954.

© 1988 American Institute of Physics

As is well known, this event happened on 1956 February 23 and is still considered the largest solar cosmic-ray event ever. In any case, by then I had long left my neutron monitor to others and had worked on several projects at the Livermore Radiation Laboratory of the University of California. Needless to say, when I reflected on how bad my timing had been, it seemed as an omen for things to come, and I concluded that my career would be rather uneventful. The report of the 1956 February 23 event as observed by the Chicago Group[2] continues to haunt me.

At Livermore, I was able to make measurements of the atmospheric neutron spectrum, with Bill Hess (now DOE) and others, on the ground and in the bubble of a B-36, flying over the United States. The most exciting work I did at Livermore was high-resolution nuclear γ-ray spectroscopy, using crystalline diffraction techniques with Hans Mark, Bob Jopson and others. For a source we used a high-current 4 MeV proton accelerator which excited nuclear rotational levels by Coulomb excitation, giving an intense flux of γ-ray lines. This allowed us to make energy measurements to a precision of 1/1000. We published several papers on our results before the accelerator was closed down for budgetary reasons. I continued the spectroscopy research, using a reactor as a source of γ-ray lines from neutron capture. We were able to make a very precise measurement of the deuteron binding energy from the capture line at 2.224 MeV to a precision of 1 part in 10^3, the best available at that time.

In 1959 I went to the Boeing Airplane Company, where I helped engineers to write proposals for Air Force space missions such as *Dyna-Soar*, a vehicle similar to the space shuttle, and later for *NASA Lunar Missions*. This was valuable training for my future academic career, which started in 1962, since a good part of a modern Professor's time is spent on writing many proposals, over the years, for funding.

A major stimulus to my own interest was the work Rich Lingenfelter, Larry Peterson and Reuven Ramaty were doing in the early and late 1960's. Soon, my small group and I (assisted by my wife and children) carried out balloon flights from the UNH football field and later from Palestine, Texas, to study atmospheric γ rays. After a few years of this work, I was fortunate enough to have obtained a NASA contract to build a γ-ray spectrometer for the *OSO-7* spacecraft.

Paul Higbie joined me to build the instrument, and later Dave Forrest joined us for the final testing before delivery. *OSO-7* was launched in 1971 September, in a less than ideal manner, with white knuckled physicists praying that GSFC engineers could save the mission – a similar situation arose in the near-missed retrieval of *SMM* by the shuttle in 1984 April. The turn on of our *OSO-7* γ-ray spectrometer quickly showed that heavy Van Allen radiation (VA) belt dosages caused the PMT gain to shift so drastically that we could only get stable spectra by turning the high-voltage off for the full VA belt orbits.

We recorded only X-ray events until early 1972 August, when a series of large flares erupted on the Sun. On August 2, I had just arrived in Munich and heard on the *American Forces Network News* that the largest flare ever had erupted on the Sun.

That evening I had telephoned UNH (upon the urging of my wife) and reached Ron Adams, a data analyst, and asked him to contact Dave Forrest, who in turn was to call GSFC, to make sure the instrument high-voltage was not turned off during the VA belt passages so we would not lose any data. On 1972 August 4 another large flare occurred. This one had our name on it. The spectra we obtained showed that during the rising phase of the flare from 0623 UT until 0632 UT γ-ray lines were present at 2.22 MeV, 0.51 MeV and possibly at 4.4 MeV

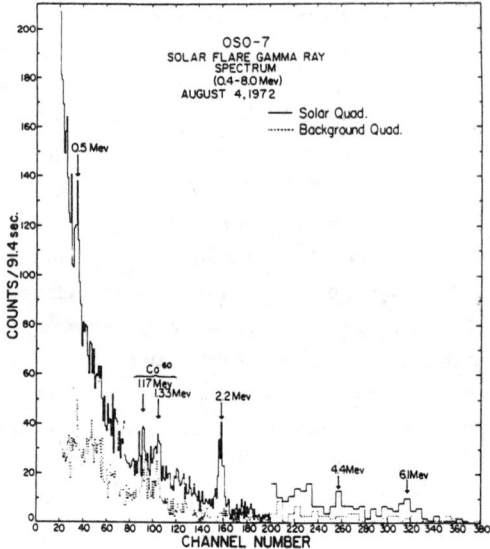

Figure 1.

Solar quadrant γ-ray spectrum (solid histogram) obtained during the rising phase of the 1982, August 4 3B flare (0623:49 to 0633:02 UT). The simultaneouysly acquired background spectrum is also shown (dotted histogram). (From E. L. Chupp et al., 1975 in S. R. Kane ed. Solar Gamma-, X- and EUV Radiation, IAU COSPAR Symp. 68, p. 341, D. Reidel Publishing Co., used by permission.

and 6.1 MeV. After 0622 UT, the Earth occulted the Sun during this flare, but during a later flare on August 7, which erupted while the Sun was eclipsed, delayed γ-ray lines at 2.2 MeV and 0.51 MeV were observed, as the satellite emerged into sunlight. The spectrum we obtained for the August 4 flare is shown in Figure 1.

This event still stands as unique, because the most intense part of solar flare γ-ray bursts seen by *SMM* occur within $\sim$ 1 minute, while the rising phase of the 1972 August 4 event was of, at least, 10 minutes duration[3]. The success of *OSO-7* firmly anchored me in γ-ray astronomy and it was the stimulus we needed to try to get a berth on *SMM*. The rest is history.

I guess I should conclude this bit of nostalgia by saying that in my 25 years in the field the intriguing questions and the challenges in γ-ray astronomy still excite me as they have in the beginning. I guess one forgets the frustrations!!

MAJOR ACCOMPLISHMENTS AND UNANSWERED QUESTIONS

The number of confirmed reports of γ-ray lines is unfortunately very small and we are all too well aware of the three problems confronting us, namely high background, low fluxes and low instrument sensitivities. It is striking however that a solid observation of a single identifiable exterrestrial γ-ray line carries an impressive amount of information. As examples, I have selected three such observations which are confirmed. Each, however, permits only limited interpretation or raises serious questions of interpretation which therefore present experimental challenges. These are: (1) γ-ray lines from solar flares, (2) the galactic ^{26}Al radioactive decay line, and (3) γ-rays from positron electron annihilation.

Table 1.

SOLAR FLARE LINES	INFERRED SOLAR CHARACTERISTICS	ARE WE THERE? OR IMPEDIMENTS	NEEDED MEASUREMENTS
0.511 MeV and associated P_s continuum; (Delayed emission) (OSO-7, SMM, Hinotori)	Temperature and density of annihilation region e.g., 10^7 K	Energy resolution of NaI inadequate; intense-instrument background line; Earth occults event	High resolution (0.1%) measurement of line; nonocculted time history; reduce upper limits
2.223 MeV and associated continuum; (Delayed emission) (OSO-7, HEA0-1, HEA0-3 SMM, Hinotori)	Abundance of ^{3}He; low-energy neutron flux and spectra; depth distribution of neutron capture; interacting particle spectrum	Need large number of nonocculted events at several solar longitudes	High sensitivity flux time histories; continuous data; explore all feasible models
Prompt nuclear deexcitation lines e.g. 4.439 MeV – 6.129 MeV	Abundances of accelerated & ambient nuclei; interaction time scale; interacting particle spectrum	Separation of bremsstrahlung continuum and nuclear excess; many unresolved lines	High resolution (0.1%) measurements of full spectrum at high sensitivity; explore all feasible models

SOLAR FLARE GAMMA-RAY LINES

I summarize in Table 1 some key results deduced from solar flare γ-ray lines, pose questions yet to be answered, and suggest an approach to further advances.

The annihilation line at 0.511 MeV has been observed in numerous flares by both the *SMM* gamma-ray spectrometer (GRS)[4] and the *Hinotori* γ-ray spectrometer[5], respectively. In both cases the spectrometers were scintillators with moderate energy resolution ($\sim$ 8% at 0.511 MeV) and only an upper limit to the temperature of the annihilation region ($\sim 10^7$ K) could be obtained from the line width. Also, there is a strong instrument background line at this energy, that is highly variable in intensity, due to radiation belt passage and cosmic ray latitude effects. Another problem in observing this line is that its emission is delayed because of the radioactive half lives of the positron emitters, and the slowing down time needed to thermalize the positrons before annihilation. This fact results in the occultation of the 0.511 MeV emission when observed from low altitude Earth orbiting satellites, which can only observe the Sun, continuously, for about 1 hour in each orbit. Future measurements of this line at least, need to be made at higher energy resolution using, for example, cooled germanium spectrometers. Obtaining nonocculted observations for over an hour after the flare and reducing the instrument background at 0.511 MeV are problems for which there is no clear solution as yet.

The solar flare γ-ray line at 2.223 MeV, due to neutron capture in photospheric hydrogen, is also delayed in emission, because of the time needed to thermalize the neutrons. Therefore, occultation of this line also occurs. However, since this line is one of the strongest lines seen in flares, it has been possible to obtain an estimate of the photospheric ^{3}He abundance by studying the time history and its intensity versus time after the flare. This possibility exists because ^{3}He has a high probability of capturing neutrons (with no γ-ray emission), therefore, its density, as well as the hydrogen density, determines the rate at which the intensity of the 2.223 MeV line falls off. Future observations of this line could be improved if weaker fluxes and more events could be detected at different solar longitudes. Interpretation of such observations is strongly model dependent, therefore obtaining more events will allow a study of the longitude dependence of the line characteristics[6].

The interpretation of the observed properties of the 2.223 MeV line and its relationship to the broader question of the acceleration of ions in solar flares depends on knowing the characteristics of other γ-ray lines produced directly or indirectly by the same ions which produce the neutrons. The total fluence of a particular γ-ray line produced by direct nuclear excitation [e.g. in an elastic scattering reaction such as C^{12} (p, p' γ)], compared to the fluence of the 2.223 MeV line, depends on the spectrum shape of the accelerated ions. Ramaty and co-workers have made extensive calculations of the yields of γ-ray lines from prompt nuclear deexcitation and the total neutron yield for Bessel function and power law incident ion spectra under thick target conditions (cf reference 7 for a recent review of this work.). By making suitable assumptions about neutron propagation and capture in the photosphere, and the location of the flare on the Sun, the fluence ratio $\phi(2.223)/\phi(4-7)$ may be calculated for different ion spectra. Here $\phi(4-7)$ is the total fluence for all prompt nuclear deexcitation lines in the (4–7) MeV energy band. For a given flare location the calculated value for this ratio depends on the spectral parameters (αT) or exponent s, for Bessel function and power-law forms, respectively[7].

There are two major problems limiting the information that can be derived from a prompt γ-ray spectrum of a solar flare. They are (1) separation of the flare electron bremsstrahlung continuum from the intrinsic flare narrow and broad line spectrum and (2) lack of knowledge about unresolved lines because of use of moderate-resolution scintillation spectrometers. It is expected that eventually flare γ-ray spectra will be obtained with high-resolution instruments such as high-purity germanium spectrometers with high sensitivity. Such information will allow extensive studies of all feasible flare particle acceleration models. Thus by a study of the observable properties of all the γ-ray lines produced in flares, it is possible to determine important characteristics of the interacting particle spectra. The γ-ray line observations taken with observations of the γ-ray continuum due to electron bremsstrahlung and meson decay above 10 MeV, and the satellite and ground level obeservation of solar neutrons, permit a fairly detailed picture to be constructed of the energetic particle characteristics for a particular flare.

GALACTIC GAMMA-RAY LINES

In Table 2, we summarize the current status concerning the galactic γ-ray lines at 0.511 MeV and 1.809 MeV, which have been discussed in detail at this workshop. Even though both lines have been solidly observed, it is interesting to note that the spatial distribution of both sources is in dispute. Consider the annihilation line at 0.511 keV. The New Hampshire 1977 result[8] using a wide field-of-view instrument was in conflict with the narrow field-of-view result of the Bell/Sandia group[9], which they tacitly assumed was due to a point source. The UNH line intensity was about a factor of 3 higher than the Bell/Sandia result and the two results could only be reconciled (Jacobson[11], Dunphy et al.[12]), if we were seeing annihilation radiation from both a point source and an extended source. An earlier French observation[13] with a wide field-of-view instrument gave also a high value for the annihilation line flux similar to the UNH result. By the end of 1979 everyone in the field felt that cosmic γ-ray line astronomy was finally on a firm basis, since HEAO-3[14] had observed, in the fall of 1979, a line flux consistent with Bell/Sandia results in 1977 and 1979. Again it was tacitly assumed a point source was responsible. The next HEAO-3 observation[14], in the spring of 1980, yielded a line flux more than half the earlier result. It was natural to invoke time variability, which implied that the source region was less than 1/2 LY in extent. Subsequent balloon experiments by the GSFC/CENS group[15] and the Bell/Sandia[16] group in 1981 and 1984, respectively, failed to detect the presence of the hypothetical "point source." Would it ever return?

Table 2.

PHOTON SPECTRUM	LOCALIZED SOURCE	EXTENDED	OBSERVATIONAL RQMT
0.511 MeV and associated P_s continuum	HEAO-3, Bell/Sandia & GSFC/CENS suggest point source turned off in 1980 (Spring)	SMM, French & UNH experiments; SMM-source steady for 8 years	Wide field-of-view imaging with increased sensitivity; high-energy resolution desirable, but not essential for source distribution
1.809 MeV	MPE consistent with "point source"	Bell/Sandia upper limit to point source; SMM supports	Wide field-of-view imaging with increased sensitivity; high-energy resolution desirable, but not essential for source distribution

The Sun observing *SMM* gamma-ray spectrometer (GRS), launched in 1980 observes the galactic disk region every fall with its wide field-of-view. Share et al.[18] have reported the results for five observations of the galactic disk through 1986, all showing clear evidence for the presence of the 0.511 MeV annihilation. Also they[18] have shown that the results are most consistent with a steady extended (diffuse) source and most importantly that the evidence for a variable point source is considerably weakened. Indeed all previous balloon and satellite observations are in good agreement if only a steady extended source is assumed and measurement errors are properly considered.

I believe that further advances in our understanding of the nature of the source of galactic annihilation radiation will require imaging instruments with greatly increased sensitivity. Spectrometers with greater energy resolution capabilities than obtainable with scintillators are useful for studying evidence for positronium formation but are not essential for determining the spatial distribution of the narrow line radiation. We can expect that advanced balloon borne instrumentation may, in the next few years, enable us to determine if an extended source alone only is sufficient to account for the observations, or if indeed there is an exotic source such as a black hole at the galactic center supplying the positrons. More naturally, an extended source could be due to the positron output of novae and supernovae events over millions of years ago, or even to red giants and massive stars. Also, positrons from galactic cosmic rays interacting with intergalactic gas must contribute.

The first solid indication that a nucleosynthetic γ-ray line at 1.809 MeV, from the decay of ^{26}Al was made by *HEAO-3* observers[19]. Theoretical calculations by many authors[20,21] showed that ^{26}Al could be produced in many astronomical sites such as novae, supernovae, and some massive stars. Since the isotope has a half life of just under a million years explosive events over millions of years of galactic history could give a distributed source of radioactive ^{26}Al. Ramaty and Lingenfelter[22] had pointed out that a narrow γ-ray line at 1.809 MeV could be detectable from the cool galactic gas distributed through the galaxy. This prediction stimulated many experimenters, and the *HEAO-3* observers were the first to find that the distribution of the radiation was consistent with a diffuse source spread over the galactic plane and furthermore the line was narrow (< 3 keV), consistent with that expected from a cool gas.

Again the *SMM* GRS could search for this galactic emission in a similar manner, as was done for the annihilation line, and Share et al.[23] confirmed the *HEAO-3* line detection at the 10 σ level. The ^{26}Al line flux has been detected by *SMM* GRS at every galactic plane transit since 1980 except for the fall of 1983 when *SMM* science observations were curtailed due to the impending repair mission. Even though the *SMM* GRS has a wide field-of-view it detects the strongest modulation in annual counting rate when the Sun passes nearest the galactic center. One cannot really determine the source distribution from *SMM* GRS data, but the presence of a source with an equivalent source intensity[23] of $(4.0 \pm 0.4) \times 10^{-4}$ γ cm^{-2} s^{-1} is consistent with the *HEAO-3* flux[19] of $(4.8 \pm 1.0) \times 10^{-4}$ γ cm^{-2} s^{-1} rad^{-1}.

A balloon observation by the Max Planck Institute (MPE) Compton telescope group in 1982 has also reported detection of the ^{26}Al line[24] from the galactic center but with a strong indication that the source is point like. This conclusion results from the image deconvolution procedure used by the MPE group. Inspired by this dramatic announcement, the Bell/Sandia group[25] reported an upper limit on a galactic center point source flux for the ^{26}Al line, over a factor of 4 below the MPE report, using data from flights made between 1977 and 1984. Furthermore, they claim to have confirmed the presence of a diffuse ^{26}Al source consistent with the *HEAO-3* and *SMM* GRS results! At this point no results are definitive!!

Of course, determining the source distribution for the ^{26}Al is essential if we are to make progress in determining the origin of the ^{26}Al. As with the annihilation line there is a critical need for wide field-of-view imaging spectrometers with greatly increased sensitivity. Since it has already been established that the ^{26}Al line has a width consistent with the cool galactic gas, it may not be necessary to use instruments with the highest energy resolution to determine the source distribution. We expect this problem can also be attacked by balloon experiments.

SN1987A

This meeting has obviously been very well planned. Our organizers have timed this event with the first public announcement of the detection of the ^{56}Co γ-ray lines from SN1987A[26,27] using the *SMM* GRS. No other γ-ray line detection could be as significant, as now we have the first real proof that nucleosynthesis takes place in supernovae events. The corresponding reports of γ-ray detections from SN1987A with balloon instruments flown by the Lockheed[28] and Caltech[29] groups, and reported here, are equally as exciting. At this workshop, I believe, there is much elation and there are only few skeptics, since these observations represent the real promise of our field.

More detailed observations of the ^{56}Co lines are greatly needed and we hope that NASA's continuing SN1987A balloon campaign will be successful.

FUTURE OF THE FIELD

The problems limiting further advances in γ-ray line astronomy have been with us for many years. Advances have been steady but slow and it seems clear that there is much that can be done in the way of development of new ways to deal with the background and low flux problems. The balloon platform offers an ideal way to test new techniques, some of which will be presented at this workshop. We can also do much with existing instruments, using balloon platforms and we hope the funding agencies supply the necessary resources.

One final thought, through the demise of Sanduleak $-69°$ 202, SN1987A has provided a strong stimulus to the balloon γ-ray astronomy program. With the knowledge gained from the use of the larger γ-ray telescopes that will study SN1987A, we should have a good basis on which to assess our future direction. While the excitement of the moment is with the supernova, a major community effort is needed to define the parameters of a major facility in the post-GRO era. Rather than wait for a NASA "opportunity" and respond individually, we should present the future needs of our field in the form of community wide planning documents. Maybe we can benefit from the experience of astronomers and high energy physicists in obtaining new major facilities. Gamma ray astronomy observations in the nuclear transition region carry unique and vitally important information, and we have only begun to realize the potential of the field. Positive steps now, will insure for future generations of γ-ray astronomers many exciting discoveries!

ACKNOWLEGEMENTS

I would like to acknowledge the many colleagues, in the US and in Europe, who have stimulated my work and have cooperated in my research. Their names are many. I would also like to thank my wife Mary for her constant support and contributions to all my work. Financial support has come primarily from NASA and I gratefully acknowledge the support, over the years of several program managers at Headquarters and GSFC.

REFERENCES

1. P. Morrison, Nuovo Cimento 7, 858 (1958).
2. P. Meyer, E.N. Parker and J.A. Simpson, Phys. Rev. 104 (3), 768 (1956).
3. E.L. Chupp et al. Nature 241, 333 (1973).
4. G.H. Share et al., *Positron-Electron Pairs in Astrophysics*,
    in ed. M. L. Burns et al., AIP Conf. Proc. 101, 15 (1983).
5. M. Yoshimori K. et al., Proc. 18th ICRC 4, 89 (1983).
6. X.-M. Hua and R.E. Lingenfelter, Ap.J. 319, 555 (1987).
7. R. Ramaty and R.J. Murphy, Space Science Rev. 45, 213 (1987).
8. B.M. Gardner et al., *The Galactic Center* in eds. G. R. Riegler
    and R. D. Blandford, (AIP Conf. Proc. 83 p. 144 (1982).
9. M. Leventhal, C.J. MacCallum and P.D. Stang, Ap.J. 225, L11 (1978).
10. M. Leventhal et al., Ap.J. 240, 338 (1980).
11. A.L. Jacobson, *The Galactic Center*, in eds. R. G. Riegler and
    R. D. Blanford, AIP Conf. Proc. 83, p. 123 (1982).
12. P.P. Dunphy, E.L. Chupp and D.J. Forrest, in ed. M. L. Burns et al.
    Positrons Electrons Pairs in Astrophysics, AIP Conf. Proc., 101, p. 237 (1983).
13. F. Albernhe et al., Astron. Astrophys. 94, 214 (1981).
14. G. R. Riegler et al., Ap.J. 248, L13 (1981).
15. W.S. Paciesas et al., Ap.J. (Letters) 260, L7 (1982).
16. M. Leventhal et al., Ap.J. 260, L1 (1982).
17. M. Leventhal et al., Ap.J. 302, 459 (1986).
18. G.H. Share et al., Ap.J. 326, 717 (1988).
19. W.A. Mahoney et al., Ap.J. 286, 578 (1984).
20. W.D. Arnett, Ap.J. 157, 1369 (1969).
21. D.D. Clayton and F. Hoyle, Ap.J. 203, 490 (1976).
22. R. Ramaty and R.E. Lingenfelter, Ap.J. (Letters) 213, L5 (1977).
23. G.H. Share et al., Ap.J. 292, L61 (1985).
24. P. von Ballmoss, R. Diehl and V. Schönfelder, Ap.J. 318, 654 (1987).
25. C.J. MacCallum et al., Ap.J. 317, 877 (1987).
26. S.M. Matz et al., These proceedings.
27. S,M. Matz et al., Nature 331, 416, 1988.
28. W. Sande et al., These proceedings.
29. T. Prince et al., These proceedings.

X-RAY OBSERVATIONS OF SN1987A WITH GINGA

Y. Tawara

Nagoya University, Furo-cho, Chikusa-ku, Nagoya 464, Japan

ABSTRACT

Continuous observations of the Large Magellanic Cloud region made with Ginga since February 25, 1987 revealed the X-rays with unusual energy spectrum from the region of 0.2 x 0.3 deg^2 including the supernova SN1987A. The first detection of a sinificant X-ray flux from SN1987A was detected about 130 days after its explosion. The energy spectrum shows the presence of unusually hard component above 10-15 keV. The light curve shows the relatively large intensity variations in the low energy band of 5.8-16.1 keV, in contrast to small variations in high energy band of 16.1-27.8 keV. About 320 days after the explosion, the SN1987A showed large increase about factor of 6 in 20 days, and they decayed to the half of the peak flux level in 30 days.

INTRODUCTION

X-ray satellite Ginga has been regularly observing the supernova SN1987A (hereafter abbreviated as SN) in the Large Magellanic Cloud since February 25, 1987 right after the optical discovery. Ginga has large effective area of 4000 cm^2 for wide energy band of 1-36 keV[1]. Because of a relatively wide field of view of 1x 2 deg^2 (FWHM) in comparison with the average source separations at the LMC, we should carefully subtract the X-ray flux from the neighboring sources to determine the X-rays from the SN1987A.

OBSERVATIONS AND RESULTS

In our first observations we use scanning mode of observations to distinguish the SN mainly from the LMC X-1, which is one of the brightest X-ray source in the LMC (~20 mCrab) and is about 0.6 deg away from the SN. From this observation, we found the presence of a hard X-ray source separated from LMC X-1[2]. The line position of the hard source is in agreement with the SN within 0.1 deg (90 % confidence).

From scans along four different paths which are separated about 0.5 deg from each other in the direction perpendicular to these paths. We could determine the 90% confidence error box of the hard X-ray source of the size 0.2 x 0.3 deg^2, which includes the SN.

In order to get precise X-ray spectrum of the SN, we use pointing observation in which the field of view of LAC does not include LMC X-1. Since there are several weak X-ray sources[3] near the SN in addition to LMC X-1; SNR's 0540-69.3, N132D, N157B, and the source 0519-69.1, we should subtract the X-ray flux of these X-ray sources from such pointing data. Thus we also have done several different pointing observations to estimate the X-ray flux from these weak X-ray sources. Furtheremore, we considered the component of very soft X-rays (E<4keV) from the LMC X-1 scattered by the collimator walls.

On the basic assumptions of the constant X-ray fluxes from these neighboring sources, the X-ray spectrum of the SN was derived from the pointing observations. Fig. 1 shows the light curve of the SN in terms of the count rate in the ranges 5.8 - 16.1 keV (a) and 16.1 - 27.8 keV (b) obtained from the pointing observations. (This figure includes the data points obtained by the most recent observation at the time when this manuscript was prepared.) The first significant X-ray flux was detected about 130 days after the supernova explosion. The intensity in terms of the photon number flux was $(6.9 \pm 2.0) \times 10^{-4}$ photons cm^{-2} sec^{-1} in the 16.1 - 27.8 keV band. The first significant detection in the lower energy band was made about 150 days after the explosion.

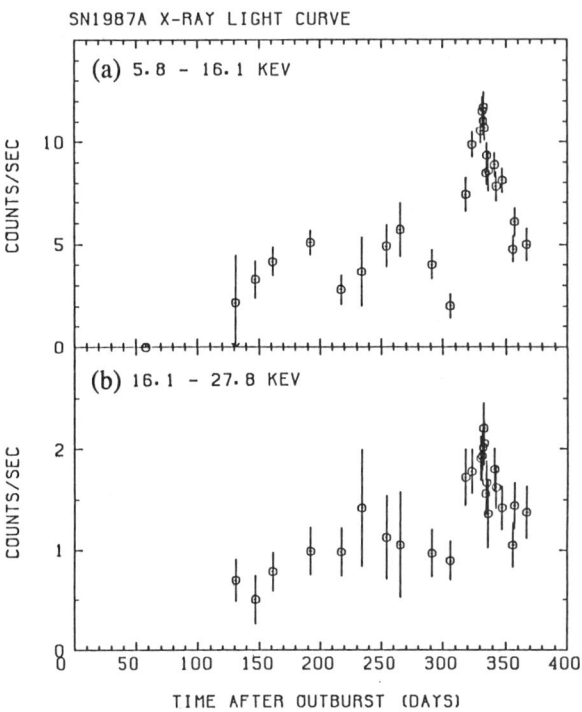

Fig.1. Light curve of SN1987A in terms of the count rate in the range 5.8-16.1 keV (a) and 16.1-27.8 keV (b) obtained from pointing observations. Background and contribution of nearby sources are subtracted. The error bars include possible fluctuation of the LMC X-1, but not the errors of other sources.

The flux in the lower energy band shows the significant variability on a time scale of ten days as shown in Fig. 1. In contrast to its variations, the flux in the higher energy band showed little variability until 306 days after the explosion. In the recent observations[4], the SN showed large flare at about 320 days after the explosion. The flux jumped up both in the lower and the higher energy bands to the maximum recorded values.

The energy spectrum of the SN1987A shows unusual shape, which is unlike any of known classes of X-ray sources. Though it may be either a power law or a thermal spectrum in the lower energy band, above 15 keV to 30 keV, the spectrum is almost flat. Actually a dip or a minimum is seen around 10-15 keV in some of the spectra. This supports the interpretation that there is the separate hard component which falls off below 15 keV. Fig. 1 shows that the hard component is much more stable with time than the soft component.

I would like to thank Drs. M. Itoh, H. Inoue and the Ginga team for permission to reproduce diagrams of their work before publication.

REFERENCES

1. F. Makino et al. Astrophys. Lett. **25**, 223 (1987)
2. T. Dotani et al. Nature, **330**, 230 (1987)
3. K. S. Long, D. J. Helfand & D. A. Grabelsky, Astrophys. J. **248**, 925 (1981)
4. M.Itoh Ph.D.Thesis, University of Tokyo, (1988)

OBSERVATIONS OF SUPERNOVA 1987A WITH INSTRUMENTATION ON
THE SOVIET SPACE STATION MIR

C. Reppin, J. Englhauser, C. Hanson, W. Pietsch, J. Trümper, W. Voges
Max-Planck-Institut für Physik und Astrophysik
Institut für Extraterrestrische Physik, 8046 Garching, FRG

E. Kendziorra, M. Bezler, R. Staubert
Astronomisches Institut der Universität Tübingen, 7400 Tübingen, FRG

R. Sunyaev, A. Kaniovskiy, V. Efremov, M. Grebenev, A. Kuznetsov,
A. Melioranskiy, D. Stepanov, I. Chulkov
Space Research Institute, Moscow, USSR

ABSTRACT

Hard X-ray emission from SN1987A in the energy range from 20-350 keV has been observed by the HEXE and Pulsar X-1 instruments on the MIR-KVANT Roentgen observatory. After the first observation of the hard X-rays on August 10, 1987 the SN became the main target of the observatory. The measured photon spectrum is extremely hard with a power law index of about 1.4. At lower energies the spectrum seems to become flatter and there is an indication of a cut off below 25 keV. The luminosity in the above energy band is 2×10^{38} erg/s. The flux does not show a significant variation between August 10 and the end of November.

INTRODUCTION

The Roentgen observatory on the KVANT module consists of four instruments covering the wide energy band from 2 to about 1300 keV and has been described by Sagdeev et al. (1986[1]). On 1987 March 31, the KVANT module was launched and docked to the MIR station on April 12. The MIR station is in a low Earth orbit with a 51° inclination. The measurements reported here have been made with the HEXE and PULSAR X-1 instruments. A more detailed comparison between the observational results and the theoretical models is given by Sunyaev et al. (1987[2a], 1987[2b]).

OBSERVATIONS

The high energy X-ray experiment HEXE was provided by the Max-Planck-Institut für Extraterrestrische Physik (MPE) and the Astronomisches Institut Tübingen (AIT). It consists of four phoswich detectors (0.32 cm NaJ(Tl), 5 cm CsJ(Tl)) with a geometric area of 200 cm^2 each. Their energy resolution is about 20 % FWHM at 60 keV and the energy range is 15-250 keV. Arrival times of photons can be determined with an accuracy of 0.3 - 25 ms, depending on the instrument operation mode. Rise time discrimination, passive graded shielding and a plastic scintillator anticoincidence are used for background reduction. The four detectors form two groups which have separate tungsten collimators with 1.6° FWHM. These collimators can be tilted by 2.3° and allow simultaneous

© 1988 American Institute of Physics

source and background observations. In the normal telemetry mode the rocking of the collimators is performed every 2 minutes. More detailed descriptions of the instrument are given by Reppin et al. (1985[3], 1987[4]).

PULSAR X-1 is also a phoswich detector system using thicker NaJ crystals (3 cm NaJ(Tl), 3 cm CsJ(Tl)) for a good high energy response and consists of four units with 250 cm^2 each. Its nominal energy range is 20-1300 keV with an energy resolution of 25 % at 12 keV (Sagdeev et al., 1986[1]). The field of view is defined by collimators with 3° FWHM. Since these collimators are fixed, background measurements are performed by tilting the MIR station in 4 minute intervals by about 10°.

Attitude measurements are done with a star tracker. The attitude drift is in the range of 1 arcmin per orbit, the absolute pointing accuracy is a few arc minutes.

Observations of cosmic X-ray sources started at the beginning of June 1987. Because of restrictions of the solar panel and the star tracker orientations, observations are normally performed during the night side of the orbit. This leads to observation times of about 25 minutes per orbit. Up to December 1987 a total of 209 orbits have been spent observing SN1987A which corresponds to a total observing time of 2,3x10^5 s for source and background.

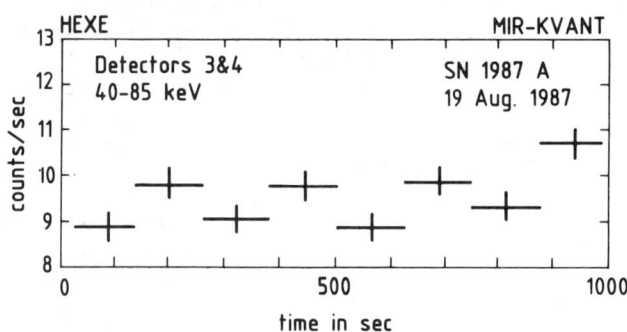

Fig. 1. Count rates obtained with HEXE in the collimator rocking mode. The data points are alternatively for off source and on source pointings. The observation began at 13:21:09 UT. Error bars are 1 σ based on count statistics. Note the suppressed zero for the vertical axis.

The first pointings towards SN1987A were made on June 8, July 16 and August 1, but were not sensitive enough to detect the SN. On August 10 the hard X-ray emission from the SN was discovered by HEXE. As illustrated in fig. 1 the source is easily detected during one orbit using the collimator rocking mode. The spectrum obtained during the August 10-21 period by HEXE is shown in fig. 2. It is based on 9600 s of data and the unfolding of the energy loss spectrum was done with a preliminary response matrix. This matrix had to be verified in detail by the calibration observations of the Crab nebula performed in September.

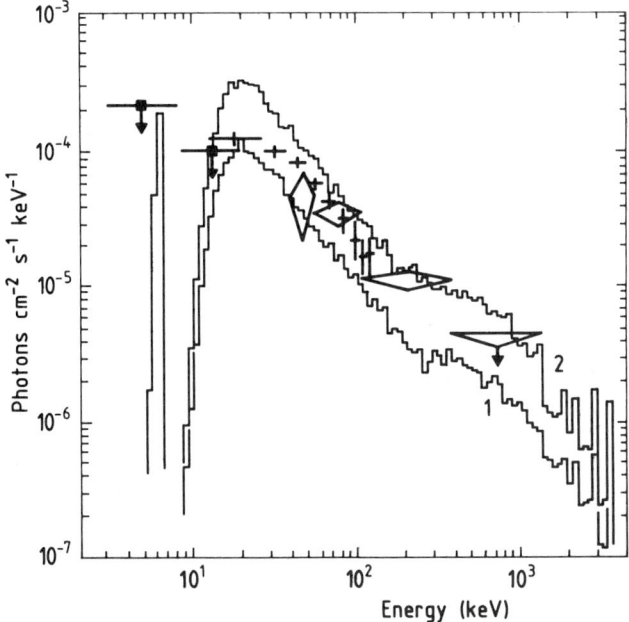

Fig. 2. The preliminary energy spectrum of the hard X-rays. Squares: 3 σ upper limits from TTM. Crosses: detections with HEXE (1 σ error bars). Diamonds: detections and 3 σ upper limits obtained with PULSAR X-1. The histograms show the results of Monte Carlo simulations by Grebenev and Sunyaev (1987) - brief details: 0.1 $M_\odot$ of ^{56}Co, expanding envelope 16 $M_\odot$ with metallicity 1/3 solar, mean expansion velocity 4150 km s^{-1}. 180 days (1) and 210 days (2) after the explosion. The histogram bins in the region of the Fe fluorescence line are 0.5 keV wide.

At higher energies up to 350 keV the source was detected by PULSAR X-1. The PULSAR X-1 broad band points are also shown in Figure 2.

The power law index of the spectrum is about 1.4 and the spectrum seems to become flatter at lower energies with an indication of a cut off below 25 keV.

Analysis of HEXE data from later observations show that there is no increase in intensity by more than 30 % until the end of November.

Since the field of view of the HEXE collimators is 1.6° FWHM there was some concern about a possible contribution from LMC X-1 which is only about 0.6° away from SN1987A. To exclude this possibility 54 offset observations were performed which are indicated in fig. 3. Using these pointings and the rocking mode of the collimators the contour lines in fig. 3 were derived for the position of the hard X-ray source. The contours are given for the 1σ, 2σ, and 3σ levels. LMC X-1 lies far outside the 3 σ contour.

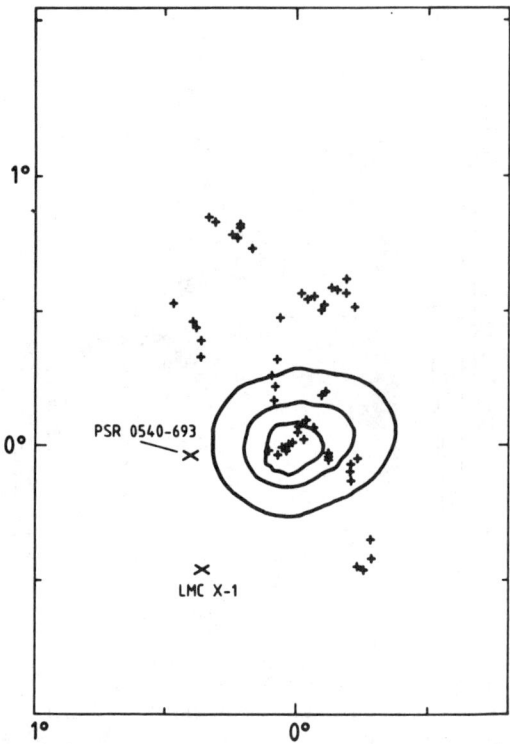

Fig. 3. The position of the source of high-energy flux obtained by analysing 54 HEXE observations at the offset positions shown (crosses). Contours are at 1 σ, 2 σ, and 3 σ confidence. The positions of LMC X-1 and 0540-693 are excluded at the 3 σ level.

DISCUSSION

At energies below 15 keV the TTM (Brinkman et al., 1983[5]), a coded mask telescope on the KVANT module, (built by the Laboratory for Space Research, Utrecht, Holland and the University of Birmingham, UK) could establish upper limits for the energy bands 2-8 and 8-20 keV from the observations in August 1987 (Skinner et al., 1987[6]). These limits are indicated in fig. 2. A spectrum from observations with the GINGA satellite on September 2-3 shows a definite flux below 10 keV which is a factor of 2-3 above the limits given by TTM (Dotani et al., 1987[7]). At energies between 20-30 keV there is good agreement between the HEXE results and the flux observed by GINGA. An observation with a balloon borne hard X-ray telescope on April 19, 1987 led to 3 σ upper limits of 3.7×10^{-4} ph/cm²skeV, resp. 1.5×10^{-5} ph/cm²skeV for the energy bands 15-45 keV, resp. 45-80 keV (Ubertini et al., 1987[8]). The upper limit for the higher energy range is about a factor of 2 below the flux observed by HEXE in August.

The X-ray luminosity in the energy band 20-350 keV is about 2×10^{38} erg/s. This is only a small fraction of the bolometric luminosity of 10^{42}

erg/s. One possibility for producing the hard X-ray emission is by Compton scattering of gamma ray photons in the shell. The source of these photons could either be a pulsar from interactions of pulsar generated cosmic rays and/or its synchrotron nebula or the decay of ^{56}Co nuclei resulting from the decay of ^{56}Ni generated during the evolution of the progenitor star. The bolometric light curve (Catchpole et al., 1987[9]) shows, after about 120 days, a decay time which is in good agreement with the decay time of ^{56}Co. Recent observations of gamma ray lines from the decay of ^{56}Co at 847 and 1238 keV with the gamma ray spectrometer on SMM (Matz et al., 1988[10]) and a balloon borne telescope with solid state detectors (Sandie et al., 1987[11]) further support the radioactive decay model. This does not exclude the existence of a pulsar inside the shell, but constrains its contribution to the total energy output.

A number of authors have performed detailed Monte Carlo calculations resulting in time dependent energy spectra (e.g. Grebenev & Sunyaev, 1987[12], Xu et al., 1987[13], Ebisuzaki & Shibazaki, 1987[14], Pinto & Woosley, 1987[15]). In fig. 2 we added the expected spectra for 6 and 7 months after the explosion obtained by Grebenev & Sunyaev (1987[12]) for the ^{56}Co decay model. They assumed a mass of 0.1 $M_\odot$ ^{56}Co, a spherically symmetric shell of 16 $M_\odot$ with an average expansion velocity of 4150 km/s and a metallicity 1/3 solar. The agreement of the observations with respect to the spectral shape and absolute flux is quite good, but the strong variation of the X-ray flux within 1 month as shown by the histograms in fig. 2 is not observed. To explain this, mixing between inner and outer layers of the shell, clumping or a nonspherical structure of the shell have to be taken into account in the models (Nomoto & Shigeyama, 1987[16], Kumagai et al., 1987[17], Woosley et al., 1987[18], Itoh et al., 1987[19], Shull & Xu, 1987[20]). Observations over the whole electromagnetic wave length band and over a long period of time will be important to refine the models.

REFERENCES

1. R.Z. Sagdeev, et al., Complex X-Ray Observatory; Academy of Sciences of the USSR, Space Research Institute, preprint 1177 (1986).
2a. R.A. Sunyaev, et al., Nature 330, 227-229, (1987).
2b. R.A. Sunyaev, et. al., Academy of Sciences of the USSR, Space Research Institute, preprint 1323, (1987).
3. C. Reppin, et al., Workshop Non-thermal and Very High Temperature Phenomena in X-ray Astronomy, Rome (1983), (eds. Perola, G.C., and Salvati, M.) 279-282 (Istituto Astronomica Roma, Rome 1985).
4. C. Reppin, et al., Proc. 20. ICRC, p. 284, Moscow (1987).
5. A.C. Brinkman, et al., Workshop Non-thermal and Very High Temperature Phenomena in X-ray Astronomy, Rome (1983), (eds. Perola, G.C., and Salvati, M.) 263-269 (Istituto Astronomica Roma, Rome 1985).
6. G.K. Skinner, Proceedings of the Fairfax workshop SN 1987A, Fairfax, (1987).
7. T. Dotani, et al., Nature 330, 230-231 (1987).

8. P. Ubertini, et al., Proceedings of the Fairfax workshop SN 1987A, Fairfax, (1987).
9. R. Catchpole, et al., M.N.R.A.S., in press, 1987.
10. S.M. Matz, et al., Nature 331, 416-418, (1988).
11. W. Sandie, et al., this conference.
12. S.A. Grebenev & R.A. Sunyaev, Soviet Astr.Lett. 13, 945-963 (1987).
13. Y. Xu, P. Sutherland, R. McCray, and R.R. Ross, submitted to APJ (1987).
14. T. Ebisuzaki & N. Shibazaki, submitted to APJ (1987).
15. P.A. Pinto & S.E. Woosley, submitted to APJ (1987).
16. K. Nomoto & T. Shigeyama, Proceedings of the Fairfax workshop SN1987A, Fairfax, (1987).
17. S. Kumagai, et al., Proceedings of the Fairfax workshop SN1987A, Fairfax, (1987).
18. S.E. Woosley, P. Pinto & L. Ensman, Ap.J. 324, in press, (1988).
19. M. Itoh, et al., Nature 330, 233-235, (1987).
20. J.M. Shull & Y. Xu, Proceedings of the Fairfax workshop SN1987A, Fairfax, (1987).

NASA'S PROGRAM FOR OBSERVING SUPERNOVA 1987A

Alan N. Bunner
NASA Headquarters, Washington, DC 20546

ABSTRACT

The National Aeronautics and Space Administration has initiated a program of balloon and sounding rocket campaigns from the southern hemisphere to study the anticipated gamma-ray, X-ray and ultraviolet emission from the supernova in the Large Magellanic Cloud. These campaigns will continue to be supported over the period during which the flux remains observable, with present plans extending through early 1989. Emphasis in the selection of payloads has been in spectroscopy, with Germanium detectors receiving a high priority among gamma-ray balloon instruments. It is expected that narrow-line sensitivities near 10^{-4} photons/cm^{2}sec at the 847 keV line of ^{56}Co will soon be available. In addition to high resolution spectrometers, other balloon payloads being flown to observe SN1987A include large area NaI imaging and spectrometry instruments and a double scatter Compton telescope. While NASA's program includes observations at radio, infrared, ultraviolet, x-ray and gamma-ray wavelengths, this paper concentrates on high energy observations.

ACTIONS TO DATE

As first word of the discovery of SN1987A reached NASA Headquarters, it was not clear whether the supernova would prove to be of Type I or Type II. However, it was clear that we needed to work quickly to be ready to field balloon payloads into the southern hemisphere to search for the predicted nucleosynthetic emission lines. So NASA initiated a series of actions to extend our joint U.S./Australian balloon agreement and to conclude an agreement for Australian rocket launches. As interest in gamma-ray and X-ray observations of this unique phenomenon grew, we soon realized that we were facing a very oversubscribed situation, particularly for balloon flight opportunities.
By 2 March, NASA's Astrophysics Division had issued a Dear Colleague letter to balloon and sounding rocket experimenters, asking for key characteristics of each available payload so that an attempt could be made to maximize the scientific return.

Meanwhile, within the first few hours of detection, both IUE and SMM had begun making observations of the supernova. The new Japanese X-ray astronomy satellite, Astro-C or "Ginga", which had fortunately just been launched on 5 February 1987, completed its turn-on sequence and was pointed towards the Large Magellanic Cloud within about 2 days of the event. These were the 3 satellites at the disposal of the community for early ultraviolet, gamma-ray and X-ray measurements of the supernova. On 5 March, a decision was made to extend the operations of an already-planned April New Zealand campaign for the Kuiper Airborne Observatory, to enable infrared supernova observations.

To help guide NASA in planning a scientific strategy for observations of this phenomenon at all wavelengths, a Workshop was convened at Goddard Space Flight Center on 6 March, which was attended by about 100 very excited scientists and several representatives from the press. A few days later, TIME magazine devoted its cover story to Supernova 1987A. By 24 March, NASA's Astrophysics Division had formed a

Supernova Science Working Group, with Bonnard Teegarden as Chairman, to provide a continuing scientific guidance to NASA on the appropriate observational response to the event. By late May, a Supernova Program Office had been formed at NASA Headquarters, with Guenter Riegler as Program Manager of all NASA supernova activities.

By early May 1987, the Kvant module had docked onto the USSR "Mir" Space Station and had begun what we all hope will be a continuous monitoring of the hard X-ray flux from the supernova. By mid-August, both Ginga and the "HEXE" instrument on Mir had detected X-rays from SN1987A.

OBSERVATIONAL STRATEGY

Our scientific strategy for observations has been driven by the theoretical predictions of (1) a 77-day half-life for the central energy source, due to a fraction of a solar mass of radioactive ^{56}Ni, (2) the emergence of the characteristic gamma-ray lines such as the 847 keV line of ^{56}Co, (3) Compton-scattered gamma-radiation producing hard X-rays and fluorescent lines such as Fe Kα, (4) the eventual emergence of a pulsar at all wavelengths, (5) the ultraviolet and infrared emission lines expected from the cooling of the expanding envelope as a function of time, and (6) the effects of the explosion on circumstellar matter and the interstellar medium.

Our hopes for long-term multi-wavelength coverage of SN1987A over the next 10 years or more include the following programs:
- SMM- The Solar Maximum Mission will provide a modest capability for spectroscopy, including coverage of the key 847 and 1238 keV lines,
- Balloons- NASA's southern hemisphere balloon campaigns will continue so long as there is sufficient flux to perform useful science,
- GRO- The Gamma Ray Observatory will be launched in mid-1990, just in time to observe the important 122 keV line of ^{57}Ni,
- Ginga- The latest Japanese X-ray astronomy mission provides X-ray photometry from 1.5 to 30 keV, and will certainly provide the necessary timing capability if an X-ray pulsar emerges,
- Sounding Rockets- are providing near-term observational coverage for both X-ray and far ultraviolet wavelengths. Periodic southern hemisphere rocket campaigns will be planned for the future as circumstances warrant.
- ROSAT- Due for launch in early 1990, ROSAT will provide high resolution X-ray imaging in the soft X-ray region (0.1 to 2 keV),
- AXAF- The Advanced X-ray Astrophysics Facility, planned for a 1995 launch, will provide high resolution imaging and spectroscopy over the 0.1 to 10 keV range,
- IUE- The International Ultraviolet Explorer has already played an important role for SN1987A in two respects: not only in the study of the uv spectrum itself, but also in providing a nicely calibrated continuous monitoring of the visual magnitude, through the use of the IUE fine error sensors. Figure 1 shows this data current as of 31 December 1987, with the clear signature of the 77 day half-life of ^{56}Co.
- HST- With the launch of the Hubble Space Telescope in mid-1989, high resolution visible and ultraviolet images and spectra will be possible.
- KAO- A continuing series of flights of the Kuiper Airborne Observatory for infrared spectroscopy has proved very important in providing key velocity, ionization state, and abundance data. These flights will continue to follow the evolution of the event in 1988 and 1989.

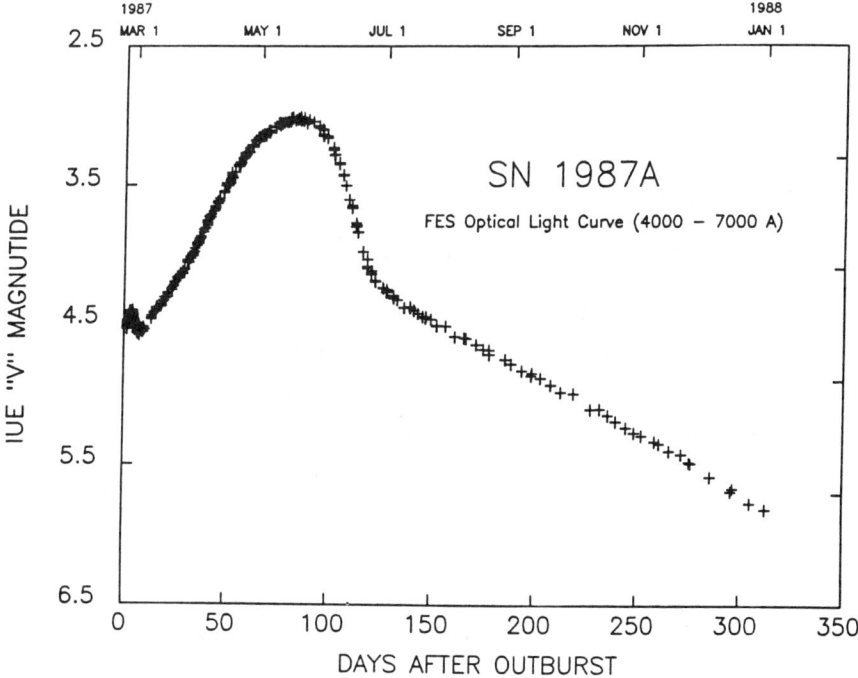

Fig. 1. The visual magnitude light curve of SN 1987A, as measured by the Fine Error Sensor on the International Ultraviolet Explorer (IUE).

Besides the observational capabilities listed above, the USSR missions GRANAT and GAMMA-1, with X-ray and gamma-ray instruments covering 2 keV to 500 MeV, are expected to be launched in 1988. The Institute of Space and Astronautical Science (ISAS) in Japan has planned Astro-D, an X-ray spectroscopy mission, for launch in 1993. The Federal Republic of Germany is making plans for Spectrosat, a high resolution X-ray spectroscopy mission which may be launched in the 1994 period. So during the next decade, the capabilities for a truly international observational program for SN1987A will be in place.

X-RAY INSTRUMENT PROPERTIES

Table I summarizes the properties of the X-ray instruments, both orbiting and sub-orbital, that are available today. All of these instruments have now flown or are in orbit except for the GSFC (Goddard Space Flight Center) sounding rocket payload, which will be flown for the first time from Woomera, Australia, in February 1988. The GSFC rocket payload is an important one because it has good energy resolution ($\Delta E = 150$ eV at 6 keV), covers a broad range of energies and has the capability to clearly separate the supernova from other nearby X-ray sources.

Table I

SUPERNOVA 1987a
PROPERTIES OF X-RAY INSTRUMENTS

ORBITING INSTRUMENTS:

Name	Type	Energy range (keV)	Spatial Resol. or field of view	Area (cm^2)
GINGA	Prop. Counter	1.5 - 30	0.8° x 1.7°	4000
HEXE/MIR	NaI Scintillator	15 - 200	1.9° x 1.9°	800
TTM/MIR	Coded-apert. ctr.	2 - 30	1.8 arc min	900
PULSAR/MIR	Scintillator	20 - 800	3° x 3°	1000
SIRENE/MIR	Gas Sc. P. C.	2 - 100	2° x 2°	314

SUBORBITAL INSTRUMENTS:

Name	Type	Energy range (keV)	Spatial Resol. or field of view	Area (cm^2)
Penn State	Mirror + CCD	0.1 - 2	1 arc min	200 (geom.)
GSFC	Mirror + Si(Li)	0.3 - 7	8 arc min	200 @ 2 keV
Columbia	Mirror + GSPC	0.1 - 2	3 arc min	40 @ 1 keV
	Collim. + GSPC	2 - 20	24 arc min	170 @ 6 keV

However, it's important to note that, unless the supernova becomes significantly brighter in soft X-rays, it will be difficult to accomplish real X-ray spectroscopy with a short sounding rocket flight. For this reason, the Supernova Science Working Group has recommended, and NASA has been working to achieve, an early flight of the Broad Band X-ray Telescope (BBXRT), a Space Shuttle attached payload. The BBXRT (Figure 2) is being built by the GSFC with the same technology as the GSFC sounding rocket payload. It uses nested conical metal foil X-ray mirrors as a "light bucket" to focus X-rays onto a cooled Si(Li) solid state detector for non-dispersive spectroscopy. We are hopeful that we can manifest the BBXRT instrument for an relatively early Shuttle flight, such as along with the ultraviolet "Astro" payload, for a late 1989 flight.

GAMMA-RAY INSTRUMENT PARAMETERS

Table II summarizes the properties of the balloon gamma-ray instruments that are in NASA's program plan for southern hemisphere SN1987A balloon campaigns. Figure 3 shows the energy range covered by both the X-ray and gamma-ray instruments and the angular resolution or field of view of each.

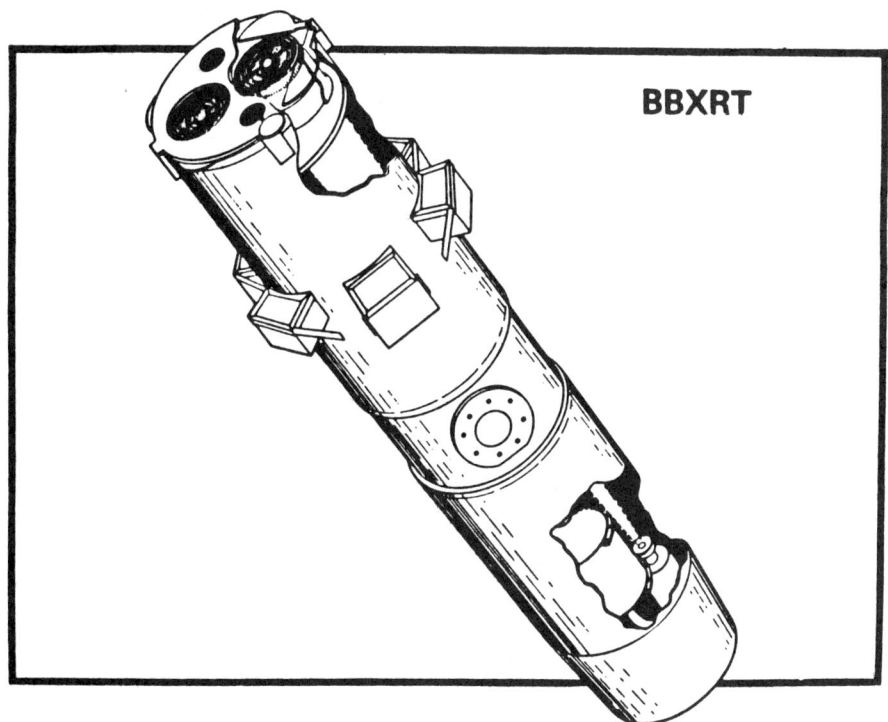

Fig. 2. Sketch of the conical foil Broad Band X-ray Telescope (BBXRT), scheduled to fly as a Shuttle-attached payload.

CAMPAIGN PLANS

The current NASA plans for southern hemisphere balloon, rocket and aircraft campaigns through the spring of 1989 are shown in Table III, prepared by Guenter Riegler. There are several noteworthy points: (a) The shorthand used on this chart identifies each instrument by a single chief U.S. investigator's name. It should be noted that all of these investigations are the work of teams of people, often involving several institutions. (b) The times of the infrared aircraft campaigns roughly coincide with the balloon campaigns. The same is not generally true with the planned times of the sounding rocket campigns, however. (c) The balloon flights are the most difficult of these three types of observation to provide. Surface winds must be very low for these 10 to 28 million cubic foot balloons to be launched, and high altitude winds must be ideal for a long flight. These conditions are satisfied at the southern hemisphere balloon launch site of Alice Springs, Australia, for only a few weeks near April and November of each year. Since this constraint prevents launching every balloon payload that might be desired, it has been necessary to prioritize the candidate flights. (d) Many of the campaign plans shown on Table III are tentative and conditional on the supernova emitting adequate gamma-ray, X-ray and ultraviolet flux levels. For this determination, we will depend on new data from SMM, Mir, Ginga, IUE and the balloon and sounding rocket flights themselves. For example, the balloon flights listed for

Table II
BALLOON GAMMA RAY INSTRUMENT PARAMETERS

Experimenter	Detector Type	Detector Area (cm2)	Energy Range (MeV)
Frontera	NaI	5400	0.02 - 0.3
Grindlay	NaI	900	0.02 - 0.3
Johnson	NaI	800	0.1 - 10
Prince	NaI	1320	0.04 - 10
Leventhal	Ge	~30	0.07 - 6.5
Lin/Matteson	Ge	24	0.02 - 10
	NaI	480	0.015 - 0.18
Mahoney	Ge	~32	0.05 - 6.5
Matteson	Ge	288	0.02 - 10
Sandie	Ge	214	0.025 - 8
Teegarden	Ge	215-250	0.02 - 10
Frye	Spark Chamber	400	>50
White	Compton Telescope	10000	1 - 30

Energy Resolution at 1 MeV: $E/\Delta E$ ~ 500 for Germanium
~ 15 for NaI

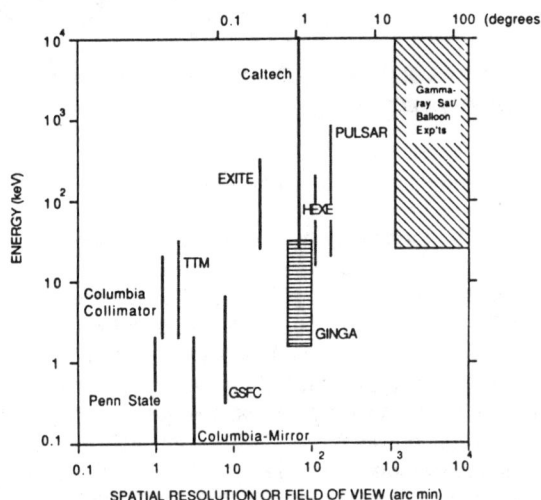

Fig. 3. The energy range and the angular resolution of the X-ray and gamma-ray instruments being used to observe the supernova. TTM refers to the coded mask imaging spectrometer from Birmingham and Utrecht on the Kvant module of the Mir space station. EXITE (Energetic X-ray Imaging Telescope Experiment) is a balloon payload from the Smithsonian Astrophysical Observatory (J. Grindlay), HEXE, the High Energy X-ray Experiment of the Max Planck Institute, and PULSAR, an instrument from the Space Research Institute (IKI), Moscow, are also on the Kvant module of Mir. GSFC refers to the NASA-Goddard Space Flight Center sounding rocket payload.

Table III NASA Supernova 1987A Campaign Plan

SUPERNOVA SN1987a CAMPAIGN PLAN FY87-FY89								12/23/87 4:47 PM	
	Apr-87	Nov-87	Feb-88 (8)	Apr-88 (8)	Jun-88	Aug-88 (9)	Nov-88 (8)	Feb-89 (10)	Apr-89(8,10)
Balloons:	Frye 4/19	Sandie 10/29	Lin/Matteson	*Cheng (3,12)		Teegarden (9)	*Cheng (3,12)		*Cheng (3,12
Gamma-Ray	Prince 5/19	Prince 11/18		*Sood (12)		Prince (9)	Matteson		Matteson
+ Microwave	Sandie 5/29	Mahoney 12/6		Sandie			Teegarden		Tee/S(4)
In order of priority	Leventhal (11)	[White (1)]		Teegarden			Sandie		Frontera
				Prince			Prince		Schönf.(5)
				White			Frontera (2)		Prince
				Johnson					
				Grindlay					
UV/X-Ray		Garmire 11/14	Serlemitsos		Serlemitsos (7)			Serlemitsos (7)	
Rockets		Cash 11/18	Garmire		Garmire (7)			Garmire (7)	
		Novick 12/4	Cash		Cash (7)			Cash (7)	
			TBD (6)		Jenkins (6)(7)			Jenkins (6)(7)	
								TBD (6)	
IR KAO	Larson 4/16	Ericksen 11/4,6		Ericksen		TBD		TBD	
Aircraft	Witteborn 4/21	Witteborn 11/10,12,14		Witteborn		TBD		TBD	
		Moseley 11/17,19		Moseley		TBD		TBD	
		Harvey 11/22		Harvey		TBD		TBD	

(1) Withdrew because of insufficient expected time at float
(2) Frontera had proposed for 4/88 flight, but withdrew after the science review
(3) non-SN; needs 2 flights 6 months apart; April 89 flight shown as placeholder if April 88 flight fails
(4) one of Teegarden or Sandie; to be decided after 4/88 flights on the basis of instrument performance
(5) non-SN flight by Schönfelder
(6) tentative; placeholders
(7) requires trigger signal
(8) Balloons: Attempt to fly as many experiments as possible, in the order shown, within the available launch window of 6 weeks per campaign.
(9) Aug 88 campaign not planned at this time; to be requested (WFF/NSBF, cosmic-ray community) if 4/88 flights show strong signal
(10) Campaign to be reviewed about 11/88
(11) Two launch attempts, no observations
(12) *Separate group of experiments on low-volume balloons - outside gamma-ray priority grouping

August 1988 will not occur unless there is a clear indication that they would yield important results. The need for an April 1989 supernova balloon expedition will be reviewed in about November 1988. All of the 1988 sounding rocket plans beyond February are contingent on a reasonable prospect of positive results.

The selection of balloon payloads for these campaigns has been made with the help of a scientific peer review panel. The following rationales have guided these selections: (a) the desire for payloads with a high senitivity for the detection of the nucleosynthetic lines of ^{56}Co, ^{57}Co, ^{22}Na and ^{44}Ti; (b) the need for payloads with high senitivity for gamma-ray line detection both for a narrow line scenario (hence the Ge detectors) and for broad line emission (which can be accomplished by the NaI instruments); (c) the need for payloads with high sensitivity for continuum detection of Compton-scattered gamma-rays, such as the Frontera payload from TESRE, Bologna; (d) the desire for some continuity from campaign to campaign, to provide some capability for monitoring the source with uniform sensitivity; (e) the need for sensitivity to a wide range of energies; (f) some imaging capability, such as is provided by the Prince (Caltech) and Grindlay (SAO) payloads, in order to discriminate against other possible sources of hard X-rays in the field, and (g) the need for some redundancy, so that the most important observation is not lost to the vagaries of the ballooning business.

Finally, Table III indicates the emphasis on the balloon campaign for the spring of 1988. This is because the SN1987A Science Working Group has stated that the scientific return from gamma-ray observations may peak in the spring of 1988, as the radioactive engine inexorably decays.

We must not overlook that fact that these ambitious balloon and rocket campaigns are causing stress elsewhere in the scientific community. We should remember to show our appreciation to our colleagues in other disciplines for their patience while we carry out these once-in-a-lifetime observations.

SMM GAMMA-RAY OBSERVATIONS OF SN 1987A

S. M. Matz* and G. H. Share
E. O. Hulburt Center for Space Research
Naval Research Laboratory
Washington, D.C. 20375   U.S.A.

E. L. Chupp
Department of Physics
University of New Hampshire
Durham, New Hampshire 03824   U.S.A.

ABSTRACT

The Gamma-Ray Spectrometer (GRS) on NASA's *Solar Maximum Mission* satellite detected gamma-ray line emission from the decay of ^{56}Co in SN 1987A beginning in 1987 August.[1,2] Balloon-borne experiments have also reported the detection of these lines in 1987 October,[3,4] December,[5] and 1988 January,[6] and excesses consistent with lines in 1987 November.[7,8] Since the initial report we have continued to monitor the emission from the supernova with the GRS, and report here observations through 1988 April 6. The average fluxes for the total observation period (1987 August 1 to 1988 April 6) are $(6.6 \pm 2.0) \times 10^{-4}$ and $(6.4 \pm 1.7) \times 10^{-4}$ photons $-$ cm^{-2} $-$ s^{-1} for the 847 and 1238 keV lines, respectively. The quoted errors represent a combination of statistical and systematic uncertainties. There has been no marked increase in the intensity of either line since the end of October, and the data are consistent with a constant flux since August 1. The best-fit energies for the two lines are 840 ± 6 and 1239 ± 10 keV.

INTRODUCTION

On 1987 February 24 a Type II supernova (SN 1987A) was discovered in the Large Magellanic Cloud, about 50 kpc from Earth.[9] Detection of a neutrino signal from this event[10,11] at about 0736 UT on February 23 established that the iron core of the star had collapsed to a neutron star or black hole. According to nucleosynthesis theory, the energy of the ensuing explosion caused the formation of heavy elements in the stellar material outside of the iron core. It has long been predicted that a major product of explosive nucleosynthesis will be radioactive ^{56}Ni.[12,13]

The ^{56}Ni decays $(\tau_{1/2} = 6.1\text{d})$ to ^{56}Co, which in turn decays more slowly

---

* Resident Research Associate at the Naval Research Laboratory under the NRC Associateship Program

($\tau_{1/2} = 77.1$d) to stable ^{56}Fe. Most of the decay energy of ^{56}Ni and ^{56}Co is in the form of gamma rays; at early times, when the overlying material is thick, the gamma rays are trapped and deposit their energy in the envelope. This trapped radiation is thought to provide much of the energy for the optical light curve.[14] This hypothesis is supported by the observation that in SN 1987A the bolometric light curve between 1987 July 23 and November 15 showed an exponential decline consistent with the decay of ^{56}Co.[15,16] From the light curve, it has been estimated that 0.07 $M_\odot$ of ^{56}Ni was produced in the explosion.[17]

As the ejecta expand and thin, more of the gamma rays escape unscattered. This decreases the amount of energy available to power the optical light curve, and allows the possibility that the gamma-ray lines can be detected directly.[18] Detailed calculations have been made of the emergence of gamma-ray lines from SN 1987A.[17,19-22] The strongest gamma-ray lines from ^{56}Co decay are at 847 keV (branching ratio = 100%) and 1238 keV (67.9%). The time history of the line intensities, along with the line ratios and line shapes, can provide a sensitive measure of the distribution of the radioactive material in the supernova envelope. We report here the observation of these gamma-ray lines from SN 1987A by the Gamma-Ray Spectrometer (GRS) on the *Solar Maximum Mission* satellite (SMM).

ANALYSIS

The GRS has been operating in low-Earth orbit since February 1980. The instrument has been described in detail elsewhere.[23] Briefly, the GRS consists of seven 3" × 3" NaI detectors, actively shielded by CsI on the sides and back. The spectrometer is usually pointed at the Sun; therefore SN 1987A must be observed through the side CsI shield. Geometrical factors and the attenuation of the shield reduce the GRS photopeak effective area for 847 keV photons from SN 1987A to ~ 42 cm^2.

Background spectra are collected when the Earth occults the supernova. To minimize systematics, source minus background differences are calculated for each orbit, and only orbits containing a sufficient amount of occulted data are accumulated in the long-term sums. Due to the position of the source and the SMM orbit, there are periods when there is little or no occultation of the supernova during an orbit. Because there are insufficient background observations at these times, there are periodic gaps in the supernova measurements. To minimize contributions from short term activation in the spacecraft and instrument, we do not use data accumulated within ~ 2.5 hours after passage through the South Atlantic Anomaly.

The selected orbital background-subtracted spectra are summed, and the summed spectra are fit to determine the intensities of the lines of interest. While the orbital subtraction removes background from slowly varying local sources, background sources which change substantially during an orbit (including the atmospheric continuum) may leave residual fluxes in the difference spectra. In order to model these residuals, the fit to the background-subtracted counting rate spectra includes, in addition to the two ^{56}Co decay lines, a power-law continuum and four lines corresponding to observed background features. Two of the background lines (at 1.173 and 1.333 MeV) are from the internal ^{60}Co calibration source, one (at 1.368 MeV) is from ^{24}Na produced in the spacecraft, and one (at $\sim$ 1 MeV) is of unknown origin. The energies and widths of all six lines are fixed. There are thus 8 parameters in the fit, which is made over 66 channels (from 0.75 to 1.48 MeV), giving 58 degrees of freedom for each spectrum. The line fluxes derived using this fit differ slightly from those given in Ref. 2, where the 847 and 1238 keV lines were fit separately.

We have tested this fitting procedure using models of supernova emission spectra which include both line and continuum contributions (Ref. 24 and M. D. Leising, private communication). These calculated photon spectra were folded through the GRS instrument response to produce counting rate spectra. These model counts spectra were then added to the summed background-subtracted spectra accumulated prior to the supernova. The average line fluxes derived from fitting these simulated supernova spectra are consistent with the photon line fluxes of the input model. This indicates that the line fluxes derived from the actual supernova data are not substantially affected by continuum radiation from either the supernova or the atmosphere.

We have determined gamma-ray line intensities for a source at the position of the supernova by fitting the background-subtracted spectra summed over each occultation period since 1981. The results are shown in Fig. 1. The line intensities prior to August 1 are consistent (at the 30% and 20% confidence level for the 847 and 1238 keV lines, respectively) with statistical fluctuations about a constant mean flux; for both lines the measured mean is consistent with zero. These results imply that there are on average no significant residual background features at 847 or 1238 keV. Since August 1 the intensities of both lines show a sustained increase. The probabilities of observing such increases from chance fluctuations above the previously observed means are $\sim 3 \times 10^{-6}$ and $\sim 3 \times 10^{-7}$ for the 847 and 1238 keV lines, respectively.

Both lines have been detected with high statistical significance, and tests decribed previously[2] have indicated that systematic background effects do not contribute substantially to the observed line intensities. The generally statistical nature of the fluctuations prior to August 1 confirms this. However, the time histories (Fig. 1) do give some indication of small residual non-statistical fluctuations. For example, the measured intensity of the 847 keV line is consistently $\sim 1\sigma$ below the mean during 1986. This is a small effect, but it may affect the derived photon fluxes for the gamma-ray lines. In order

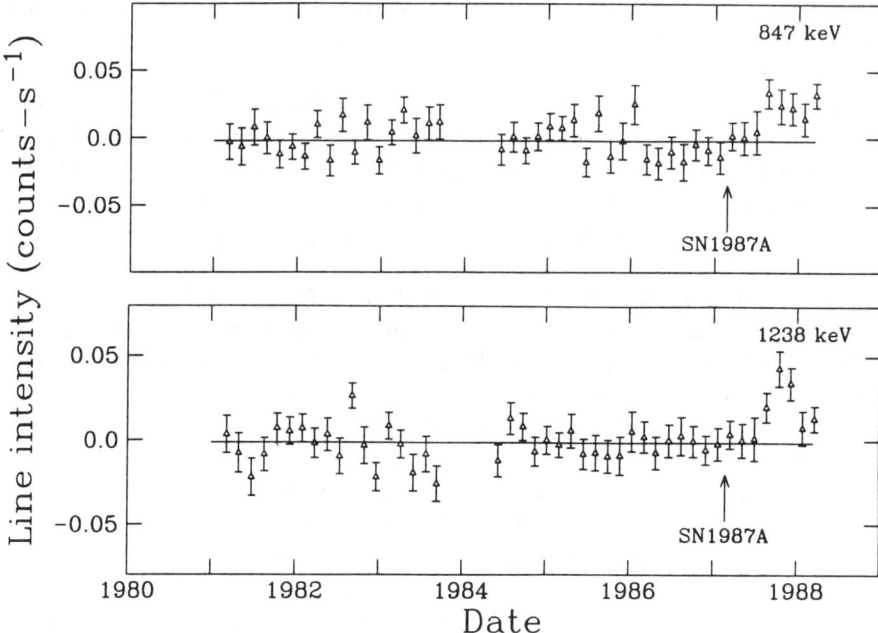

Fig. 1: The fitted intensities for the 847 and 1238 keV lines in the background subtracted spectra for SN 1987A. The data have been summed over periods when the source was significantly occulted by the Earth. Each point represents about 35 days of observation; each period of observation is followed by a gap of about 18 days. The mean fitted intensities for the data prior to 1987 August 1 are indicated by the solid lines.

to account for the possibility of such low-level systematics, we have increased the size of the error bars on the photon fluxes for both lines based on an estimate of possible systematic effects.

DISCUSSION

The resulting fluxes and errors for both lines during the eight observation periods since the supernova explosion are given in Table 1, and the values are plotted vs. time in Fig. 2. Note that we determine an average line flux from a fit to the sum of orbital difference spectra over the stated observation interval. Because usable observations are not available uniformly throughout the intervals, all times in the intervals are not equally weighted. The average fluxes for the period since first detection (1987 August 1 to 1988 April 6) are

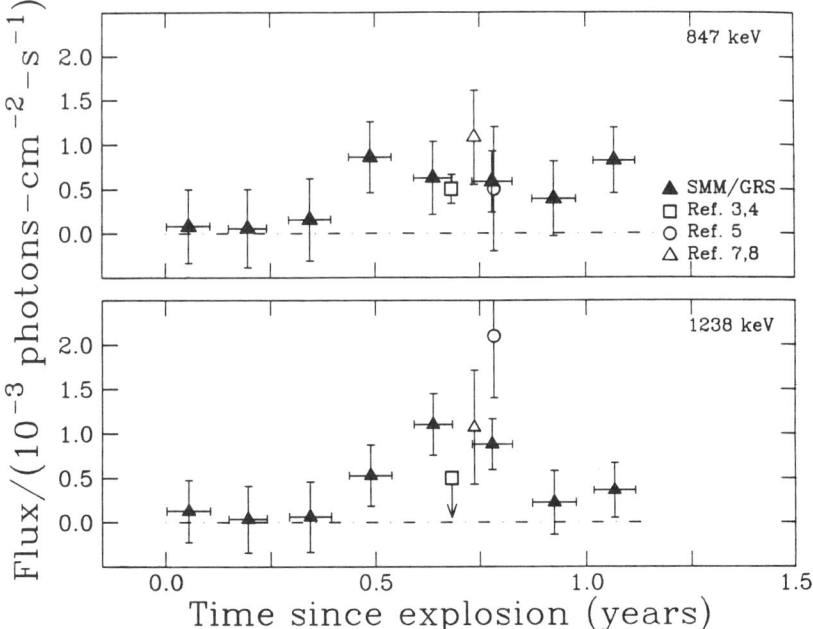

Fig. 2: Photon fluxes derived from the fitted intensities in Figure 1, for accumulations since the supernova explosion. Also shown are line fluxes from the balloon observations of Sandie et al.[3,4] and Mahoney et al.,[5] and the integrated continuum plus line fluxes of Cook et al.[7,8] The 1238 keV point of Sandie et al. represents their quoted 3σ upper limit for a narrow line.

$\sim (6.6 \pm 2.0) \times 10^{-4}$ and $\sim (6.4 \pm 1.7) \times 10^{-4}$ photons$-$cm$^{-2}-$s^{-1} at 847 and 1238 keV, respectively. Allowing the mean of the peaks to vary gives best-fit line energies of 840 ± 6 and 1239 ± 10 keV, consistent with the laboratory values. The best-fit line widths are consistent with the instrument resolution; the 2σ upper limit on the FWHM of the 847 kev line is ~ 65 keV during this period.

Also shown in Fig. 2 are results from three recent balloon flights. The Ge detector flown by Sandie et al.[3,4] measured a narrow line flux at ~ 847 keV consistent with that observed by GRS. They report a 3σ upper limit for a narrow line at 1238 keV which is below the GRS value; however, the values may be consistent if the line is broad. Mahoney et al.[5] flew a Ge detector on 1987 December 6; they detected a slightly broadened line near 1238 keV at 3σ significance, and insignificant flux ($< 1\sigma$) in the 847 keV line. Their 1238 keV line flux is marginally ($< 2\sigma$) higher than the average flux measured by

TABLE 1

GRS gamma-ray line fluxes for SN 1987A
in units of $10^{-4} \gamma - cm^{-2} - s^{-1}$

		847 keV	1238 keV
1987	Feb 23 – Apr 2	0.8 ± 4.2	1.3 ± 3.5
	Apr 18 – May 21	0.6 ± 4.4	0.4 ± 3.8
	Jun 10 – Jul 16	1.6 ± 4.6	0.6 ± 4.0
	Aug 1 – Sep 7	8.6 ± 4.0	5.3 ± 3.5
	Sep 27 – Oct 30	6.2 ± 4.1	11.0 ± 3.5
	Nov 16 – Dec 22	5.8 ± 3.4	8.8 ± 2.9
1988	Jan 8 – Feb 14	3.9 ± 4.2	2.2 ± 3.6
	Mar 1 – Apr 6	8.2 ± 3.7	3.6 ± 3.1

GRS from 1987 November 16 to December 22. In 1988 January, Rester et al. flew a Ge detector in the Antarctic and observed positive fluxes in both the 847 and 1238 keV line regions.[6] However, their reported line fluxes were only preliminary, so we make no direct comparison to the GRS results. Cook et al.[7,8] did not fit lines, but reported total fluxes (continuum plus line) in broad energy bands around 847 and 1238 keV. As seen in Fig. 2 these integrated fluxes are in good agreement with the GRS fitted line fluxes.

The relatively large errors in the GRS fluxes prevent us from placing strong constraints on the time evolution of the line intensities. The time histories can be adequately fit with a large variety of models. For each line, the measured fluxes since August 1 are consistent with a constant flux; therefore, the apparent variations are not significant.

The measured line fluxes can be compared with the maximum fluxes which could have been observed during the observation intervals, taking the ^{56}Co decay into account and assuming no attenuation of gamma rays in the source. We have assumed that 0.07 $M_\odot$ of ^{56}Ni was produced in the explosion, and that the distance to SN 1987A is 50 kpc. The results are shown in Figure 3. Though the line fluxes are not rapidly rising, Fig. 3 shows an indication (not conclusive) of a general thinning of the envelope overlying the ^{56}Co, with about 4% of the gamma rays at these energies escaping in our last observation interval. The energy contained in the measured gamma-ray lines is consistent

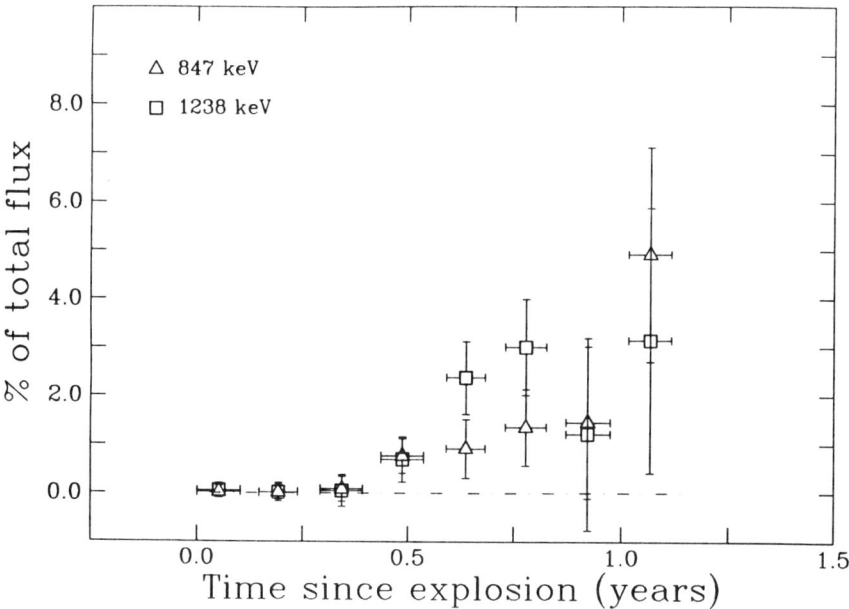

Fig. 3: The observed gamma-ray line fluxes as a fraction of the flux which would have been observed assuming no attenuation in the source.

with the amount of decay energy escaping from the supernova, based on the light curve.[16]

In Ref. 2 we showed a comparison of spectra taken from 1987 August 1 to October 31 with a spectrum taken during a time of similar background in 1985. In order to show the gamma-ray line profiles more clearly, we have plotted in Fig. 4 the sum of the supernova spectra from 1987 August 1 to 1988 April 6, minus spectra accumulated under similar background conditions in 1985 and 1986. This subtraction reduces the residual background lines and continua. Recall that there are no significant residual background features at 847 or 1238 keV, so this difference will not generate features at these energies. Both gamma-ray lines show up clearly as peaks in this spectrum, and there are no other obvious features in this energy range. For comparison, we have also plotted gaussian peaks at 847 and 1238 keV added to a power-law continuum. The widths of the lines are equal to the instrument resolution at those energies. The continuum seen in the spectrum may be at least partially due to residual atmospheric background; it is not currently possible to separate the supernova continuum from the local background.

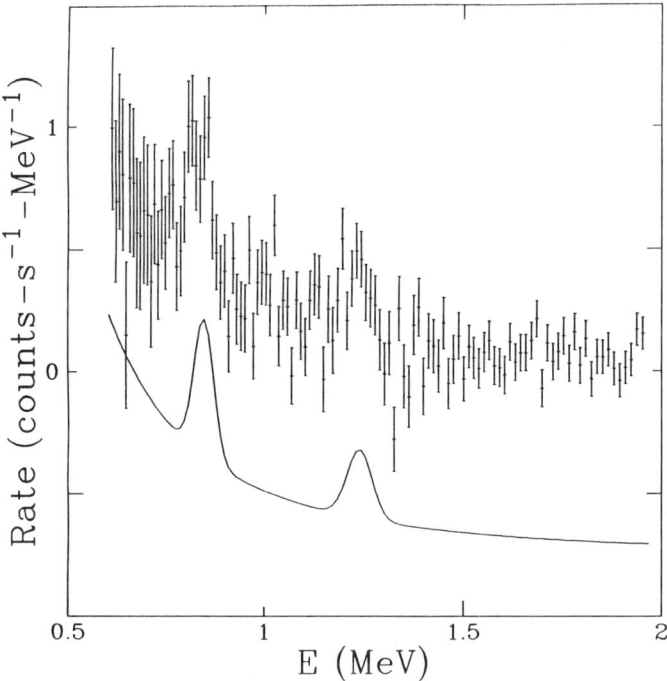

Fig. 4: The background-subtracted spectrum for SN 1987A, accumulated from 1987 August 1 to 1988 April 6, minus the sum of corresponding spectra from 1985 and 1986. The residual continuum is at least partially atmospheric. The solid line illustrates the expected GRS response to gamma-ray lines at 847 and 1238 keV, added to a power-law continuum.

CONCLUSIONS

The GRS has detected gamma-ray line emission from the direction of SN 1987A from the decay of ^{56}Co. The presence of lines has been confirmed by balloon-borne experiments. Fluxes are roughly at predicted levels, although both the X rays[25,26] and gamma rays appeared earlier than expected. Possible explanations for this include mixing of the radioactive ^{56}Co closer to the surface or departures from spherical symmetry in the ejecta (blobs, jets, etc.). Further observations by the GRS and balloon experiments may help establish the distribution of ^{56}Co in the ejecta.

ACKNOWLEDGEMENTS

We are indebted to Mark Leising for assistance in analysis and helpful discussions. We would also like to thank Kent Wood, Tom Vestrand, and Jim Kurfess for important contributions to this work. We thank Dan Messina for his direction of the production analysis. This work was supported by NASA contract S-14513-D at NRL and grant NAG 5-720 at UNH.

REFERENCES

1. Matz, S. M., et al. *IAU Circ.* 4510 (1987).
2. Matz, S. M., et al. *Nature* **331**, 416 (1988).
3. Sandie, W., et al. *IAU Circ.* 4526 (1988).
4. Sandie, W., et al., in these proceedings.
5. Mahoney, W.A., et al. *IAU Circ.* 4584 (1988).
6. Rester, A. C., et al. *IAU Circ.* 4535 (1988).
7. Cook, W. R., et al. *IAU Circ.* 4527 (1988).
8. Cook, W. R., et al., in these proceedings.
9. Kunkel, W. and Madore, B. *IAU Circ.* 4316 (1987).
10. Hirata, K., et al. *Phys. Rev. Lett.* **58**, 1490 (1987).
11. Bionta, R.M., et al. *Phys. Rev. Lett.* **58**, 1494 (1987).
12. Truran, J.W., Arnett, W.D., and Cameron, A.G.W. *Canadian J. Phys.* **45**, 2315 (1967).
13. Bodansky, D., Clayton, D.D., and Fowler, W.A. *Ap. J. Suppl.*, **16**, 299 (1968).
14. Colgate, S.E., and McKee, C. *Ap. J.* **157**, 623–643 (1969).
15. Catchpole, R. M., et al. *Mon. Not. R. A. S.* **231**, 75P (1988).
16. Whitelock, P.A., et al. *Mon. Not. R. A. S.* (submitted).
17. Woosley, S.E., Pinto, P.A., and Ensman, L. *Ap. J.* **324**, 466 (1988).
18. Clayton, D. D., Colgate, S. A., and Fishman, G. J. *Ap. J.* **155**, 75 (1969).
19. McCray, R., Shull, J.M., and Sutherland, P. *Ap. J. (Letters)* **317**, L73 (1987).
20. Chan, K.W. and Lingenfelter, R.E. *Ap. J. (Letters)* **318**, L51 (1987).
21. Gehrels, N., McCallum, C.J., and Leventhal, M. *Ap. J. (Letters)* **320**, L19 (1987).
22. Shibazaki, N. and Ebisuzaki, T. *Ap. J. (Letters)* **327**, L9 (1988).
23. Forrest, D. J., et al. *Solar Phys.* **65**, 15 (1980).
24. Leising, M. D. *Nature* **332**, 516 (1988).
25. Sunyaev, R., et al. *Nature* **330**, 227 (1987).
26. Dotani, T., et al. *Nature* **330**, 230 (1987).

IMAGING OBSERVATIONS OF SN1987A AT GAMMA-RAY ENERGIES

W. R. Cook, D. M. Palmer, T. A. Prince,
S. M. Schindler, C. H. Starr & E. C. Stone
California Institute of Technology, Pasadena, California 91125 USA

ABSTRACT

The Caltech imaging γ-ray telescope was launched by balloon from Alice Springs, NT, Australia for observations of SN1987A during the period 18.60-18.87 November 1987 UT. The preliminary results presented here are derived from 8200 seconds of instrument livetime on the supernova and 2500 seconds on the Crab Nebula and pulsar at a float altitude of 37 km. We have obtained the first images of the SN1987A region at γ-ray energies confirming that the bulk of the γ-ray emission comes from the supernova and not from LMC X-1. A count excess is detected between 300 and 1300 keV from the direction of the supernova, one third of which comes from energy bands of width 80 and 92 keV centered on 847 and 1238 keV, respectively. The excess can be interpreted as a line photon flux plus scattered photon continuum from the radioactive decay of ^{56}Co synthesized in the supernova explosion. We compare our data to recent predictions and find it to be consistent with models invoking moderate mixing of core material into the envelope.

INTRODUCTION

The observations reported here were performed with the Caltech Gamma-Ray Imaging Payload (GRIP), a balloon-borne coded-aperture telescope sensitive to radiation in the energy range from 30 keV to 10 MeV[1]. The instrument in its current configuration employs a rotating hexagonal-celled uniformly redundant array (URA) and a 5 cm thick by 41 cm diameter position sensitive NaI scintillation detector to image a 14° diameter field of view with 1.1° angular resolution. The mask cell size in the November flight was 4.8 cm and the separation of the mask and detector was 2.5 m. The instrument has flown twice previously[2] including a flight in May 1987 from Australia to observe SN1987A. No γ-ray or hard X-ray emission was detected in this initial observation of SN1987A and preliminary upper limits have been reported elsewhere[3]

IMAGES

Images of the Crab and SN1987A regions from the flight on 18 November 1987 are shown in Fig. 1 and 2. The Crab region image is the sum of 4 different observation periods. The source peak is clearly evident, as is the ring structure which results from use of a rotating mask having multiple repetitions of the basic URA pattern[4]. The ring can be removed in later analysis. The locations of the image peaks (each greater than 8σ) from individual Crab observations varied over a range of ±0.25° in the azimuthal direction and ±0.1° in the elevation direction. We attribute the larger azimuthal variations to incomplete correction for deviations between the actual and the assumed magnetic field direction used as a pointing reference during flight. The image of Fig. 1 was obtained by aligning the centroids of the peaks from the individual Crab observations.

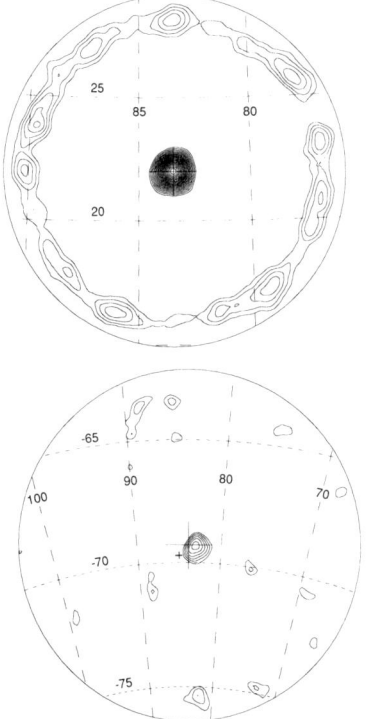

Fig. 1 Image of the Crab region from 30 to 2700 keV covering a 14° field of view. Right ascension (vertical lines) and declination (horizontal lines) are indicated. The contours indicate the number of excess counts in a given direction calibrated in units of the statistical significance of the excess, with contours beginning at the 2σ level and spaced by 1σ intervals.

Fig. 2 Image of the SN1987A region from 40 to 1300 keV. Parameters as in Fig. 1 except for the contour spacing which is 0.5σ. The large cross indicates the expected position of SN1987A. The small cross indicates the expected position of LMC X-1.

Figure 2 shows an image of the supernova region for energies between 40 and 1300 keV. The image is a composite of data from 4 individual observation periods with the individual fields aligned by using the pointing corrections determined from the Crab observations. Flux is detected at a nominal 4.9σ level from within 0.28° of the expected position of SN1987A. The 1σ statistical error in the peak location is estimated to be 0.16° in both azimuth and elevation directions. The residual deviation is in the azimuthal direction and may be due to incomplete magnetic field offset correction. In the elevation direction, which is approximately aligned with the declination axis in these observations, the peak centroid is consistent with the position of the supernova. This indicates that the observed flux above 40 keV does not have a strong contribution from LMC X-1 which is 0.47° away in declination.

SPECTRA

Measurements of the γ-ray flux from SN1987A for selected energy intervals are listed in Table I. The differential flux measurements were derived from the number of excess counts in the image at the position of the supernova determined from the data of Fig. 2. Because many of the factors that determine photon detection efficiency are energy dependent, a calculation of the expected excess counts was made by integrating a power-law differential continuum flux, $F = K(E/100 keV)^{-\alpha}$, folded with the estimated photon detection efficiency. The predicted and measured image counts were compared, iteratively adjusting the factors K and α in the power law until a consistent result was obtained. This analysis for the Crab yielded a flux of 4.4×10^{-4} $(E/100 \text{ keV})^{-2}$ $(\text{cm}^2 \text{ s keV})^{-1}$ between 40 and 500 keV.

		Table I	
Energy Interval (keV)	Weighted Mean Energy (keV)	SN1987A Flux Measurement	Statistical Significance
		Differential Flux $(\text{cm}^2 \text{ s keV})^{-1}$	
40-303	152	$1.95 \pm 0.48 \times 10^{-5}$	4.1σ
303-1286	625	$0.56 \pm 0.22 \times 10^{-5}$	2.5σ
1608-3752	2471	$< 3.71 \times 10^{-6}$	3.0σ
		Integral Flux $(\text{cm}^2 \text{ s})^{-1}$	
807-887	847	$1.08 \pm 0.53 \times 10^{-3}$	2.0σ
1192-1284	1238	$1.07 \pm 0.64 \times 10^{-3}$	1.7σ
807-887 + 1192-1284	-	$2.15 \pm 0.83 \times 10^{-3}$	2.6σ

The first two differential flux estimates in the table were obtained using a power law spectral index, $\alpha = -0.9$, and flux normalization, $K = 2.8 \times 10^{-5}$ $(\text{cm}^2 \text{ s keV})^{-1}$. The second energy interval was chosen to include possible 1238 keV emission, and the gap between 1286 and 1608 keV was chosen to eliminate the strong 1460 keV background line contribution.

It is of interest to estimate the flux near the energies of the prominent γ-ray lines expected from ^{56}Co decay. Table I gives integral flux values for intervals centered on 847 keV and 1238 keV and a sum for the combined intervals. The widths of the intervals are chosen to be approximately 1.2 times the FWHM instrumental resolution. Because the energy dependence of the detection efficiency is not strong in these narrow intervals, no spectral index assumption was required.

We stress that the results reported here are of a preliminary nature. The present analysis incorporates only the preliminary calibration data for the new coded-aperture mask and low-energy collimator used in the 18 November flight. Further, we have recently obtained more detailed magnetic field data which will be used in later analyses and which may remove the pointing uncertainties mentioned above. We estimate the possible systematic errors in the absolute flux measurements to be less than 50 percent. Systematic errors in the relative flux measurements should be smaller and generally negligible compared to the statistical errors in Table I. Most conceivable systematic errors, in particular pointing uncertainties, will tend to yield an absolute measured flux lower than the true flux. We note that the estimate of the Crab spectrum given above is ~20-40% below other measurements at 100 keV[5] and may indicate an overall systematic underestimate of the flux.

DISCUSSION

The results given in Table I should be considered in the context of other measurements of the hard X-ray and γ-ray flux made during the period of August through November 1987. These include results using the Mir-Kvant observatory[6], the Ginga satellite [7], and the Solar Maximum Mission (SMM) [8]. Our continuum flux measurement on 18 November at 152 keV is consistent with the flux of approximately 1.5×10^{-5} (cm^2 s keV)$^{-1}$ at 150 keV measured by the Mir-Kvant observatory during the period 10-21 August 1987, although a factor of two increase in flux is marginally acceptable, differing by approximately 2σ from the measured value. Our finite flux detection in the 303 to 1286 keV range lies just above the 3σ upper limit reported from the Mir-Kvant Pulsar X-1 instrument for the 10-21 August period, indicating a probable increase in flux at γ-ray energies near 1 MeV. While our integral fluxes measured in the region of 847 keV and 1238 keV should be interpreted as continuum plus line emission, the values given in Table I are consistent with the γ-ray line fluxes reported by the SMM instrument for the August through October period: $1.0 \pm 0.25 \times 10^{-3}$ cm^{-2} s^{-1} at 847 keV and $6 \pm 2 \times 10^{-4}$ cm^{-2} s^{-1} at 1238 keV. Our measurements should also be compared to the results obtained from three other balloon-borne instruments. These include the sensitive hard X-ray measurements of the NASA Marshall Space Flight Center group on 29-30 October [9] which cover the energy range 45 to 200 keV, and the high-resolution γ-ray line measurements of the Lockheed [10] and NASA Jet Propulsion Laboratory[11] groups on 29-30 October and 6 December respectively.

A likely origin of the observed hard X-ray and γ-ray flux is the decay of ^{56}Co which is a product of the decay of ^{56}Ni synthesized in the supernova explosion [12-14]. The ^{56}Co decays are expected to yield both line emission (in particular at 847 and 1238 keV) and a continuum of Compton degraded photons. Approximately 0.075 solar mass of ^{56}Ni was synthesized in the initial explosion as is well determined from the late time decay of the bolometric luminosity of the supernova [15]. The results given in Table I are compared with recent theoretical predictions in Fig. 3. The continuous histogram is based on a Monte Carlo calculation of the expected photon flux on day 250, taken from Pinto and Woosley[16]. Similar calculations have been carried out by other authors [17-21]; the calculation in Fig. 3 is a recent example of models which incorporate mixing to explain the early turn-on of X-rays and γ-rays seen by the Mir-Kvant, Ginga, and SMM instruments. The model assumes a 6 solar mass He core, a 10 solar mass hydrogen envelope, 1.3×10^{51} erg explosion kinetic energy, and a moderate degree of mixing of ^{56}Co out into the helium envelope.

In general, our differential flux measurements shown in Fig. 3 are consistent with the calculations. To compare the measured and calculated values more directly in the 40 to 303 keV interval where energy dependent efficiency factors change rapidly, we have folded the Monte Carlo spectrum in Fig. 3 with the estimated photon detection efficiency. We find that the Monte Carlo calculation predicts an excess of 3480 photons in the image compared to the measured value of 2251±559. This is a nominal 2.2σ discrepancy, but given the range of possible systematic errors discussed above, we do not view the difference as serious at this time. A 30% underestimate of the absolute flux would reduce the difference to less than 1σ.

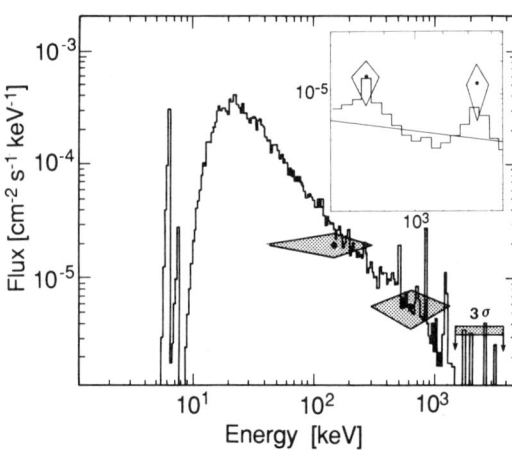

Fig. 3 Flux measurements and calculations for SN1987A. The continuous histogram is a Monte Carlo calculation of the expected flux on day 250 of the supernova from Pinto and Woosley[16] convolved with a gaussian representation of the instrumental energy resolution. The measurements are indicated by diamonds with centers plotted at the weighted mean energy and the width indicating the energy interval of the measurement. The inset shows an expanded view of the 750 to 1350 keV interval. The straight line shown in the inset indicates a power law continuum flux with spectral index $\alpha = -0.9$

The insert in Fig. 3 is an expanded view of the energy region between 750 and 1350 keV and includes our integral flux measurements for the 847 and 1238 keV regions plotted as differential fluxes by dividing by the width of the energy intervals. Our results are consistent with the Monte Carlo calculations which have been convolved with the instrumental energy resolution. In particular, when taken together with our continuum flux estimate at 625 keV, the results are consistent with a flux excess in the region of the expected γ-ray lines from ^{56}Co decay. The chance probability for obtaining excesses as large as those observed at 847 and 1238 keV is 1% assuming a power law continuum flux with spectral index $\alpha = -0.9$ and using a flux normalization (with errors) determined from the measured flux of $0.56\pm0.22 \times 10^{-5}$ $(cm^2 \text{ s keV})^{-1}$ at 625 keV. An excess in the region of the 847 and 1238 keV lines will include both line photons and a contribution from forward scattered Compton photons. To estimate the unscattered line photon intensity it is thus necessary to subtract off the forward scattered contribution assuming a specific model for the depth distribution of ^{56}Co in the supernova remnant.

ACKNOWLEDGEMENTS

We gratefully acknowledge the important contributions of W. Althouse, D. Burke, A. Cummings, and M. Finger to the development of the GRIP telescope. We thank the personnel of the National Scientific Balloon Facility and the NASA Wallops Flight Facility for their excellent balloon launch support. This work was supported by NASA grant NGR 05-002-160.

References

1. Althouse, W. E., Cook, W. R., Cummings, A. C., Finger, M. H., Prince, T. A., Schindler, S. M., Starr, C. H., and Stone, E. C., Proc. 19th Int. Cosmic Ray Conf., La Jolla, CA, 3, 299 (1985).

2. Althouse, W. E., Cook, W. R., Cummings, A. C., Finger, M. H., Palmer, D. M., Prince, T. A., Schindler, S. M., Starr, C. H., and Stone, E. C., Proc. 20th Int. Cosmic Ray Conf., Moscow, USSR, 1, 84 (1987).

3. Cook, W., Palmer, D., Prince, T., Schindler, S., and Stone, E., IAU Circ. 4400 (1987).

4. Cook, W. R, Finger, M., Prince, T. A., and Stone, E. C., IEEE Trans. Nucl. Sci., Vol. NS-31, 771-775 (1984).

5. Jung, G. V., thesis, Univ. Calif. San Diego (1986).

6. Sunyaev, R. et al., Nature, 330, 227-229 (1987).

7. Dotani, T. et al., Nature, 330, 230-231 (1987).

8. Matz, S. M., Share, G. H., Leising, M. D., Chupp, E. L., Vestrand, W. T., Purcell, W. R., Strickman, M. S., and Reppin, C., Nature, 331, 416-418 (1988).

9. Wilson, R., Fishman, G., Meegan, C., Paciesas, W., and Pendleton, G., submitted to Ap. J. Lett. (1988).

10. Sandie, W. G., Nakano, G. H., Chase, L. F. Jr., Meegan, C. A., Wilson, R. B., Paciesas, W. S., and Lasche, G. P., submitted to Ap. J. Lett. (1988).

11. Mahoney, W. A., Varnell, L. S., Jacobson, A. S., Ling, J. D., Radocinski, R. G., and Wheaton, Wm. A., submitted to Ap. J. Lett. (1988).

12. Bodansky, D., Clayton, D. D., and Fowler, W. A., Ap. J. Suppl., 16, 299-371 (1968).

13. Clayton, D. D., Colgate, S. A. and Fishman, G. J., Ap. J., 155, 75-82 (1969).

14. Colgate, S. A. and McKee, C., Ap. J., 157, 623-643 (1969).

15. Woosley, S., Ap. J., in press (1988).

16. Pinto, P. A. and Woosley, S. E., Nature, in press (1988).

17. Itoh, M., Kumagai, S., Shigeyama, T., Nomoto, K., and Nishimura, J., Nature, 330, 233-235 (1987).

18. McCray, R., Shull, J. M., and Sutherland, P., Ap. J., L73 (1987).

19. Ebisuzaki, T. and Shibazaki, N., Ap. J. Lett., in press (1988).

20. Shibazaki, N. and Ebisuzaki, T., Ap. J. Lett., in press (1988).

21. Xu, Y., Sutherland, P., McCray, R., and Ross, R. R., submitted to Ap. J. Lett. (1988).

HIGH-RESOLUTION OBSERVATIONS OF GAMMA-RAY LINE EMISSION FROM SN 1987A

W. G. Sandie, G. H. Nakano, and L. F. Chase, Jr.
Lockheed Palo Alto Research Laboratory
3251 Hanover St., Palo Alto, Ca. 94304

G. J. Fishman, C. A. Meegan, and R. B. Wilson
High Energy Astrophysics Laboratory
NASA Marshall Space Flight Center, Huntsville, Al. 35812

W. S. Paciesas
University of Alabama in Huntsville
Huntsville, Al. 35899

G. P. Lashe
DARPA, Arlington, Va. 22209

ABSTRACT

A balloon-borne gamma-ray spectrometer comprising an array of high-purity germanium (HPGE) detectors was flown from Alice Springs, Australia, on October 29-30-31, nominally 250 days after the observed neutrino pulse. High-resolution data, typically 2.5 keV at 1.33 MeV, for two transists of the supernova SN 1987A were obtained along with interspersed background data. A significant net flux of gamma rays with energy 847 keV was observed from the direction of SN 1987A on each transit. No prominent gamma-ray features were seen at other energies, although data analysis is still in progress. A preliminary estimate of the line flux at 847 keV is $\sim 5\times 10^{-4}$ photons cm^{-2} s^{-1} with statistical significance greater than three sigma. This line may be interpreted as emission from the first excited state of ^{56}Fe due to the radioactive decay of ^{56}Co, providing strong evidence for nucleosynthesis in the supernova. No emission was seen from the second excited state of ^{56}Fe at 1238 keV. A preliminary upper limit for the 1238-keV line is $\sim 5\times 10^{-4}$ photons cm^{-2} s^{-1}. The flux estimates may be considerably altered as the systematics of the experiment are better understood.

INTRODUCTION

Gamma-ray line emissions provide the most valuable characterization of nucleosynthetic and inelastic processes occurring in supernova development. The supernova SN 1987A, identified with the blue supergiant progenitor, Sanduleak -69 202, has been extensively modeled by theoreticians.[1-10] For a progenitor with a mass of 19 solar masses, the models synthesize a "neutronized core" of about 1.4 solar masses. Only about 7.5×10^{-2} solar masses of the total amount of ^{56}Ni synthesized forms in a thin shell outside the core and is potentially visible. This thin shell is ejected with velocity ~ 2000 km s^{-1} and underlies the residual mass of hydrogen ejecta, which is moving radially outward at much greater speed, $\sim 17{,}000$ km s^{-1}. Initially the mass of ejecta concentrated about the core is opaque to the gamma rays produced by the radioactive decay of short-lived ^{56}Ni to ^{56}Co, half-life 6.1 days. Comptonized gamma radiation from the decay of longer-lived ^{56}Co is apparently the primary energy source currently driving the light curve, since the e-folding decay time of the bolometric luminosity, 114 days, corresponds precisely to that of ^{56}Co.

As the mass of ejecta expands and becomes more tenuous, the opacity decreases, ultimately revealing gamma-ray line emission of 847 keV and 1238 keV from the decay ^{56}Co to the first and second excited states of ^{56}Fe in the underlying shell. The rise time of the gamma-ray line flux depends upon the amount of ^{56}Ni synthesized in the explosion, which decayed into ^{56}Co, the exponential decay of ^{56}Co, the opacity of the ejecta, which is diminishing with time, and the amount of turbulent mixing of ^{56}Co in the ejecta. Models predict peak intensities for the 847-keV line of 0.002 to 0.008 photons cm^{-2} s^{-1} occurring 200 to 400 days after flare-up.

Detection of gamma-ray line emission from the nickel-cobalt-iron chain is evidence for the process of nucleosynthesis in SN 1987A. The time-dependent intensity and spectral shape of the gamma-ray line emission observed constrains the amount of ^{56}Co produced and its turbulent distribution in the dynamically expanding envelope. The measured flux intensity 250 days after flare-up bodes well for high-resolution balloon-borne observations utilizing a large array of germanium detectors.

SPECTROMETER

The observations reported here utilized the same gamma-ray spectrometer flown in May 1987 and described previously.[11] The instrument comprises a three-by three array of nine HPGE detectors, cooled to cryogenic temperature using a cooler with capacity 15 liters of liquid nitrogen. Each germanium detector is a coaxial diode nominally 5.5 cm in diameter by 5.5 cm in length. The HPGE detectors have combined total area of 214 cm^2. The resolution of each detector is nominally better than 2.5 keV at 1.33 MeV. The spectrometer has a useful efficiency over the energy range 25 to 8000 keV. To reduce background, the detector array is surrounded by a live NaI(Tl) antishield with a nominal thickness of 10 cm. The live antishield serves to reject out-of-aperture gamma rays and in-aperture Compton-scattered events. Each detector has a 22° full-angle field of view defined by collimator holes in the upper NaI(Tl) shield. In-aperture charged particles are rejected by a 4-mm-thick plastic anticoincidence shield. The NaI(Tl) antishield is viewed by 16 3-in.-diameter photomultiplier tubes and the plastic scintillator is viewed by 4 1-in.-diameter photomultiplier tubes. Data are recorded in list mode, 13 bits for each HPGE pulse analyzed. Two eight-bit addresses for antishield pulse heights are allocated for events in anticoincidence with the HPGE pulse. Data are formatted into a 128-kbs Miller-DMM serial data stream and transmitted to the ground.

OBSERVATIONS

A balloon-borne payload containing the spectrometer was launched from Alice Springs, Northern Territory, Australia, at 09:20:37 UT on October 29, 1987 by the field crew of the National Scientific Balloon Facility. The payload reached a nominal float altitude of 130,000 ft. Observations were made on the first transit of the SN 1987A by alternately pointing the spectrometer at the supernova for ~40 min. and then slewing away for background determination for ∼20 min. The average slant column depth of residual atmosphere in the direction of the supernova was ∼5 g cm^{-2}. Data from the first transit in the region around 847 keV are shown in the three panels of Figure 1. An excess of gamma rays at 847 keV appears initially as a shoulder to an instrumental background line from ^{27}Al (844 keV) obtained with the spectrometer pointed at the supernova. The excess does not appear in data obtained while pointed away from the supernova.

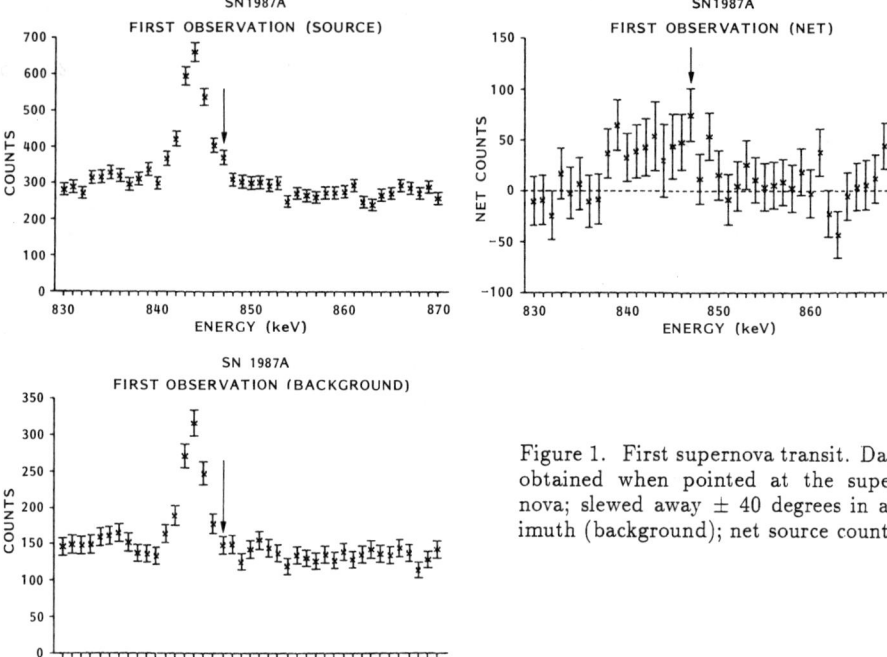

Figure 1. First supernova transit. Data obtained when pointed at the supernova; slewed away ± 40 degrees in azimuth (background); net source counts.

Similarly, data taken during the second transit of the supernova are shown in the three panels of Figure 2. The excess again appears in the data obtained while pointed at the supernova and does not appear in the background data. The fact that the excess appears in both transists is compelling. The composite data for both transits are shown in Figure 3. Subtraction of the background to obtain net source counts is done by normalizing the spectra to the same total counts in energy regions above and below the region of interest. For comparison, the same analysis was done on a neighboring background line at 911 keV due to ^{228}Th with the results shown in Figure 4. Again there is no evidence for a shoulder at high or low energy. Finally, data obtained in the energy region around 1238 keV are shown in Figure 5. No prominent feature appears at or near 1238 keV.

As a further demonstration of the absence of an instrumental line at 847 keV, data obtained at float altitude during flight N-219, May 29, 1987 (∼ 7-h of data), are shown in Figure 6. These data, mostly background, exhibit no excess flux in the vicinity of 847 keV.

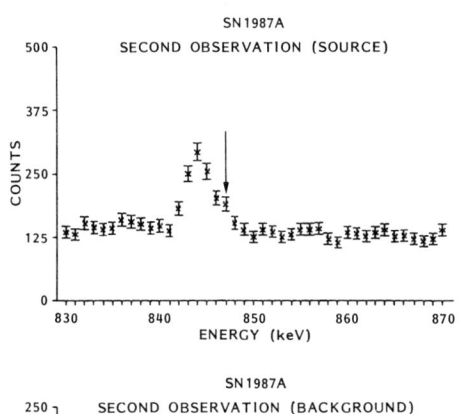
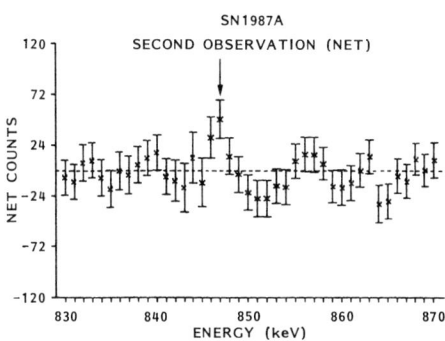
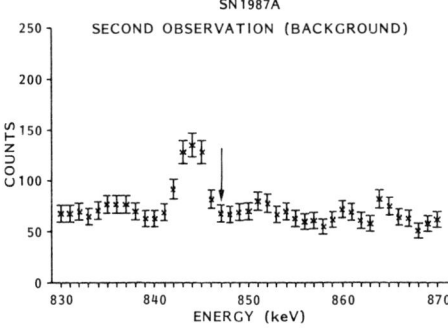

Figure 2. Second supernova transit. Data obtained when pointed at the supernova; slewed away ± 40 degrees in azimuth (background); net source counts.

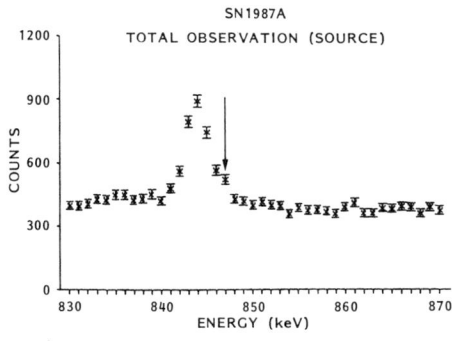
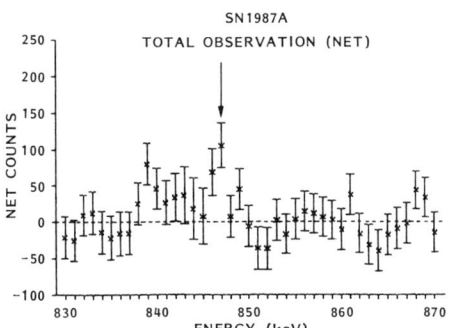
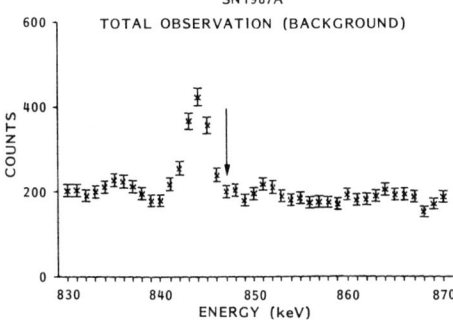

Figure 3. The two supernova transits summed together.

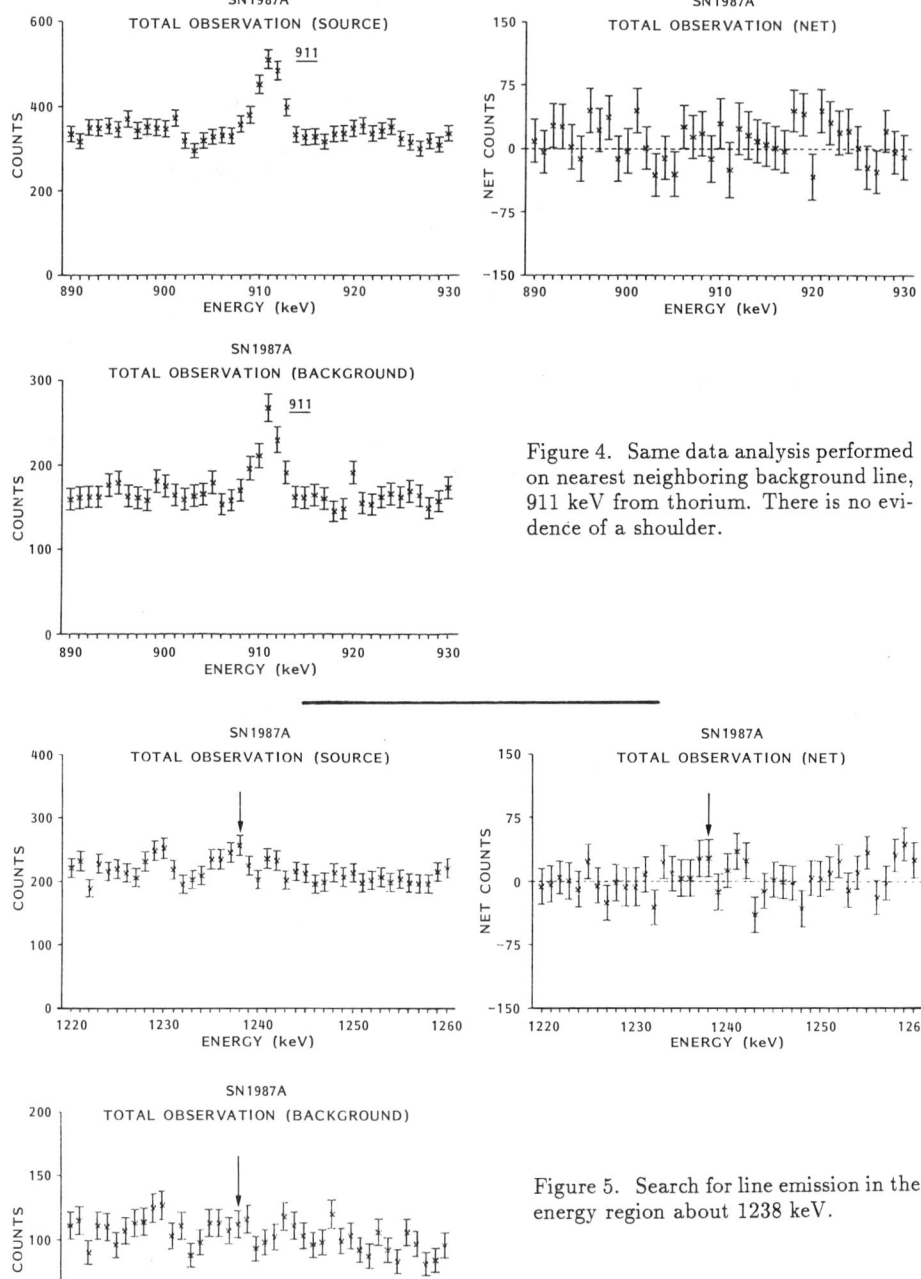

Figure 4. Same data analysis performed on nearest neighboring background line, 911 keV from thorium. There is no evidence of a shoulder.

Figure 5. Search for line emission in the energy region about 1238 keV.

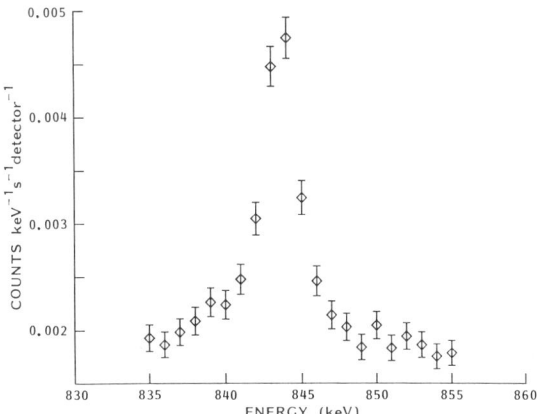

Figure 6. Aluminum line spectrum as measured in the previous flight, N-219, from Alice Springs, May 29, 1987. The spectrum contains all data, on and off the supernova, acquired during ~7 h at float altitude. As expected, there is no evidence of excess flux at 847 keV.

Pointing truth was verified by observations of the Crab during each supernova transit and by observing planets with an onboard TV camera. The preliminary flux estimate in the direction of the supernova is $\sim 5 \times 10^{-4}$ photons cm^{-2} s^{-1} for the spectral region 6-keV-wide surrounding the 847-keV line. A three-sigma upper bound on the flux from the 1238-keV line is also $\sim 5 \times 10^{-4}$ photons cm^{-2} s^{-1}. These data may be significantly altered when systematics are taken into account. In particular, sporadic noise in the germanium detector electronics, involving multiple detector events, which are unambiguously discarded, necessitate large dead-time corrections. Preliminary dead-time corrections have been applied in the flux determinations. One of the nine detectors exhibited excessive leakage current and was considered inoperative for the flight duration. The total geometric area of the instrument (eight operational detectors) was reduced to 190 cm^2.

CONCLUSIONS

A three-sigma net excess of gamma-ray line emission at 847 keV from the direction of the supernova was observed with a high-resolution germanium spectrometer. A preliminary estimate of the line flux is $\sim 5 \times 10^{-4}$ photons cm^{-2} s^{-1}. The 847-keV line emission is expected from the decay of ^{56}Co to the first excited state of ^{56}Fe, which may be taken to confirm the process of nucleosynthesis in the supernova. No other prominent spectral features were observed, although the data are still being analyzed; notably the line emission at 1238 keV was not prominent and a preliminary three-sigma upper limit on the 1238-keV flux from the supernova is $\sim 5 \times 10^{-4}$ photons cm^{-2} s^{-1}. The upper bound on the 1238-keV line is not inconsistent with the 847-keV flux arising from ^{56}Co decay. The absence of the predicted blue shift of $\sim$3-keV for the 847-keV line may be of significance, suggesting that the material may be emitting in a direction perpendicular to the line of sight. The 847-kev line could be enhanced through inelastic scattering of neutrons exciting the 847-keV line in ^{56}Fe, however this effect is constrained by the light curve which is clearly dominated by ^{56}Co decay.

The line intensities are consistent with the results obtained at lower resolution with a scintillation detector reported by SMM investigators[12].

ACKNOWLEDGMENTS

The balloon-borne experiment is a considerable technical feat. We are deeply indebted to those who have tirelessly performed these complex engineering tasks: at LPARL; Francis Pang, Robert Fujimoto, and Bruce Imai (Electronics), Joseph Kilner and Stephen Geller (Software), Gregg Bell (Thermal), Donald Isaac and Jon Hamilton (Mechanical), Harry Mann (Detectors) at MSFC; Robert Austin, Fred Berry, and W. Thomas Sutherland (Electronics), Stan Dothard, Max Love, William Hammon, and Joe Ozbolt (Mechanical), M. Polites (Pointing System), and Martin Brock (Software). We are also grateful for the support of R. Kubara and crew of the National Scientific Balloon Facility.

REFERENCES

1. S. E. Woosley, submitted to Astrophys. J., 5 October 1987.
2. P. A. Pinto, and S. E. Woosley, submitted to Astrophys. J., 8 October (1987).
3. W. D. Arnett, (1987). Astrophys. J., **319**, 136.
4. S. Kumagai, M. Itoh, T. Shigeyama, K. Nomoto and J. Nishimura, (1987). Proceedings of the GMU Workshop. To be published.
5. K. W. Chan and R. E. Lingenfelter, (1987). Astrophys. J. (Letters), **318**, L51.
6. Y. Xu, P. Sutherland, and R. McCray, (1987). Submitted to Astrophys. J. (Letters), July 1987.
7. N. Gehrels, C. J. MacCallum, and M. Leventhal, (1987). Astrophys. J. **322**, 215.
8. T. Ebisuzaki, and N. Shibazaki 1987, (private communication).
9. R. McCray, J. M. Shull and P. Sutherland, Astrophys. J. (Letters), **317**, L73, (1987).
10. T. Shigeyama, K. Nomoto, M. Hashimoto, and D. Sugimoto, Nature, **328**, 320 (1987).
11. W. Sandie, G. H. Nakano, L. F. Chase, Jr., G. J. Fishman, C. A. Meegan, R. B. Wilson, W. Paciesas, and G. P. Lasche, (1987) Proceedings of the GMU Workshop. To be published.
12. S. M. Matz, G. H. Share, M. D. Leising, E. L. Chupp, and W. T. Vestrand, (1987) IAU Circular No. 4510.

HARD X-RAY OBSERVATIONS OF SN1987A

R. B. Wilson, G. J. Fishman, and C. A. Meegan
Space Science Laboratory, NASA Marshall Space Flight Center
Huntsville, Alabama 35812

W. S. Paciesas and G. N. Pendleton
Department of Physics, The University of Alabama in Huntsville
Huntsville, Alabama 35899

ABSTRACT

We have detected continuum emission of hard x-rays from a source in the LMC region which includes SN1987A. The detection reported here, if interpreted as coming entirely from SN1987A, indicates that during the interval from day 180 to day 249 the source increased significantly in intensity and its spectrum hardened. This is qualitatively consistent with theoretical models.

INTRODUCTION

Observations of SN1987A and the LMC region in hard x-rays and low-energy gamma-rays were performed on October 29-30, 1987, during a flight of a balloon-borne payload launched from Alice Springs, Australia. The payload included two types of detectors: an actively-shielded array of germanium detectors designed to perform high-resolution measurements of gamma-ray line emission, and two large-area, passively-shielded scintillation detectors designed to perform sensitive observations of the hard x-ray continuum and to search for possible emission from a remnant pulsar in the supernova. We report herein on continuum spectral observations obtained with the latter detectors. Results from the germanium detectors are reported in a companion paper.[1]

X-ray emission from SN1987A was first detected around day 180.[2,3] The subsequent temporal evolution of the emission has been inconsistent with most early theoretical expectations, but more recent calculations[4-6] that invoke substantial mixing of ^{56}Ni in the expanding gas cloud have been more successful in this regard. We compare our results with the model calculations for the epoch of our observation.

INSTRUMENTATION

The scintillation detectors are adaptations of a detector designed for the Burst and Transient Source Experiment[7,8] to fly on the Gamma Ray Observatory. Each consists of a 50.8-cm-diameter, 1.27-cm-thick NaI(Tℓ) scintillator, viewed by three 5-inch photomultiplier tubes mounted on a truncated cone, which forms a diffuse reflecting light collector. The cone is shielded from behind by a layer of lead of thickness 0.86 g cm^{-2}. The energy resolution of these detectors is typically 30% at 88 keV.

The detectors were mounted at a fixed elevation of 45° on the balloon gondola. Azimuth was controlled by a pointing system using a high sensitivity nulling magnetometer offset from the pointing direction by a stepper-motor drive. The field of view of the detectors was defined by one-dimensional passive slat collimators placed in front of each detector, aligned to restrict the field of view only in azimuth. The slats were constructed of 0.76 mm lead strips, with 0.51 mm layers of tin on each side, to reduce lead fluorescence emission. For one detector the slats were uniformly spaced, resulting in a triangular response with a FWHM of 10.3°. For the other, alternate slats were removed, so that half the detector area has a 10.3° FWHM response, while the other half has an 18.4° FWHM response. The detector response parallel to the slats is approximately a cosine function, typical of detectors with a large area compared to their thickness. Atmospheric attenuation resulted in an effective elevation cutoff at about 25°. Charged particle veto counters were not used.

The data telemetered from the instrument consist of 128-channel spectra, read out separately from each detector every 5.5 s. Energy bins are nominally 2.5 keV per channel from 18 to 160 keV, 5 keV per channel from 160 to 320 keV, and 20 keV per channel from 320 keV to 960 keV. Data binned into eight larger energy channels having 2.1 ms time resolution were also transmitted, permitting a search for a rapid pulsar. These data have not been used in the present analysis.

OBSERVATIONS

The experiment was launched on October 29, 1987, reaching a float altitude of 4 mb at 1330 UT. The flight was terminated at 0330 UT on October 31, after 38 hours at float altitude. Observations of the region of the LMC containing SN1987A were performed on October 29-30 during a period of ~10 hours each day. The gondola was typically pointed in the direction of SN1987A for 40 minutes at a time, with azimuth updated every 20 minutes. Background data were obtained offset by 40° in azimuth, alternating in direction in successive intervals. They were each about 20 minutes in duration.

Altitude was stable at 3.0 ± 0.15 mb from 1400 to 2330 UT, during the first transit of the source. During the second observation, the altitude changed from 3.8 to 2.6 mb in the 2 hours after meridian passage of the LMC. The trajectory of the balloon covered 1.7° in latitude during the second observation, compared to 1.0° during the first. As the background counting rate is observed to vary significantly with these parameters, analysis has been restricted to the first observation of the source region for the results presented here.

The pointing direction was confirmed to be within 2° of the desired pointing direction by a CCD camera observation of Venus as it set. Pointed and scanning observations of the Crab Nebula near meridian passage also confirm that pointing was within 5° of the predicted location of that source. Pointing was stable to within 0.5° peak-to-peak.

DATA ANALYSIS

Data analysis was performed by first summing the 128-channel spectra over intervals of approximately 5 minutes within each of the on-source and off-source intervals. Gain variations were monitored by observing the atmospheric background line at 511 keV, which allows gain determination to within 0.1% in 5 minutes. Since gain changes are primarily due to temperature variations which occur on time scales of hours, the 5-minute gain data were smoothed with a polynomial fit and the results were then used to rebin counts in the pulse height spectra into 17 energy bands. Gain varied by ~8% during the source passage on October 29.

The data were corrected to constant atmospheric depth using curves of count rate versus altitude constructed from data collected between 4.1 and 2.9 mb. These background data were collected as float altitude was reached, when the LMC region was at a low elevation angle, and were thus substantially cut off by the atmosphere.

In order to measure the background variation as a function of azimuth, the gondola was rotated at constant angular velocity during three intervals at float altitudes. The data collected at the time of LMC transit are insufficient to determine the variation versus angle, with an upper limit of about ±0.4% change in counting rate (5% of the Crab Nebula) for a 40° offset. Data taken at other times are contaminated by point source contributions and are not suitable to address this systematic uncertainty. In the present analysis we have assumed no azimuth background variation between 120° and 240°, which is a satisfactory fit to data collected near source transit.

No correction has been made to the counting rates to account for the 1.0° change in latitude which occurred during the observation. The latitude change was monotonic, thus affecting background and source pointing intervals in the same manner. Counting rate variations versus latitude appear to be present at about 1% per degree, and could contribute significantly to the uncertainty in our flux measurements.

The corrected 45-200 keV counting rates during the source transit are shown in Figure 1. The measured counting rate in both detectors from the source intervals is significantly higher than that obtained from background intervals. Several of the source and background pointing intervals were found to contain Vela X-1 within the field of view of the detectors, as identified by the presence of periodicity at its pulse period. Those intervals were not used in subsequent analysis and are not shown in the figure. The higher background rate in detector 1 is due to its wider collimator response.

For each energy interval, the excess count rate in each ~5-minute source pointing interval was determined by subtracting the average rates of the appropriate background intervals, corrected to the altitude of the source interval, from the on-source counting rate. These were converted to fluxes using detection efficiencies that include an atmospheric scattering

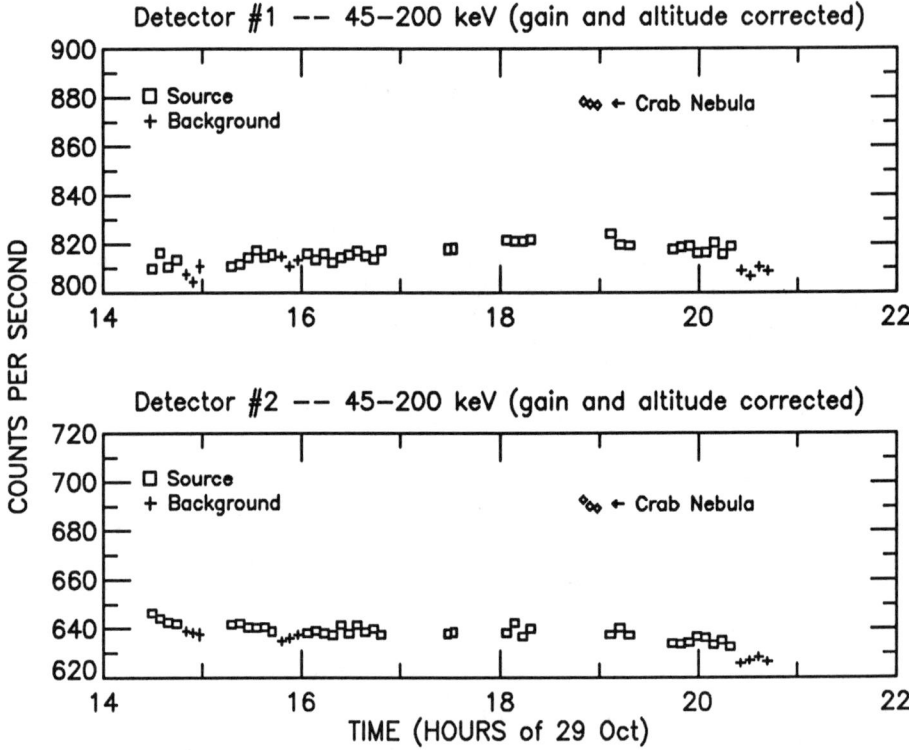

Figure 1. Detector counting rates (45-200 keV) vs. time during the LMC transit on October 29. Rates have been adjusted for gain and altitude variations (see text). The Crab Nebula pointing intervals are indicated. The vertical extent of each data point represents its ±1σ statistical error. Intervals which are contaminated by Vela X-1 are not shown.

contribution, due to photons that Compton scatter but are still collected. The fraction of photons detected after such scattering was determined by a Monte Carlo simulation, which assumed a power-law input spectrum with a spectral index of -1, and calculated the angular and energy distribution of source photons reaching the depth at which the source was observed. The simulated collimator and detector response functions were then used to estimate the fraction of these photons which are detected. A major portion of the response at energies of 35-70 keV is due to scattered photons.

The photon spectrum, as measured separately by each detector, is shown in Figure 2. Significant excess source count rates were detected in all channels between 18 and 730 keV. Observed differences in the flux measurements at energies above 200 keV indicate, however, that systematic errors are substantially larger than the statistical errors. The azimuthal dependence of the background counting rate is clearly different in the two detectors above 700 keV and, though less dramatic, a similar effect may also account for discrepancies at energies as low as 200 keV. Detailed treatment of these systematic effects is beyond the scope of the the present analysis and we restrict further discussion to the region below 200 keV.

The calculation of detector response in the present analysis does not fully account for partial energy deposition from higher energy photons, an effect which is most significant below 45 keV. Systematic corrections due to atmospheric effects are also large in this energy region, so that the soft (power law spectral index ~-3) component evident in Figure 2 below 45 keV is subject to large systematic uncertainties. If real, it is likely to originate in other sources within the field of view of both detectors. GINGA observations[3] of the LMC region show emission in 15.9 to 20.4 keV from LMC X-1 which was almost as intense as emission from the supernova ~190 days after the event. Similarly, the average fluxes listed for LMC X-4 in the HEAO A-4 catalog[9] imply that it alone

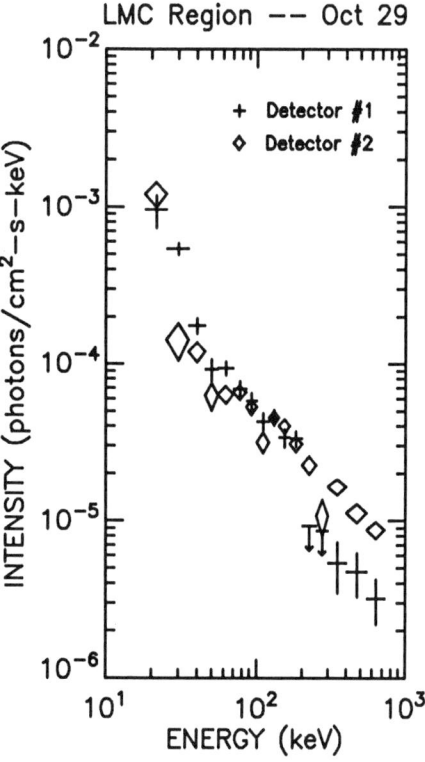

Figure 2. Net flux from the LMC region as determined separately by the two detectors.

would contribute enough flux to be responsible for the soft component we observe if it were near the peak phase of its 30.5 day period. However, the latter is insufficiently well known[10] to be extrapolated to the present epoch.

Another strong source which is a possible contaminant is SMC X-1. In the detector with the more restrictive collimator, the exposure to this source was between 10% and 50% in both on-source and clockwise-offset background intervals, resulting in its partial subtraction in the net flux calculations. Atmospheric scattering from this source has the effect of increasing the detected incident flux, while reducing the difference between source and background intervals. A cursory analysis indicates that the maximum net exposure to this source is ~15%. Using the SMC X-1 spectrum observed by Coe et al.,[11] we find that contamination from this source could account for the enhancement we measure below 45 keV if the net exposure to this source is near our estimated maximum.

Although other sources in the field of view during the supernova observation are likely to contaminate our measurements at energies of less than 45 keV, we know of no sources with a sufficiently high intensity or hard spectrum to be significant at energies greater than 45 keV.

CONCLUSIONS

Neither previously known sources nor systematic effects can account for the relatively hard spectrum which we have detected in the 45-200 keV band. We therefore attribute the flux we have detected to SN1987A. Our results from the combined detectors are shown in Figure 3, together with a representative model calculation (K. Nomoto, private communication; also see Ref. 4). In this model, the production mechanism for the hard spectrum is entirely due to Compton scattering of gamma-ray line emission from the decay of ^{56}Co in an expanding cloud surrounding the supernova. Significant mixing of the parent ^{56}Ni with the expanding gas must be assumed in order to obtain hard x-rays as early as observed after the event (e.g., Ref. 2). Although no detailed spectral fitting has been performed, we can state qualitatively that the spectrum which we observe is more intense and slightly harder than that seen by Sunyaev et al.[2] This behavior is also consistent with the model of Kumagai et al.[4]

Similar models have been constructed by Pinto and Woosley[5] and by Shibazaki et al.[6] The former calculate a spectrum for the time of our observation which also agrees very well with our measurements. Their spectrum may be approximated by two power laws, with spectral indices of -1.6 below 120 keV and

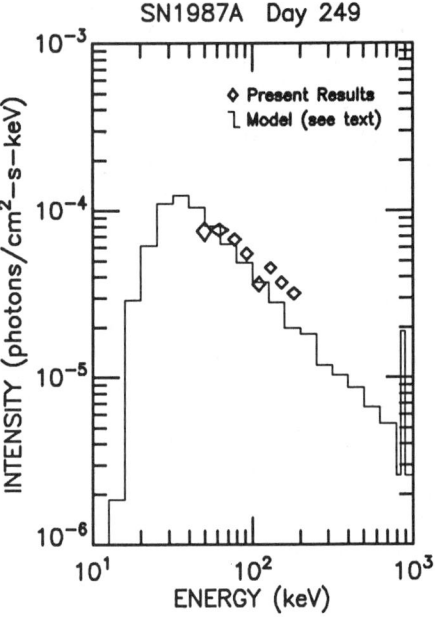

Figure 3. The hard x-ray spectrum of SN1987A. The present results (day 249) are shown as diamonds. The histogram shows the results of a model calculation (K. Nomoto, private communication; also see Ref. 4) for day 250.

-1.0 at higher energies. Our measurements indicate a slightly harder spectrum than this. Shibazaki et al.[6] have constructed a model which distributes the ^{56}Co produced in the supernova in a radial bar in order to simulate filamentary structure. At energies below 75 keV, our data are in better agreement with the fully mixed, non-filamentary case which they consider, although systematic uncertainties in our values near 50 keV are sufficient to prevent exclusion of either case.

ACKNOWLEDGMENTS

We are grateful to J. Odom for his support of this effort. We wish to express our appreciation to the technical personnel at MSFC, led by R. W. Austin. They include W. T. Sutherland, M. Brock, M. Polites, F. Berry, S. Dothard, W. Hammon, M. Love, J. Ozbolt, and E. Roberts. Balloon flight operations were expertly conducted by NSBF. R. Sood provided valuable assistance in Australia. We also thank K. Nomoto, S. Woosley, and T. Ebisuzaki for providing results in advance of publications.

REFERENCES

1. Sandie, W. et al. (these proceedings).
2. Sunyaev, R. et al., Nature 330, 227-229 (1987).
3. Dotani, T. et al., Nature 330, 230-231 (1987).
4. Kumagai, S., Itoh, M., Shigeyama, T., Nomoto, K., and Nishimura, J., in Proc. 4th George Mason Workshop: Supernova 1987A in the LMC (ed. Kafatos, M.) (Cambridge University Press, in press).
5. Pinto, P. A. and Woosley, S. E., Ap. J. (in press).
6. Shibazaki, N., Ebisuzaki, T., and Takahashi, Y., in Proceedings of the Symposium "The Physics of Neutron Stars and Black Holes" (ed. Tanaka, Y.) (in press).
7. Fishman, G. J. et al., Proc. 19th Int. Cosmic Ray Conf., Vol. 3, 343-346 (1985).
8. Fishman, G. J., Meegan, C. A., Parnell, T. A., Wilson, R. B., and Paciesas, W., in High Energy Transients in Astrophysics (ed. Woosley, S.), 651-664 (New York: AIP, 1984).
9. Levine, A. M. et al., Ap. J. Suppl. 54, 581-617 (1984).
10. Lang, F. L. et al., Ap. J. Lett. 246, L21-L25 (1981).
11. Coe, M. J., Bell Burnell, S. J., Engel, A. R., Evans, A. J., and Quenby, J. J., M.N.R.A.S. 197, 247-251 (1981).

50 - 500 MeV γ-RAY EMISSION IN THE EARLY PHASE OF SN1987A

R.K. Sood, J.A. Thomas, L. Waldron and R.K. Manchanda
University College, ADFA, Univ. of NSW, Canberra, Australia.

P. Ubertini, A. Bazzano and C.D. La Padula
Instituto di Astrofisica Spaziale, CNR, Frascati, Italy.

G.K. Rochester and T.J. Sumner
Imperial College, London SW7 2BZ, England.

G. Frye, T. Jenkins, R. Koga and P. Albats
Case Western Reserve University, Cleveland, USA.

ABSTRACT

SN1987A was observed on 19th April 1987 with a combined high energy γ-ray and hard x-ray payload, flown on a stratospheric balloon from Alice Springs, Australia. The γ-ray detector, sensitive in the energy range 50-500 MeV, was an optical spark chamber with 400cm^2 area, a field of view of 60° FWHM and a time resolution of 10 μs. The counting rate profile at ∼2.2 mb float altitude has lead to a 3σ upper limit to the steady γ-ray flux of 7×10^{-4} ph cm^{-2}s^{-1} in the 50-500 MeV range. This upper limit is compared to our predictions for the time profile of γ-ray emission from SN1987A resulting from pulsar acceleration of particles to cosmic ray energies.

INTRODUCTION

Supernovae have long been considered to be a likely source of both the nucleons and electrons in cosmic rays.[1,2] Shock acceleration of particles in the outer layers of the ejected shell can result in these particles attaining relativistic energies behind the shock[3,5]. Alternatively a rotating magnetised neutron star in the centre of a young supernova can provide the acceleration of particles up to TeV energies with a particle luminosity of ∼10^{42} erg s^{-1} [6,8]. Detection of the presence of relativistic nucleonic species relies upon the observation of the emission of high energy (E > 50MeV) γ-rays produced by way of Π° decay as a consequence of the interaction of the relativistic charged particles with material surrounding the site of the explosion. If a pulsar is responsible for the acceleration of these particles then the overlying material must be sufficiently thin for the γ-rays to escape if they are to be observed.[9] Typically it is predicted that these conditions should prevail within several weeks of the occurrence of a Type II supernova. However it is now clear that SN1987A was somewhat atypical and we present here results of our calculations of the time profile expected for γ-ray emission using recent data on computed density and velocity distributions in the envelope together with current estimates of mass loss. A factor describing the expected pulsar spin down rate is included.

OBSERVATIONS

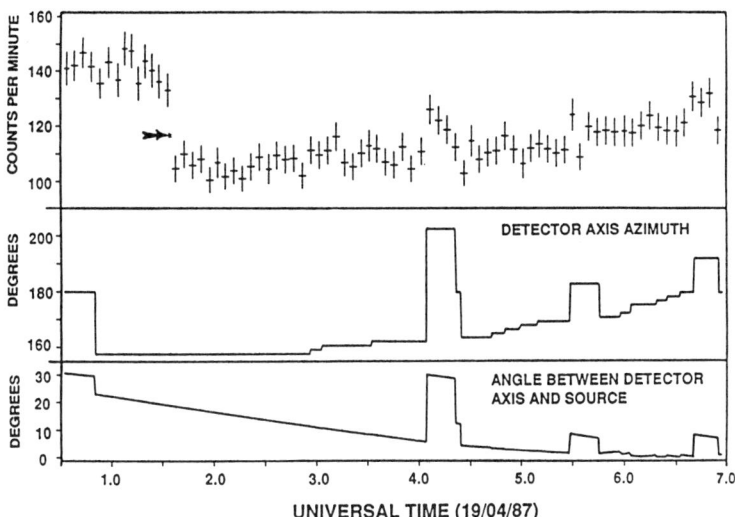

Figure 1. (a) The counting rate profile during the flight. The discontinuity at 0130 UT is due to a commanded change in the discriminator threshold of the Cerenkov counter.
       (b) The orientation profile during the flight.
       (c) The angle between the detector axis and SN1987A during the flight.

The high energy γ-ray observations were made with the 30-gap spark chamber and scintillator/Cerenkov telescope belonging to University College, ADFA, which has been flown several times previously[10,11]. The detector was launched on a 21 Mcf balloon from Alice Springs, Australia, on the 19th April 1987. The counting rate profile obtained during the flight is shown in fig. 1, corrected to a residual pressure of 2.2 mb. All the observed counting rate variations can be attributed to changes in cut-off rigidity during the flight (correlation coefficient of 0.82). A preliminary upper limit of $\sim 7 \times 10^{-4}$ ph $cm^{-2} s^{-1}$ was obtained from the counting rates by assuming the source flux is less than the statistical variation of the observed detector background[12]. A preliminary scan of the spark chamber data in which 7900 events out of a total recorded number of

48000 were visually identified as γ-ray conversion pairs has since shown that there was no γ-ray emission seen from SN1987A to a 3σ confidence level above background of 3×10^{-4} ph cm^{-2}s^{-1} [13]. More detailed analysis of the measured track angles and their projected location on a celestial map will improve the sensitivity of the

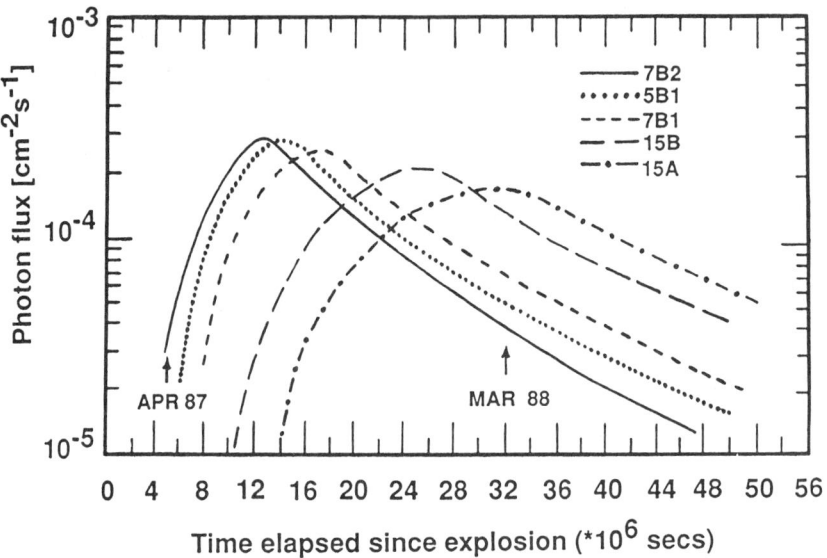

Figure 2. Predicted time dependence of the high energy γ-ray flux at the Earth from SN1987A assuming the time dependent cosmic ray power within the shell given by Ginzburg and Ptuskin [26] and based on the models 15A and 15B of Woosley et al. [23] This prediction is derived from data on the production of γ-rays in the Earth's atmosphere by cosmic rays. Model 7B1 differs from 15B in that 7.5 solar masses are ejected instead of 15, but at the same velocity, i.e. with the same kinetic energy release as 15A. Model 7B2 differs from 15B in that 7.5 solar masses are ejected instead of 15, but at √2 times the velocity, i.e. with the same kinetic energy release as in 15B. Model 5B1 differs from 15B in that 5 solar masses are ejected. All models are consistent with an initial photospheric velocity of the order 10^9 cm s^{-1}.

observation still further to at least an upper limit of 10^{-5} ph $cm^{-2}s^{-1}$. When all the events have been read into the computers a timing analysis will be carried out in an attempt to reveal the existence of a pulsar at the site of the supernova.

DISCUSSION

In order to make a theoretical estimate of the γ-ray flux we follow the model in which cosmic rays are accelerated within the shell by a rapidly rotating magnetised neutron star left at the site of the supernova explosion. The progenitor star for SN1987A is believed to be the blue B3 I supergiant Sk 69-202 rather than the more usual M-supergiant for type II supernovae. It is also inferred that the star may have been metal deficient[14,16] and that the kinetic energy involved in the explosion was much less than the canonically expected value for type II events[16,17]. The theory of core collapse and post burst evolution of type II supernovae is fairly well developed and discussed in the literature[18,20]. A detailed discussion with a variety of progenitor star masses has been presented by Woosley and Weaver[20]. Recently several model calculations of the supernova explosion in SN1987A have been made[21,23] to estimate the flux of nuclear lines from the nebula. These assume the ejected mass to be between 8 and 15 solar masses and the time dependent density distribution to vary as $\sim t^{-3}$. Using the Woosley et al.[23] model for the quantitative density distribution we have computed the high energy γ-ray flux above 50 MeV expected from the source.

It is assumed in the first instance that the radially outward moving intensity of cosmic rays within the shell does not vary with time and that the spectrum of these cosmic rays is similar to that observed at the Earth during solar minimum. In order to avoid the need for having to consider a series of nuclear interactions at different energies in which the multiplicity of Π^o production is not too well defined, a simpler method of carrying out the necessary calculation is employed here which depends upon the results obtained by Sood et al.[24] for the dependence of the flux of cosmic ray induced high energy γ-rays on depth in the Earth's atmosphere. These γ-rays are produced by Π^o decay and secondary bremsstrahlung[25] in exactly the same way as the γ-rays from the supernova in the neutron star acceleration model. If the cosmic ray spectra are the same in the two cases[26] then the only major difference is the composition of the material with which the energetic particles are interacting, which is air around the Earth and a mixture of hydrogen and helium around the neutron star. An examination of the cross sections and multiplicities for Π^o production of the elements involved and also of the pair production cross sections responsible for the absorption of the γ-rays reveals that the 80/20 hydrogen/helium mixture which is presumed to surround the site of the supernova is up to twice as efficient as air at converting cosmic rays to γ-rays, depending on its thickness. A simple extrapolation of the atmospheric production of γ-rays, bearing in mind that Π^0 production from hydrogen cannot occur if the energy of the incoming particle is less that about 340 MeV, leads to the prediction of the time dependence of the γ-ray flux at the Earth from SN1987A under the assumption that the internal cosmic ray luminosity is constant. If this luminosity is assumed to be 10^{42} erg s^{-1}, at the lower end of the range suggested by Berezinsky et al.[27] as being necessary to explain the observed cosmic ray intensity at the Earth if supernovae are indeed the prime source, then, using the Woosley et al.[23] 15A and 15B models, we estimate the high energy γ-ray flux from SN1987A on day 55 to have been somewhere between 10^{-20} and 10^{-14} photons cm^{-2} s^{-1}, far below the level at which it was possible to detect it.

If it is a neutron star which is responsible for accelerating the cosmic rays then the cosmic ray luminosity must decline as the pulsar decelerates.[6] Ginsburg and Ptuskin[26] have shown that if excess antiprotons observed in the cosmic rays[28,29] are generated by young supernovae then these objects, which will also be bright in high energy γ-rays, are likely to have a shell which becomes transparent to high energy protons after about 10^7 seconds. This prediction agrees with the conclusions reported here. Their model also predicts that the time dependence and flux of the cosmic ray power within the shell surrounding the neutron star will typically be given by

$$L_{cr} = L_0/(1+t/T)^2 \text{ erg s}^{-1}$$

where $T=5\times 10^7$ is the initial characteristic time of neuton star luminosity decrease due to magnetic dipole braking and t is the time after the supernova explosion. Folding this power function with the time dependence caused by the decreasing thickness of the envelope gives the γ-ray time dependence shown in the fig. 2. A value of 9.2×10^{41} ergs s^{-1} has been used for L_0; this is consistent with that estimated by Ginsburg and Ptuskin[26] for the cosmic ray power output required from a typical supernova if supernovae are to account for all the cosmic rays in the Galaxy. It can be seen that the maximum γ-ray flux occurs $\sim 1.2\times 10^7$ seconds after the explosion for model 5B1 and has a maximum value of $\sim 3.0\times 10^{-4}$ photons cm^{-2} s^{-1}. Our model does not yet include magnetically induced turbulent cosmic ray motion within the expanding envelope[26]. However, addition of this refinement to the theory could not increase the flux beyond $\sim 6\times 10^{-4}$ photons cm^{-2} s^{-1} for model 5B1. It is interesting to note that Berezinsky and Ginsburg[30] have recently predicted a very similar maximum γ-ray flux from SN1987A, after due allowance has been made for differing assumptions about the cosmic ray power in the source, by a totally different method.

CONCLUSION

It is clear from the above account that the sensitivity of the γ-ray detector flown on a balloon over central Australia is sufficiently high to obtain confirmation of those theories of the origin of cosmic rays which postulate a neutron star within the expanding shell of material around a supernova. Unfortunately the development of the shell of expanding material around SN1987A appears, according to the best possible estimates available at this time, to be such as to have made it impossible to observe γ-rays from it on 19th April, 1987. However fig. 2 suggests that the γ-ray emission should not have declined too far below its peak by the time the next planned balloon flight takes place in March 1988. We should therefore be able to obtain a clear indication of the validity of the models shown in fig. 2. Moreover, by that time the shell of overlying material should have been reduced to about 30 gm cm^{-2} which is thin enough to allow pulsed γ-rays produced directly by the neutron star itself to escape and be observed.

ACKNOWLEDGEMENTS

These observations would not have been possible without the expertise and dedicated efforts of J. Panetierri and A. Meilak of University College, ADFA and M. Mastropietro of IAS. We also thank Prof. V.D. Hopper for his many words of wisdom.

REFERENCES

1. W. Baade and F. Zwicky, Proc. Nat. Acad. Sci. Amer. 20, 259, (1934).
2. V.L. Ginzburg and S.I. Syrovatskii, *Origin of Cosmic Rays*, Pergamon Press, (1964).
3. S.A. Colgate, *Origin of Cosmic Rays*, Reidel Publ. Co., (1975).
4. S.A. Colgate and M.H. Johnson, Phys. Rev. Lett. 5, 235, (1960).
5. R.A. Chevalier, Ap. J. 272, 765, (1983).
6. J.E. Gunn and J.P. Ostriker, Phys. Rev. Lett. 22, 728, (1969).
7. T.K. Gaisser et al., Nature 329, 314, (1987).
8. G. Cavallo and F. Pacini, Astron. Astrophys. 88, 367, (1980).
9. V.S. Berezinsky and O.F. Prilutsky, Astron. Astrophys., 66, 325, (1978).
10. V.D. Hopper et al., Ap. J. Lett. 186, L552, (1973).
11. G.M. Frye and C.P. Wang, Ap. J., 158, 925, (1969).
12. R.K. Sood et al., I.A.U. circular 4405, (1987).
13. R.K. Sood et al., Supernova 1987a in the LMC *Fairfax, Virginia*, (1987).
14. V.M. Blanco et al., Cerro Tololo Observatory, *preprint*, (1987).
15. W.D. Arnett, Ap. J. Lett., *in press*, (1987).
16. W. Hillebrandt et al., Nature 327, 597, (1987).
17. R.M. Catchpole et al., *preprint*, (1987).

18. A. Burrows, Ap. J. 318, L57, (1987).
19. S. Miyaji and K. Nomoto, Ap. J. 318, 307, (1987).
20. S.E. Woosley and T.A. Weaver, Ann. Rev. of Astron. Astrophys. 377, (1981).
21. P.R. Wood and D.J. Faulkner, Proc. Astr. Soc. Australia, *in press*, (1987).
22. K.W. Chan and R.E. Lingenfelter, Ap. J. 318, L51, (1987).
23. S.E. Woosley et al., Ap. J., *in press*, (1988).
24. R.K. Sood et al., J. Phys. E., 15, 462, (1982).
25. K.P. Beuermann, J. Geophys. Res., 76, 4291, (1971).
26. V.L. Ginzburg and V.S. Ptuskin, J. Astrophys. Astr. (Ind) 5, 99, (1984).
27. V.S. Berezinsky et al., 19th ICRC 1, 305, (1985).
28. A. Buffington et al., Ap. J., 248, 1179, (1981).
29. R.L. Golden et al., Phys. Rev. Lett. 43, 1196, (1979).
30. V.S. Berezinsky and V.L. Ginzburg, Nature 329, 807, (1987).

SUMMARY OF GAMMA-RAY OBSERVATIONS OF SN 1987A

Neil Gehrels
Goddard Space Flight Center, Greenbelt, MD 20771

Marvin Leventhal
AT&T Bell Laboratories, Murray Hill, NJ 07974

Crawford J. MacCallum
Sandia National Laboratories, Albuquerque, NM 87185

ABSTRACT

Gamma-ray line and continuum observations of SN 1987A through May 1988 are reviewed and discussed. It is shown that the measurements are reasonably consistent with each other if the latest values are used and if line width and systematic uncertainties are taken into account. The light curve of the line and continuum fluxes, the line ratios, and the line widths all indicate that the freshly synthesized ^{56}Co producing the emission is not confined to a thin shell at the inner edge of the ejecta as early models had assumed.

INTRODUCTION

This paper was updated after other papers in this volume. It summarizes the available gamma-ray data on SN 1987A at the time this Proceedings goes to press in May 1988. It was written for the benefit of the gamma-ray community, with the permission (and at the urging) of the observers whose results are included.

The gamma-ray data on SN 1987A available in May 1988 are still preliminary, but are beginning to give a consistent observational picture. At the time of the Workshop and shortly thereafter, the data on the 847 and 1238 keV lines from ^{56}Co decay were new and exciting, but some discrepancies existed. In particular, the preliminary flux measurements reported for the balloon-borne instruments flown from Australia and Antarctica were inconsistent with the data from the Gamma Ray Spectrometer (GRS) on the Solar Maximum Mission (SMM) and were inconsistent with each other. It should be noted, however, that even in this early period, the X-ray and gamma-ray data were driving the theoretical models of the velocity structure and envelope mass of the ejecta. For example, the early detection and relatively flat light curves for both the line and continuum fluxes were indicating a mixing of the inner ejecta material[1,2,3,4,5]. The situation in May 1988 is improved. There are new observations available from high-sensitivity balloon spectrometers, from SMM, and from the Soviet Mir space station. In addition, the earlier data have been further analyzed. The data are now reasonably consistent if one takes care to compare similar quantities.

© 1988 American Institute of Physics

OBSERVATION SUMMARY

There have been four balloon campaigns from which observations of SN 1987A were obtained. They were in May-June 1987 from Australia, November-December 1987 from Australia, January 1988 from Antarctica, and April-May 1988 from Australia. A mix of sodium iodide (NaI) and germanium (Ge) detectors have been flown, with a gamma-ray imaging instrument included in each of the Australian campaigns. An instrument list is given in a companion paper by Bunner[6]. The actual achieved sensitivities (3σ) for line detection at 1 MeV (3 keV line width assumed for this comparison) have improved from 1.5×10^{-3} ph cm^{-2} s^{-1} in May-June 1987 to 2×10^{-4} ph cm^{-2} s^{-1} in April-May 1988.

Two satellite missions have provided virtually continuous monitoring of the hard X-ray and gamma-ray emission from the supernova. These are SMM with its GRS and the Mir with its High Energy X-ray Experiment (HEXE) and Pulsar X-1. The data sets from these two missions are complementary, with SMM primarily measuring the line emission and Mir primarily measuring the continuum. The line sensitivity (3σ) SMM has achieved is between 5×10^{-4} and 10^{-3} ph cm^{-2} s^{-1} depending on integration time.

Gamma-Ray Line Measurements

All reported measurements of gamma-ray lines from SN 1987A are listed in Table 1. The three highest-branching-ratio lines at 847, 1238, and 2599 keV from ^{56}Co decay have all been observed. The measurements of the 847 and 1238 keV lines are shown as functions of time in Figure 1.

The Caltech flux values for November 1987 and April 1988 are listed as upper limits because they include continuum in addition to line flux. For the Lockheed/MSFC observation of the 847 keV line in October 1987, fluxes for both narrow (5 keV) and broad (12 keV) bins have been published[10]. We list the number for the broad bin in order to include all the flux. The 1238 keV limit was determined by scaling ($\sqrt{18/6}$) their limit[17] for a 6 keV line width to an 18 keV width (the expected width if the 847 keV line is 12 keV wide). For the Florida/GSFC observation, we list in Table 1 the statistical uncertainty, but plot in Figure 1 the statistical and 30% systematic uncertainty added in quadrature. For the JPL observation, the 847 keV line was detected with less than 1σ significance and is therefore plotted in Figure 1 as a 3σ upper limit.

The measurements for each line are seen to be in reasonably good agreement with each other, with the only 2σ disagreement being between the 1238 keV measurements by Florida/GSFC and SMM in early 1988 (day 300-350). The data for both the 847 and 1238 keV lines are consistent with a flat light curve at $\sim 10^{-3}$ ph cm^{-2} s^{-1} since day 170 (August 1987). They are also consistent with various "mixed" models[1,2,3,4,5]. In particular, we have plotted the Pinto and Woosley model[1] 10 HMM which has a 6 M$_\odot$ He core, a 10 M$_\odot$ H-He envelope, 1.3×10^{51} erg expansion energy, and 0.075 M$_\odot$ of ^{56}Ni synthesized in the explosion. The model assumes that a small fraction of the ^{56}Co is accelerated by radioactive heating (or possibly not fully decelerated by the original reverse shock) to velocities as high as 4000 km s^{-1}, and thereby mixes into the core and envelope. Although the quantity mixed up in such models is only a few percent of the total ^{56}Co, the effect on the early light curve is dramatic since the inner slow-moving ^{56}Co is highly attenuated at early times. This is demonstrated by comparing with model 5 L, which is one of the better non-mixed models for matching the gamma-ray line data[18].

Table 1 Gamma-Ray Line Observations of SN 1987A

Instrument	E_0 (keV)	Date	Day #	Line Width (keV)	Flux (10^{-4} ph cm^{-2} s^{-1})	Ref.	Comments
SMM	847	2/23/87 - 4/2/87	0 - 38	-	0.8 ± 4.2	7	
"	1238	"	"	-	1.3 ± 3.5	"	
"	847	4/18/87 - 5/21/87	54 - 87	-	0.6 ± 4.4	"	
"	1238	"	"	-	0.4 ± 3.8	"	
Caltech	847	5/20/87	86	-	< 15.	8	3 sigma limit.
"	1238	"	"	-	< 23.	"	"
"	2599	"	"	-	< 14.	"	"
Lockheed/MSFC	847	5/29/87 - 5/30/87	95 - 96	3 (assumed)	< 17.	9	"
"	1238	"	"	3 (assumed)	< 13.	"	"
SMM	847	6/10/87 - 7/16/87	107 - 143	-	1.6 ± 4.6	7	
"	1238	"	"	-	0.6 ± 4.0	"	
"	847	8/1/87 - 9/7/87	159 - 196	-	8.6 ± 4.0	"	
"	1238	"	"	-	5.3 ± 3.5	"	
"	847	9/27/87 - 10/30/87	216 - 249	-	6.2 ± 4.1	"	
"	1238	"	"	-	11. ± 3.5	"	
Lockheed/MSFC	847	10/29/87 - 10/31/87	248 - 250	12	10. ± 2.	10	
"	1238	"	"	18 (assumed)	< 8.5	"	3 sigma limit.
Caltech	847	11/19/87	269	81 (assumed)	≤ 11. ± 5.	11	Includes continuum.
"	1238	"	"	93 (assumed)	≤ 11. ± 6.	"	"
JPL	847	12/7/87	287	8 (assumed)	5. ± 7.	12	
"	1238	"	"	8.2 ± 3.4	21. ± 7.	"	Line centroid at 1240.8±1.7 keV.
SMM	847	11/16/87 - 12/22/87	267 - 303	-	5.8 ± 3.4	7	
"	1238	"	"	-	8.8 ± 2.9	"	
Florida/GSFC	847	1/8/88 - 1/11/88	319 - 322	7.5	11. ± 4.	13,14	30% systematics, all 3 lines.
"	1238	"	"	~14	19. ± 4.	"	Peaks at 1226 & 1240 keV.
"	2599	"	"	~21	8.6 ± 3.2	"	Peaks at 2582 & 2603 keV.
SMM	847	1/8/88 - 2/14/88	319 - 356	-	3.9 ± 4.2	7	
"	1238	"	"	-	2.2 ± 3.6	"	
"	847	3/1/88 - 4/6/88	372 - 408	-	8.2 ± 3.7	"	
"	1238	"	"	-	3.6 ± 3.1	"	
Caltech	847	4/12/88 - 4/13/88	414 - 415	94 (assumed)	≤ 8. ± 4.	15	Includes continuum.
"	1238	"	"	93 (assumed)	< 14.	"	3 sigma limit. Includes continuum.
GSFC/Bell/Sandia	847	5/2/88	434	-		16	847 keV not yet reported.
"	1238	"	"	15 - 25	8.1 ± 1.7	"	

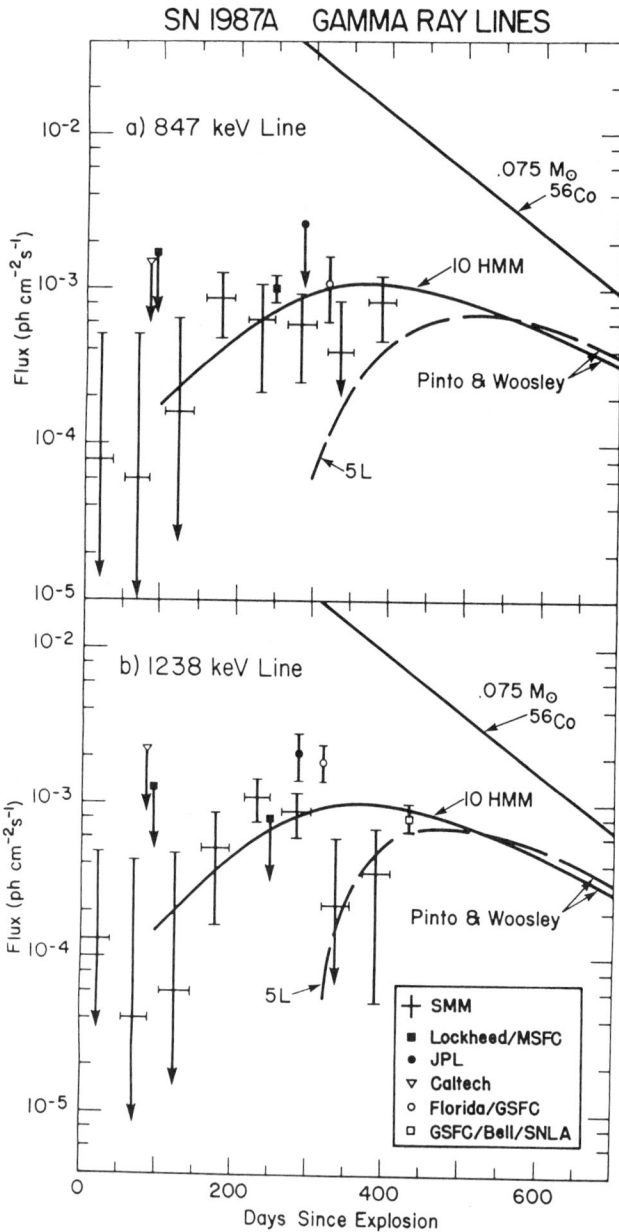

Figure 1 - Measurements of the 847 and 1238 keV line fluxes from ^{56}Co decay. Also shown is the line for the limiting case of 0.075 $M_\odot$ (see ref. 1 and references therein) of unobscured ^{56}Co, and curves for models 10 HMM[1] and 5 L[18].

Gamma-Ray Continuum Measurements

The hard X-ray and gamma-ray continuum data on SN 1987A available to date are from the following: the Mir HEXE and Pulsar X-1 phoswich instruments[19,20] with spectra for August 1987 and January 1988; the Marshall Space Flight Center (MSFC) NaI instrument[21] flown in conjunction with the Lockheed/MSFC spectrometer, with a spectrum for 29-31 October 1987; the Caltech NaI instrument[11,15] with spectra for 19 November 1987 and 12-13 April 1988; the Jet Propulsion Laboratory (JPL) Ge instrument[12] with a spectrum for 7 December 1987; and the Frascati xenon proportional counter[22] flown in collaboration with groups from Canberra University, Case Western Reserve University, Imperial College, and Tubingen University, with a measurement for 5 April 1988. The data are shown in Figures 2 and 3.

Figure 2 shows the development of the spectra as a function of time. Also plotted are curves for the 10 HMM and 5 L models of Pinto and Woosley[1,18]. The data are reasonably consistent with the 10 HMM model which includes ^{56}Co mixing, but are significantly higher at early times than the unmixed model 5 L. This again indicates that the ^{56}Co is not all in a thin shell at the inner edge of the ejecta.

Figure 3 shows all continuum data plotted together. It is seen that the spectrum has not changed much since its first measurement in August 1987. The total variation is about a factor of 2. A representative spectral shape between 50 keV and 1 MeV is 8×10^{-3} $E^{-1.2}$ ph cm^{-2} s^{-1} keV^{-1} (E in keV).

DISCUSSION

The early appearance of gamma-ray lines from ^{56}Co and X-ray and gamma-ray continuum presumably from Compton down-scattering of line photons[23,24] implies that some fraction of the ^{56}Co, originally produced as ^{56}Ni in the inner layers of the ejecta, has found its way closer to the surface of the ejecta. The time scale for gamma-ray escape from the inner layers without some upward mixing is typically a year or more[24] for resonable models of this event.

There are many geometries for this mixing, the two extreme cases being 1) a single jet or clump of material that acquires a high velocity that propels it near the surface, and 2) a more uniform mixing of some fraction of the ^{56}Co upward into the mantle and envelope. For both cases, even a small quantity of material moved nearer the surface can produce a large change in the early light curve of the lines since the emission from the inner slow-moving ^{56}Co is highly attenuated at early times. For example, comparing the flux from unattenuated ^{56}Co, shown by the straight lines in Figure 1, with the measured line fluxes, we see that only about 1%[25] of the total 0.075 M$_\odot$ (ref. 1 and references therein) of ^{56}Co is required near the surface to explain the line fluxes in August 1987 (~ day 160).

Although the data available to date are not detailed enough to decide definitively between case 1 and 2 above, there is some suggestion that a more uniform mixing geometry (case 2) is a better model. There are four pieces of evidence in this regard:
1) There is no obvious factor of 2 decrease in the line fluxes between August and November 1987 as would be expected for the decay of unobscured ^{56}Co. The data do not rule out such a decrease, but seem more consistent with a constant flux in this time period.
2) The simultaneous early appearance in August 1987 of line and continuum emission would not be explained by unobscured ^{56}Co, since there would be very little scattering of the line photons from the unobscured material.

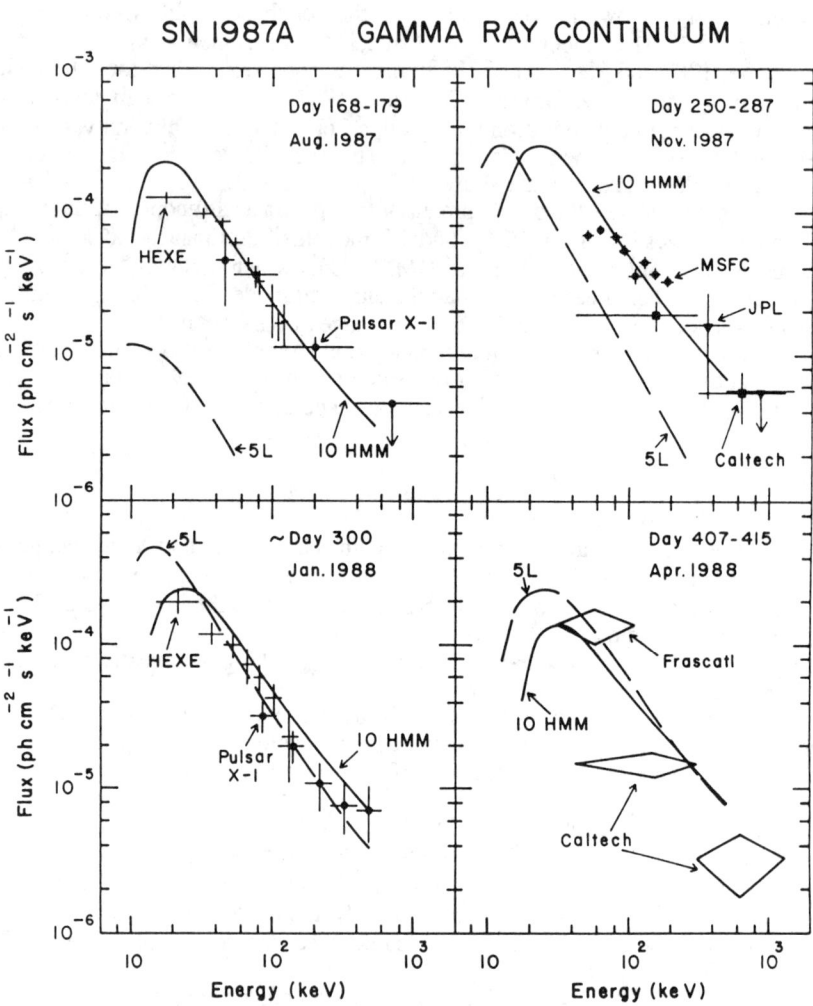

Figure 2 - Measurements of the continuum compared with models 10 HMM[1] and 5 L[18].

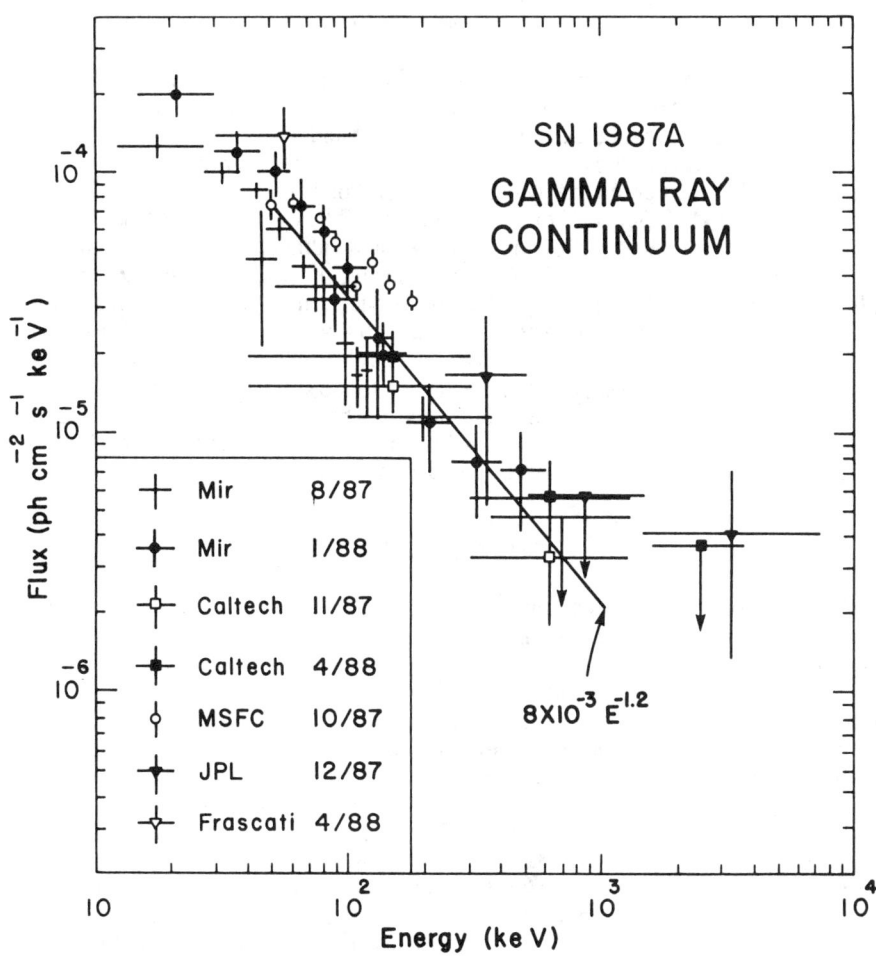

Figure 3 - All continuum measurements. A representative spectral shape is given by the line.

3) Summing the SMM data from August to December 1987 (day 159 to 303), the ratio of the 1238 keV line flux to that of the 847 keV line is 1.2±0.5. This is to be compared with expected values of 0.7 for unobscured ^{56}Co, ~ 1.0 for mixed models, and > 2.0 for unmixed models. The data are marginally consistent with unobscured emission, but are in better agreement with deeper emission such as for the mixed models. The data are inconsistent with unmixed models which have the ^{56}Co confined to the inner-most layers of the ejecta

4) The lines from a single clump of material would be narrow. The observed line emission appears to be broad[10,12,13,16] in disagreement with the single-clump model. Some line broadening would be expected for both a jet and a mixed model.

One of the results of the ^{56}Co mixing is that the lines are stronger relative to the continuum emission than predicted by early models. For example, the ratio of the flux in the continuum band 50-500 keV to the flux in the 847 keV line has remained constant (to within ~ 50%) from August 1987 to April 1988 at a value of about 8. Calculations[24] for models with ^{56}Co at a constant depth give values for this ratio that decrease with time, but with a typical value of 30 near the peak of the gamma-ray light curves. The low value of the continuum to line ratio is also apparent when compared with a general treatment of photon scattering[26]. The reason for the relatively strong lines in the mixed models is that the line emission at early times is due to the small fraction of ^{56}Co that has acquired a high velocity and mixed near the surface. The continuum emission from this line-producing material is low due to its small amount and small scattering depth. The continuum that is observed is due primarily to deeper material.

THE FUTURE

We first derive estimates and limits on the gamma-ray emission at the times of the next two NASA balloon campaigns[6] in November 1988 and April 1989. For the 1988 campaign, choosing 15 November (day 631) as a representative date, the 847 keV line flux is constrained to be less than 1.6×10^{-3} ph cm^{-2} s^{-1} based on 0.075 $M_\odot$ of unobscured ^{56}Co. This high a value is unlikely. The flux could possibly remain at 1.0×10^{-3} ph cm^{-2} s^{-1} based on the constant light curve since August 1987, but the most likely value in our opinion is 5×10^{-4} ph cm^{-2} s^{-1} predicted by model 10 HMM[1]. This is a factor of 2 below the flux during the last campaign in April 1988. The 1238 keV line flux will be 20-30% lower than that of the 847 keV line. For the possible 1989 campaign, choosing April 15 (day 782) as a representative date, the 847 keV flux will be between 2 and 4×10^{-4} ph cm^{-2} s^{-1}, with the lowest value being most likely in our opinion. For both dates, only the most sensitive balloon spectrometers will be able to detect the lines. The continuum emission should drop off more rapidly than the line emission as the envelope thins.

The gamma-ray lines should become narrower as times progresses. This will occur as the deeper material (with lower total velocity) and material on the limbs of the ^{56}Co inner shell (with lower velocity perpendicular to the line of sight) become visible[27]. If the Fe II fine structure lines observed in the infrared[28,29] are produced only by Fe formed from ^{56}Co decay, then the gamma-ray line widths when the envelope becomes optically thin will be ~ 8 keV FWHM for the 847 keV line and ~ 12 keV FWHM for the 1238 keV line.

Finally we note the possibility that a buried pulsar could be detected in the ongoing balloon campaigns. Because of the neutrino pulse, it is generally believed that SN 1987A formed a neutron star which is very likely to be an infant pulsar based on supernova and pulsar counting statistics. A present day limit on the luminosity L of such a pulsar can be obtained from the late time optical light curve which is still falling with nearly a 77 day half-life. This requires that $L \lesssim 10^{40}$ ergs s^{-1} which is about 100 L_{Crab}. It is reasonable to assume that the SN 1987A pulsar might radiate a Crab-like X-ray and gamma-ray spectrum with 100 times the intensity[30]. Taking the distance to the LMC into account, the apparent luminosity of the SN 1987A pulsar could be as large as 1/10 the Crab pulsar neglecting opacity. While the opacity of the remnant is clearly energy and model dependent, it is fair to say that by mid 1988, the remnant is ~ 10% transparent to low energy gamma rays. Since pulsations can only be seen in unscattered radiation, this means that the observable pulsar luminosity must currently be less than ~ 1/100 the observed Crab pulsar luminostity. Such a pulsar could be detected by the most sensitive balloon instruments. Since $L \propto \mu^2 \Omega^4$ and it is thought that most pulsars have similar magnetic dipole moments μ, it should be possible to place a limit of $\Omega \lesssim 3\Omega_{Crab}$ on the spin frequency if the pulsar is not detected in current data sets.

Acknowledgements We are grateful to the Florida/GSFC, Lockheed/MSFC, and JPL groups for permission to quote results from preprints. We thank James Ling and Thomas Prince for useful comments.

REFERENCES

1. P.A. Pinto and S. E. Woosley, Nature 333, 534 (1988).
2. S. Kumagai et al., Astronomy and Astrophys. Lett., preprint (1988).
3. N. Shibazaki and T. Ebisuzaki, Astrophys. J. Lett. 327, L5 and L9 (1988).
4. S. Grebenev and R. Sunyaev, Soviet Astron. Lett. 13, 945 (1987).
5. A. Fu and W. D. Arnett, Astrophys. J., preprint (1988).
6. A. N. Bunner, this volume (1988).
7. S. M. Matz et al., this volume (1988).
8. W. Cook et al., IAU Circular 4400 (1987).
9. W. Sandie et al., IAU Circular 4463 (1987).
10. W. G. Sandie et al., Astrophys. J. Lett., preprint (1988).
11. W. R. Cook et al., this volume (1988).
12. W. A. Mahoney et al., Astrophys. J. Lett., preprint (1988).
13. A. C. Rester et al., Astrophys. J. Lett., preprint (1988).
14. J. I. Trombka, private communication (1988).
15. W. Cook et al., IAU Circular 4584 (1988).
16. S. Barthelmy et al., IAU Circular 4593 (1988).
17. W. G. Sandie et al., this volume (1988).
18. P. A. Pinto and S. E. Woosley, Astrophys. J., preprint (1988).
19. C. Reppin et al., this volume (1987).
20. R. Z. Sagdeev, Physics Today, May, p. 30 (1988).
21. R. B. Wilson et al., this volume (1988).
22. P. Ubertini et al., IAU Circular 4590 (1988).
23. R. McCray et al., Astrophys. J. Lett. 317, L73 (1987).
24. N. Gehrels et al., Astrophys. J. Lett. 320, L19 (1987).
25. S. M. Matz et al., Nature 331, 416 (1988).
26. M. D. Leising, Nature 332, 516 (1988).
27. R. E. Lingenfelter, private communication (1988).
28. H. Moseley et al., IAU Circular 4576 (1988).
29. M. R. Haas et al., IAU Circular 4578 (1988).
30. Y. Xu et al., Astrophys. J. 327, 197 (1988).

Comments on the Reported Detection of ^{56}Co Gamma Ray Lines From SN 1987A

M. Leventhal

AT&T Bell Laboratories
Murray Hill, New Jersey 07974

Now that our workshop is drawing to a close I would like to make some comments of an editorial nature. While we have discussed many important and exciting topics over the last three days, in my mind they were all overshadowed by the reported historical detection of 847 keV and 1238 keV ^{56}Co line radiation from SN 1987A. To begin with, I would like to respectfully disagree with a comment that Stan Woosley made on the first day of the meeting. He suggested that the simple detection of the ^{56}Co gamma ray lines from SN 1987A was, by itself, unimportant because of the exponential behavior of the late time bolometric light curve which is decaying with about the ^{56}Co halflife (77 days). The theoretical implication is that .07 solar masses of radioactive ^{56}Co is unavoidably powering the light curve and a gamma ray line confirmation is unnecessary. However, similar exponentially decaying light curves have been seen from a number of extragalactic Type I supernovae for years yet considerable skepticism remains within the astrophysical community about the radioactive origin of the energy source. Unfortunately there are many ways to make an exponential light curve without radioactivity and the mere coincidence in halflives has not been convincing enough. Already at this conference Ken Brecher has suggested a pulsar powered exponential for SN 1987A. The detection of gamma ray lines will forever end this controversy.

Having said that, I would like to express my belief that the ^{56}Co lines have now been seen by the SMM satellite group and the Lockheed/Marshall and Caltech balloon groups and offer them my congratulations on this exciting "discovery". Crawford MacCallum and I have been trying to make this discovery for 15 years. Apparently it was not meant to be for us. The line detections, reported for the first time at this workshop, seem to be coming earlier than theoretically expected and, surprisingly, seem to indicate that relatively unshielded ^{56}Co is being detected. The discovery groups should not be concerned about this. As is usually the case, observation will lead theory in the right direction. Theoretical models will be modified quickly to accommodate the new results. It appears that significant departures from a spherically symmetric onion skin type post explosion are called for.

What is most disturbing for our field is the fact that once again important new observations are being made right at the sensitivity threshold of the instruments (2σ to 5σ). The fact that three groups are reporting consistent results is encouraging. Nevertheless many of us have been burnt in the past with results of a somewhat marginal statistical nature. I urge the discovery groups to take an extremely cautious and conservative attitude in the analysis and publication of their results. It would be prudent to resist the great pressure to publish prematurely in the media before a careful analysis is complete. Perhaps the three groups might agree amongest themselves to submitt simultaneous papers to a journal and thus avoid a race to beat each other into print.

The proper attitude of the community at this time should be one of skepticism. All of us should study these new exciting, but statistically modest, results to see that mistakes are not being made. The systematic effects are serious in these experiments and it is still possible that we are fooling ourselves. It would be wise of the discovery groups to seek out the opinions of some "devils advocates" within the community prior to publication. If we "screw up" on this one our field will suffer a grievous setback.

The observational task is clear for those of us fortunate enough to be involved in the ongoing SN 1987A program. It is extremely important for us to produce repeated gamma ray line observations of high energy resolution and high statistical quality. Such observations will remove any doubts about the reality of the results and provide a diagnostic of the explosion which is so desperately needed by our theoretical colleagues. The gamma ray lineshapes and light curves contain a great wealth of information about the spectacular supernova phenomenon which can be obtained in no other way.

Finally let me say than it is natural for those of us who have not participated in the initial "discovery" of the lines to feel a bit letdown. We must overcome these feelings. The larger task remains before us. When we look back upon these years what will be remembered is the contribution that the gamma ray community made to unravelling the supernova mystery. We have yet to insure that our contribution is a central one. Additional discoveries and surprises are to be expected in the months ahead.

GAMMA-PRODUCING RADIOACTIVITIES FROM SUPERNOVAE

S. E. Woosley

and

Philip A. Pinto

Board of Studies in Astronomy and Astrophysics
Lick Observatory, University of California at Santa Cruz
Santa Cruz CA 95064

Abstract

The production of three isotopes critical to astronomical γ-ray spectroscopy, ^{44}Ti, ^{56}Co, and ^{57}Co, is briefly reviewed along with the information each contains. Emphasis is placed on SN 1987A, the only Type II supernova likely to be seen in γ-lines for decades to come. The 847 keV line from ^{56}Co decay in this supernova should peak approximately 400 days after its explosion with a flux of about 1×10^{-3} cm^{-2} s^{-1}. For comparison, the second best candidate, a Type Ia in the Virgo cluster (20 Mpc) gives a peak flux 100 times smaller than this 100 days after the explosion. ^{57}Co decay in SN 1987A will also present a potentially detectable signal $\sim 1 \times 10^{-4}$ cm^{-2} s^{-1} in 1989 through 1991. ^{44}Ti, chiefly from Type I supernovae, is a wild card, but may be responsible for the diffuse pair background.

I. TYPE II SUPERNOVAE: SN 1987A

Because of its proximity Supernova 1987A (SN 1987A) has presented an unanticipated and unprecedented opportunity to study the dynamics and nucleosynthesis of a Type II supernova using γ-ray spectroscopy. Good agreement of the bolometric luminosity of the supernova on the tail of its light curve as determined consistently by groups in South Africa[1,2], Chile[3-5], and Australia[6] with the known mean lifetime of ^{56}Co, 111.3 d[7] (but note an approximate 2% uncertainty), shows conclusively that 0.07 M$_\odot$ (plus or minus about 20% for uncertainty in the distance) of ^{56}Ni was synthesized in the explosion[8] and that radioactive decay is now powering the light curve.

Based upon the identity of the presupernova star, Sk -69 202, now almost universally accepted, the progenitor star had a mass on the main sequence[8] of 19 ± 3 M$_\odot$ and a helium core mass of 6 ± 1 M$_\odot$ when it died. At death, its luminosity was near 100,000 L$_\odot$, its temperature near 16,000 K, and its radius near 50 R$_\odot$. Two critical unknowns are the explosion energy and the mass of the

© 1988 American Institute of Physics

hydrogen envelope at the time of explosion. Based upon constraints imposed by the light curve, the photospheric velocity and temperature histories, the speed of the slowest moving hydrogen[9] yet detected (2100 km s^{-1}), which sets an upper bound on the expansion rate of the helium core), and the appearance of hard x-rays, presumably Comptonized γ-rays from ^{56}Co decay, in August[10,11], Woosley[8] concludes that the explosion energy was 8×10^{50} ($M_{env}/5$ M$_\odot$) with M_{env} between 3 and 14 M$_\odot$ and probably close to 10 M$_\odot$. Similar conclusions have been reached by Arnett[12,13], Nomoto and Shigeyama[14], Shigeyama et al.[15], and Itoh et al.[16].

An interesting dilemma has unfolded in that the models which agree best with other constraints, basically from the light curve, would not have given a detectable x-ray flux in mid-August when it was in fact discovered, but would have turned on about 75 days later. On the other hand, more rapidly expanding models[17] having lower mass envelopes are capable of turning on in the x-ray band at the proper time, but having done so, continue to increase in x-ray luminosity well beyond what has been observed during the last few months of 1987. They also agree less well with other constraints on the model (light curve, *etc.*) and are disallowed by the upper limits on the 847 keV line placed by SMM and other γ-ray experiments prior to August, 1987.

The solution that has been proposed[10,16-18] to this dilemma is to invoke mixing of the radioactive material out from the center of the core, perhaps to the edge of the helium core, perhaps even into the hydrogen layer. This breaking of the chemical stratification of the ejecta is alleged to occur as a consequence of Rayleigh-Taylor instabilities that exist[19] in the supernova following passage of the deceleration front ("reverse shock") into the core or because the energy released by radioactive decay itself creates a hot, low density "bubble" of radiation that accelerates the overlying layers of oxygen and helium[8,20]. So far the alleged mixing has been artificially parametrized, with most groups simply homogenizing a region of the supernova out to some arbitrary mass (or, equivalently, velocity). We have taken a different, though still artificial tack in our papers, allowing chemical segregation to persist in the mixed region at some level, but smoothing the abundance contours so that a decreasing abundance of heavy elements is mixed farther out. The particular model employed here, Model 10HM[8,17,21,22], mixed ^{56}Co out to velocities as great as 3000 km s^{-1}, but the mass fraction at the edge of the helium core (1700 km s^{-1}), for example, was only 2% of the central value.

In models of this sort an increasing fraction of ^{56}Co is revealed to compensate for the exponential decay in its overall abundance. The hard x-ray flux thus remains relatively constant for about 200 days (Fig. 1). Using this same model, one can also estimate the anticipated flux of ^{56}Co decay lines,

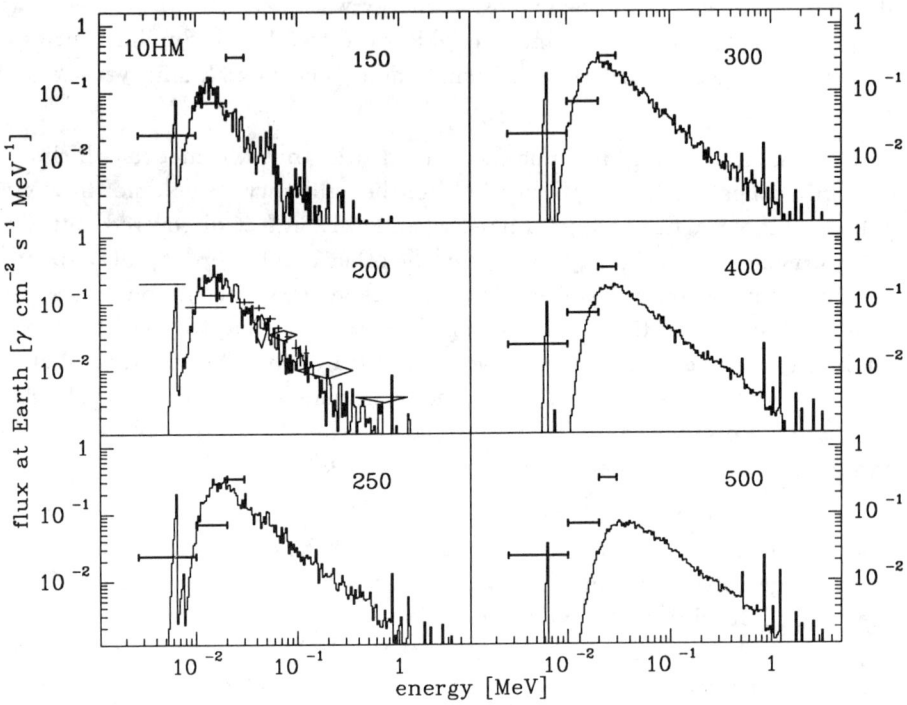

Fig. 1 - Time history of the hard x-ray flux from Model 10HM[17], a 1.4×10^{51} erg explosion of a 6 $M_\odot$ helium core with an overlying hydrogen envelope mass of 10 $M_\odot$. The ejected composition has been mixed, though not thoroughly. Data points near day 200 are from Dotani et al.[11] and Sunyaev et al.[10]. The flux in the iron line at 6.8 keV, day 200 is 1×10^{-4} cm^{-2} s^{-1}. Subsequent line fluxes in units of 10^{-4} photons cm^{-2} s^{-1} are 0.86 on day 250; 0.67 on day 300; 0.3 on day 400; and 0.09 on day 500. An iron abundance equal to 1/2 solar has been assumed in the atmosphere.

e.g. at 847 keV, as shown in Figure 2. [Note added in proof: After this paper was presented at the meeting, γ-line observations were reported by a number of groups, beginning with SMM. These measurements showed an earlier onset to the γ-ray line emission than anticipated in this paper, though the peak values were about the same. The observations thus indicate even more mixing than the model used for Figure 2. Revised models and Monte Carlo calculations in good agreement with all observations up to April, 1988 were subsequently published by Pinto and Woosley[23]. In the interest of scientific honesty, this summary which was written before the meeting and circulated there was not extensively revised.]

The ^{56}Co should be confined to a region moving slower than a few thousand kilometers per second at the very most, and more likely about 1000 km s^{-1} for most of the ^{56}Co. This implies a relatively narrow line (Fig. 3) that will require germanium detectors for its resolution. Owing to the mixing and the fact that the supernova still has an optical depth of several at the time the 847 keV line peaks in intensity, the line will be asymmetric about its zero energy (as redshifted for the recession of the LMC, 270 km s^{-1}; 0.76 keV; energies given in Figure 3 should be decremented by this value). Material moving towards us has a smaller column depth and thus the line centroid is blue-shifted.

Following the demise of the signal from ^{56}Co decay, the species ^{57}Co will present a target of opportunity (Fig. 4) during 1989 through 1991 as challenging and perhaps even more interesting[24-26]. Because of its longer lifetime (e-folding lifetime 391.91 days), ^{57}Co emissions at 122.06 keV (81%) and 136.47 keV (11%) persist as the supernova becomes optically thin, resulting in a very broad peak. While the mass of ^{56}Co in SN 1987A is known from the bolometric luminosity of the tail of the light curve, the mass of ^{57}Co must be inferred from theory (which means that its detection will provide unique and important information). If we assume, as in Figure 4, that ^{57}Co exists in that proportion to mass 56 required to give a final solar abundance ratio to ^{57}Fe/^{56}Fe, 0.0243, there should be 1.7×10^{-3} M$_\odot$ of ^{57}Co produced in SN 1987A. Actually iron is mostly due to Type I supernovae, so there is no good reason to assume a solar ratio for the iron isotopes produced in a Type II. The iron to oxygen ratio produced in SN 1987A, for example, is estimated to be one-third solar[8]. The process responsible for producing both ^{56}Ni and ^{57}Ni is explosive nucleosynthesis, specifically nuclear statistical equilibrium and incomplete explosive silicon burning in those silicon zones which experience temperatures in excess of five billion degrees. The ratio ^{57}Ni/^{56}Ni depends linearly upon the neutron density and thus is proportional to the *neutron excess*, $\eta = \Sigma(N_i - Z_i)(X_i/A_i)$, where N and Z are the neutron and proton numbers of the nucleus having mass fraction X and nucleon number A. The abundance of ^{57}Ni may be further modified during "freeze-out", especially

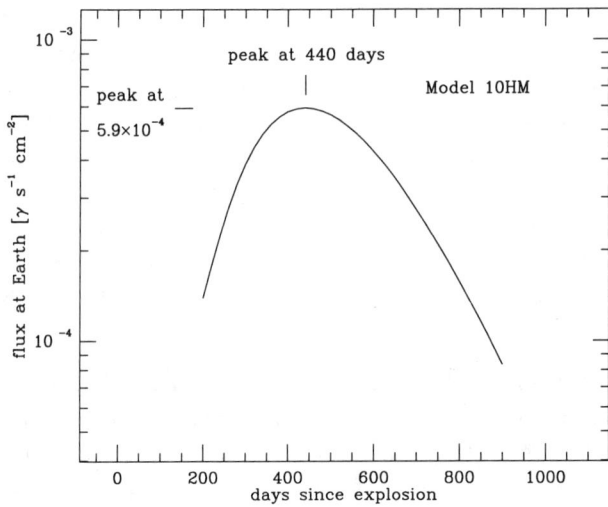

Fig. 2 - Light curve of SN 1987A in the 847 keV line of ^{56}Co decay based upon Model 10HM [see also ref. 23 for revisions].

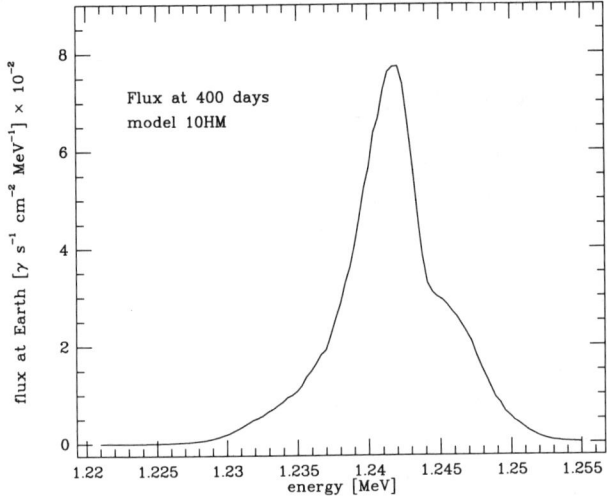

Fig. 3 - Line profile at day 400 of the 1238 keV line from Model 10HM. The overall energy scale should be shifted downward by 1.11 keV to correct for the recession of the LMC (see also ref. 23).

if a large abundance of free α-particles is present[27]. Its final value could easily be a factor of two larger, but probably not much smaller than solar. Thus the ^{57}Co flux in Figure 4, aside from uncertainties in the supernova model itself, might be twice as great. Because this species is produced by the same process as ^{56}Co, the line profile should in theory be like that of the 847 keV line (Fig. 3). Early on, the 122 keV line would be obscured by Comptonized photons from ^{56}Co decay (Fig. 1), thus prospects for detection would be best at late times.

Supernova 1987A has also produced a poorly determined mass of ^{44}Ti. If ^{44}Ca were produced as ^{44}Ti in solar proportions to iron, the ^{44}Ti mass would be 8.6×10^{-5} M$_\odot$. Actually ^{44}Ca may come from a rare class of Type I supernovae, helium detonations (see below), and is produced in considerably smaller proportions in SN 1987A. Explosive nucleosynthesis calculations in actual 15 and 25 M$_\odot$ stars[28,29] show that the ratio of the mass fractions of ^{44}Ti and ^{56}Ni synthesized in explosive silicon burning range from 0.2 to 0.4 times solar. We thus tentatively adopt a mass of ^{44}Ti in SN 1987A of 3×10^{-5} M$_\odot$. The mean life of ^{44}Ti is 68 years and its decay to ^{44}Sc produces two γ-rays of energy 78.4 and 67.85 keV. ^{44}Sc decays immediately to ^{44}Ca almost always producing a γ-ray of 1.157 MeV. The flux of any of these lines at sufficiently late times that the supernova is optically thin should be given by $1.3 \times 10^{-6} e^{-t/68y}$ cm^{-2} s^{-1}, a goal perhaps for our graduate students' graduate students.

II. TYPE Ia SUPERNOVAE

In recent years there has been a convergence upon a single favored model for the common Type Ia supernova event[30]: a thermonuclear explosion of an accreting CO-white dwarf that has approached the Chandrasekhar mass. The explosion is propagated by a subsonic flame front, presumably highly turbulent, known as a "deflagration wave"[31,32]. Comparison to the known and highly homogeneous spectra and light curves of Type Ia supernovae suggests that the successful model will produce about 0.5 M$_\odot$ of ^{56}Ni. Here we specifically consider deflagration Model 2 of Woosley and Weaver[29], which has been shown[30] to reproduce well the observed light curve and late time spectrum, and is similar to W7 of Nomoto et al. which has been shown[33] to produce a good early time spectrum as well. This model ejects 0.85 M$_\odot$ of iron group elements of which 0.51 M$_\odot$ are in the form of ^{56}Ni. The explosion energy is 1.04×10^{51} erg and the velocity at the outer edge of the iron-nickel core, about 9500 km s^{-1}. The x- and γ-ray spectrum of this model were calculated using the same techniques[17] as for Figures 1 – 4 and the results are shown in Figures 5 – 7.

Apparently, at peak, a Type Ia supernova in the Virgo cluster, where one might hope to have one to several events per year[34], will have a γ-line flux (847 keV) about 100 times less than that of SN 1987A. In order to detect

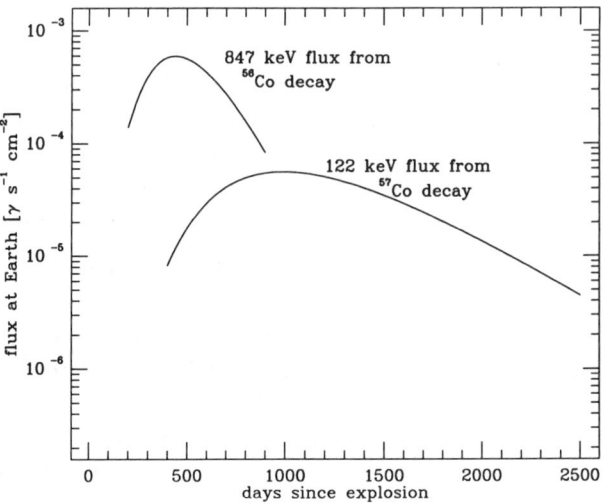

Fig. 4 - Light curve of Model 10HM in the 122 keV line of ^{57}Co decay.

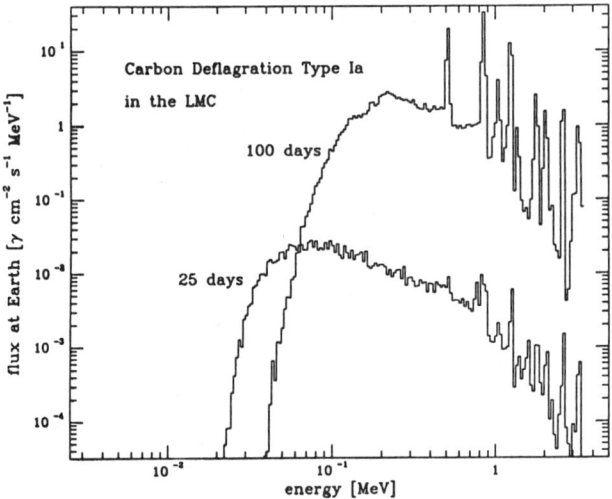

Fig. 5 - Hard x-ray flux from Comptonized photons generated by the decay of radioactive ^{56}Co in a carbon deflagration model for a Type Ia supernova (Model 2 of ref. 29). Note that, unlike Figs. 6 and 7, a distance of 50 kpc has been assumed.

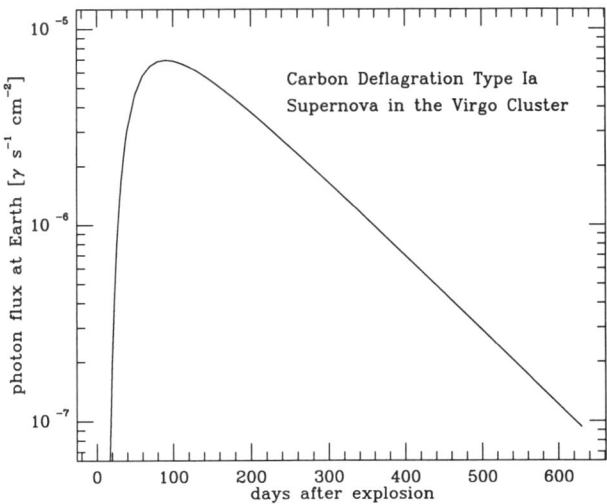

Fig. 6 - Light curve of a carbon deflagration supernova in the 847 keV decay line of ^{56}Co. A distance of 20 Mpc, typical of the Virgo cluster has been assumed.

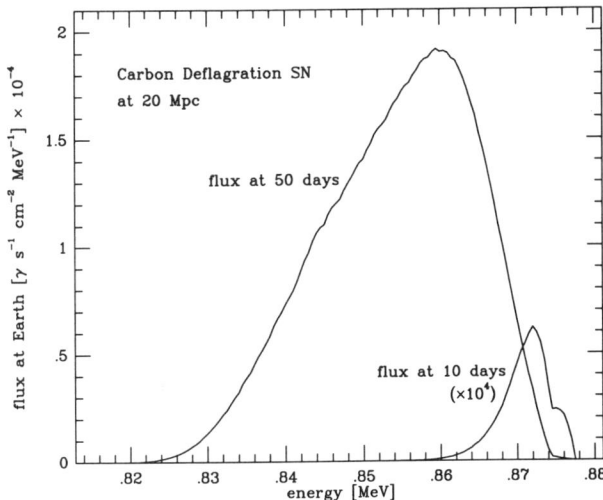

Fig. 7 - Line profiles at 10 and 50 days for a carbon deflagration, Type Ia supernova at 20 Mpc.

such faint events using the best currently available detectors (sensitivity about 1×10^{-4} cm^{-2} s^{-1}), the Type Ia supernova would need to be closer to the Earth than 5 Mpc. If the absolute magnitude of a Type Ia supernova at peak is -19, then such a supernova would need have an apparent magnitude at peak brighter than 9.5. To *study* the supernova, and obtain a few data points within the line one would probably need another magnitude or so and a detector having a resolution of $\sim$15 keV. Since supernovae as bright as 8th magnitude are very rare (about once per 20 years) one needs much better sensitivity than current balloons are capable of and better energy resolution than anticipated at 847 keV for the OSSE experiment on GRO ($\sim$8%).

Type Ia supernovae may also eject quantities of ^{44}Ti of interest to γ-line astronomy. A typical deflagration – Ia event will produce about 0.2 to 0.5 $\times 10^{-4}$ M$_\odot$ of it[31,35], comparable to Type II supernovae. Helium detonations, however, which have yet to be directly correlated with a particular class of Type I's (Ib perhaps[36,37] or Ip[30]), produce much more. Woosley, Taam, and Weaver[38] found that a pure helium dwarf of 0.64 M$_\odot$, when detonated, produced 0.009 M$_\odot$ of ^{44}Ti, so much that, solely upon the basis of its ^{44}Ca nucleosynthesis, this type of supernova must be a rare event. Detonations in the helium layers overlying a degenerate carbon oxygen core produce from 0.6 to 4×10^{-4} M$_\odot$ of ^{44}Ti. Such is the decay time of ^{44}Ti that, save for SN 1987A, we are unlikely to see its γ-line emissions from any extragalactic supernova.

However, in our own Galaxy the remnants of recent supernovae may glow for centuries in potentially detectable[39] γ-lines of ^{44}Ti. The rare Type I's responsible for making ^{44}Ca would be the stronger emitters but the fainter Type Ia's and Type II's might also be detectable §I). For normalization, at a distance of 1 kpc, the flux from a supernova remnant initially containing 10^{-4} M$_\odot$ of ^{44}Ti would be $7.4\times 10^{-3}\,e^{-t/68y}$ cm^{-2} s^{-1}. At a fiducial age of 400 years this implies a flux of 2×10^{-5} cm^{-2} s^{-1}. Most known remnants of interest are either very old or several kiloparsecs away. Thus detection of ^{44}Ti from a single Galactic supernova would require the fortuitous occurrence (in the past few centuries) of one nearby or major advances in detector sensitivity.

III. ^{44}Ti AND THE DIFFUSE PAIR BACKGROUND

The nucleus ^{44}Ti may contribute in another fashion to γ-ray spectroscopy, however. The decay of its daughter, ^{44}Sc, to ^{44}Ca is accomplished by positron emission 95% of the time. Thus ^{44}Ti, like ^{26}Al, contributes[40] to the diffuse background from pair annihilation[41,42]. To proceed in as model independent a fashion as possible let us assume that ^{56}Fe and ^{44}Ca are currently being produced in the Galaxy in solar proportions, 0.0012. Further, each Type I supernova produces, *on the average*, 0.5 M$_\odot$ of ^{56}Fe and the rate of Type I supernovae is (to a factor of 2) one per century[34,43,44]. Then, again on the average, Type I and II supernovae must make about 6×10^{-4} $M_\odot$ of ^{44}Ti for every Type Ia supernova that occurs, i.e., every century. Note that this is about 10 to 25 times the fiducial amount estimated in the previous section for a deflagration model or for SN 1987A. Presumably most of the ^{44}Ti is being produced in rare events (helium dwarf detonation?). Anyway, continuing the argument, which must be globally true so long as ^{44}Ca is made as ^{44}Ti and no one has ever found any other way, this implies a ^{44}Ti synthesis rate in the Galaxy of 6 $M_\odot$ every million years. Since, as Ramaty and Lingenfelter have pointed out[40], positrons produced by the decay of the 3 $M_\odot$ of ^{26}Al believed to inhabit the interstellar medium and be renewed every million years, can produce $\sim$20% of the diffuse signal from pair annihilation, it follows that 6 $M_\odot$ (with an uncertainty of at least a factor of three) of ^{44}Ti might produce the remainder. Depending upon the lifetime of the positrons in the interstellar medium there could be fluctuations both in space and in time of the diffuse signal.

This work has been supported by the National Science Foundation (AST-84-18185) and by the NASA Theory Program. We are also grateful for a grant of time at the San Diego Supercomputer Center where a portion of these calculations were carried out.

BIBLIOGRAPHY

1. J. W. Menzies, R. M. Catchpole, G. van Vuuren, H. Winkler, and 12 others, *MNRAS*, **227**, 39p, (1987).

2. R. M. Catchpole, J. W. Menzies, A. S. Monk, W. F. Wargau, and 16 others, *MNRAS*, **229**, 15p, (1987).

3. M. M. Phillips, S. R. Heathcote, M. Hamuy, and M. Navarrete, CTIO preprint submitted to *Astron. J.*, (1987).

4. M. Hamuy, N. B. Suntzeff, R. Gonzalez, and G. Martin, *A. J.*, **95**, 63, (1988).

5. J. Danziger, P. Bouchet, R. A. E. Fosbury, C. Gouiffes, L. B. Lucy, A. F. M. Moorwood, E. Oliva, and F. Rufener, paper presented at George Mason Fall Workshop on Supernova 1987A, ed. D. Kazansas, (Cambridge Univ. Press: Cambridge), in press, (1988).

6. M. A. Dopita, J. A. Dawe, N. Achilleos, R. J. V. Brissenden, and 9 others, preprint Mt. Stromlo and Siding Spring Observatories, (1987).

7. *Nuclear Data Sheets*, M. J. Martin, editor, **51**, 67, (1987).

8. S. E. Woosley, *Ap. J.*, in press, July, (1988).

9. J. H. Elias, and S. C. Gregory, in preparation for *Ap. J.*, (1988).

10. R. Sunyaev and 33 others, *Nature*, **330**, 227, (1987).

11. T. Dotani and 36 others, *Nature*, **330**, 230. (1987).

12. W. D. Arnett, *Ap. J.*, 319, 136, (1987).

13. W. D. Arnett, paper presented at George Mason Fall Workshop on Supernova 1987A, ed. D. Kazansas, (Cambridge Univ. Press: Cambridge), in press, (1988).

14. K. Nomoto and T. Shigeyama, paper presented at George Mason Fall Workshop on Supernova 1987A, ed. D. Kazansas, (Cambridge Univ. Press: Cambridge), in press, (1988).

15. T. Shigeyama, K. Nomoto, K. Hashimoto, and D. Sugimoto, *Nature*, **328**, 320, (1987).

16. M. Itoh, S. Kumagai, T. Shigeyama, K. Nomoto, and J. Nishimura, *Nature*, **330**, 233, (1987).

17. P. A. Pinto and S. E. Woosley, *Ap. J.*, in press, June, (1988).

18. N. Shibazaki and T. Ebisuzaki, T., *Ap. J. Lettr.*, in press, (1988).

19. R. Chevalier and R. I. Klein, *Ap. J.*, **219**, 994, (1978).

20. S. E. Woosley, P. A. Pinto, and L. Ensman, *Ap. J.*, **324**, 466, (1988).

21. S. E. Woosley, in *Proc. IAU Colloq. No. 108: Atmospheric Diagnostics of Stellar Evolution*, ed. K. Nomoto, (D. Reidel: Dordrecht), in press, (1988).

22. S. E. Woosley, paper presented at George Mason Fall Workshop on Supernova 1987A, ed. D. Kazansas, (Cambridge Univ. Press: Cambridge), in press, (1988).

23. P. A. Pinto and S. E. Woosley, *Nature*, in press, (1988).

24. D. D. Clayton, *Ap. J.*, **188**, 155, (1974).

25. S. E. Woosley, P. A. Pinto, P. Martin, T. A. Weaver, *Ap. J.*, **318**, 664, (1987).

26. K. W. Chan and R. E. Lingenfelter, *Ap. J. Lettr.*, **318**, L51, (1987).

27. S. E. Woosley, in *Nucleosynthesis and Chemical Evolution*, 16th Advanced Course Swiss Society of Astronomy and Astrophysics, *Saas-Fee Lecture Notes*, ed. B. Hauck, A. Maeder, and G. Meynet, (Geneva Observatory: Geneva), p. 1, (1986).

28. S. E. Woosley and T. A. Weaver, in *Essays in Nuclear Astrophysics*, ed. C. A. Barnes, D. D. Clayton, and D. N. Schramm, (Cambridge Univ. Press: Cambridge), p. 377, (1982).

29. S. E. Woosley and T. A. Weaver, in *Radiation Hydrodynamics in Stars and Compact Objects*, ed. D. Mihalas and K.-H. A. Winkler, (Springer Verlag: Berlin), p. 91, (1986), and unpublished.

30. S. E. Woosley and T. A. Weaver, *Ann. Rev. Astron. and Ap.*, **24**, 205, (1986) and references therein.

31. K. Nomoto, F. -K. Theilemann, and K. Yokoi, *Ap. J.*, **286**, 644, (1984).

32. S. E. Woosley, T. S. Axelrod, and T. A. Weaver, in *Stellar Nucleosynthesis*, ed. C. Chiosi and A. Renzini, (D. Reidel: Dordrecht), p. 263, (1984).

33. D. Branch, J. B. Doggett, K. Nomoto, and F. -K. Thielemann, *Ap. J.*, **294**, 619, (1985).

34. G. Tammann, in *Supernovae: A Survey of Current Research*, ed. M. J. Rees and R. J. Stoneham, (D. Reidel: Dordrecht), p. 371, (1982).

35. F. -K. Thielemann, K. Nomoto, and K. Yokoi, *Astron. and Ap.*, **158**, 17, (1986) and private communication.

36. D. Branch and K. Nomoto, *Astron. and Ap.*, **164**, L13, (1984).

37. D. Branch, in *Proc. IAU Colloq. No. 108: Atmospheric Diagnostics of Stellar Evolution*, ed. K. Nomoto, (D. Reidel: Dordrecht), in press, (1988).

38. S. E. Woosley, R. E. Taam, and T. A. Weaver, *Ap. J.*, **301**, 601, (1986).

39. D. D. Clayton, S. A. Colgate, and G. J. Fishman, *Ap. J.*, **155**, 75, (1969).

40. R. Ramaty and R. E. Lingenfelter, in *The Galactic Center*, ed. D. C. Backer, (Am. Inst. of Phys.: New York), in press, (1987).

41. D. D. Clayton, *Nature Phys. Sci.*, **244**, 137, (1973).

42. S. E. Woosley, *Ap. J. Lettr.*, to be submitted, (1988).

43. S. Van den Bergh, R. D. McClure, and R. Evans, *Ap. J.*, **323**, 44, (1987).

44. E. Cappellaro and M. Turatto, M., *Astron. Ap.*, **190**, 10, (1988).

GAMMA-RAY LINES FROM SUPERNOVAE

K. W. Chan and R. E. Lingenfelter
Center for Astrophysics and Space Sciences
University of California, San Diego, CA 92093

ABSTRACT

We have calculated the time-dependent fluxes and shapes of the gamma-ray lines expected from decay of ^{56}Ni and other radionuclei in both Type I and Type II supernovae. The lines from different models of these supernovae have very distinct shapes. The widths, line centroids, shapes, intensity ratios, and light curves of the line emission can be used to distinguish between the different models and constrain the parameters in these models. We discuss gamma-ray lines from SN 1987A in the Large Magellanic Cloud and also demonstrate other features of gamma-ray line emission from supernovae.

INTRODUCTION

It has been recognized for a long time that detectable gamma-rays can be expected from decays of explosively synthesized nuclei in supernovae[1,2]. The most intense gamma-ray lines are expected to be those from ^{56}Ni→^{56}Co→^{56}Fe, and the first detection of two of these lines has just been reported[3-5]. Decays of ^{57}Co→^{57}Fe and ^{44}Ti→^{44}Sc→^{44}Ca, however, should also be important sources of gamma rays in supernovae.

Such radionuclei, carried outward in the expanding supernova nebula, decay and then deexcite by gamma-ray line emission. At early times, the gamma rays interact with the material in the supernova and are compton-scattered down to x-ray energies which are photoelectrically absorbed and their energy is eventually released as optical emission. However, as the supernova expands, some of the gamma-rays begin to escape without scattering.

These gamma-ray lines are doppler-broadened by the velocity spread of the radionuclei in the expanding nebula. The gamma-ray line shapes thus reflect the velocity distribution within the supernova, modified by the opacity along the line of sight, and their measurement with high resolution can give us information on this distribution.

We are making systematic calculations of the time-dependent fluxes and shapes of these gamma-ray lines from both Type I and Type II supernovae generally, as well as the recent Type II supernova 1987A specifically. The details of these calculations are discussed elsewhere[6-8]. In the following, we shall only highlight some of the features that might be expected from these lines.

GAMMA-RAY LINES FROM TYPE I SUPERNOVAE

The principal gamma-ray lines expected[9,10] from Type I supernovae are those from freshly synthesized ^{56}Ni. It decays into ^{56}Co in a mean life of 8.8 days with major lines at 158 keV (100%), 812 keV(75%), and 750 keV (48%). ^{56}Co in turn decay into ^{56}Fe in a mean life of 113.6 days with major lines at 847 keV (100%), 1238 keV (68%), and 2599 keV (17%). Also ^{57}Ni, produced by neutron capture, decays rapidly to ^{57}Co which subsequently decays into ^{57}Fe in a mean life of 392 days, giving major lines at

© 1988 American Institute of Physics

122 keV (89%) and 136 keV (11%). Although these lower energy lines are more strongly attenuated, they should still be detectable because of the longer ^{57}Co life time.

The expected time-dependent gamma-ray line fluxes from a Type I supernova at a distance of 10 Mpc, which we calculated for models of thermonuclear deflagration[9] and detonation[10] on carbon-oxygen white dwarfs, are shown in Figure 1. The fluxes from such a supernova should be detectable by the Gamma-Ray Observatory which has a line sensitivity of about 2×10^{-5} photon cm^{-2} s^{-1} around 847 keV. Moreover, the proposed Nuclear Astrophysics Explorer with a sensitivity of 2×10^{-6} photon cm^{-2} s^{-1} should be able to detect the 847 keV line from supernovae as far away as 40 Mpc, which will include all of those that occur in the Virgo Cluster.

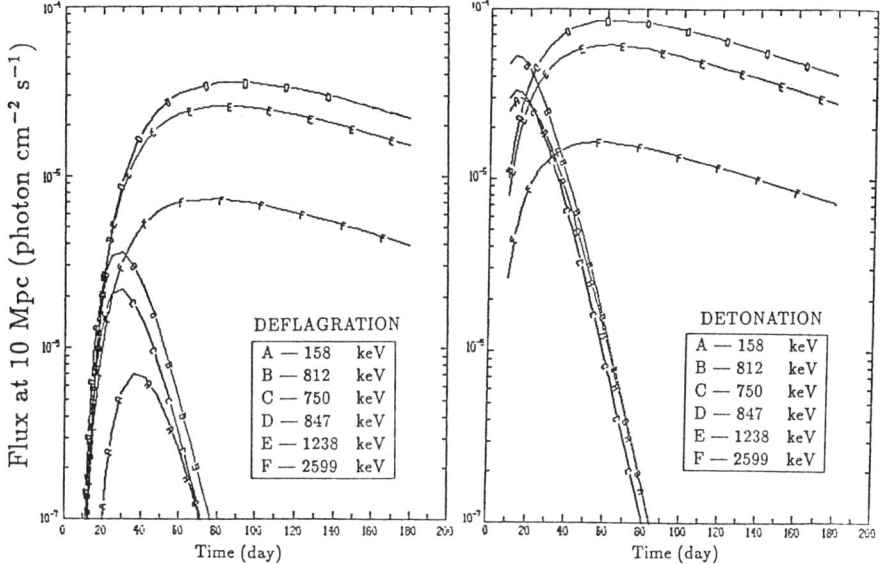

Fig. 1. Gamma-ray line fluxes from ^{56}Ni and ^{56}Co decay for deflagration[9] and detonation[10] models of a Type I supernova at 10 Mpc.

Detection of these lines and measurements of the time-varying line ratios and shapes can provide a direct determination not only of the mass of ^{56}Ni produced in the supernova and its distribution within the expanding nebula, but also the total mass and energy of the ejected nebula and the velocity versus mass distribution in that ejecta. Such observations should clearly distinguish between the proposed models of Type I supernova.

Measurements of the line widths and profiles are essential, since they contain more detailed information than the fluxes alone. In the detonation model[10] about one solar mass of ^{56}Ni is produced with relatively little overlying material and the ejecta expands with a speed of the order of 10^4 km s^{-1}, reaching $\sim 3 \times 10^4$ km s^{-1} at the edge of the supernova. In the deflagration model[9] only about a half solar mass of ^{56}Ni is synthesized with a comparable mass of lighter elements lying on top of it and the expansion velocity is also only about half of that of the detonation model. The resulting differences in the expected widths and profiles of the 847 keV line are shown in Figure 2. In both cases the emerging lines are strongly blueshifted as a result of the greater attenuation of the red wing of the line which comes from the deeper backward moving hemisphere of the ejecta. The deflagration line width of about 5% or less requires an energy resolution better than that of NaI detectors (> 6% at this energy).

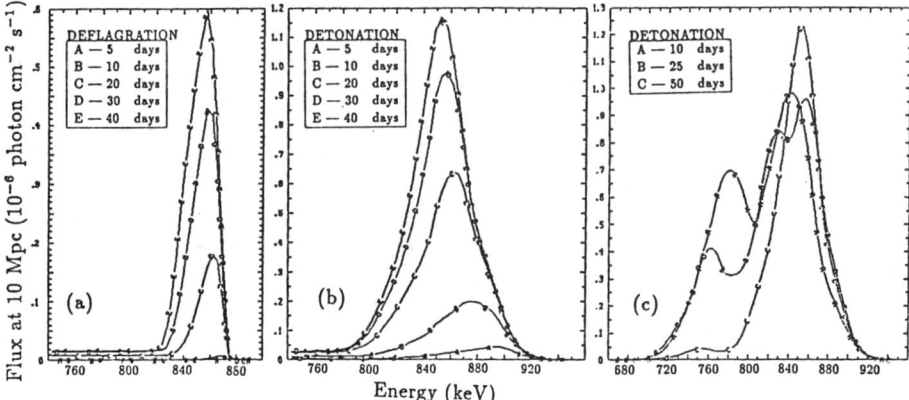

Fig. 2. Evolution of the ^{56}Co 847 keV line profile for detonation (a) and deflagration (b) models of Type I supernova; and the added contribution of 750 and 812 keV from ^{56}Ni for the detonation model (c).

Moreover, even in the detonation model, at early times the lines are so strongly blueshifted that the 750 keV and 812 keV lines from ^{56}Ni decay blend together with the 847 keV line from ^{56}Co decay. As can be seen in Figure 2c, the rapidly decaying ^{56}Ni line at 812 keV first appears in the 847 keV range before the longer lived 847 keV line from ^{56}Co decay actually shows up, causing confusion in line identification. This kind of line blending would be particularly important for fast expanding supernovae like those modeled by the detonation. High resolution spectroscopy is clearly needed to measure these lines.

In our own Galaxy, the 1157 keV line from ^{44}Ti→^{44}Sc→^{44}Ca decay should be detectable from Type I supernovae on a much longer time scale. Since ^{44}Ti decays with a mean life of 78 years, it can produce detectable gamma-ray line emission for a few hundred years. Thus observations of this line can reveal the locations of the most recent Type I supernovae in our Galaxy[1].

Moreover, as we have shown[6], measurement of the line profile can give a unique determination of the age of each supernova. This is possible because the gamma rays observed at different doppler-shifted energies, E, at any particular time, t, were emitted at different times, t' = $(E/E_o)t$, where E_o is the rest energy of the line. These time differences reflect different light travel times across the nebula. Thus we look back at gamma rays emitted at earlier times when we observe the red-shifted half of the line from the more distant half of the nebula and look forward at later times when we observe the blue-shifted half of the line from the nearer half. This leads to a net red-shift of the observed line that increases with time, since the more red-shifted photons from the far side were emitted at earlier times when less of the the ^{44}Ti had decayed and the emission rate was thus higher.

An example of the expected time-dependent line profile of the 1157 keV line from a supernova at a nominal distance of 7 kpc, calculated[6] for a helium dwarf detonation model[10], is shown in Figure 3. As can be seen, the peak of the line shifts by ~ 7 keV/100 yr over the first 500 years and measurement of this red shift can thus give a direct measure of the age of the supernova remnant. Although the rate of the shift depends on the expansion velocity, this can be directly determined from the measured line width. Such line fluxes from remnants as old as 500 years should still be detectable with GRO. But the predicted gamma-ray line fluxes depend on the expected yield of ^{44}Ti/^{56}Ni, which ranges from about 10 times the solar ratio of their decay products ^{44}Ca/^{56}Fe for helium dwarf detonations[10] to only about 0.1 times that ratio for carbon-oxygen dwarf deflagration[9]. Even the latter, however, should be detectable by GRO for about 200 years.

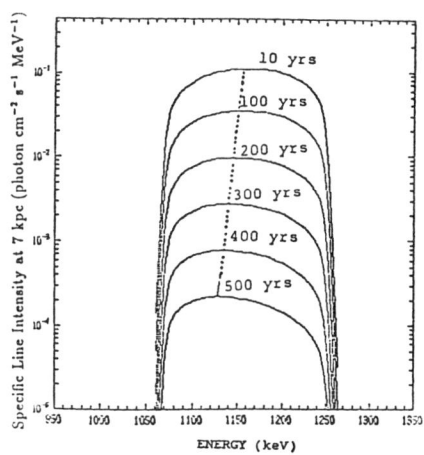

Fig. 3 1157 keV line from ^{44}Sc decay in a Type I supernova at 7 kpc.

GAMMA-RAY LINES FROM TYPE II SUPERNOVAE

The principal gamma-ray lines expected from Type II supernovae are also those from ^{56}Co decay, but, as we have shown[7], they may have very different shapes from those of Type I supernovae. We have calculated the time-dependent gamma-ray line fluxes and profiles expected from the core collapse model[11,12] of a 15 M$_\odot$ star. Although more ^{56}Ni is produced in the collapsing core of the star, nearly all of it is trapped to form a neutron star. Only a small amount of synthesized ^{56}Ni, ~ 0.1 M$_\odot$, is ejected and it is buried deep under the much more massive (> 10 M$_\odot$) ejecta. The expansion velocity of the ^{56}Ni layer is also only about 1,000 km s^{-1}, much slower than that expected from Type I supernovae. If there is a rarefaction behind the ^{56}Ni layer, created by the core collapse forming a compact remnant, then the line shape can be quite distinctive, as can be seen from Figure 4. The low velocity of the radioactive ^{56}Ni layer gives narrow gamma-ray lines with a width of only a few keV. Greater attenuation, or limb darkening, of the gamma rays from radionuclei moving transverse to the line of sight strongly suppresses the flux at the line center, but the rarefaction in the center of the nebula allows the most red-shifted gamma

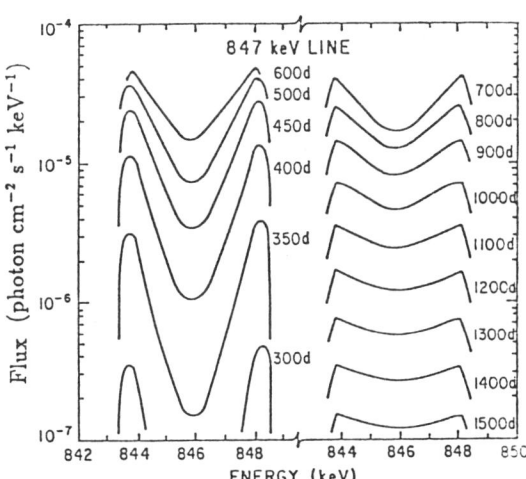

Fig. 4. Evolving 847 keV line from a Type II supernova in the Large Magellanic Cloud.

rays from radionuclei moving directly away from us to be no more attenuated than the most blue-shifted gamma rays from radionuclei moving directly toward us. Thus the resulting line is doubled-peaked with a deep central trough at early times.

If there is no rarefaction, however, and the radionuclei fill the inner part of the remnant with an appreciable fraction moving at much lower velocities, then the trough in the line profiles can be filled, as is assumed in one recent model[13] of supernova 1987A.

Because of the obscuration by the massive ejecta and the small amount of ^{56}Ni synthesized, the gamma-ray line fluxes expected from Type II supernovae are much lower than for Type I. Nonetheless, the detectable fluxes were expected[7,13,14] from the recent supernova 1987A in the Large Magellanic Cloud. Figure 5 shows the time-dependent fluxes of the ^{56}Co and ^{57}Co lines, expected[7] from the core collapse model[11,12] of a 15 $M_\odot$ star at the distance of the Large Magellanic Cloud. The peak fluxes are about 2 x 10^{-4} photon cm^{-2} s^{-1} occurring about 500 to 600 days after the explosion. Because the gamma-ray opacity decreases with increasing energy, the 1238 keV line should peak earlier and have a larger peak flux than the 847 keV line, even though its emissivity is less. The two low energy lines from ^{57}Co decay eventually dominate the gamma-ray emission at later times because of the longer life time of that species

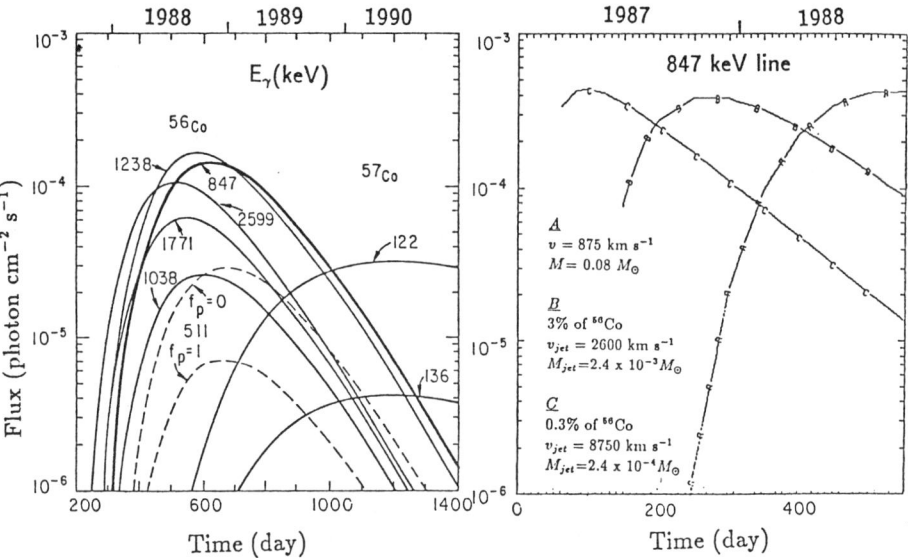

Fig. 5. Calculated[7] time-dependent line fluxes from a Type II supernova in the Large Magellanic Cloud.

Fig. 6. Hypothetical[17] time-dependent 847 keV line fluxes from random jets, or plumes, ejecting 0.3% and 3% of the ^{56}Co at higher velocity.

Since the lines are are still strongly attenuated even at maximum, a different ejecta mass or even a different velocity distribution with the same ejecta mass can significantly change the maximum line flux, as can be seen from a comparison of the 847 keV line fluxes in Figures 5 and 6 (curve A). Detectable line fluxes could also be emitted much earlier, if even a small fraction of the ^{56}Co were ejected at much higher velocities. The occurrence of jets, or plumes, of high velocity ejecta has been suggested[15,16] in model calculations exploring the effects of rotation and magnetic fields on supernova explosions. These calculations suggest that jets of ejecta may carry as much as 10^{-3} $M_\odot$ at velocities of several times 10^3 km s^{-1}. We have calculated[17] the expected time-dependent 847 keV line emission, shown in Figure 6 (curves B and C), for two hypothetical systems of randomly emitted jets carrying 3% of ^{56}Co at 3 times the velocity and 0.3% of ^{56}Co at 10 times the velocity expected in the 15 $M_\odot$ model discussed above. As can be seen, such jets could produce observable gamma-ray lines within months of the explosion. The profiles of such lines should be wider and more blue-shifted than the lines from the lower velocity ^{56}Co layer.

The recent detection[3-5] of 847 and 1238 keV lines after only 160 to 250 days may very well result from such jets or plumes. Since the observed ratio of the 1238 to 847 keV line fluxes is consistent with the branching ratio of 68%, the line fluxes cannot be greatly attenuated. Thus the reported[3] 847 keV flux of $(1.0 \pm 0.25) \times 10^{-3}$ photon cm^{-2} s^{-1} could be produced by only about 0.0015 $M_\odot$ of exposed ^{56}Co — only about 2% of the 0.07 $M_\odot$ of ^{56}Co required[18] to account for the bolometric luminosity. The lack of a significant blue shift in the observed[4] lines would not be surprising if the observed ^{56}Co is only in a few jets, or plumes, since in a random distribution their most probable directions would be roughly transverse to the line of sight, thus greatly reducing any red or blue shift. If this is the explanation of the observed line emission, then that from the bulk of the ^{56}Co is still to come. Moreover, if there is mixing of the ejecta, then these line fluxes would also be expected[13] to reach both higher and earlier maxima than those shown in Figure 5.

Acknowledgements. We thank Phil Pinto and Stan Woosley for helpful discussions and NASA for financial support of this work under grant NSG-7541.

REFERENCES

1. D. D. Clayton, S. A. Colgate and G. J. Fishman, *Ap. J.*, **155**, 75 (1969).
2. D. D. Clayton, *Ap. J.*, **188**, 155 (1974).
3. S. M. Matz *et al.*, this volume (1987).
4. W. Sandie *et al.*, this volume (1987).
5. W. R. Cook, D. M. Palmer and T. A. Prince, this volume (1987).
6. K. W. Chan and R. E. Lingenfelter, *20th Internat. Cosmic Ray Conf. Papers*, **1**, p. 164. (1987).
7. K. W. Chan and R. E. Lingenfelter, *Ap. J.*, **318**, L51. (1987).
8. K. W. Chan and R. E. Lingenfelter, in preparation (1987).
9. K. Nomoto, F.-K. Thielemann and K. Yokoi, *Ap. J.*, **286**, 644 (1984).
10. S. E. Woosley, R. E. Taam and T. A. Weaver, *Ap. J.*, **301**, 601 (1986).
11. T. A. Weaver and S. E. Woosley, in *Supernova Spectra*, (A.I.P., N.Y., 1980) p. 15.
12. T. A. Weaver and S. E. Woosley, *Ann. N. Y. Acad. Sci.*, **336**, 335 (1980).
13. P. Pinto and S. E. Woosley. *Ap. J.*, submitted. (1987)
14. N. Gehrels, C. J. MacCallum and M. Leventhal. *Ap. J.*, **320**, L19. (1987)
15. P. Bodenheimer and S. E. Woosley, *Ap. J.*, **269**, 281 (1983).
16. E. M. D. Symbalisty, D. N. Schramm and J. R. Wilson, *Ap. J.*, **291**, L11 (1985).
17. R. E. Lingenfelter and K. W. Chan, Workshop on SN 1987A, GSFC, March 6, 1987.
18. S. E. Woosley, this volume (1987).

PARTICLE ACCELERATION AND GAMMA-RAY PRODUCTION IN SN1987A

A. K. Harding
Code 665, NASA/GSFC, Greenbelt, MD 20771

T. K. Gaisser and T. Stanev
Bartol Research Institute, University of Delaware, Newark, DE 19716

ABSTRACT

A pulsar wind model which has been successful in accounting for the observed radiation from the Crab nebula is used to estimate the power and energy of accelerated particles in SN1987A and the resulting gamma-ray line and continuum fluxes in terms of the pulsar period and surface magnetic field strength. For a pulsar period less than ~ 10 ms (or spin-down power greater than 10^{39} erg/s) pion-decay gamma rays at TeV and possibly PeV energies should be observable with ground-based detectors in the Southern hemisphere. Nuclear excitation line fluxes of C, O and N are small compared to the 1 - 10 MeV continuum flux of synchrotron radiation from TeV electrons. This electron continuum flux should be detectable and may be comparable to the expected line fluxes from radioactive decays.

INTRODUCTION

Young supernova remnants are expected to be bright sources of energetic gamma rays and neutrinos from collisions of accelerated particles in the expanding envelope[1-3]. High energy photons will result from the decay of neutral pions produced in collisions of protons with material in the envelope and from synchrotron radiation by accelerated electrons and positrons. Neutrinos from the decay of charged mesons are also expected[4]. The relatively high density and magnetic field in the remnant during its first few years of expansion enable these processes to take place much more efficiently than in older remnants such as the Crab nebula, where proton collisions are not expected to be important.

Observations of high energy radiation from SN1987A could provide important clues to the nature of particle acceleration in young supernova remnants. If $> 10^{39}$ erg/s were available to accelerate protons above 10 TeV then the resulting TeV gamma rays would be observable with present ground-based detectors in the Southern hemisphere[5]. The accelerated particles themselves will not be directly observable for two reasons. First, they will be contained in the expanding envelope by diffusion in turbulent magnetic fields. Second, even those particles that escape would be deflected by galactic and intergalactic magnetic fields and become part of the general cosmic ray pool. We must thus depend on neutral secondaries for a signal.

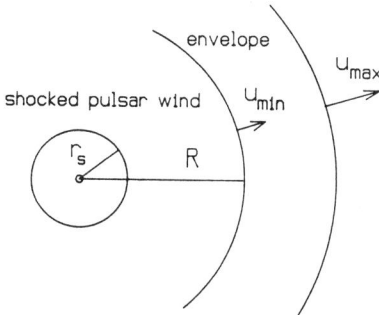

Fig. 1. Schematic illustration of pulsar wind model.

ACCELERATION MODEL

In order to calculate the expected gamma-ray spectrum and flux, we must know the spectrum and power of radiating particles, the density, magnetic field and size of the region in which they propagate and the optical depth to the photons they produce. The gamma-ray flux depends primarily on the power in accelerated protons and electrons, and different acceleration models may give observable signals. We have examined a model[5] in which protons and electrons are accelerated at a pulsar wind shock inside the expanding shell. The pulsar wind model, originally due to Rees and Gunn[6], has been very successful in accounting for the optical radiation from the Crab nebula[7]. The model is illustrated schematically in Fig. 1. If the central engine is a rapidly rotating pulsar then the spin-down power available for energetic particles is

$$L_p = 4 \times 10^{39} \text{erg s}^{-1} \epsilon \, B_{12}^2 \, P_{10}^{-4} \qquad (1)$$

where $B_{12} = B/10^{12}$ G is the pulsar surface magnetic field, $P_{10} = P/10$ ms is the pulsar period and ϵ is the efficiency for particle acceleration. In vacuum, the spin-down power would appear as magnetic dipole radiation at the pulsar rotation frequency. Because the wave frequency is probably below the ambient plasma frequency, this power goes instead into driving a relativistic MHD wind carrying electrons and positrons. The wind is confined by the inner edge of the expanding envelope and a shock forms at r_s where the wind ram pressure is balanced by the pressure in the nebula due to the accumulated energy outflow from the pulsar:

$$r_s \simeq 10^{14} \text{cm} \, \tau_{yr} u_8^{3/2} \simeq 0.1 \, R \qquad (2)$$

where τ_{yr} is the age of the nebula in years, $u_8 = u_{min}/10^8 \text{cm s}^{-1}$ is the expansion velocity of the inner edge of the envelope and R =

$u_{min}\tau$ is the radius of the expanding wind bubble. The wind magnetic field at the shock is

$$B_s \simeq 3 \text{ Gauss } B_{12} P_{10}^{-2} \tau_{yr}^{-1} u_8^{-3/2} \qquad (3)$$

We postulate that acceleration of charged particles (both ions and electrons) occurs by a first order Fermi mechanism at the shock. The acceleration time is inversely proportional to the magnetic field in Eqn (3) which is large. The maximum energy of protons is obtained by equating the acceleration time to the time for protons to diffuse away from the shock:

$$E_p^{max} \simeq 1.9 \times 10^4 \text{ TeV } B_{12} P_{10}^{-2} \qquad (4)$$

For electrons, the maximum energy occurs where the acceleration time equals the synchrotron loss time:

$$E_e^{max} \simeq 8.6 \text{ TeV } P_{10} B_{12}^{-1/2} \tau_{yr}^{1/2} u_8^{3/4} \qquad (5)$$

As shown in Ref. 8, particles with energy less than

$$E_{esc} \simeq 10^5 \text{ TeV } \xi^{-1} B_{12} u_9^{1/2} P_{10}^{-2} \qquad (6)$$

will be contained in the expanding envelope for times longer than the age of the supernova. Here, $u_9 = u_{max}/10^9 \text{cm s}^{-1}$ is the outer expansion velocity of the envelope and $\xi > 1$ is the ratio of the diffusion coefficient to its minimum value, which is one-third the Larmor radius times the particle velocity. Comparing the energy in Eqn (6) with the maximum acceleration energies in Eqns (4) and (5), it is evident that all accelerated particles will be confined within the remnant. When the accelerated particle energy is limited by escape, as in the case of protons above, the spectrum from shock acceleration is generally a power law, $N(E_p) \propto E_p^{-\gamma}$. However, when the radiative loss time is much less than the escape time, as is the case with the electrons, there will be a pile-up at the maximum energy[9], resulting in a nearly monoenergetic spectrum at E_e^{max}.

CONTINUUM GAMMA RAYS

After the first few days of expansion, the density is low enough that all the secondary mesons from proton collisions decay before interacting. Only neutral particles such as photons and neutrinos escape from the envelope; protons are contained and have an effectively infinite pathlength for interaction. The photon and neutrino spectra are then linearly proportional to the power L_p in accelerated protons. Table 1 shows calculated pion-decay photon fluxes for SN1987A normalized to $L_p = 10^{40}$ erg/s in protons having a power law number spectrum with index γ between 1 and 10^8 GeV injected into the nebula 50 kpc away. Since in this model the high energy protons result from a diffusive process, we would not expect the gamma-ray signal to be pulsed.

Table 1: Photon fluxes at Earth ($cm^{-2}s^{-1}$) for $L_p = 10^{40}$ ergs/sec

	$\gamma=2$	2.2	2.4	2.6	2.8
>100 MeV	2.1(-6)	5.0(-6)	6.7(-6)	7.3(-6)	7.4(-6)
>1 TeV	3.5(-10)	1.3(-10)	3.0(-11)	5.9(-12)	1.1(-12)
>200 TeV	5(-13)	8.9(-14)	7.5(-15)	5.6(-16)	3.9(-17)

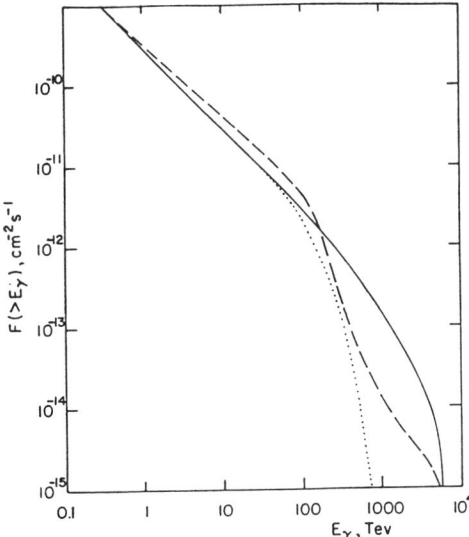

Fig. 2. Integral spectrum of photons from proton interactions for $L_p = 10^{40}$ erg/s at 50 kpc.

Figure 2 shows the integral spectrum of pion-decay gamma rays for an E^{-2} proton spectrum with a cutoff at 10^8 GeV, normalized to $L_p = 10^{40}$ erg/s. The solid line is the production spectrum; the dotted and dashed lines include absorption due to pair production by the high energy photons on the intervening microwave background with and without cascading from electron inverse Compton scattering.

Because of the large opacity of the shell at early times, a photon signal would be suppressed at first even if the accelerator turned on immediately. After a time, $t_a \simeq 0.5$ yr $u_9^{-3/2}(M_s/M_\odot)^{1/2}$, when the proton interaction rate falls below the expansion rate of the nebula, adiabatic losses of the magnetically trapped particles in the expanding envelope of mass M_s will cause the photon signal to decrease quadradically in time. The fraction of accelerated protons which interact to produce photons and neutrinos is

$$f = \min[1, \frac{\rho \, \sigma_N c \, r}{m_H \, u}] \qquad (7)$$

where ρ is the density at radius r and σ_N is the nuclear interaction cross section. There will thus be a window during which the flux of secondary photons is at its maximum. Since the time and duration of this window is sensitively dependent on the density and velocity profile of the envelope, we have calculated the pion-decay photon light curve for several different envelope models, shown in Fig. 3. In calculating these light curves, we assumed a constant proton density throughout the envelope normalized to a total power of $L_p = 10^{40}$ erg/s . For the constant density model, the shell becomes optically thin after a few months but the flux decreases rapidly after about a year. In the more realistic models where interior regions of higher density expand with smaller velocities, the window occurs later (at about one year) but lasts much longer (about ten years) due to smaller adiabatic losses.

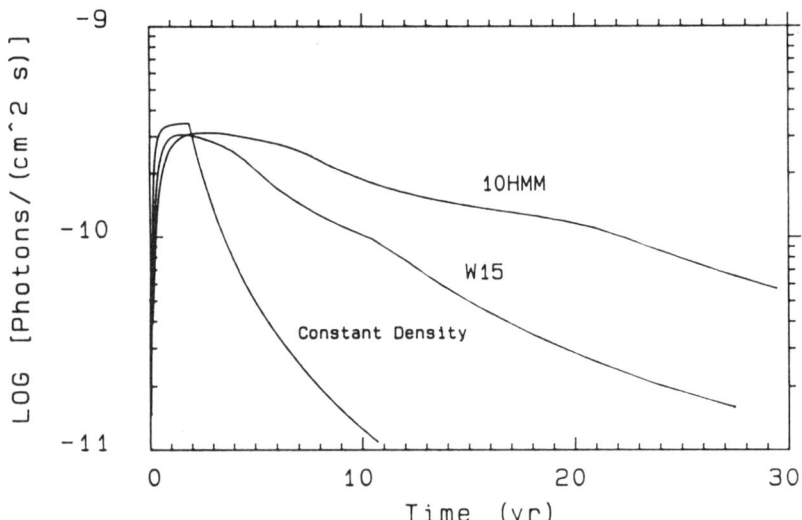

Fig. 3. Pion-decay gamma ray light curve for $L_p = 10^{40}$ erg/s, $\gamma = 2$ and different models for the envelope: a constant density model for a 15 $M_\odot$ shell with outer expansion velocity of 10^9cm/s, model W15B of Ref. 10 and model 10HMM of Ref. 11.

Accelerated electrons will also produce secondary photons from the synchrotron radiation that limits their acceleration to E_e^{max} as given in Eqn (5). We estimate the flux of synchrotron photons from the pile-up peak, assuming that all energy loss comes in photons at the synchrotron peak frequency of 6.4 MeV (= $\nu_c/3$), which in this model is independent of all parameters and constant in time. The flux of such photons at Earth is proportional to the total pulsar spin-down power and is:

$$\phi_e(6.4 \text{ MeV}) = 1.2 \times 10^{-3} \epsilon_e B_{12}^2 P_{10}^{-4} \text{ photons cm}^{-2} \text{ s}^{-1} \quad (8)$$

where ϵ_e is the fraction of spin-down power going into pile-up electrons. The photon production spectrum is the single particle emissivity, which is proportional to $\nu^{1/3}$ below ν_c. Such fluxes could be easily detectable with balloon experiments in the Southern hemisphere when the shell becomes optically thin to 6.4 MeV gamma rays.

NUCLEAR DEEXCITATION LINES

Inelastic collisions of accelerated protons with nuclei in the expanding envelope can produce gamma-ray lines from nuclear deexcitation. The strongest of these lines for a target of solar composition are the 4.44 MeV ^{12}C line, the 6.13 MeV ^{16}O line and the 2.313 MeV ^{14}N line (Ref. 12). Since these lines result only from nuclear deexcitation and not from radioactive decay, their detection in any source is a sure sign of the presence of accelerated particles.

The envelope of SN1987A is very enriched in heavy elements and thus is much different from solar composition targets for which most calculations of gamma-ray line fluxes have been done. As an upper limit to the expected 4.44 MeV deexcitation line flux from SN1987A, we assume that the accelerated protons interact with a pure CNO thick target, in which case the yield will be[12,13]

$$Q_{4.44}(E_p) \simeq 2.53 \times 10^{-3} \left(\frac{E_p}{10 \text{ MeV}}\right)^{0.3} \text{ photons/proton} \quad (9)$$

Using the above yield, the expected flux from a power law proton spectrum, $N_p(E_p) = E_o E_p^{-2}$ with lower cutoff energy $E_o > 10$ MeV, is

$$\phi_{4.4} \simeq 2.25 \times 10^{-6} \epsilon_p B_{12}^2 P_{10}^{-4} \left(\frac{E_o}{10 \text{ MeV}}\right)^{-0.7} \text{ ph. cm}^{-2} \text{ s}^{-1} \quad (10)$$

where ϵ_p is the fraction of the pulsar spin-down power going into accelerated protons. Even if $\epsilon_p = 1$, the above flux will not be detectable with present instruments unless $P \geq 4$ ms. Furthermore, the ratio of the 4.44 MeV line flux to the synchrotron continuum radiation from Eqn (8) is

$$\frac{\phi_{4.4}}{\phi_e(6.4 \text{ MeV})} \simeq 1.87 \times 10^{-3} \frac{\epsilon_p}{\epsilon_e} \left(\frac{E_o}{10 \text{ MeV}}\right)^{-0.7} \quad (11)$$

Therefore, unless $\varepsilon_e \ll \varepsilon_p$ the continuum flux from accelerated electrons will always overwhelm the 4.44 MeV line flux. Considering the high efficiency of electron acceleration in the Crab nebula, it does not seem likely that nuclear deexcitation lines will be observable in SN1987A. In fact, the electron synchrotron flux in Eqn (8) may be comparable to the expected line fluxes from radioactive decay[14,15].

DISCUSSION

The most easily detectable and unambiguous signature of particle acceleration in SN1987A is the TeV gamma-ray flux from pion decay. At present, three air Cherenkov detectors in the Southern hemisphere have the capability of searching for ~ TeV signals from SN1987A: the University of Durham (UK) detector at Narrabri[16], the Potchefstroom detector in South Africa[17], and the Adelaide experiment[18]. Their threshold for detection of a d.c. signal is roughly 3×10^{-11} photons cm^{-2} s^{-1} above 1 TeV. The observability window of > 5 years could make SN1987A detectable also around 100 MeV with EGRET on GRO. From our predicted fluxes in Table 1, detection of a TeV signal from SN1987A would require $L_p \gtrsim 10^{39}$ erg s^{-1} in protons with a -2 spectral index while EGRET would require only $L_p \gtrsim 5 \times 10^{38}$ erg s^{-1}.

REFERENCES

1. V.S. Berezinsky and O.F. Prilutsky, Astr. Ap., **66**, 325 (1978).
2. M.M. Shapiro and R. Siberberg, in Relativity, Quanta and Cosmology, ed. F. deFinis (Johnson Reprint Corp, New York, 1979), **2**, 745.
3. H. Sato, Prog. Theor. Phys., **58**, 549 (1977).
4. T.K. Gaisser and T. Stanev, Phys. Rev. Lett., **58**, 1695 (1987).
5. T.K. Gaisser, A.K. Harding and T. Stanev, Nature, **329**, 314 (1987).
6. M.J. Rees and J.E. Gunn, Mon. Not. Roy. Astr. Soc., **167**, 1 (1974).
7. C.F. Kennel and F.V. Coroniti, Ap. J., **283**, 694 (1984).
8. G. Auriemma, T.K. Gaisser and P. Lipari, preprint (1987).
9. R. Schlickeiser, Astron. Astrophys., **136**, 227 (1984).
10. S.E. Woosley, P.A. Pinto and L. Ensman, Ap. J., **324**, 000 (1988).
11. P.A. Pinto and S.E. Woosley, preprint (1988).
12. R. Ramaty, B. Kozlovsky and R.E. Lingenfelter, Ap. J. Supp., **40**, 487 (1979).
13. R. Ramaty, G. Borner and J.M. Cohen, Ap. J., **181**, 891 (1973).
14. K.W. Chan and R.E. Lingenfelter, Ap. J., **318**, L51 (1987).
15. N. Gehrels, C.J. McCallum and M. Leventhal, Ap. J., **320**, L19 (1987).
16. K.E. Turver, Proc. Durham ARW on VHE Gamma Ray Astronomy (in press).
17. H.I. deJager et al., S. African J. of Physics, **9**, 107 (1986).
18. R.W. Clay et al., Proc. Astron. Soc. Australia, **6**, 338 (1986).

IRON STORY

Michel CASSE

Service d'Astrophysique

Institut de Recherche Fondamentale

CEN Saclay, France

IRON NUCLEOSYNTHESIS

The sky harbours a huge number of gravitationally confined fusion reactors, the stars. Sometimes one of them explodes, enriching into iron the galactic environment. Indeed, one of the major conclusions of the monumental article of B^{2}FH on the formation of nuclei in stars was that the iron-peak abundances observed in the solar system could be univocally explained in terms of an equilibrium (or "e") process, adjusting only the temperature and the ratio of free protons to electrons. The exceptional binding energy of the ^{56}Fe nucleus causes it to be the dominant species in a thermal bath where nuclear statistical equilibrium (or quasiequilibrium) prevails, i.e. at temperatures of a few billions degrees, provided that the n/p ratio is close to 30/26, its own n/p ratio, (Hoyle 1946, Burbidge et al 1957, Hoyle and Fowler 1964). More generally, if temperature is high enough, the abundance of a given nuclear species peaks at or near its own n/p ratio. This trend has been confirmed by a whole series of calculations (e.g. Hainebach et al, 1974, Woosley 1986 and references therein) and can be now considered as a safe rule. Since the early times it was clear that ^{56}Fe was synthesized, quickly after photodesintegration of ^{28}Si (n/p=1) as ^{56}Ni in stellar zones adjacent to the infalling core. However this state of affairs seemed odd to the proponents and they imagined that a ^{56}Ni-^{56}Fe transformation takes place deep inside massive stars, before explosion, under the effect of weak interactions principally electron captures (^{56}Ni + 2e→^{56}Fe + 2ν). The decay of ^{56}Ni, changing the n/p ratio, in turn shifted the equilibrium abundance toward ^{56}Fe.

Later on, in 1968, explosive nucleosynthesis makes a clashing entry, invoking nuclear build-up and matter ejection so prompt that essentially no weak interaction can occur in the stellar material, but has to proceed subsequently in the expanding envelope to stabilize the nuclides produced. The idea that "the abundance of the iron isotopes A=56 and A=57 reflects instead the nuclear properties of those isotopes of nickel" (as phrased by Hainebach et al, 1974) gained wide acceptance (Bodansky, Clayton and Fowler 1968, Truran, Arnett and Cameron 1967, Clayton and Woosley 1969, Arnett, Truran and Woosley 1971, Hainebach et al 1974, Woosley 1986). Indeed, a single zone with slight neutron excess, reflecting the past history of weak interaction in the material, can reproduce the solar abundances of many nuclei (^{52,53}Cr, ^{55}Mn, ^{54,56,57}Fe, and ^{58}Ni). Subsequently, Woosley, Arnett and Clayton (1973) and Hainebach et al (1974) refining the "freeze-out" corrections, which take care of the deviations to the statistical equilibrium in the course of the ejection due to the rapid decrease of temperature and density, arrived at the conclusion that the solution of Fowler and Hoyle (i.e. production of ^{56}Fe as ^{56}Fe) overproduces ^{52}Cr, ^{53}Cr and ^{60}Ni by a factor of $\simeq$ 10. These excesses, not apparent in the work of the pioneers of nucleosynthesis, arose because the photospheric Fe abundance had been revised upwards in the mean time (Garz and Kock 1969), bringing it in agreement with the meteoritic one.

Then, stellar evolutionary calculations confirmed that stellar explosions could inseminate the surrounding space with radioactive nickel (for a review see Woosley 1986 and Nomoto and Hashimoto 1987), exploding binary white dwarfs (type I supernovae) being much more prolific than massive imploding-exploding stars (type II supernovae).

Incidently ^{56,57}Ni were used by Cassé and Soutoul 1975 and Soutoul, Cassé and Juliusson (1978) to set constraints on the time delay between nucleosynthesis and acceleration of cosmic ray nuclei by supernovae.

THE MAGELLANIC SUPERNOVA

The production of iron as nickel is one of the main predictions of the parametrized models of explosive nucleosynthesis, and fortunatelly it is open to definite empirical tests. The profound implications of explosive nucleosynthesis for gamma ray line astronomy were put forward by Clayton (1973a, b) and continuously repeated since (see Clayton 1982, Woosley, Axelrod and Weaver 1981 and Lingenfelter, this volume, for historical perspectives, and Ramaty and Lingenfelter 1982) for a broader scope of gamma ray-line astronomy).

Briefly, Clayton, Colgate and Fishman (1969) were the first to make a quantitative evaluation of the flux expected in the lines of nickel and cobalt, originating from remote type I supernovae, by far the most generous sources. Numerous other attempts followed (Van Hise 1974, Arnett 1979, Colgate, Petschek and Kriese 1980, Axelrod 1980, Sutherland and Wheeler 1984 Woosley, Taam and Weaver 1986, Gehrels, Leventhal and McCallum, 1987, Ambwami and Sutherland 1988). Unfortunately Gamma-ray line observations of recent supernovae led only to upper limits (Matz et al 1987) and astrophysicists of the eighties resigned themselves to leave to the next generations the task of revealing the signature of the decay chain ^{56}Ni ^{56}Co ^{56}Fe.

And then came the light of the magellanic supernova. Probably, SN 1987 A provides us with a once in a lifetime opportunity to study at close distance the genesis of iron.

Three empirical arguments speak in favour of the formation and ejection of iron in the form of its proton-rich ancestor :

i) the exponential decay of the late light curve of the magellanic supernova with a characteristic time strickingly close to that of ^{56}Co, the daughter of ^{56}Ni, e.g. 110 days (Woosley, this conference and references therein), the early light curve being dominated by the energy imparted by the shock (e.g. Schaeffer et al 1987, Arnett 1987, Woosley et al 1987, Woosley, Pinto and Ensman 1988, Shigeyama et al 1987, Wampler et al 1987).

ii) the discovery by the japanese GINGA satellite and the soviet KVANT experiment on board of the MIR space platform of hard X-rays with an exceptionally flat spectrum, compatible with the comptonisation of high energy photons arising from the cobalt desintegration (Sunyaev et al 1987, Itoh et al 1987, Woosley and Pinto 1987, Pinto and Woosley 1988, Woosley 1988, Ghehrels McCallum and Leventhal 1987, McRay, Shull and Sutherland 1987, Xu et al, 1988).

iii) but the clearest proof of this scenario is the direct detection by the SMM satellite of the two main nuclear gamma ray lines resulting from the decay of ^{56}Co at 847 and 1238 keV respectively, and its probable confirmation by two independent balloon observations.

More specifically, the gamma ray spectrometer on board of the NASA's Solar Maximum Mission satellite has delivered a spectrum of SN 1987 A, which after background rejection, shows evidence of a pattern in the vicinity of 847 keV. This is a convincing signature of the radioactive cobalt produced by the supernova (Matz et al, this volume and 1988). The identification is strenghthened by the presence of an accompanying line at 1238 keV, arising also from cobalt decay, though marginal compared to the first one (3 confidence level instead of 5). The average flux inferred from data acquired between August 1 and October 31 is (1 ± 0.25) 10^{-3} photon cm^{-2} s^{-1} for the 847 keV line. The measured line intensity ratio is close to the natural one (0.68). Thus the absence of differential absorption on the two lines in spite of the energy dependence of the Klein-Nishina cross section suggests that the emitting region has a very low optical depth.

Two balloon flights, sponsored by NASA, have given positive results at the time of the workshop, confirming the findings of SMM. Those are the Gamma-Ray Imaging Telescope developed at the California Institute of Technology (Cook, Palmer and Prince) and the detector operated by the Lookheed-Marshall collaboration (Sandie et al, this conference).

These detections are one of the most fundamental results acquired in this decade in the field of nuclear astrophysics; they were awaited impatiently. Predictions of the cobalt line emission of SN 1987 A have been presented by Woosley and coworkers (see Woosley 1988, Chan and Lingenfelter 1987, and Gehrels, McCallum and Leventhal 1987). The internal density and velocity profiles, calculated on the basis of presupernova models were assumed to be "frozen" and used as inputs to the codes of gamma-ray transport. To the general surprise, all expectations seem rather far from reality, but it is premature to conclude since they apply to the bulk of the gamma line emission, whereas the data concern a blob, an excrescence of cobalt-rich material, of very small mass ($2 \ 10^{-4}$ Mo), which has apparently been brought near the surface of the ejecta by some sort of hydrodynamical instability (Chevalier, this conference). This is a tiny fraction of the mass of ^{56}Co required to sustain the late light curve, which is 0.07 Mo (i.e. Woosley 1988, Itoh et al 1987).

PROSPECTS

The future will tell if such a mass of radioactive material is really present in the debris since SMM and other experiments will continue to monitor the supernova (e.g. the NASA programm, Bunner, this meeting). The early and quasi-simultaneous appearance of the X (Sunyaev et al 1987, Itoh et al 1987) and gamma fluxes demonstrates that the beautiful symmetry and stratification of the theoretical models have been broken. Now theorists have to face the difficulties of developing three dimensional modeling of supernova explosion and expansion. In the meantime observers will probably succeed to measure the width and shape of the prominent gamma lines with balloon-borne detectors of sensitivity better than 10^{-4} photon $cm^{-2} \ s^{-1}$ and excellent energy resolution (see Bunner for a planning of NASA experiments). The gathering of line profiles should allow a significant test of the distribution of radioactive products of nucleosynthesis in the deepest supernova layers (Gehrels, this conference).

According to current theory not only ^{56}Fe, but also ^{57}Fe is produced under the form of Ni. Hopefully the line arising from ^{57}Co decay at 122 keV (Clayton, 1974, Pinto and Woosley 1988, Chan and Lingelfelter 1987) will enrich the panorama for, if theory is right, one experiment or another (SMM, Sigma, GRO,

Jupiter III, the American-French experiment, NASA balloons etc...) will detect it. This line will bring information on the neutron excess and freezing conditions of the nuclear abundances in the ^{57}Co formation zone (Woosley, this conference). Thus the future of gamma-ray line astronomy appears brilliant.

Distinct observations at other wavelengths might reveal a fast rotating neutron star among the debris of the explosion, realizing the prophecy of Zwicky (1934) : "with all reserve, we advance the view that a supernova represents the transition of an ordinary star into a neutron star". Harding (this conference, see also Gaisser, Harding and Stanev 1987), and other pulsar supporters, such as Berezinsky and Ginzburg 1987, are looking forward to the detection of very high energy gamma rays induced by the decay of neutral pions, themselves generated in collisions between hypothetically accelerated protons and the intervining matter. A less exotic suggestion, presented by Brecher at this meeting and Michel, Kennel and Fowler 1988, is to follow the long term decline of the light curve since it could reveal the underlying pulsar, once radioactivity will be extinguished.

Finally, it is worth mentioning the consequences on galactic evolution of the mass of ^{56}Ni inferred from the light curve i.e. 0.07 Mo, which is rather meager. Since SN 1987 A is, despite its aspect, a type II supernova (e.g. Shaeffer et al 1987 Woosley 1988 and references therein), one can infer that this amount is typical of the objects of its class. Therefore the thorough incineration of C-O white dwarfs into ^{56}Ni, associated to type I events, seems to be the main source of iron in the galaxy (Andreani, Vangioni-Flam and Andouze, 1988).

CONCLUSION

Today the rather strange idea that iron, the iron of our hemoglobin, the iron of railways, the father of rust, is not made as iron, is leaving the realm of pure speculation, thanks to the conspiration of an opportune supernova, the skill of the observers and the enormous technological power put at their disposal. Therefore, it is not exaggerated to conclude that the theorists of the stellar evolution and nucleosynthesis have been knighted by the magellanic supernova.

Aknowledgements

I am endebted to Gerald Share and Neil Gherels for their great hospitality and to all other organizers for the quality of the Washington meeting.

REFERENCES

Arnett, W.D. 1987, Ap. J., 319, 136
Axerod, T.S. 1980, in type I supernovae, ed. J.C. Wheeler. Austin : University of Texas Press, p. 80.
Ambwani, K., and Sutherland, P. 1988, Ap. J. 325, 820.
Andreani, P., Vangini-Flam, E., and Audouze, J. 1988 submitted to Ap. J.
Arnett, W.D. 1979, Ap. J. Letters, 230 L 37.
Berezinsky, V.S., and Ginzburg, V.L. 1987, Nature, 329, 807.
Bodansky, D., Clayton D.D., and Fowler, W.A. 1968, Phys. Rev. Lett. 20, 161.
Burbidge, E.M., Burbidge, G.R., Fowler, W.A. and Hoyle, F. 1957, Rev. Mod. Phys., 29, 547.
Cassé M. and Soutoul A. 1975, Ap. J. (Letters) 200, L76.
Chan, K.W., and Lingenfelter, R.E. 1987, Ap. J., 318, L55.
Chevalier, R.A., and Francson, C. 1987, Nature, 328, 44.
Clayton, D.D. 1973a, in Explosive Nucleosynthesis, ed. D.N. Schramm and W.D. Arnett. Austin : University of texas Press, p. 264.
Clayton, D.D. 1973b, in Gamma-ray astrophysics, ed. F.W. Stecker and J.I. Trombka, Washington : NASA, p. 263.
Clayton, D.D. 1974, Ap. J. 188, 155.
Clayton, D.D. 1982 in Essays in Nuclear Astrophysics.
Clayton, D.D., Colgate, S.A., and Fishman, G.J. 1969, Ap. J., 155, 75.
Clayton, D.D., and Woosley, S.E. 1969, Ap. J., 157, 1381.
Clayton, D.D. 1987, Nature, 330, 423.
Colgate, S.A., and McKee, C. 1969, Ap. J., 157, 623.
Colgate, S.A., Petschek, A.G. and Kriese, J.T. 1980, Ap. J. Letters, 237, L81.
Donati et al. 1987, Nature, 330, 230.
Gaisser, T.K., Harding, A., and Stanev, T. 1987, Nature, 329, 314.
Garz, G., and Kock, M. 1969, Astr. and Ap., 2, 274.
Gehrels, N., Mac Callum, C.J., and Leventhal, M. 1987, Ap. J. 320, L19.
Gehrels, N. Mac Callum, C.J., and Leventhal, M. 1987, Ap. J. (Letters) 320, L19.
Haineback, K.L., Clayton, D.D., Arnett, W.D. and Woosley, S.E. 1974, Ap. J., 193, 157.
Hoyle, F. M.N.R.A.S., 1946, 106, 343.
Itoh, M. et al. 1987, Nature, 330, 233.
Maeder, A., 1988, to appear in Proc. of ESO Workshop on SN 1987A, ed. J. Danziger, in press.
Matz, S.M., Share, G.H., Leising, M.D., Chupp, E.L., Vestrand, W.T., Purcell, W.R., Strickman, M.S. and Reppin, C. 1988, Nature, 331, 416.
Matz, S.M., Share, G.H., Kinzer, E.O., Chupp, E.L., Forrest, D.J., and Reppin C., 1987, 20th International Cosmic Ray Conference.

McCray, R., Shull, J.M., and Sutherland, P. 1987, Ap. J. (Letters), 317, L73.
Michel, F.C., Kennel, C.F., and Fowler W.A. 1988, Science, 238, 938.
Nomoto, K., and Hashimoto, 1987, in the Early Universe and its Evolution, Erice, Progress in Particle and Nuclear Physics, 18.
Pinto, P.A., and Woosley, S.E. 1988, Ap. J., in press.
Ramaty, R., and Lingenfelter, R. 1982, Ann. Rev. of Nuclear and Particle Sci., 32, 235.
Schaeffer, R., Cassé, M., Mochkovitch, R., and Cahen, S. 1987, Astron. and Ap., Letter, 184, L1.
Shigeyama, T., Nomoto, K., Hashimoto, and Sugimoto, D. 1987, Nature, 328, 320.
Soutoul, A., Cassé, M., and Juliusson, E. 1978, Ap.J. 219, 753.
Sutherland, P., and Wheeler, J.C., 1984, Ap. J., 280, 282.
Truran, J.W., Arnett, W.D., and Cameron A.G.W. 1967, Can. J. Phys. 45, 2315.
Sunyaev, R. et al. 1987, Nature, 330, 227.
Van Hise, J.R. 1974, Ap. J., 193, 657.
Wampler, E.J., Truran, J.W., Lucy, L.B., Höfflich, P., and Hillebrandt, W. 1987, Astron. and Ap., Letters, 182, L51.
Woosley, S.E. 1986, in Nucleosynthesis and Chemical Evolution, 16th Advanced Course of Astrophysics and Astronomy, ed. B. Hauck, A. Maeder, and G. Meynet. Sauverny : Geneva Observatory, p.129.
Woosley, S.E. 1988, Ap. J. in press.47.
Woosley, S.E., Pinto, P.A., and Ensman, L. 1988, Ap. J. 324, 466.
Woosley, S.E., Arnett, W.D., and Clayton, D.D. 1973, Ap. J. Suppl. 231, 26.
Woosley, S.E., Pinto, P.A., Martin, P.G., and Weaver, T.A., 1987, Ap. J. 318, 664.
Woosley, S.E., Taam, R.E., and Weaver, T.A. 1986, Ap. J. 301, 601.
Woosley, S.E., and Weaver, T.A. 1986, Ann. Rev. Astron. and Ap., 24, 205.
Xu, Y., Sutherland, P., Mc Ray, R., and Ross, R.R. 1988, to appear in Ap. J.

GAMMA-RAY LINE DIAGNOSTICS OF NOVAE

Mark D. Leising*
E.O. Hulburt Center for Space Research
Naval Research Lab, Washington DC 20375

ABSTRACT

The nuclear processing thought to power classical nova outbursts produces substantial quantities of unstable nuclei. These nuclei emit γ-rays and β^+'s which might be detectable to present and future γ-ray spectrometers. We discuss the importance of detection of γ-rays for improving our understanding of the nova phenomenon and we review the observations made to date.

INTRODUCTION

Both theory and observation point to a thermonuclear runaway in the degenerate envelope of a white dwarf as the cause of the classical nova outburst[1,2]. However, neither the theory nor observations have been refined enough to provide much confidence in our understanding of the details of the physical processes involved. Gamma-ray lines offer an opportunity to directly probe the conditions within the burning regions and the dynamic processes initiated by the violent runaway. In general, the γ-ray luminosities depend only on the abundances of the emitters and the characteristics of any attenuating medium, not on the ambient temperatures or densities after the runaway (although β^+ annihilation is sensitive to the physical conditions). Detection of γ-ray lines would provide firm observational constraints on theoretical models of the outbursts.

BASIC THEORY

When a white dwarf in a binary accretes matter slowly enough, a substantial envelope (of order 10^{-4} $M_\odot$) can be accumulated. The base of the envelope, which is mainly supported by electron degeneracy pressure, is heated by the luminosity of the white dwarf and steady hydrogen burning until a runaway ensues. The temperature then rises rapidly, exceeds 10^8 K (when the increased thermal pressure causes the envelope to begin expanding), and peaks shortly thereafter. The peak temperature is determined by the mass of the white dwarf and its envelope, and the available energy. For the most massive dwarfs and envelopes the temperature might reach 5×10^8 K, but the typical value is probably half of that. The high temperatures transform nuclei capable of capturing protons into unstable proton-rich nuclei. Further processing and energy generation must then await the weak decays of those nuclei. The available energy thus depends on the abundances of the catalysts such as C, O, and Ne. If enough of

*National Research Council Resident Research Associate at NRL

these nuclei from the white dwarf can be exposed to the hot bath of protons from the envelope, the energy generated can exceed the gravitational binding of the envelope, which can then be ejected. These ideas have been largely developed by Starrfield, Sparks, and Truran, and coworkers (see references in 1,2).

The extreme temperature gradient across the envelope at the peak of the burning produces rapid convective energy transport which can thoroughly mix the envelope material. Very abundant unstable nuclei with lifetimes longer than the convective time scale could appear at the surface where they are in principle detectable from their nuclear decay or positron annihilation γ-rays. Unstable nuclei with even longer lifetimes (greater than a few days) could survive the ejection and thinning of the envelope. Then their decay could be observed in γ-rays even if their abundances are relatively small. Those with lifetimes exceeding the time between nova events which eject them could accumulate in the interstellar medium and be detected as (nearly) constant sources of γ-rays. We discuss the potentially interesting nuclei in each lifetime regime separately.

^{26}Al (mean lifetime = 10^6 years)

This is the only isotope discussed here which has been definitely detected in an astrophysical source, the Galactic plane, with an apparent concentration toward the Galactic center direction. Aluminum-26 was first detected by the HEAO 3 γ-ray spectrometer[3,4] and its presence was confirmed by that on the Solar Maximum Mission[5]. Novae represent a plausible source of the observed ^{26}Al, but there are several other viable sources (see ref. 6 and references therein). No single nova would alone be detectable in the 1.809 MeV line of ^{26}Al, because of its slow decay. However the roughly 4×10^7 novae which are thought to occur in the Galaxy during the ^{26}Al lifetime could produce the observed flux of 4×10^{-4} cm^{-2} s^{-1} if on the average they eject 10^{-7} $M_\odot$ of ^{26}Al.

Calculations of nova nucleosynthesis suggest a production of mass fraction $X(^{26}Al) \sim 10^{-4}$ [7,8] and observed novae (and calculated nova models) typically eject a total mass of $\sim 10^{-4}$ $M_\odot$[1], but these numbers can both vary by a factor of 10 in either direction. Some novae have been observed to eject material extremely rich in neon and heavier elements. If these enrichments simply reflect enhanced pre-outburst abundances of nuclei in this mass range, the production of ^{26}Al should also be increased proportionately. Together, the uncertainties in total Galactic ^{26}Al production of novae allow for their being the sole source, but the same can be said for other potential sources. The determination of the origin of the ^{26}Al might await a definitive measurement of its distribution in the Galaxy. Because novae are thought to occur predominantly near the Galactic center, the γ-ray distribution would be strongly peaked in that direction.

Two subsequent detections of Galactic ^{26}Al by detectors flown on balloons have not yet clarified this situation. One experiment indicated a distribution of ^{26}Al strongly concentrated toward the

Galactic center direction[9], while another suggested a more broadly distributed source[10], although neither measured the distribution with high degree of confidence. All of the ^{26}Al observations and their interpretations are discussed elsewhere in these proceedings.

^{22}Na (3.75 years)

With its 3.75 year lifetime, ^{22}Na decays fast enough to be detected in the ejectae of individual novae and yet lives long enough to accumulate from the outbursts of many novae. This nucleus was suggested by Clayton and Hoyle[11] as a potentially interesting diagnostic of novae. Upper limits to the γ-ray flux at 1.275 MeV from ^{22}Na have been obtained for Nova Cygni 1975 [12], for a diffuse Galactic center component[3,13], and for several individual novae with ejecta rich in neon[13]. If 40 novae occur in the central Galaxy per year and eject 10^{-4} $M_\odot$ on average, the production of ^{22}Na is limited to $X(^{22}Na) \leq 2 \times 10^{-4}$ in those novae, based on the lowest upper limit for the Galactic center 1.275 MeV line flux, 1.2×10^{-4} cm^{-2} s^{-1} [13]. This mass fraction is about the largest predicted by calculations of nucleosynthesis in material which starts with solar abundances. This upper limit does constrain the frequency of novae near the Galactic center which eject material 100 times overabundant in neon (such as N CrA 81 [14] and N Aql 82 [15]), and convert neon to ^{22}Na with the highest calculated efficiency, to about six per year[13]. It has been suggested, based on rather uncertain statistical arguments, that several previously undiscovered neon-rich novae in the nearby Galactic disk should contribute significantly to the Galactic plane 1.275 MeV flux[16]. Scaling these arguments by the SMM upper limit, the average ^{22}Na mass ejected per neon-rich disk nova is limited to roughly 1.5×10^{-7} $M_\odot$.

The upper limits to the mass of ^{22}Na ejected by each of four observed individual neon-rich novae are $\sim 10^{-6}$ $M_\odot$[13]. These novae typically eject only $\sim 10^{-5}$ $M_\odot$ of material so this does not greatly constrain theoretical models. Nucleosynthesis in this mass range in novae has in general been investigated with fairly crude simulations. There are indications that the net production of very fragile species (such as ^{22}Na and, to a lesser extent, ^{26}Al) might be dramatically increased over those of standard parameterized (isothermal, in mass) calculations because of their removal from the burning regions by rapid convection[8]. One hydrodynamic calculation of a nova outburst on a Ne-O white dwarf, including nuclear reactions beyond ^{22}Na, ejected 1.6×10^{-7} $M_\odot$ of that isotope[8]. However the only other hydrodynamic nova calculation (on a C-O white dwarf) including ^{22}Na synthesis known to this author, ejected only 10^{-13} $M_\odot$ of ^{22}Na (Prialnik 1987 private communication). Improved theoretical estimates of the synthesis of species such as this are clearly needed, but detection of ^{22}Na in novae would go a long way toward establishing the validity of theoretical models of the entire nova phenomenon.

^{7}Be (77 days)

The synthesis of ^{7}Li (the daughter of ^{7}Be) is important for more than confirmation of nova theory. Calculations of big bang nucleosynthesis suggest a production of (Li/H) ~ 10^{-10} by number[17], in very good agreement with observed atmospheric abundances in hot, low mass, metal-poor halo stars[18]. Population I objects show striking uniformity in lithium abundance, near (Li/H) ~ 10^{-9}, except in some stars which presumably have convective zones deep enough to destroy surface lithium. Possible explanations are then either: 1) Big bang estimates of Li production are correct and there is a Galactic source which has produced 90% of the Population I Li, or 2) The big bang somehow produced (Li/H) ~ 10^{-9} which has been uniformly destroyed in hot halo subdwarfs, and there has been no Galactic production of Li.

As pointed out by Arnould and Nørgaard[19], and Starrfield et al.[20], novae are possibly significant contributors to the Galactic Li abundance. It would be ejected as ^{7}Be which, as Clayton[21] suggested, could possibly be observed from its 478 keV β^- capture γ-ray line (which is emitted in 10% of its captures). The net production of ^{7}Be depends sensitively on the thermal history of the ejected material, and therefore on the treatment of convection (it is easily destroyed at high temperatures), and on the pre-outburst abundance of ^{3}He, which would presumably be accreted from the giant companion of the white dwarf. A nova at 1 kpc would have to eject ~ 4×10^{-9} $M_\odot$ of ^{7}Be to produce a flux of 10^{-4} cm^{-2} s^{-1} initially. According to calculations[20], this mass could be achieved in a total ejected mass of 10^{-4} $M_\odot$ if the red giant atmosphere accreted onto the white dwarf contained 20 times the solar abundance of ^{3}He.

No limits on this line from novae have been reported to date. The SMM γ-ray spectrometer can reach a sensitivity of ~ 4×10^{-4} cm^{-2} s^{-1} in 77 days for an object which happens to be in its field of view, but there have been no novae closer than 3 kpc observed since its launch. It is also difficult in a low resolution spectrometer to separate a line at 478 keV from the line at 511 keV (which is produced in the Earth's atmosphere and the Galactic plane) and associated continua (Compton scattered, from the atmosphere, and [possibly] triplet positronium decay, from the Galaxy). Nevertheless, future instruments and the luck of a very close nova could produce a most interesting detection or upper limit.

^{13}N (862 s), ^{14}O (102 s), ^{15}O (176 s), ^{18}F (158 m)

As pointed out by Clayton and Hoyle[11], short-lived positron emitters might be carried quickly to the nova surface where the decay positrons would annihilate with electrons. The resulting 511 keV photons might be detectable from nearby Galactic novae, yielding important information on the burning conditions and convective transport. These ideas were refined somewhat and the ^{18}F nucleus was added as a possible candidate[22].

In the hot hydrogen burning, many of the CNO nuclei initially present are transformed into these unstable species. The relative abundances among them depend on the peak temperature reached. Clearly the longer-lived nuclei offer the best prospects for detection, because the nova envelope can expand further before they decay and their surface abundances depend less on the rapidity of convective transport from the burning region to the surface. It is not clear that these unstable nuclei can reach relatively thin regions at all before decaying. Self-consistent calculations of their transport are not yet available. However, it may be that significant energy production very near the photosphere is required to explain the apparent super-Eddington luminosities of some novae[2]. Beta-unstable nuclei carried there by rapid convection are a natural (perhaps necessary; A. Shankar, private communication 1988) explanation in the context of the thermonuclear runaway model.

For typical nova models with ^{13}N abundances of a few percent by mass, fluxes at 1 kpc at 511 keV from that nucleus might reach 10^{-2} cm^{-2} s^{-1} over about 30 minutes following runaway[22]. Some models with ten times more ^{13}N suggest fluxes exceeding 0.1 cm^{-2} s^{-1}. Abundances of ^{18}F of 10^{-3} by mass in the outer envelope lead to 511 keV fluxes well over 10^{-3} cm^{-2} s^{-1} for periods of several hours (at 1 kpc). Explosive hydrogen burning, possibly in novae, is a likely source of the bulk of Galactic ^{18}O [23], which results from the decay of ^{18}F. For novae to be significant contributors, ejected mass fractions of at least $X(^{18}O) = 2 \times 10^{-3}$ (after decay of ^{18}F) are required.

As one has no advance knowledge that novae, even relatively nearby ones, are about to occur, detecting them requires large field of view detectors or extremely good luck. Also, we may wait for a long time for a nova as close as 1 kpc. Still, these fluxes are potentially detectable in existing and planned detectors (such as the SMM γ-ray spectrometer and GRO OSSE and BATSE) to distances of a few kiloparsecs. The most extreme estimated fluxes from ^{13}N positrons might be detectable to OSSE even from novae up to the distance of the Galactic center, where that detector may be pointed for long periods and where novae are thought to be very frequent. Gamma rays from the positrons emitted by shorter-lived nuclei do not appear to be currently detectable, except from very nearby novae. Note that the ^{14}O decay also produces a line at 2.31 MeV at a comparable intensity to its annihilation line. Also a significant continuum from Compton scattering of the lines would be produced and might be detectable in lower energy detectors.

CONCLUSION

Clearly, there is the potential for learning a tremendous amount about a fascinating physical situation by observing γ-ray lines from novae. The theory has not yet evolved to the point where upper limits on γ-ray fluxes are terribly interesting, but it soon should. The next generation of γ-ray spectrometers will greatly improve the possibility of detecting emission from novae, especially in the fortuitous event of a very close nova. A detection would place important constraints on models of novae and on general theories of convection, and possibly even address larger issues.

I would like to thank Don Clayton, who originated so many ideas of importance to astrophysical nuclear spectroscopy, for his encouragement and enthusiasm for this subject.

REFERENCES

1. Gallagher, J. S., and Starrfield, S., Ann. Rev. Astr. Ap., **16**, 171 (1978).
2. Truran, J. W., in Essays in Nuclear Astrophysics, ed. C. A. Barnes, D. D. Clayton, and D. N. Schramm (Cambridge: Cambridge University Press), p. 467 (1982).
3. Mahoney, W. A., Ling, J. C., Jacobson, A. S., and Lingenfelter, R. E., Ap. J., **262**, 742 (1982).
4. Mahoney, W. A., Ling, J. C., Wheaton, W. A., and Jacobson, A. S., Ap. J., **286**, 578 (1984).
5. Share, G. H., Kinzer, R. L., Kurfess, J. D., Forrest, D. J., Chupp, E. L., and Rieger, E. 1985, Ap. J. (Letters), **292**, L61.
6. Clayton, D. D., and Leising, M. D., Physics Reports, **144**, 1 (1987).
7. Wiescher, M., Görres, J., Thielemann, F.-K., and Ritter, H., Astr. Ap., **160**, 56 (1986).
8. Woosley, S. E., in Nucleosynthesis and Chemical Evolution, ed. B. Hauck, A. Maeder, and G Meynet (Geneva Observatory: Geneva) p. 85 (1986).
9. von Ballmoos, P., Diehl, R., and Schönfelder, V., Ap. J., **318**, 654 (1987).
10. MacCallum, C. J., Huters, A. F., Stang, P. D., and Leventhal, M., Ap. J., **317**, 877 (1987).
11. Clayton, D. D., and Hoyle, F., Ap. J. (Letters), **187**, L101 (1974).
12. Leventhal, M., MacCallum, C., and Watts, A., Ap. J., **216**, 491 (1977).
13. Leising, M. D., Share, G. H., Chupp, E. L., and Kanbach, G., Ap. J., **328**, 755 (1988).
14. Williams, R. E., Ney, E. P., Sparks, W. M., Starrfield, S. G., Wyckoff, S., and Truran, J. W., M. N. R. A. S., **212**, 753 (1985).
15. Snijders, M. A. J., Batt, T. J., Roche, P. F., Seaton, M. J., Morton, D. C., Spoelstra, T. A. T., and Blades, J. C., M. N. R. A. S., **228**, 329 (1987).
16. Higdon, J. C., and Fowler, W. A., Ap. J., **317**, 710 (1987).
17. Yang, J. et al., Ap J., **281**, 493 (1984).
18. Spite, F., and Spite, M., Astr. Ap., **115**, 357 (1982).
19. Arnould, M., and Norgaard, H., Astr. Ap., **42**, 55 (1975).
20. Starrfield, S., Truran, J. W., Sparks, W. M., and Arnould, M., Ap. J., **222**, 600 (1978).
21. Clayton, D. D., Ap. J. (Letters), **294**, L97 (1981).
22. Leising, M. D., and Clayton, D. D., Ap. J., **323**, 159 (1987).
23. Wiescher, M., Görres, J., and Thielemann, F.-K., Ap. J., **326**, 384 (1988).

NUCLEOSYNTHESIS OF GAMMA-RAY EMITTERS IN NOVAE

Michael T. Wolff
Universities Space Research Association, Columbia, MD 21044

Mark D. Leising[1]
Naval Research Laboratory, Washington, D.C. 20375

ABSTRACT

A program is underway at the Naval Research Laboratory to develop a numerical hydrodynamic capability to model thermonuclear outbursts on compact objects. As part of this program, a nuclear reaction network comprised of 67 nuclei, including all the important isotopes of the elements from hydrogen to phosphorus, and the important reactions among them, has been developed. The nuclear processing expected to occur in the proton-rich nova environment can be followed with this network, which employs a fast numerical algorithm developed at NRL for the solution of the coupled system of non-linear differential equations. Results of a one-zone calculation with a simple parameterization of temperature and density are compared with similar, previous studies. The validity of such parameterized models is tested by comparison with another model with follows the temperature and density histories of a published hydrodynamic nova model.

The abundance of several unstable proton-rich isotopes which live long enough to be potentially observable via decay or positron annihilation gamma rays are reported. These include ^{7}Be, ^{13}N, ^{14}O, ^{15}O, ^{18}F, ^{22}Na, and ^{26}Al8. The sensitivities of the resulting abundances to the different models and the potential for observing the gamma-ray lines are discussed.

I. INTRODUCTION

Theoretical models of thermonuclear outbursts in the envelopes of white dwarfs have been relatively successful in explaining the gross features of observed novae (see, for example, refs. 1 and 2 and references therein). The models generally include the essential hydrodynamics with nuclear reaction networks sufficient to describe the nuclear energy generation that drives the explosion. However, these networks are not complete enough to address the nucleosynthesis of all interesting nuclear species, including some of particular interest to gamma-ray astronomy. So far these questions have been examined with parameterized studies by evolving more complete reaction networks along prescribed paths in temperature and density space for one or two zones in crude attempts to simulate the conditions at the peak burning region of the envelope (e.g., ref. 3)

[1]Resident Research Associate at the Naval Research Laboratory, Under the NRC Associateship Program

As part of a program to incorporate an extensive nuclear reaction network into a hydrodynamic code, we have developed a limited network which is adequate to describe the production of gamma-ray producing species in nova outbursts. Here we report on the application of that network to the nova problem in parameterized studies. We first verify our code by duplicating a published study. We check the validity of such one-zone models by comparing the results of the most extensive network coupled to a hydrodynamic model published to date, with our results for a one-zone model which follows the same temperature and density profile. We calculate a parameterized model designed to simulate the nucleosynthesis in a nova outburst on a massive O-Ne-Mg rich white dwarf. We also check the effect of greatly reducing the time scale of the temperature decline on the production of fragile nuclear species. This is a naive attempt to estimate the importance of the removal of matter from the burning regions by rapid convection.

II. COMPUTATIONAL DETAILS

We compute the abundances of the nuclear species with a nuclear reaction network that includes all the important nuclear species between hydrogen and phosphorus. Table 1 shows the nuclei included in our network for the present calculations. We include all (p,γ), (p,α), (p,n), (α,γ), (α,n), (α,p), (n,α), (n,γ), (n,p) reactions and β-decays involving the reactants in our network for which we could find rates. Wherever possible we use the analytic rate expression,[3-10] as suggested by Woosley and Hoffman.[4] We include both $^{26}Al^m$ and $^{26}Al^g$ in our network as separate species. We also include the electron capture on ^7Be and the (γ,p) reactions on ^8B and ^{17}F.

In reaction networks one must solve a system of coupled non-linear first order differential equations, one for each nuclear species. We use a numerical scheme developed for such reaction systems called CHEMEQ.[11] CHEMEQ is designed to solve large sets of strongly coupled equations where the characteristic time scales over which the species abundances vary may be widely divergent. CHEMEQ applies a specialized numerical technique called the Selected Asymptotic Integration Method to this set of equations. Note that this is a semi-explicit method in contrast to many calculations of this sort done with implicit methods (see, for example, ref. 12). By semi-explicit we mean that the equation set is divided into those "normal" equations that can be integrated explicitly and those equations that are "stiff" and must be integrated using the asymptotic predictor-corrector scheme. The computer realization of our numerical algorithm runs very quickly on the NRL CRAY X-MP because it is easily vectorized. We have tested our solution technique against a published result[13] and against an implicit network of 21 species. We find the solutions obtained are indistinguishable and take about the same amount of computer time to run for this small test network. For larger networks, we have chosen the CHEMEQ routine both because of the ease with which it may be coupled to any of a number of hydrodynamic solution techniques and the ease with which new reactive species may be inserted in the

TABLE 1.

Proton Rich Nova Nucleosynthesis Network

Z	A_{min}	A_{max}	Z	A_{min}	A_{max}	Z	A_{min}	A_{max}
H	1	2	C	11	14	Na	20	23
He	3	4	N	13	15	Mg	21	26
Li	6	7	O	14	18	Al*	22	27
Be	7	9	F	17	19	Si	24	30
B**	8	10	Ne	18	22	P	27	32

* $^{26}Al^g$ and $^{26}Al^m$ are treated as two separate species.

** ^9B is not in the network.

network. Finally, we have also done test runs that allow us to estimate our errors in the final abundances to be no more than approximately 5%, much smaller than the uncertainty in many of the reaction rates.

III. RESULTS AND DISCUSSION

We usually approximate the conditions in the peak burning region of a nova in a white dwarf envelope by assuming a one-zone model where the temperature and density are initially set at some high temperature and density T_0 and ρ_0, and then allowed to decrease exponentially with time with e-folding times of τ_T and τ_ρ, where $\tau_\rho = 1/3 \ \tau_T$ (corresponding to adiabatic expansion). The expansion is allowed to continue until the temperature T_f is reached. The model parameters are given in Table 2 along with the nucleosynthetic yields. For our final model (model 4) we have used a fit to the temperature and density profile given by Prialnik.[2]

Model 1 should be considered a test of our solution technique. Here we use the same T_0, ρ_0, τ_T, and initial abundances as Woosley and Hoffman.[4] The initial abundances are solar with an admixture of

TABLE 2.

Parameterized Nova Nucleosynthesis Models

Model	WH	1	2	3	4
T_0	2.52(8)	2.52(8)	2.52(8)	2.52(8)	--
ρ_0	1.4(4)	1.4(4)	1.4(4)	1.4(4)	--
τ_T	260.0	260.0	86.7	260.0	--
T_f	9.5(6)	9.9(6)	9.9(6)	9.9(6)	--
$X(^1H)$	0.684	0.693	0.705	0.539	0.488
$X(^4He)$	0.228	0.226	0.211	0.238	0.243
$X(^7Be)$	--	1.45(-14)	8.22(-12)	1.20(-14)	3.34(-8)
$X(^{12}C)$	1.80(-2)	1.62(-2)	7.38(-3)	2.08(-2)	3.58(-2)
$X(^{13}C)$	1.65(-2)	7.37(-3)	1.02(-3)	9.61(-3)	4.72(-2)
$X(^{13}N)$	--	7.53(-3)	3.84(-3)	9.70(-3)	5.67(-4)
$X(^{14}N)$	2.97(-2)	2.67(-2)	2.94(-2)	2.73(-2)	0.147
$X(^{15}N)$	1.91(-2)	1.57(-2)	1.80(-2)	2.01(-2)	5.63(-5)
$X(^{14}O)$	--	3.67(-5)	3.45(-3)	3.59(-5)	5.8(-10)
$X(^{15}O)$	--	1.11(-3)	1.15(-2)	1.37(-3)	7.08(-7)
$X(^{16}O)$	3.56(-5)	3.31(-5)	3.09(-5)	4.53(-5)	1.95(-4)
$X(^{17}O)$	1.96(-5)	4.24(-5)	3.80(-3)	2.47(-4)	2.99(-2)
$X(^{18}O)$	4.49(-6)	6.17(-8)	1.11(-6)	4.04(-7)	1.64(-4)
$X(^{18}F)$	--	1.17(-6)	6.18(-5)	7.61(-6)	1.99(-3)
$X(^{19}F)$	2.90(-8)	1.90(-8)	7.11(-6)	7.41(-8)	3.93(-6)
$X(^{20}Ne)$	3.83(-4)	3.70(-4)	7.14(-4)	1.54(-2)	8.36(-4)
$X(^{21}Ne)$	8.74(-8)	7.62(-8)	4.47(-7)	3.04(-6)	1.05(-8)
$X(^{22}Na)$	3.05(-7)	2.89(-7)	8.05(-7)	1.10(-5)	8.57(-8)
$X(^{23}Na)$	7.22(-6)	6.49(-6)	1.95(-4)	2.55(-4)	1.58(-5)
$X(^{24}Mg)$	1.94(-7)	3.17(-7)	2.89(-6)	1.39(-5)	1.67(-6)
$X(^{25}Mg)$	6.67(-5)	8.77(-5)	2.53(-4)	3.74(-3)	1.82(-4)
$X(^{26}Mg)$	--	5.66(-9)	1.31(-6)	2.46(-7)	4.12(-9)
$X(^{26}Al^g)$	3.48(-5)	6.78(-6)	1.34(-5)	3.15(-4)	3.81(-5)
$X(^{27}Al)$	1.93(-6)	4.37(-6)	3.57(-5)	1.94(-4)	1.98(-5)

4% by mass for both ^{12}C and ^{16}O. Our resulting abundances are shown and for the most part compare well with the Woosley and Hoffman results which we label as model "WH". Note that for several nuclei the WH abundances contain the abundances of the radioactive progenitors (^{13}C, ^{14}N, ^{15}N, and ^{18}O). Our results differ from the WH results for ^{26}Al. This may be due to our including the ground state and metastable isomer of ^{26}Al as two separate species; in contrast Woosley and Hoffman treated the metastable state as equilibrated with the ground state. The resulting abundances in our model 1 mostly are uninteresting from an observational point of view.

In model 2 we crudely simulate the effects of convection by dramatically shortening the time scale over which the burning material expands and cools. As the convective region recedes through the envelope, material at a given point can be decoupled from the peak temperature region, and thus experience a rapid decrease in temperature (decreasing on the convective time scale) from the peak temperature, T_0. The effect is to remove from the peak burning regions the ^{7}Be, ^{18}F and ^{22}Na nuclei which can be easily destroyed at high temperatures. This results in larger abundance yields for each of these species over those of model 1. Whether such an effect is realized (or possibly magnified) in a nova envelope will be determined with improved hydrodynamic models or by observations of the gamma-ray lines. Finally, note that the increases in the yields of ^{14}O and ^{15}O in model 2 are due mainly to the earlier termination time of this model. These nuclei have not had as much time to decay; the differences from model 1 are not relevant to their detectability. What is actually important for these short-lived species is their abundances in the outermost regions of the envelope and therefore their abundances in the burning region one convective turnover time before the convective zone recedes from the surface.

Model 3 appears to be the most interesting from the standpoint of the production of gamma-ray emitters. Here we simulate the nucleosynthesis expected to occur in a nova on a white dwarf rich in oxygen, neon and magnesium. The time scale for cooling is the same as for model 1 but the initial mixture consists of solar abundances except for ^{16}O which is 10% by mass and 5% by mass for both ^{20}Ne and ^{24}Mg. The effect is quite dramatic. The resulting abundance of ^{22}Na is more than a factor of ten higher than either of the two previous models and only a factor of ten lower than the lower production limits placed on the ^{22}Na mass fraction from neon rich novae near the galactic center.[14] The production of ^{26}Al^g is also interesting. Ten galactic novae per year, each ejecting 10^{-4} $M_\odot$ of material with this mass fraction of ^{26}Al^g could account for 0.3 $M_\odot$ of ^{26}Al in its lifetime. This is ~10% of that actually detected.[15]

For our final model (model 4) we have departed from the simple exponential model and carefully set up a one-zone simulation of the peak burning shell from the hydrodynamic nova simulation of Prialnik.[2] We determine from her model a temperature and density at each time. The initial abundances are solar with an admixture of 13% for both ^{12}C and ^{16}O, approximately what she found was mixed into the dwarf envelope by species diffusion and convection by the time the thermonuclear runaway had begun. We allow the simulation to proceed until the model cools to 7×10^7K. Note that this is significantly different than an exponential model. The results are given in the last column of Table 2. Our principal interest here is to compare the results for the nucleosynthetic products from our one-zone model with the results of her hydrodynamic model. Our results are generally close to hers with a couple of notable exceptions. For ^{22}Na we find $X(^{22}Na) = 8.6 \times 10^{-8}$ whereas she finds 1.5×10^{-8}. For ^{18}F we obtain $X(^{18}F) = 2.0 \times 10^{-3}$ whereas she obtains 7.0×10^{-6}, a difference of roughly a factor of 300.[16] We also find that ^{16}O is effectively destroyed while in her calculation it largely survives the burning.

IV. CONCLUSIONS

Let us now summarize our principal conclusions. First, although the <u>final</u> abundances of ^{13}N, ^{14}O, and ^{15}O are relatively uninteresting for gamma-ray astronomy, during the burn the <u>peak</u> abundances for each approaches the initial carbon abundance (<u>i.e.</u>, several percent) in those models where carbon was initially enhanced. If material this rich in β^+ emitters can reach the photosphere, a very close nova (distance ~ 1 kpc) represents a promising target for a gamma-ray telescope.[17] Second, the prompt removal of fragile nuclei such as ^{18}F and ^{22}Na can enhance the resulting nucleosynthetic yields and may lead to the possibility of observing their respective gamma-ray lines with instruments to be launched in the near future such as those on the Gamma Ray Observatory. Third, enhancing the initial abundances of neon and magnesium may result in the production of observable abundances of ^{22}Na and ^{26}Al^g. In particular, our one-zone neon rich model leads to a nucleosynthetic yield for ^{26}Al^g that indicates neon-rich novae may account for a significant fraction of the observed diffuse 1.809 MeV flux (but the frequency and distribution of such novae are very uncertain). Finally, the confrontation between our one-zone model and the more complete hydrodynamic calculation of Prialnik[2] shows fair agreement in the predicted abundances. It is clear that new hydrodynamic calculations with a complete reaction network adequate to describe the nucleosynthesis beyond phosphorous are very much needed.

This work was done while Michael Wolff was a Universities Space Research Association Visiting Scientist at the Naval Research Laboratory and Mark Leising was a National Research Council-Naval Research Laboratory Research Associate. This work was supported by the Office of Naval Research. We also thank Dina Prialnik for providing us with unpublished abundance results from her nova models.

REFERENCES

1. Starrfield, S., in <u>The Classical Nova</u>, ed. M. F. Bode and A. Evans (New York: Wiley) (1985).
2. Prialnik, D., <u>Ap. J.</u>, **310**, 222 (1986).
3. Wiescher, M., Gorres, J., Thielemann, F.-K., and Ritter, H., <u>Astron. Astrophys.</u>, **160**, 56 (1986).
4. Woosley, S. E. and Hoffman, R. D., preprint (1987).
5. Fowler, W. A., Caughlan, G. R., and Zimmermann, B. A., <u>Ann. Rev. Astr. and Astrophys.</u>, **13**, 69 (1975).

6. Woosley, S. E., Fowler, W. A., Holmes, J. A., and Zimmermann, B. A., Atomic Data and Nuclear Data Tables, **22**, 371 (1978).
7. Wallace R. K. and Woosley, S. E., Ap. J. Suppl., **45**, 389 (1981).
8. Wiescher, M., and Kettner, K.-U., Ap. J., **263**, 891 (1982).
9. Harris, M. J., Fowler, W. A., Caughlan, G. R., and Zimmermann, B. A., Ann. Rev. Astron. and Astrophys., **21**, 165 (1983).
10. Caughlan, G. R., Fowler, W. A., Harris, M. J., and Zimmermann, B. A., Atomic Data and Nuclear Data Tables, **32**, 197 (1985).
11. Young, T. R., NRL Memorandum Report 4091 (1980).
12. Woosley, S. E., Arnett, W. D., and Clayton, D. D., Ap. J. Suppl., **26**, 231 (1973).
13. Cowan, J. J., and Rose, W. K., Ap. J., **217**, 51 (1977).
14. Leising, M. D., Share, G. H., Chupp, E. L., and Kanbach, G., Ap. J., **328**, in press (1988).
15. Mahoney, W. A., Ling, J. C., Wheaton, Wm. A., and Jacobson, A. S., Ap. J., **286**, 578 (1984).
16. Prialnik, D., private communication (1987).
17. Leising, M. D., and Clayton, D. D., Ap. J., **323**, 159 (1987).

MEASUREMENTS OF ASTROPHYSICAL REACTION RATES FOR RADIOACTIVE SAMPLES[*]

P.E. Koehler, H.A. O'Brien and C.D. Bowman
Physics Division, Los Alamos National Laboratory, MS-D449,
Los Alamos, New Mexico 87545, USA

ABSTRACT

Reaction rates for both big-bang and stellar nucleosynthesis can be obtained from the measurement of (n,p) and (n,γ) cross sections for radioactive nuclei. In the past, large backgrounds associated with the sample activity limited these types of measurements to radioisotopes with very long half lives. The advent of the low-energy, high-intensity neutron source at the Los Alamos Neutron Scattering CEnter (LANSCE) has greatly increased the number of nuclei which can be studied. Results of (n,p) measurements on samples with half lives as short as fifty-three days will be given. The astrophysics to be learned from these data will be discussed. Additional difficulties are encountered when making (n,γ) rather than (n,p) measurements. However, with a properly designed detector, and with the high peak neutron intensities now available, (n,γ) measurements can be made for nuclei with half lives as short as several weeks. Progress on the Los Alamos (n,γ) cross-section measurement program for radioactive samples will be discussed.

INTRODUCTION

Over the years, many reaction rates of importance to both big-bang and stellar nucleosynthesis calculations have been measured in the laboratory. These rates have then been integrated into the reaction network used in the calculations and have improved the general understanding of several types of nucleosynthesis events. At present, the rates for several important reactions have not been measured, necessitating the use of theoretical estimates[1]. This may lead to large uncertainties in the isotopic yields from nucleosynthesis calculations. Many of the unmeasured rates involve neutron-induced reactions on radioactive nuclei. Previous measurements[2,3] of this type have been limited to nuclei with very long half lives due to potentially large backgrounds associated with the sample activity. The advent of pulsed spallation neutron sources, such as LANSCE[4], has opened up the possibility of making cross-section measurements for neutron-induced reactions on nuclei with short half lives. In this paper, we will give some examples of recent measurements of this type and briefly discuss the nuclear astrophysics to be learned. We will then discuss our plans for additional measurements, briefly outlining the techniques involved and the expected results. We expect that these measurements will greatly aid in the understanding of the astrophysical environments in which nucleosynthesis occurs, and will aid in the calculation of the expected nucleosynthesis yield of isotopes of interest to gamma-ray astronomy.

[*]Work supported by the U.S. Department of Energy

EXPERIMENTAL TECHNIQUE

The experimental technique[5] for these measurements requires a large peak neutron intensity and a properly designed detector so that the detected rate for the reaction of interest is larger than the background rate associated with the decay of the sample under study. At LANSCE, the high peak neutron intensity is obtained by bombarding a tungsten target with an intense burst of protons. The protons are accelerated by LAMPF, and compressed into an intense pulse by the newly commissioned Proton Storage Ring (PSR). At the design intensity of the PSR (100 μA, at 12 Hz), the water moderated neutron intensity at 1 eV for a flight path of 7 m is 4×10^6 neutrons/(eV cm^2 sec), and the neutron spectrum of this "white" source is approximately proportional to $1/E_n$. The relatively long pulse width (250 ns) from the PSR limits the useful upper energy to about 50 keV at which point the energy resolution is about 25% for the 7 m flight path used in our measurements. The "white" nature of the neutron source means that measurements at all neutron energies are obtained simultaneously. This high neutron intensity allows measurements to be made with sample sizes in the 100 ng to a few hundred μg range. The small sample size means that the necessary radioactive samples are easier to produce and that only relatively modest activities must be handled. Even these small samples can still present some rather large background problems, but a properly designed detector can reduce the sample-related backgrounds to acceptable levels. Because the requirements for the detectors differ, we will discuss (n,p) and (n,γ) measurements separately below.

A*(n,p) MEASUREMENTS

For A*(n,p) measurements, where A* is a radioactive nucleus, the sample-related backgrounds can be reduced to manageable levels by choosing a charged-particle detector of thickness no greater than that needed to stop the protons from the reaction of interest. The detection efficiency for radioactive decay emissions from the sample can thus be reduced to order 10^{-6} of the proton detection efficiency. Also, very few nuclei emit charged particles under bombardment by slow neutrons. Hence, the sample can be of relatively low specific activity, and can even be a chemical compound. Several unstable nuclei (^{22}Na, ^{26}Al, ^{56}Co) have been considered as candidates for observation by gamma-ray telescopes[6], and at least one[7] (^{26}Al) has been detected. Perhaps at this workshop we will hear of other successful observations. In certain astrophysical scenarios, A*(n,p) reactions play an important role in the calculations of the production of these isotopes. For example, the relatively large ^{26}Al(n,p)^{26}Mg cross section is given as evidence that ^{26}Al may not be produced in sites where neutrons are important.[8] Also, some A*(n,p) reactions may play an important role in the nucleosynthesis in explosive environments of very rare stable isotopes.[9]

We will discuss ^{7}Be(n,p)^{7}Li ($t_{1/2}$=53 days) as a first example of an important A*(n,p) reaction rate. In this case, the reaction is of importance to the big-bang nucleosynthesis of ^{7}Li. Our measurements were made with 90 ng ($\approx$30 mCi) of ^{7}Be, in less than one week of beam time with the PSR operating at only one-tenth of its design intensity. Prior to our measurements[5], only the thermal cross sections had been measured directly[10]. The rate[11] for this reaction used in calculations was based on this rather imprecise thermal value and on some also rather imprecise ^{7}Li(p,n)^{7}Be measurements[12] converted to (n,p) using detailed balance. Our new measurements have substantially reduced (by a factor of

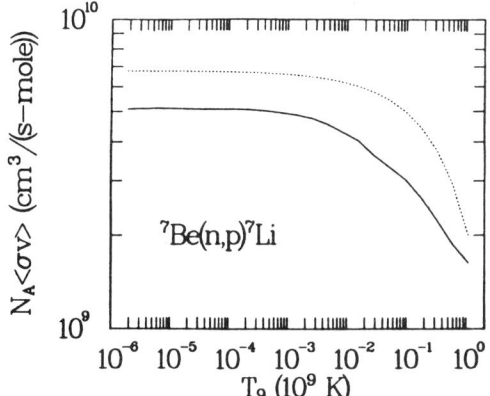

Fig. 1. The ^{7}Be(n,p)^{7}Li reactivity verses temperature. The solid curve is the rate calculated from our data, while the dashed curve is the theoretical rate of ref. 11.

almost ten at thermal energy) the uncertainty in the reaction rate. The reaction rate determined primarily from our new data is compared to the old rate in fig. 1. The rate based on our data is only 60% to 80% of the old rate in the temperature range of interest[13] in the big-bang calculations ($T_9 \approx 0.3 - 1$). Calculations[14-16] have shown that this difference can lead to as much as a 20% increase in the amount of ^{7}Li produced in the big bang.

A second example of our recent A*(n,p) measurements[17] is ^{22}Na(n,p)^{22}Ne ($t_{1/2}$=2.6 years). In this case, the measurements were made with 360 ng ($\approx$2.25 mCi) of ^{22}Na. This reaction may play a role in the nucleosynthesis of ^{22}Na and ^{22}Ne, and may aid in the interpretation of the neon-E anomaly[18] in meteorites. The astrophysical reaction rate calculated from our data is compared to the theoretical rate[1] in fig. 2. At our energies, most of the rate is due to protons emitted to the first excited state of ^{22}Ne. The theoretical rate is about a factor of ten lower than the experimentally determined one at very low temperatures. However, due to a resonance at E_n=170 eV, the two rates cross at T_9=0.05. We are currently exploring the astrophysical implications of this result.

A final example of A*(n,p) measurements is our recent data on the ^{36}Cl(n,p)^{36}S reaction[17]. These measurements were made with 410 μg (9 μCi) of ^{36}Cl. Because the half life for this sample is long ($t_{1/2}$=3x10^5 years), a high peak neutron intensity is not essential to the measurements. However, the relatively high average neutron intensity available from LANSCE is still important to measuring this comparatively small cross section within a reasonable time. Our preliminary measurements are displayed in fig. 3. Because the thermal cross section has not yet been measured, we display yields rather than cross sections. The data reveal several resonances for E_n>900 eV. This reaction is denoted by a * in Howard et al[9], a mark which they reserve for rates important to the nucleosynthesis of rare nuclei (^{36}S in this case) in explosive carbon burning. It remains to be seen how our measurements will affect the

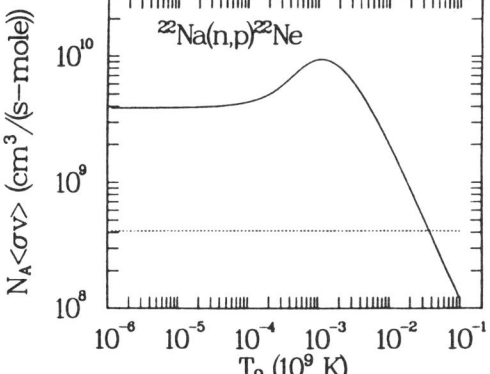

Fig. 2. The ^{22}Na(n,p)^{22}Ne reactivity verses temperature. The solid curve is the rate calculated from our measurements, while the dashed curve is the theoretical rate of ref. 1.

results of future calculations.

A*(n,γ) MEASUREMENTS

Because many nuclei have sizable (n,γ) cross sections at low neutron energy, A*(n,γ) measurements require a separated isotope on a low mass backing as a sample. This requirement should not be too difficult to meet for the very small samples sizes required. For these types of measurements the decay radiation from the sample is often of the same type as that from the reaction of interest. Pileup of the low-energy decay γ-rays can result in a signal the same size as that from a neutron-capture event. To overcome this potentially large background one can make use of the fact that the gamma-ray decay energy, E_d, is almost always much lower in energy than the total energy, E_c, of the neutron capture cascade. Hence, a detector which registers all of the energy from the capture cascade, and which has a very short output pulse width, τ, can effectively overcome this background. Of course, the size of this background is a very strong function of the ratio, E_d/E_c, and of τ, and one can always think of very difficult cases for which measurements are still not possible. Our calculations show that measurements on many interesting samples with half lives as short as a few weeks can be made. A few of these cases[19] are summarized in table 1. One other source of potentially serious background is from neutrons scattered into the detector and subsequently captured by one of the detector nuclei. This could conceivably result in a signal indistinguishable from the reaction of interest.

One obvious choice for a detector is a large tank of liquid scintillator such as that used in previous measurements on stable samples. The tank would be almost 100% efficient for the capture cascade, and by loading the liquid with ^{10}B it can be made much less sensitive to background neutrons. However, we feel that the relatively new scintillator barium fluoride (BaF_2) is a better choice. Monte Carlo calculations we have made using the computer code CYLTRAN[20] indicate that a thickness of roughly 15 cm of BaF_2 is adequate to make an approximately 100% efficient detector in essential agreement with the calculations of Wisshak et al.[21] In our application, the small sample size allows us to have a very small neutron beam (0.5 cm diameter). Hence, a 30 cm cube of BaF_2 with a 4 cm beam hole is sufficient to make a detector which is approximately 100% efficient. A detector of this size, composed of eight 15 cm cubes is relatively inexpensive to build compared to more complex designs such as the Karlsruhe sphere[22]. The fast component of BaF_2 ($\tau \approx 20$ ns for 15 cm crystals) provides for effective pileup rejection. Pileup of the slow component of BaF_2 contributes essentially only a constant offset signal which can be cancelled with the use of the proper electronics.

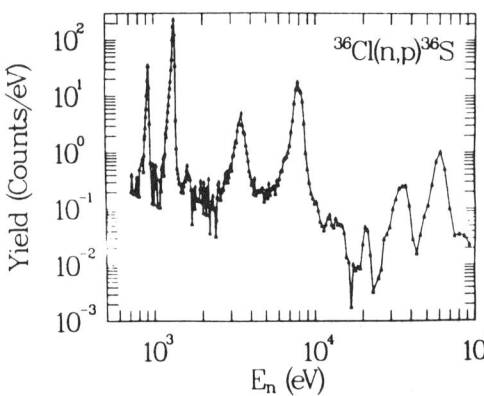

Fig. 3. Preliminary yield verses neutron energy from our $^{36}Cl(n,p)^{36}S$ measurements. The yield has not been corrected for the variation with energy of the neutron flux.

TABLE 1

A list of a few of the many isotopes of importance to the s-process of nucleosynthesis for which $A^*(n,\gamma)$ measurements should be possible at LANSCE.

Isotope	Half life	Motivation
^{85}Kr	10.7 y	Very important branch point.
^{115}Cd^m	44.8 d	Causes some of mass flow to bypass s-only ^{116}Cd.
^{151}Sm	93 y	Branch responsible for ^{152}Gd. Half life depends on temperature.
^{152}Eu	13 y	Same branching as ^{151}Sm.
^{154}Eu	8.8 y	Part of branch causing mass flow to bypass s-only ^{154}Gd. Half life depends on temperature.
^{155}Eu	5 y	Important branch point.
^{153}Gd	241 d	Same branching as ^{151}Sm.
^{163}Ho	33 y	Branch at ^{163}Dy, ^{164}Er abundance, and yields estimate for matter density.
^{170}Tm	129 d	Causes mass flow to bypass ^{170}Yb.
^{171}Tm	1.9 y	Continuation fo branch at ^{170}Tm.
^{179}Ta	665 d	Origin of ^{180}Ta, nature's rarest stable isotope.
^{185}W	75 d	Important brach point.
^{204}Tl	3.8 y	Important to ^{205}Pb - ^{205}Tl clock for early solar system.

A BaF$_2$ scintillator has several advantages over a large liquid scintillator tank. First, the total detector size is much smaller so that shielding from room background is much easier to accomplish. Second, the output pulse width can probably be made shorter for BaF$_2$ than for a tank of liquid scintillator of the required size ($\approx$1.5 m diameter). Third, the pulse-height resolution of BaF$_2$ can be varied by accepting more or less of the signal from the slow decay component. When using only the fast component, the expected resolution should be about equal to that obtained with a large tank of liquid scintillator (20-30%). However, for those stable samples which are available only in very small quantities as separated isotopes, or for radioactive samples of very low specific γ-ray activity, the opportunity exists with BaF$_2$ for improving the pulse-height resolution dramatically and, hence, for improving the accuracy of the measurement, perhaps to as good as 1% (ref. 21). The main disadvantage of BaF$_2$ is its greater sensitivity to the background from scattered neutrons. We have made Monte Carlo calculations using the code MCNP[23] to estimate the size of this background. The calculations indicate that our detector would be about 1% efficient to scattered neutrons at 10 keV. This can be reduced even further (to 0.3%) by lining the central beam hole with a 1 cm thick layer of ^{10}B$_4$C. Hence, for almost all cases the background due to scattered neutrons can be reduced to acceptable levels for the range of energies possible in our measurements.

The resulting $A^*(n,\gamma)$ measurements will have a large impact on our understanding of the dynamics of s-process nucleosynthesis - a point where current models fail to reproduce the observations.[24] The first few measurements at selected s-process branch points should help reveal more about the mean properties of the s-process environment, the competition between radioactive decay and neutron capture helping to place more stringent limits on quantities such as the mean neutron density.[25] A larger series of measurements will allow an examination of the dynamics of the s-process. For example, the measurements will allow for a better understanding of the time dependence of the neutron flux and the peak neutron intensity. We hope that this in turn will spur the development of better models of the s-process environment.

REFERENCES

1. See for example, S.E. Woosley et al., At. Data Nucl. Data Tables 22, 371 (1978).
2. A. Emsallem et al., Nucl. Phys. A368, 108 (1981); H. Weigmann et al., Nucl. Phys. A368, 117 (1981).
3. Yu.M. Popov et al., Z. Phys. A 322, 685 (1985).
4. R.N. Silver, Physica 137B, 359 (1986).
5. P.E. Koehler et al., Submitted for publication to Phys. Rev. C.
6. See for example, D.D. Clayton, Space Sci. Rev. 24, 147 (1982).
7. W.A. Mahoney, J.C. Ling and A.S. Jacobson, Ap. J. 262, 742 (1982).
8. W.A. Fowler, Rev. Mod. Phys. 56, 149 (1984).
9. W.M. Howard et al., Ap. J. 175, 201 (1972).
10. R.C. Hanna, Phil. Mag. 46, 381 (1955).
11. N.E. Bahcall and W.A. Fowler, Ap. J. 157, 659 (1969).
12. R.L. Macklin and J.H. Gibbons, Phys. Rev. 109, 105 (1958).
13. A.M. Boesgaard and G. Steigman, Ann. Rev. Astron. Astrophys. 23, 319 (1985).
14. J. Yang et al., Ap. J. 281, 493 (1984).
15. G. Beaudet and H. Reeves, Astron. Astrophys. 134, 240 (1984).
16. P. Delbourgo-Salvador et al., Astron. Astrophys. 150, 53 (1985).
17. P.E. Koehler et al., to be published.
18. D.C. Black, Geochim. Cosmochim. Acta., 36, 377 (1972).
19. F. Kaeppeler, private communication (1986).
20. J.A. Halblieb, Sr. and W.H. Vandevender, "CYLTRAN", Sandia National Laboratories Report, SAND 74-0030 (1974); H.H. Hsu et al., IEEE Trans. Nucl. Sci. NS-31, 390 (1984).
21. K. Wisshak, F. Kaeppeler and G. Schatz, Nucl. Instr. and Meth 221, 385 (1984).
22. K. Wisshak and F. Kaeppeler, Nucl. Instr. and Meth. 227, 91 (1984); and in Nuclear Data for Basic and Applied Science, Vol. 2", (Gordon and Breach, 1986) p. 1319.
23. J.F. Breismeister, ed., "MCNP - A general Monte Carlo Code for Neutron and Photon Transport", Los Alamos National Laboratory Report LA-7396-M, Rev. 2 (1986).
24. See for example, W.M. Howard et al., Ap. J. 309, 633 (1986).
25. F. Kaeppeler et al., Ap. J. 257, 821 (1982).

GAMMA-RAY SPECTROSCOPY OF RADIONUCLEI FROM GALACTIC NUCLEOSYNTHESIS

William A. Mahoney
Jet Propulsion Laboratory 169-327, Pasadena, CA 91109

ABSTRACT

A variety of radionuclei are produced during stellar evolution, especially during explosive events such as novae and supernovae. The detection of gamma-ray line emission from the decay of these radionuclei was one of the prime objectives of the HEAO 3 high-resolution gamma-ray spectrometer. A number of exciting results have been obtained to date from the analysis of the HEAO 3 data and they are summarized here. The most important discovery was that of a narrow cosmic line at 1809 keV from the decay of approximately 3 $M_\odot$ of ^{26}Al in the present interstellar medium. In addition, searches have been performed for galactic sources of ^{60}Fe, ^{44}Ti, and ^{22}Na, all produced in explosive events. While no positive detections were made, interesting new limits have been obtained. The HEAO 3 data have also been searched for the extended source of 511 keV line emission which is expected to result from galactic nucleosynthesis.

INTRODUCTION

It is widely believed that the synthesis of the bulk of the elements heavier than helium occurs in stars[1]. During the processes of stellar evolution, especially in the explosive phases of novae and supernovae, a number of radioisotopes are created. The gamma-ray line emission from the decay of these radionuclei and from the annihilation of positrons therefore provides a direct probe of the sites, rates, and models of nucleosynthesis. Table 1 summarizes the predicted yields of the most abundant radionuclei believed to be produced in both Type I and Type II supernovae. Most abundant is ^{56}Ni produced primarily in Type I supernovae, but with a substantial yield in Type II supernovae. However, in the absence of rapid mixing, attenuation in the expanding envelope will probably prevent the direct observation of gamma rays from the short lived ^{56}Ni decay. Even gamma rays from the decay of the daughter nucleus, ^{56}Co, which has a 111-day lifetime before decaying to ^{56}Fe, will be strongly attenuated for a time dependent on the expansion velocity of the shell. Until the recent detection of line emission from SN1987A[2-5], there had not been a supernova in the vicinity of the Galaxy recent enough for the resulting gamma rays to be visible. On the other hand, the sensitivity of gamma-ray spectrometers has been insufficient to detect extragalactic supernovae.

Two radionuclei with lifetimes comparable to the time between synthesizing

© 1988 American Institute of Physics

TABLE 1
ISOTOPIC DECAY CHAINS AND PREDICTED SUPERNOVA YIELDS

DECAY CHAIN	MEAN LIFE (YEARS)	TYPE I[A] NUCLEI / SUPERNOVA	TYPE II[B] NUCLEI / SUPERNOVA	PHOTON ENERGY (MeV)	PHOTONS / DISINTEGRATION
$^{56}Ni \rightarrow {}^{56}Co$	0.024	1.3×10^{55}	4.5×10^{54}	0.158	0.99
				0.812	0.86
				0.750	0.50
				0.270	0.37
				0.480	0.37
$^{56}Co \rightarrow {}^{56}Fe$	0.31	6.3×10^{51}		0.847	1.0
				1.238	0.68
				2.598	0.17
				1.771	0.16
				1.038	0.14
$^{57}Co \rightarrow {}^{57}Fe$	1.07	5.5×10^{52}	2.2×10^{53}	0.122	0.86
				0.136	0.10
				0.014	0.10
$^{44}Ti \rightarrow {}^{44}Sc$	68	2.3×10^{51}	2.8×10^{51}	0.068	1.0
				0.078	0.98
$^{44}Sc \rightarrow {}^{44}Ca$	6.5×10^{-4}			1.157	1.0
$^{22}Na \rightarrow {}^{22}Ne$	3.75	9.1×10^{48}	1.4×10^{51}	1.275	1.0
$^{26}Al \rightarrow {}^{26}Mg$	1.04×10^{6}	1.8×10^{50}	1.1×10^{51}	1.809	1.0
$^{60}Fe \rightarrow {}^{60}Co$	2.15×10^{6}	4.8×10^{46}	4.2×10^{50}	0.059	1.0
$^{60}Co \rightarrow {}^{60}Ni$	7.61	4.6×10^{47}		1.173	1.0
				1.333	1.0

[A] NOMOTO, K., ET AL., 1984, AP. J., 286, 644.

[B] WOOSLEY, S. E., ET AL., 1981, COMMENTS NUCL. PART. PHYS., 9, 185.

events in the Galaxy are ^{44}Ti, believed to be produced primarily in Type II supernovae, and ^{22}Na, produced mainly in novae. Because of the short lifetime, and because of the dominance of nearby events[6], the distribution of gamma-ray emission from their decay should, at any given time, be dominated by several point sources. In the case of ^{22}Na, there will probably be an underlying extended emission. Unfortunately, the predicted intensities fall near or below the sensitivity of the HEAO 3 gamma-ray spectrometer.

Both ^{26}Al and ^{60}Fe have lifetimes of a million years, four orders of magnitude longer than the average time between galactic supernovae. Thus gamma-ray emission from their decay should represent the cumulative remains of approximately 10^4 supernovae and should form an extended distribution in the equatorial plane of the Galaxy corresponding to the stellar population responsible for their origin. The principal source of ^{26}Al is controversial, the most likely candidates being supernovae[7,8], novae[9-11], Wolf-Rayet stars[12,13], and red giants[14]. Following the discovery[15] and confirmation[16] of ^{26}Al in the Galaxy, interest has moved from the question of its existence to a determination of its spatial distribution and origin. The second long-lived radioisotope, ^{60}Fe, is likely produced by explosive helium burning in Type II supernovae[17].

Study of positron annihilation radiation will also enhance our understanding of galactic nucleosynthesis. Given the existence of the ^{26}Al in the interstellar medium (ISM) and the fact that ^{26}Al decays mainly via positron emission, there must also exist an extended galactic source of 511 keV emission from the annihilation of these positrons. In addition to ^{26}Al, two likely contributions to the positrons in the ISM are ^{44}Ti which decays to ^{44}Sc and then to ^{44}Ca via positron emission[18], and the escape of positrons created by the decay of ^{56}Co in Type I supernovae[19]. The ^{44}Ti should be dispersed directly into the ISM while some fraction of the ^{56}Co positrons may escape from the expanding shell into the ISM where they would have a lifetime of order 10^6 years[20].

Intense emission in a narrow line at 511 keV from the direction of the galactic center has been known for over a decade[21]. The original analysis of the HEAO 3 data for a point source model demonstrated a statistically significant decrease in the line flux between the fall of 1979 and the spring of 1980[22]. Since then, the measurement of an extended component has been reported[23] based on data from the gamma-ray experiment on the Solar Maximum Mission (SMM). In an attempt to separate the point and extended components, and to directly compare the HEAO results with those from SMM, a reanalysis of the HEAO 3 data was undertaken.

DIFFUSE SOURCES

HEAO 3 was a scanning mission with a spin period of 20 minutes. Conventional analysis techniques involve the accumulation of counts from many scans followed by a source-background fitting. Because the HEAO 3 shield has non-zero transmission above a few hundred keV, even at large angles from the viewing direction, this technique does not work for an extended source where a global fit to the data must be performed. Additionally, for both point and extended sources, the older technique is prone to systematic errors resulting from an intense, variable background. Thus we have developed a new technique[24] which returns the amplitude of an assumed source function computed for each ten minute stretch of data. The calculated fluxes from hundreds of such scans are then averaged together to give the final result. Since 10 minutes is short compared to the spacecraft orbital period, systematic effects from background variations are nearly eliminated. While this method results in a significant reduction in both the systematic and statistical errors, it is not yet possible to measure the spatial distribution of a source flux. Rather a source distribution must first be assumed, either a point source at a specified location, or a planar source with a specified distribution. This function is then convolved with the instrument transmission as a function of scan angle to derive the source flux for each scan.

This analysis technique was used to search for extended cosmic gamma-ray line emission at 1809 keV from the decay of ^{26}Al, and indeed a net cosmic line was discovered[15]. Two galactic distributions were studied[25] which we believe are representative of stellar populations suggested as the source of the ^{26}Al, both of which are peaked in the direction of the galactic center. For the older disk population containing novae and red giants, we chose the total visual luminosity of the Galaxy[26]. The measured galactic CO distribution[27] was assumed to model the distribution of Type II supernovae and massive main sequence stars. For both models, a statistically significant net cosmic excess was measured. The peak of the emission was found to fall within about 20° of the galactic center[25]. This was subsequently confirmed by results from the SMM gamma-ray experiment[16]. Assuming the gamma-ray emission originated from the decay of ^{26}Al distributed throughout the ISM, the implied galactic mass content is about 3 $M_\odot$[15]. The measured limit of < 3 keV FWHM on the intrinsic line width indicates that the ^{26}Al has come to rest in the ISM before decaying. While both extended distributions give acceptable fits to the data, the model of a point source at the galactic center gives a fit of significantly lower confidence. Thus, while the exact distribution of the emission has not yet been measured with the HEAO 3 data, the results clearly favor an extended source model. A similar conclusion was reached by the Bell/Sandia group[28], however, the Max-Planck group has suggested the emission results from a very compact source at the galactic center[29].

Both ^{60}Fe and ^{26}Al have similar lifetimes and the predicted supernova yields are comparable (Table 1). While no evidence for galactic emission has been found, we have been able to place an interesting limit of 1.8 x 10^{-4} photons/cm^2-s-rad on the flux in the 1173 and 1333 keV lines which result from the ^{60}Fe-decay chain[30]. This limit refers to the galactic center region and is based on an assumed extreme population I distribution. The result limits the galactic ^{60}Fe mass content to <6 $M_\odot$.

^{22}Na is unique in that its lifetime is about 100 times the average time between galactic novae. However, because of the dominance of novae that occur near the Earth, the distribution of the resulting gamma-ray line emission is expected to be dominated at any given time by several point sources on top of an extended component peaked toward the galactic center[6]. Flux limits on the emission from ^{22}Na for the extreme population I distribution are included in Table 2. The limit of 4.4 x 10^{-4} photons/cm^2-s-rad from the galactic center[30] constrains the galactic mass content of ^{22}Na to < 10^{-5} $M_\odot$. The search for possible point sources of ^{22}Na is the subject of a future study.

TABLE 2

DIFFUSE GALACTIC LINE EMISSION[a]

Isotope	Line Energy (keV)	Assumed Line Width (keV)	Intensity [10^{-4} photons cm^{-2} s^{-1} rad^{-1}]	
			Net Line	3σ Upper Limit
^{22}Na	1275	8	1.7±0.9	4.4
^{26}Al	1809	-	4.3±0.8	-
^{60}Co	1173	3	0.50±0.57	2.21
^{60}Co	1333	3	0.57±0.65	2.52
^{60}Fe,^{60}Co[b]	-	-	0.53±0.43	1.82

[a] From the vicinity of the galactic center.
[b] Weighted average.

POINT SOURCES

Gamma-ray emission from nucleosynthetic radioisotopes with shorter lifetimes will appear as point sources. Searches of the HEAO 3 data have not yielded any sources of gamma-ray emission from the decay of ^{56}Ni from either galactic or extragalactic supernovae.

The gamma-ray emission from ^{44}Ti should appear as a few point sources, much like the emission from the decay of ^{22}Na produced in galactic novae[6]. Since it is unknown where these point sources might be located, we have placed limits on a series of assumed point sources uniformly distributed around the galactic plane. For an expected ejection velocity of order 7000 km/s for the ^{44}Ti produced in Type I supernovae, the gamma-ray line at 1157 keV would be broadened to order 100 keV FWHM, rendering our instrument less sensitive to its detection. The 68 keV and 78 keV lines are more promising, especially the latter where there is no line in the instrumental background spectrum. Thus we chose to investigate the energy band from 58 to 90 keV. Fits to the data at both line energies provided no evidence for a cosmic point source anywhere in the galactic plane. The limit obtained for an assumed point source at the galactic center is 3×10^{-4} photons/cm^2-s, which is typical of all point sources in the galactic plane. This limit corresponds to a ^{44}Ti mass at the galactic center of $< 2 \times 10^{-4}$ M$_\odot$. The flux limit combined with Monte Carlo calculations provides a limit on the ^{44}Ti yield in a typical Type I supernova of the order of 5×10^{-4} M$_\odot$.

POSITRON ANNIHILATION

The search for an extended galactic annihilation line proceeded in much the same manner as the 1809 keV analysis except for two complicating factors: a) a variable cosmic point source at the galactic center[22], and b) a strong atmospheric line[31]. The effects of the atmospheric annihilation line were minimized by eliminating periods when the instrument was pointing toward the Earth, and by including a model of the atmospheric emission in the fitting procedure. An attempt to separate the point and extended components by including both as free parameters in the fit was not successful because their strong correlation increased the errors to the point where no useful conclusions could be drawn. Thus the analysis was limited to the CO distribution used for the ^{26}Al studies. This is the same model as that used for the SMM analysis. Since the SMM investigators attributed all the 511 line emission to an extended source, the resulting HEAO 3 diffuse fluxes can be directly compared to those reported by SMM.

The net line fluxes obtained by HEAO 3 for the CO model during both the fall of 1979 and the spring of 1980 are summarized in Table 3. Figure 1 shows the corresponding cosmic spectra near 511 keV. Owing to radiation damage of the germanium detectors[32], the instrument energy resolution at 511 keV had degraded from less than 3 keV FWHM initially to about 7 keV by the spring of 1980. The curves fitted to the data represent a line at the expected energy with a width corresponding to the energy resolution of the instrument. From Table 3 it is clear that the fluxes obtained during the fall of 1979 and during the spring of 1980 are both consistent with a constant value of $(1.35 \pm 0.22) \times 10^{-3}$ photons/cm^2-s-rad.

TABLE 3

511 keV LINE FLUX FOR GALACTIC CO DISTRIBUTION
[10^{-3} photons cm^{-2} s^{-1} rad^{-1}]

Fall 1979	Spring 1980	Weighted Average
1.50±0.33	1.23±0.29	1.35±0.22

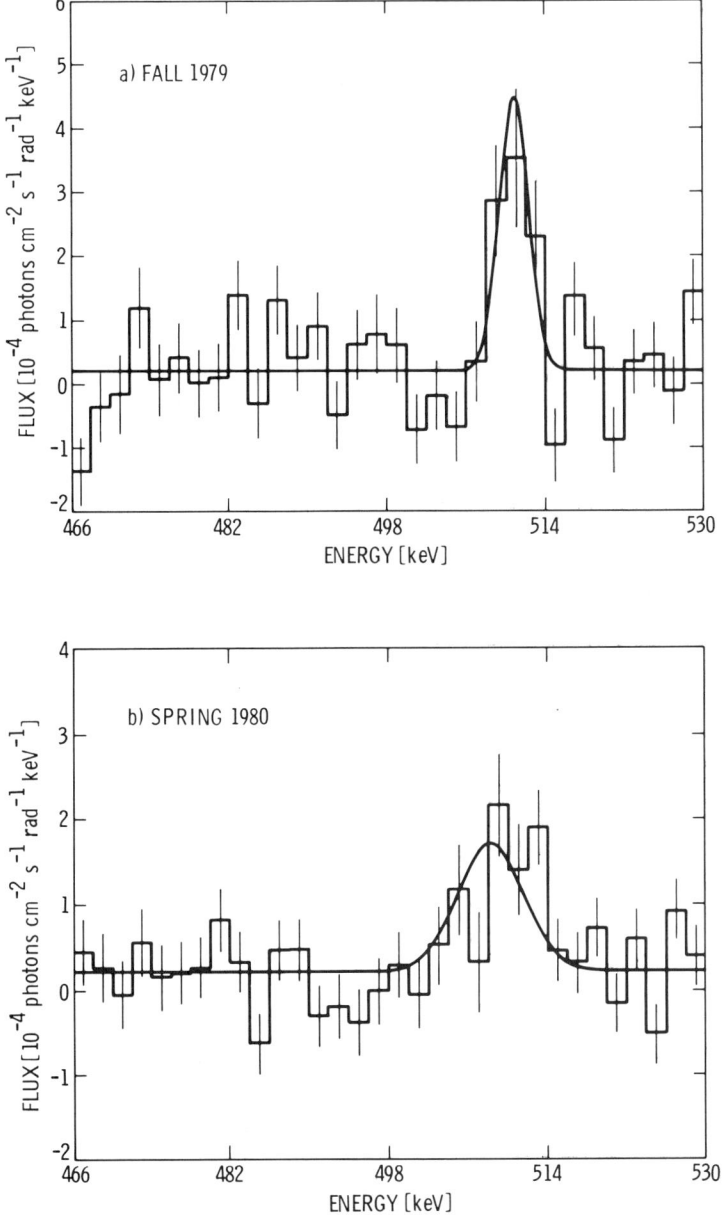

Figure 1. Net gamma-ray spectra from the galactic center based on an assumed CO model, measured a) in the fall of 1979, and b) in the spring of 1980.

Similarly, this flux is consistent with the value of $(1.6 \pm 0.3) \times 10^{-3}$ photons/cm^2-s-rad reported from SMM observations. Finally, during both HEAO observations, the intrinsic width of the 511 keV line was not resolved, with upper limits of 3 keV and 9 keV for the fall of 1979 and spring of 1980, respectively.

During the study of the extended galactic 511 keV line emission, the point source model with more complicated background components was also investigated. The new results support time variability of the emission, but at a reduced level of confidence. The net line flux measured in the fall of 1979 is lower than originally reported[22] while that observed in the spring of 1980 is higher. We are presently extending the capabilities of our analysis programs to specifically study variations in the cosmic emission as the galactic center is occulted by the Earth. A sharp change in intensity would correspond to a point source which could then be separated from the extended component.

The origin of the extended 511 keV line is not completely understood. The annihilation of positrons from the decay of ^{26}Al accounts for 10% - 20% of the observed intensity, depending on the fraction of positrons forming positronium prior to annihilation. A likely contributor suggested recently[18] is ^{44}Ti. If one assumes that all the present galactic ^{44}Ca was produced originally as ^{44}Ti in Type I supernovae, the annihilation of the resulting positrons can account for the remainder of the observed 511 keV line emission, depending on the frequency of galactic supernovae. Two other potential contributors are ^{56}Ni and ^{22}Na. However, positrons emitted during ^{56}Co-decay may not escape from the supernova shell, and it appears that far too little ^{22}Na is produced.

SUMMARY

The results from the HEAO 3 high-resolution gamma-ray spectrometer have made a number of important contributions to the understanding of galactic nucleosynthesis and to the understanding of the formation of the solar system. The discovery and confirmation of approximately 3 $M_\odot$ of ^{26}Al in the present ISM demonstrate conclusively that intermediate-mass nuclei are presently being synthesized in the Galaxy. The implied interstellar ^{26}Al/^{27}Al isotopic ratio is 1×10^{-5}. This is comparable to the corresponding isotopic ratio measured is meteoritic material[33] implying that, contrary to previous belief, such a high ^{26}Al fraction was probably typical of the ISM at the time of formation of the solar system. While the exact distribution of the interstellar emission is presently unknown, the HEAO results clearly demonstrate the source is extended. Galactic novae presently appear to be the most likely source of the ^{26}Al.

Unlike ^{26}Al, gamma-ray emission from the decay of ^{44}Ti should appear as one or a few point sources, but at unknown locations. No evidence for point source emission from the galactic plane was found. The flux limit for a point source at the galactic center is about 3 x 10^{-4} photons/cm^2-s, corresponding to a limit of 2 x 10^{-4} M$_\odot$ of ^{44}Ti. This constrains the ^{44}Ti yield of a typical Type I supernova to approximately 5 x 10^{-4} M$_\odot$ or less.

Fitted to an extended source model, the HEAO 3 data for the 511 keV annihilation radiation yield an average flux of (1.35 ± 0.22) x 10^{-3} photons/cm^2-s-rad, consistent with the value of (1.6 ± 0.3) x 10^{-3} photons/cm^2-s-rad deduced from SMM data. Furthermore, the intrinsic line width is unresolved in both the fall of 1979 and the spring of 1980, with upper limits of 3 keV and 9 keV FWHM respectively. It appears that about 10% - 20% of the annihilation radiation line can be explained by the decay of ^{26}Al, with the remainder likely resulting from the decay of ^{44}Ti. Contributions from ^{56}Ni or ^{22}Na appear less likely.

ACKNOWLEDGEMENT

I would like to acknowledge A. S. Jacobson for carrying the HEAO 3 gamma-ray spectrometer through all phases from conception through data analysis. I also thank my colleagues A. Dunklee, J. Ling, R. Radocinski and W. Wheaton for assisting in all phases of the data reduction and analysis, and J. Higdon for stimulating discussions and comments. The research described in this paper was carried out at the Jet Propulsion Laboratory, California Institute of Technology, under contract with the National Aeronautics and Space Administration.

REFERENCES

1. Burbidge, E. M., G. R. Burbidge, W. A. Fowler, and F. Hoyle, Rev. Mod. Phys., 29, 547 (1957).
2. Cook, W. R., D. Palmer, T. Prince, S. Schindler, C. Starr, and E. Stone, IAU Circ. 4527 (1988 January 12).
3. Mahoney, W. A., L. S. Varnell, A. S.Jacobson, J. C. Ling, R. G. Radocinski, and Wm. A. Wheaton, Nature, to be submitted (1988).
4. Matz, S. M., G. H. Share, M. D. Leising, E. L. Chupp, W. T. Vestrand, W. R. Purcell, M. S. Strickman, and C. Reppin, Nature, 331, 416 (1988).
5. Sandie, W., G. Nakano, and L. Chase, IAU Circ. 4526, (1988 January 11).
6. Higdon, J. C., and W. A. Fowler, Ap. J., 317, 710 (1987).
7. Arnett, W. D., Ann. N.Y. Acad. Sci., 302, 90 (1977).
8. Ramaty, R., and R. E. Lingenfelter, Ap. J. (Letters), 213, L5 (1977).

9. Wallace, R. K., and S. E. Woosley, Ap. J. (Suppl.), 45, 389 (1981).
10. Hillebrandt, W., and F. K. Thielemann, Ap. J., 255, 617 (1982).
11. Clayton, D. D., Ap. J. 280, 144 (1984).
12. Dearborn, D. S. P., and J. B. Blake, Ap. J. (Letters), 288, L21 (1985).
13. Prantzos, N. and M. Cassé, Ap. J., 307, 324 (1986).
14. Norgaard, H., Ap. J., 236, 895 (1980).
15. Mahoney, W. A., J. C. Ling, W. A. Wheaton, and A. S. Jacobson, Ap. J., 286, 578 (1984).
16. Share, G. H., R. L. Kinzer, J. D. Kurfess, D. J. Forrest, E. L. Chupp, and E. Rieger, Ap. J. (Letters), 292, L61 (1985).
17. Woosley, S. E., T. S. Axelrod, and T. A. Weaver, Comments Nucl. Part. Phys., 9, 185 (1981).
18. Woosley, S., private communication.
19. Ramaty, R., and R. E. Lingenfelter, Nature, 278, 127 (1979).
20. Bussard, R. W., R. Ramaty, and R. J. Drachman, Ap. J., 228, 928 (1979).
21. Leventhal, M., C. J. MacCallum, and P. D. Stang, Ap. J. (Letters), 225, L11 (1978).
22. Riegler, G. R., J. C. Ling, W. A. Mahoney, W. A. Wheaton, J. B. Willett, A. S. Jacobson, and T. A. Prince, Ap. J. (Letters), 248, L13 (1981).
23. Share, G. H., R. L. Kinzer, J. D. Kurfess, D. C. Messina, W. R. Purcell, E. L. Chupp, D. J. Forrest, and C. Reppin, Ap. J., 326, 717 (1988).
24. Wheaton, Wm. A., et al., Ap. J., to be submitted (1988).
25. Mahoney, W. A. , J. C. Higdon, J. C. Ling, W. A. Wheaton, and A. S. Jacobson, Proceedings, XIX ICRC, 1, 357 (1985).
26. Bahcall, J. N., and R. M. Soneira, Ap. J. (Suppl.), 44, 73 (1980).
27. Burton, W. B., and M. A. Gordon, Astr. Ap., 63, 7 (1978).
28. MacCallum, C. J., A. F. Huters, P. D. Stang, and M. Leventhal, Ap. J., 317, 877 (1987).
29. v. Ballmoos, P., R. Diehl, and V. Schonfelder, Ap. J., 318, 654 (1987).
30. Mahoney, W. A., J. C. Ling, A. S. Jacobson, and R. E. Lingenfelter, Ap. J., 262, 742 (1982).
31. Mahoney, W. A., J. C. Ling and A. S. Jacobson, JGR, 86, 11098 (1981).
32. Mahoney, W. A., J. C. Ling, and A. S. Jacobson, Nucl. Instr. Meth., 185, 449 (1981).
33. Lee, T., D. A. Papanastassiou, and G. J. Wasserburg, Ap. J. (Letters), 211, L107 (1977).

OBSERVATIONS OF DIFFUSE GALACTIC GAMMA RADIATION

Gerald H. Share
E.O. Hulburt Center for Space Research
Naval Research Laboratory
Washington, D.C. 20375

ABSTRACT

A new window for studying our Galaxy has emerged after over two decades of concerted effort with satellite and balloon-borne gamma-ray experiments. These experiments have discovered what appears to be a diffuse glow along the Galactic plane in lines at 0.511 and 1.809 MeV. I summarize these observations, with emphasis on measurements made with the Gamma-Ray Spectrometer (GRS) on NASA's Solar Maximum Mission satellite (SMM). The GRS is currently being used to study other diffuse Galactic emissions including the positronium continuum, radioactive lines from recent supernovae and novae, and de-excitation lines from interactions of cosmic-ray protons with interstellar carbon and oxygen.

INTRODUCTION

Lingenfelter summarizes the early theoretical expectations for gamma-ray astronomy in these Proceedings.[1] He points out that diffuse emissions from the Galactic plane were expected to be a significant part of the celestial gamma-ray sky. This has indeed been the case (see reference 1 for citations). The 1952 predictions of the presence of $\gtrsim$20 MeV photons from π^0-decays and from bremsstrahlung were borne out in observations by the OSO-7 detector. In the same manner, the presence of a diffuse Galactic source of line emission from ^{26}Al decay was predicted in 1977 and observed by HEAO-3 in 1979[2] with an intensity about an order of magnitude higher than the estimates. In 1969 Stecker made an estimate of the Galactic flux of 0.511 MeV radiation resulting from cosmic-ray interactions on interstellar material.[3] This estimate was also significantly below the initial observations,[4] leading Clayton[5] in 1973 to suggest that a significant flux of 0.511 MeV photons is expected from positrons emitted by ^{56}Co and ^{44}Sc from decay of ^{56}Ni and ^{44}Ti produced in supernovae. In 1964 Hayakawa predicted the presence of nuclear line emission from de-excitation of interstellar nuclei after collision with cosmic-ray protons.

The presence of unexpectedly intense fluxes of diffuse Galactic gamma-ray line emission has been a pleasant surprise for astrophysicists who have spent their careers in search of these elusive high-energy photons. Over the past ten years I have been fortunate to be associated with a team of investigators from

© 1988 American Institute of Physics

the University of New Hampshire (Principal Investigator Edward
Chupp), Max Planck Institute for Extraterrestrial Physics
(Garching, FRG), and the Naval Research Laboratory analyzing data
from the Gamma-Ray Spectrometer (GRS) on NASA's Solar Maximum
Mission satellite (SMM). This venerable instrument was designed
for solar studies, but its performance has been so remarkably
stable that it has been exceptionally well suited for studying
diffuse Galactic emissions. In this paper I shall summarize how
these measurements have been made, what we have learned to date,
and what plans we have for future analyses.

INSTRUMENT AND TECHNIQUE

A drawing of the essential elements of the GRS[6] is shown in
Figure 1. It consists of seven 7.5 x 7.5 cm NaI detectors enclosed

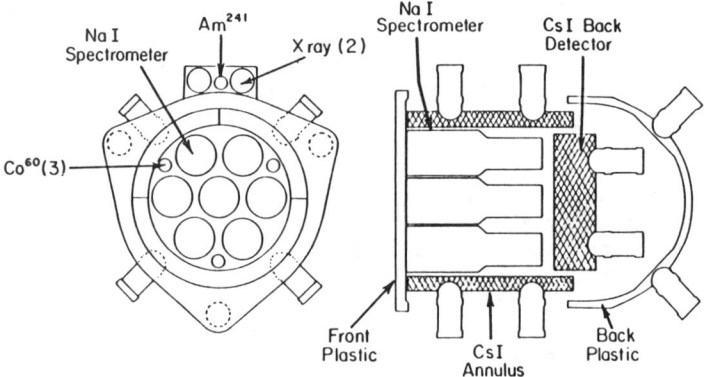

Fig. 1 Schematic drawing of the SMM Gamma-Ray Spectrometer.

in an anti-coincidence shield consisting of a CsI annulus, a CsI
backplate, and plastic scintillation counters. This shield
provides excellent rejection of charged particles and moderate
rejection of photons; it also defines a broad aperture of ~130° at
511 keV.[7] The instrument has performed flawlessly since its launch
in 1980. Its excellent gain stabilization is illustrated in Figure
2 which displays a background spectrum accumulated in 2×10^7 s
from 1980 to 1986; a detailed discussion of the features in this
spectrum can be found elsewhere[7,8]. The energy resolution of the
GRS is 7% at 662 keV.

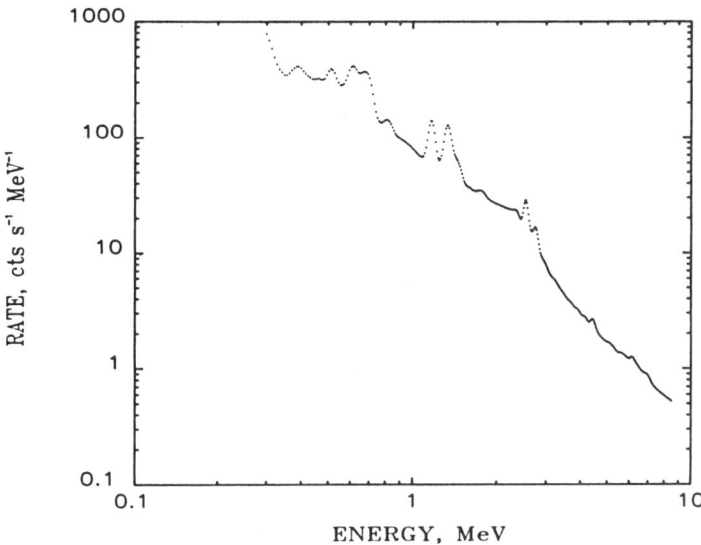

Fig. 2 Spectrum accumulated in the GRS from 1980 to 1986.

The basic technique which we have used to study Galactic gamma radiation involves fitting a given range of the GRS spectrum with a model containing a continuum and one or more Gaussian-shaped lines[7,9], on timescales as short as 3 days, and comparing the measured line intensities with what would be expected for a Galactic source transiting the aperture. Figure 3 illustrates such a measurement. SMM is viewed in its orbit around the Earth from above the North pole. Because SMM is solar oriented, a celestial source transits its aperture in a clockwise direction and would exhibit the basic time profile shown when SMM is pointed away from the Earth. When SMM is in solar night, or near dawn or dusk, the Earth blocks part of the transit, producing modified time profiles shown in the Figure. We have used such profiles to confirm the celestial origin of line radiation detected by the GRS.[7]

SUMMARY OF OBSERVATIONS

This technique for detecting Galactic emissions with the GRS was first applied to the 1.809 MeV line from ^{26}Al which was discovered[2] in data from HEAO-3. We chose this line, not only because of the early suggestion of a Galactic source[10], but because there are no strong lines from the Earth's atmosphere near that energy. This simplified the analysis of the background and enabled us to use the Earth to block the incident radiation and confirm its celestial origin.[9] We have continued our observations of the

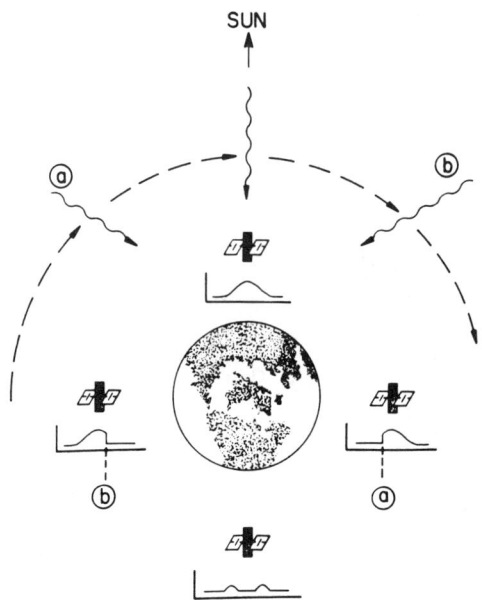

Fig. 3 Schematic representation of how a Galactic source is detected by the GRS. Reprinted courtesy of the Astrophysical Journal published by the Univ. of Chicago Press (c) 1988 (Amer. Ast. Soc.).[7]

Galactic ^{26}Al line subsequent to this earlier report. Shown in Figure 4 are the measured intensities of the line feature near 1.8 MeV observed from 1980 to 1987 in spectra accumulated when SMM was pointed away from the Earth. The peaks observed each December coincide with the transit of the Galactic center through the GRS' aperture. The gradual increase in rate is due to the increase in a nearby line (1.786 MeV) attributed to the sum of the 1.275 MeV and positron annihilation lines from the buildup of ^{22}Na, produced by spallation in the detector's aluminum housing[9]. The curve fit to the data points is a model which is comprised of a constant, a function representing ^{22}Na production, and the calculated response of the GRS to a constant Galactic source of ^{26}Al. The last three transits have peak intensities consistent with the earlier observations.[9] Higdon provides an excellent review of likely sources of diffuse Galactic ^{26}Al in another paper in these Proceedings.[11]

The high statistical significance (>16σ) of the yearly increase in the 1.81 MeV line intensity exhibited in Fig. 4 has led us to initiate a study of the Galactic longitude distribution of ^{26}Al. This analysis utilizes the Earth as an occulting disk to define a narrower effective field of view. A detailed description of this technique and some preliminary results are also discussed in these Proceedings.[12]

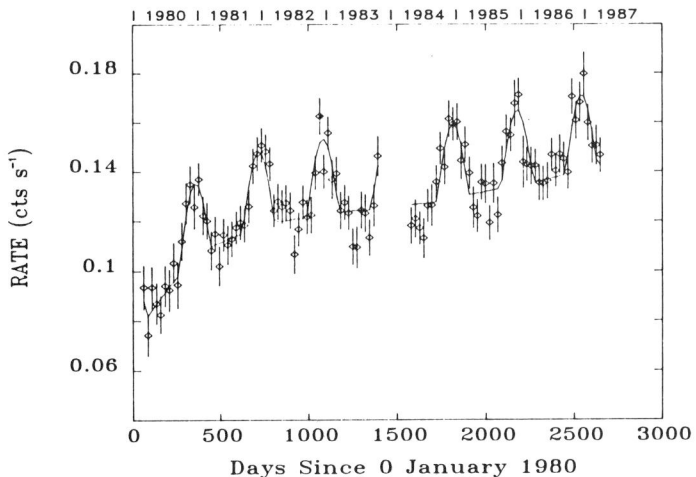

Fig. 4 Intensities of 1.81 MeV line radiation observed with the GRS pointed away from the Earth.

Following the successful utilization of the GRS to observe the Galactic ^{26}Al line, we began work on the Galactic positron annihilation line. This analysis was complicated by the intense background at this energy both from radioactive materials produced within the instrument and spacecraft and from the Earth's atmosphere. After correcting for these background contributions we were able to demonstrate that the GRS had detected a relatively intense emission of 511 keV line radiation from the general vicinity of the Galactic center.[7] This was done by demonstrating that the line's intensity increased coincidently with the annual transit of the Galactic center and was modified as expected for a Galactic source when occulted by the Earth. Since that time we have accumulated data from the 1986/87 transit of the Galactic center. Figure 5 shows this expanded time profile of the 511 keV line intensity measured by the GRS when SMM was pointed away from the Earth. The curve drawn through the data points is a model which consists of a constant, a function representing ^{22}Na (decays via emission of a positron) production, and the calculated response of the GRS to a constant Galactic source of annihilation radiation. There is no evidence for significant year-to-year variability in the peak intensity observed each December.

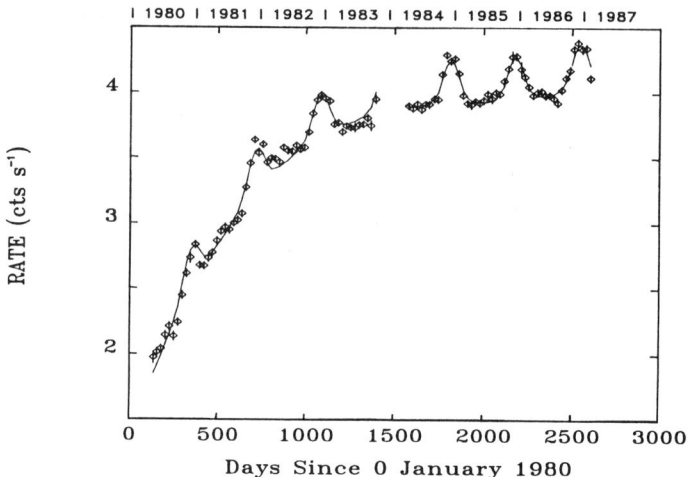

Fig. 5 Time history of corrected 511 keV line measurements with the GRS viewing the sky.

The time-averaged flux, if attributed to a point source at the Galactic center, is $(2.1 \pm 0.4) \times 10^{-3}$ photon cm^{-2} s^{-1}, where the error is dominated by systematic uncertainties. The SMM observations are compared with previous measurements of the 511 keV line in Figure 6 taken from reference 7. Three balloon-borne observations were performed using germanium spectrometers with 15° apertures while the GRS was viewing the Galactic center region. No evidence was found in any of these experiments for a strong Galactic center source. We explain the disagreement between the upper limits from the balloon experiments and the SMM observations as being due to a distributed Galactic source of positron annihilation radiation. For a distribution proportional to that of Galactic CO (see for example ref. 13), the GRS measurements require a flux of $(1.6 \pm 0.3) \times 10^{-3}$ photon cm^{-2} s^{-1} rad^{-1} at the Galactic center.

The dashed horizontal lines in Figure 6 give our estimates of the intensities that the different experiments would have reported for such an extended Galactic source. These estimates are consistent with most of the observations, with the exception of the flux originially reported for the HEAO-3 measurement in the Spring of 1980.[14] Results from a recent reanalysis of the HEAO-3 data, discussed by Mahoney in these Proceedings,[15] are consistent with the presence of an extended Galactic source of 511 keV line radiation near the intensity observed by SMM. Within the uncertainties, this diffuse source appears to account for fluxes reported from most of the experiments, with the possible exception of the Bell/Sandia observations in 1977 and 1979 which suggest the presence of an additional variable source.[16]

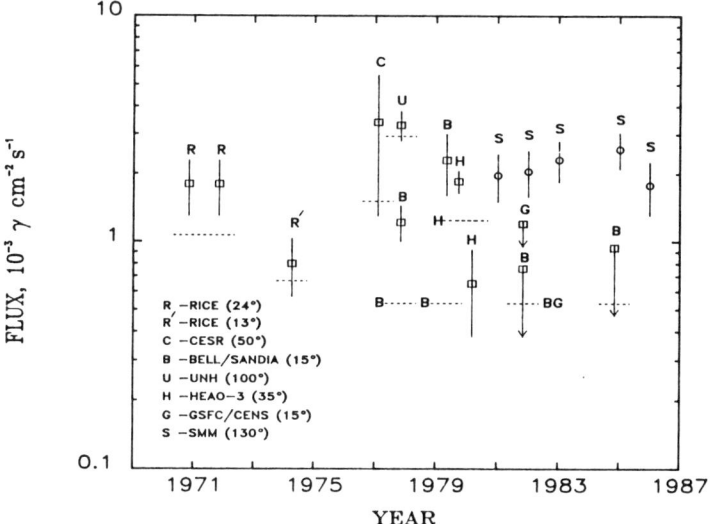

Fig. 6 History of Galactic center 511 keV line. Reprinted courtesy of the Astrophysical Journal published by the Univ. of Chicago Press (c) 1988 (Amer. Ast. Soc.).[7]

A variety of sources can account for the diffuse Galactic annihilation line.[7] The most likely are the positrons emitted in the decays of ^{56}Co and ^{44}Sc, produced in supernovae, and ^{26}Al. Detailed discussions of these sources can be found in these Proceedings.[1,11,17,18]

CURRENT AND PLANNED STUDIES

Searches for diffuse Galactic line emission from other radioactive species have already begun using the GRS. Leising discusses limits set on the production of ^{22}Na produced in novae.[19] A similar search for line emission from ^{60}Co (1.17 and 1.33 MeV) and ^{44}Sc (1.157 MeV) is now in progress. Work on the Galactic 511 keV source has been extended to a search for the positronium continuum expected when annihilation takes place in the interstellar medium.[20] Preliminary results from this analysis indicate that such a continuum has been detected by the GRS at a level consistent with such a low temperature and low density medium.[21] Work is also in progress to set limits on emission of 4.43 MeV and 6.13 MeV lines from direct excitation of interstellar carbon and oxygen by low energy cosmic radiation. Higdon is not optimistic that such emissions will be detectable soon;[22] he estimates that the intensities are not likely to exceed about 5 x 10^{-6} photon cm^{-2} s^{-1} rad^{-1}. Preliminary limits from the GRS are over a factor of 3 better than any other previous observation, but fall short of testing these estimates by over an order of magnitude.[23]

From our studies of the annihilation line and the Galactic carbon and oxygen de-excitation lines, we have gained an understanding of how the GRS data may be utilized to study diffuse Galactic continuum radiation from 0.3 to about 8.5 MeV. Preliminary results from these analyses indicate that the Galactic continuum from 4 to 8.5 MeV has been detected by the GRS at a level consistent with other observations within and near this energy range.

ACKNOWLEDGMENTS

I am indebted to my colleagues at NRL: Neil Johnson, Bob Kinzer, Jim Kurfess, Mark Leising, Steve Matz, Danny Messina, Bill Purcell, and Mark Strickman, without whom the full capabilities for utilizing the GRS for Galactic studies would not have been realized. Of course none of this work would have been possible without such an excellent gamma ray spectrometer; our colleagues at the University of New Hampshire and the Max Planck Institute for Extraterrestrial Physics in Garching, are to be congratulated for designing and developing this instrument. This work has been supported at NRL by NASA contract S-14513-D.

REFERENCES

1. R.E. Lingenfelter, this volume (1988).
2. W.A. Mahoney, et al., Ap. J. 286, 578 (1984).
3. F.W. Stecker, Ap. Sp. Sci. 3, 579 (1969).
4. W.N. Johnson, III, and R.C. Haymes, Ap. J. 184, 103 (1973).
5. D.C. Clayton, Nature Phy. Sci. 244, 137 (1973).
6. D.J. Forrest, et al., Solar Phys. 65, 15 (1980).
7. G.H. Share, et al., Ap. J. 326, 717 (1988).
8. G.H. Share, et al., in High Energy Radiation Background in Space (AIP Proceedings), Am. Inst. Phys., New York (1988).
9. G.H. Share, et al., Ap. J. Lett. 292, L61 (1985).
10. W.A. Mahoney, et al., Ap. J. 262, 742 (1982).
11. J.C. Higdon, this volume (1988).
12. W.R. Purcell, et al., this volume (1988).
13. M.D. Leising and D.D. Clayton, Ap. J. 294, 591 (1985).
14. G.R. Riegler, et al., Ap. J. Lett. 248, L13 (1981).
15. W.A. Mahoney, this volume (1988).
16. M. Leventhal, et al., Ap. J. 302, 459 (1986).
17. M. Signore and G. Vedrenne, this volume (1988).
18. S.E. Woosley and P.A. Pinto, this volume (1988).
19. M.D. Leising, et al., Ap. J. 328, in press (1988).
20. B.L. Brown and M. Leventhal, this volume (1988).
21. G.H. Share, et al., Bull. APS 32, 1112 (1987).
22. J.C. Higdon, 20th Int. Cos. Ray Conf. 1, 160 (1987).
23. G.H. Share, et al., BAAS 19, 694 (1987).

THEORETICAL ESTIMATES OF DIFFUSE GALACTIC γ-RAY EMISSIONS

J. C. Higdon
Joint Science Department
The Claremont Colleges
11th Street and Dartmouth Avenue
Claremont, CA 91711

ABSTRACT

Theoretical investigations of line emissions at 1.809 and 0.511 MeV, generated respectively by the decay of unstable ^{26}Al and e^+ annihilation, are discussed in detail.

1.809 MEV γ-RAY LINE EMISSION

The decay of ^{26}Al with a mean life of 1.04×10^6 yr generates a γ-ray line at 1.809 MeV. The search for this nucleosynthetic line from type II supernovae was suggested independently by Ramaty and Lingenfelter[1] and by Arnett[2]. But the discovery of diffuse 1.809 MeV line emission by HEAO 3[3] at ten times the expected intensity[1,2], confirmed by SMM measurements[4], has stimulated new theoretical work to try to understand its origin.

At the present time a controversy exists in the interpretation of the 1.809 MeV γ-ray data. The pioneering HEAO 3 and SMM detections[3,4] were performed with low angular resolutions of 42° and 120° FWHM respectively. Subsequent analyses[6] of the HEAO 3 data have demonstrated that a wide range of spatial distributions require a ^{26}Al mass of 2.5 $M_\odot$ to reproduce the observed 1.809 MeV line emissions. Recently von Ballmoos, Diehl, and Schonfelder[7] have performed a balloon-flight measurement of the galactic-center region with a Compton telescope with a resolution of ~ 10° FWHM at 1.809 MeV. They reported that their 1.809 MeV emission was consistent with a single point source with a flux of $(6.7 \pm 2.7) \times 10^{-4}$ cm^{-2}s^{-1}, located at a longitude $l \cong 0°$, but their accuracy was insufficient to exclude models of diffuse emission sharply peaked at $l \cong 0°$.

Since the report of von Ballmoos et al.[7], Mahoney et al.[8] analyzed their HEAO 3 data assuming a point source at the galactic center; their reported point source flux was $(1.4 \pm 0.9) \times 10^{-4}$ cm^{-2}s^{-1} in disagreement with von Ballmoos et al.[7] at a 2 σ level. Further, MacCallum et al.[9] have analyzed data from four balloon flights for evidence of galactic-center 1.809 MeV line emission. They found for $|l| \leq 7.5°$, an equivalent source flux of $(1.3 \pm 0.9) \times 10^{-4}$ cm^{-2}s^{-1}, again inconsistent at a 2 σ level with von Ballmoos et al.[7], but consistent with analyses of Mahoney et al.[3] and Share et al.[4] Unlike the HEAO analysis[8], the result of MacCallum et al.[9] is independent of the modeling of the position of any hypothetical point source.

In view of the consistent analyses of Mahoney et al.[8] and MacCallum et al.[9] I have assumed that the bulk of the galactic 1.809 MeV line emission is not produced by a single source located in the galactic nucleus, but is generated by sources distributed throughout the Galaxy. Based on ^{26}Al production models a variety of sources have been suggested for such diffuse emission, in addition to type II supernovae: asymptotic giant branch stars[10,11,12], Wolf-Rayet stars[13,14,15], and novae[16,17,18,19,20].

1. ASYMPTOTIC GIANT BRANCH STARS

Main-sequence stars in the range 1 to ~ 9 M_o evolve onto the asymptotic giant branch (AGB) of the Hertzsprung Russell diagram[21]. Such AGB stars consist of dense carbon oxygen (CO) cores surrounded by very extended hydrogen envelopes[21]. In an AGB star convective dredging brings to the surface some of the isotopes synthesized by helium burning at the base of the stellar envelope[21]. The efficiency of this dredging process is not well determined[23]. These AGB stars lose their stellar envelopes both by stellar winds and by the ejection of planetary nebular shells (mass ~ 0.4 M_o[22]) on times ranging from 2 x 10^6 yr for a 1 M_o star to 10^5 yr for a star of 8 M_o[21,12].

Norgaard[10] has suggested that ^{26}Al is synthesized at the base of the convective envelope of an AGB star, and that subsequently such material is brought to the surface on times comparable to the ^{26}Al mean life time, τ_{26}. Such ^{26}Al is produced by ^{25}Mg(p,γ)^{26}Al. He found that the production ratio of ^{26}Al/^{27}Al approaches values of ~ 1 in the convective helium burning regions. Cameron investigated further ^{26}Al synthesis in AGB stars[12]. At ~ 5 x 10^7 K, typical of the hydrogen-burning shells, he found that ^{26}Al is produced by ^{24}Mg(p,γ)^{25}Al(β^+)^{25}Mg(p,γ)^{26}Al. He determined that, first, the ^{26}Al/^{27}Al production ratio is between 3 to 33, and, second, ^{27}Al is increased by 0.3 over its initial value in the stellar envelope.

The maximum possible contribution of AGB stars to the interstellar ^{26}Al occurs when, first, all the envelope material undergoes nuclear processing, and, second, the time between the last significant nucleosynthetic episode and the loss of nebula is short compared to τ_{26}. In this extreme case a nebular mass of ~ 0.4 M_o[22] could be expected to contain a maximum of ~ 2.4 x 10^{-5} M_o of freshly synthesized ^{26}Al, M_s(^{26}Al), if the ^{26}Al/^{27}Al production ratio $\approx$ 3 and an enhancement of stable ^{27}Al $\approx$ 0.3 for a nominal mass fraction of unprocessed ^{27}Al of 6.6 x 10^{-5}, representative of that in the Sun[24]. With a galactic frequency for the planetary nebula formation, ν_{pn}, of ~ 1 yr^{-1} [22], the interstellar mass of ejected ^{26}Al, M_s(^{26}Al)$\tau_{26}\nu_{pn}$, potentially can be as large as ~ 25 M_o. This model yield overestimates the AGB star contribution because nebula formation is dominated[23] by the contribution of low-mass stars, ≤ 2 M_o, where the time scale[12] for stellar envelope loss, τ_L, is ~ 2 x 10^6 yr, somewhat greater than τ_{26}. The effect of this time lag between ^{26}Al production and the loss of a stellar envelope is approximated crudely here by the factor $\exp(-\tau_L/\tau_{26})$. Thus the interstellar mass of ^{26}Al, which can be generated by AGB stars, based on Cameron's[12] yields, is estimated to be $\exp(-\tau_L/\tau_{26})M_s(^{26}Al)\tau_{26}\nu_{pn}$ ~ 3 M_o. Such a result suggests that the ^{26}Al mass derived by Mahoney et al.[3] and Share et al.[4] can be generated solely by AGB stars. However, model ^{26}Al yields are plagued by large uncertainties in nuclear reaction rates. Clearly a more accurate estimate of the contribution of AGB stars to interstellar ^{26}Al awaits detailed nucleosynthetic models applicable to low-mass AGB stars.

Planetary nebulae in the solar vicinity belong[22] to the disk population with ages ~ (3 to 5) x 10^9 yr, which correspond to main-sequence masses ~ 1.4 to 2 M_o. The galactic distribution of such a population is well known[19,22,25]. I have calculated the 1.809 MeV intensity, produced solely by such a disk population[19], normalized to a interstellar ^{26}Al mass of 2.5 M_o. In Figure 1 I have plotted the resultant 1.809 MeV γ-ray intensity (dashes) as a function of galactic longitude for 0° ≤ l ≤ 180°.

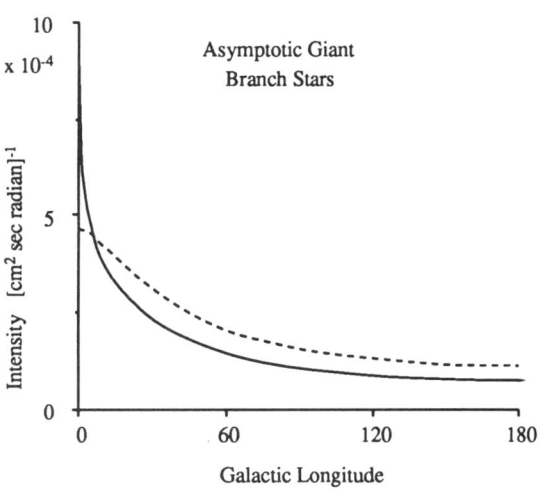

Fig. 1. 1.809 MeV intensities from the decay of 2.5 M_o of ^{26}Al for a model galactic disk (dashes) and for a model composite disk and spheroid (solid).

Planetary nebulae, which are not members of the disk population, have been detected[22] in the vicinity of the galactic center. These stars are members[22] of the much older (~ 1.5 x 10^{10} yr) galactic spheroid population. This older population contributes[25] as much as 0.34 to the visual luminosity of the Galaxy. I have assumed that the density of spheroid planetary nebulae scales with the spheroid visual luminosity. Employing the density[26] of the spheroid population, as a function of galactocentric distance (normalized to ~ 0.8 M_o of ^{26}Al), I have derived the 1.809 MeV γ-ray intensity, produced by such spheroid planetary nebulae, and by disk planetary nebulae (normalized to ~ 1.7 M_o of ^{26}Al) for the decay of a total ^{26}Al mass of 2.5 M_o. In Figure 1 I have plotted the total γ-ray intensity (shown as a solid line) for such a spheroid-disk distribution as a function of longitude. However, I suggest that the contribution of the spheroid planetary nebulae is less than that indicated in Figure 1, because the abundances of the Mg seed isotopes seem to be significantly lower[27] in the spheroid than in the disk.

2. NOVAE

A variety of investigators[16,17,18,19,20] have suggested that novae are a major source of ^{26}Al. The recent identification[29] of a new class of nova outbursts, accreting oxygen-neon-magnesium (ONeMg) white dwarfs, has increased[20,29] the possibility of nova production of ^{26}Al.

Most nova outbursts[30] are thermonuclear runaways in gas envelopes of CO white dwarfs; these envelopes are accreted from stellar companions. Such outbursts[30] are powered by carbon-nitrogen-oxygen (CNO) cycle hydrogen burning with peak temperatures ~ (1.5 to 3) x 10^9 K. Nucleosynthesis in typical nova outbursts is limited[31] primarily to proton-induced reactions. However, novae have been detected[31] with extreme overabundances (~ 10 to 100) relative to solar in CNO; further, several novae have been observed with large overabundances relative to solar of neon, sodium, magnesium, and aluminum. Significant production of such heavy nuclei seems to be unlikely[31] at the temperatures typical of nova shells. The source[31] of these overabundances seems to be the mixing of underlying white-dwarf core material with the H-rich accreted envelope. Starrfield et al[29] have suggested that Ne-rich ejecta is produced by the mixing of accreted gas with dredged-up core material from a ONeMg white dwarf.

Nova ^{26}Al production is difficult to model in part due to large uncertainties[31] in the temperature history of nova ejecta. However, Hoffman and Woosley[20] have calculated ^{26}Al production in accreting ONeMg white dwarfs via $^{24}Mg(p,\gamma)^{25}Al(\beta^+)^{25}Mg(p,\gamma)^{26}Al$. They determined that the mass fraction of ^{26}Al in ejecta could be as large as 3.6 x 10^{-3}; for accreting CO white dwarfs the ^{26}Al mass fraction was ~ 30 to 100 less. Further, they found that convection increases ^{26}Al production by ~ 3 in accreting ONeMg white dwarfs because ^{26}Al can be convected away from the nuclear burning region before it is destroyed. Clearly, model ^{26}Al yields are sensitive to the details of the poorly-understood convective process.

^{26}Al produced by accreting ONeMg white dwarfs can be a major source of ^{26}Al, when some model[20] yields are used. Employing a maximum[20] ^{26}Al mass fraction of 3.6 x 10^{-3} and a total[29] ejecta mass of ~ 2 x 10^{-5} M_o, a typical accreting

ONeMg white dwarf ejects a ^{26}Al mass, $M_e(^{26}Al)$, $\approx 7 \times 10^{-8}$ $M_\odot$. Such accreting ONeMg white dwarfs are expected to produce[29] $f_{ne} \sim 0.25$ of all nova outbursts. The galactic frequency of novae of all classes[19], ν_n, is 45 yr^{-1}. Thus the nova contribution to interstellar ^{26}Al is estimated here to be $M_e(^{26}Al)f_{ne}\nu_n\tau_{26} = 0.85$ $M_\odot$. More accurate estimates of nova ^{26}Al require better determinations of the frequency of accreting ONeMg white dwarfs and nuclear production rates as well as better models of convection.

3. WOLF-RAYET STARS

Significant amounts[13,14,15] of ^{26}Al can be synthesized in the cores of massive (≈ 50 $M_\odot$) stars and subsequently be ejected via intense stellar winds. Wolf-Rayet stars (WRS) are massive post main-sequence stars powered by helium core burning[15]. ^{26}Al is produced[14] during the short lived[14] ($\approx 2 \times 10^6$ yr) main-sequence phase by H burning via $^{25}Mg(p,\gamma)^{26}Al$. When the stellar envelope is blown off during the post main-sequence WR phase by the stellar wind (mass loss rates $\sim 3 \times 10^{-5}$ $M_\odot$ yr^{-1}), the previously synthesized ^{26}Al is ejected[14] into the interstellar medium. For initially 50 $M_\odot$ stars, in solar neighborhood, the average WRS ^{26}Al yield[15] is $\sim 3 \times 10^{-5}$ $M_\odot$.

The galactic ^{26}Al yield depends on the galactic WRS distribution which is poorly known in the inner Galaxy due to visual obscuration and uncertain metallicity effects. Pranztos, Casse, and Arnould[15] have constructed two models for the WRS spatial distribution. In their conservative case A the total number of galactic WRS is 1800, and they eject 0.12 $M_\odot$ of ^{26}Al during the last τ_{26}. In their rather extreme case B the total number of galactic WRS is 5500, and they eject 0.5 $M_\odot$ of ^{26}Al during the last τ_{26}. Employing these yields suggests that Wolf-Rayet stars are not the dominant source of galactic ^{26}Al, and at best contribute ~ 0.2 of the interstellar ^{26}Al required by the analyses[3,4] of the HEAO 3 and SMM observations.

0.511 MEV γ-RAY LINE EMISSION

The SMM γ-ray spectrometer (GRS) has detected[32] diffuse γ-ray line emission at 0.511 MeV, generated by e^+ annihilation. The SMM γ-ray spectrometer has a large 130° aperture and, consequently, the unique separation of any diffuse emission source, peaked at $l \approx 0°$, from the variable[33] emission of the galactic nucleus is difficult. However, the maximum year-to-year variation of the emission measured by the GRS is ≤ 0.3, and the comparison[32] of the SMM results with contemporaneous, narrow-aperture, balloon-flight measurements suggests that the bulk of the 0.511 MeV emission detected by the GRS is truly diffuse. If the 0.511 MeV γ-ray emissivity follows the interstellar molecular cloud distribution, Share et al. found that the intensity at $l = 0°$ is $(1.6 \pm 0.3) \times 10^{-3}$ cm^{-2}s^{-1} radian^{-1}. A galactic luminosity, Q_{511}, of 1.7×10^{43} s^{-1} is required to produce such an assumed emissivity distribution.

The relationship between Q_{511} and Q_{e^+}, the galactic e^+ production rate, depends on the nature of the ambient medium in which the e^+ annihilate. In cold ($\leq 10^2$ K) and warm ($\approx 10^4$ K) interstellar phases, once decelerated, only ~ 0.1 of e^+ annihilate directly into two 0.511 MeV γ-rays, and ~ 0.9 of e^+ form[34] positronium atoms (Ps). 0.25 of the time a Ps forms[34] in the singlet spin state (^{1}Ps), which decays emitting two 0.511 MeV γ-rays. 0.75 of the time a Ps forms[34] in the triplet spin state (^{3}Ps), which decays producing a three γ-ray continuum. Thus $γ_{511}/e^+ = (0.1 \times 2) + (0.9 \times 0.25 \times 2) = 0.65$ in the cold and warm phases. In hot ($\geq 10^6$ K) plasmas e^+ behavior is dominated[35] by Ps formation in dust grains where a ^{3}Ps is broken down to a ^{1}Ps, thus the formation of the 3 γ continuuum is suppressed, and $γ_{511}/e^+$ approaches 2. For media at intermediate temperatures ($\approx 10^4$ to 10^6 K) $γ_{511}/e^+$ is intermediate between 0.65 and 2, and depends[34] sensitively on the Ps formation rate in the gas phase by charge exchange. GRS observations have detected[32] a low-energy γ-ray continuum and a Ps fraction of ~ 0.9 is indicated, requiring that the bulk of the e^+ annihilate in either the warm or cold phases. Thus $Q_{e^+} \sim Q_{511}/0.65 \approx 2.5 \times 10^{43}$ s^{-1}.

The rates of e^+ production from possible stellar sources have been summarized by Lingenfelter[36] and are listed in Table 1. He assumed a rate of one type I supernova (SN I) per 100 yr with a ^{56}Ni yield of 0.5 M$_\odot$ and a e^+ escape fraction ε. From Table 1 the most probable source of galactic e^+ is SN I by $β^+$ decay[37,38] from either ^{56}Co or ^{44}Sc. The contribution of ^{26}Al decay is only 0.14 of that required to produce the GRS result. The relative contributions of ^{56}Co and ^{44}Sc depend on the fraction of e^+ that can escape into the interstellar medium from the dense SN I ejecta. Such escape is the dominant e^+ source if $ε \geq 0.008$. The e^+ escape fraction is difficult to calculate; it is dominated by large uncertainties in our knowledge of both the magnetic field topology and the intensity of plasma turbulence in the ejecta. Theoretical estimates for ε remain controversial and range[38,39] from ~ 10^{-5} to 0.1.

Table 1

Positron Production Rates[36]

Source	Process	Rate (10^{43} e$^+$/s)
SN I	^{56}Co($β^+$)^{56}Fe	$70ε/τ_{100}$
SN I	^{44}Sc($β^+$)^{44}Ca	$0.6/τ_{100}$
AGBS & Novae	^{26}Al($β^+$)^{26}Mg	0.35
Novae & WR Stars	^{22}Na($β^+$)^{22}Ne	<0.7
Cosmic Rays	pp ⇒ $π^+$	≤0.1
γ-Ray Bursters	γγ ⇒ $e^+ e^-$	<0.3
Pulsars	e^+ Cascade	<0.1

However, e^+ escape will produce the intensity of diffuse annihilation radiation detected by the GRS if ε is only $\approx$ 0.04, for a conservative SN I birthrate of 1 every 100 yr. If e^+ escape proves untenable, the decay of ^{44}Ti, entrained in the expanding SN I ejecta, to ^{44}Ca via ^{44}Sc, could be the primary source[38] of diffuse galactic e^+ if the SN I birthrate is closer to 1 every 25 yr. The galactic SN I birthrate has been estimated[40] to be 1 every 36 yr with large uncertainties.

SUMMARY

Galactic sites of ^{26}Al and e^+ have been examined. The major source of ^{26}Al by default seems to be asymptotic giant branch stars[10,11,12]. Secondary ^{26}Al sources are novae[16,17,18,19,20] and Wolf-Rayet stars[13,14,15]. The relative contributions of these sources depend on a variety of poorly-understood phenomena: the relevant nuclear reactions, convective dredging, and post-main-sequence stellar mass loss. More research is needed to identify irrefutably all major sources of galactic ^{26}Al. The identity of the source of diffuse galactic e^+, whose annihilation radiation was detected by SMM[32], is more hopeful. It is most likely type I supernovae from the decay of either ^{56}Co [37] or ^{44}Sc [38]. The relative contributions of ^{56}Co and ^{44}Sc are uncertain and depend on the efficiency of e^+ escape and the frequency of galactic type I supernovae.

This work has been supported in part by the National Aeronautics and Space Administration (NAG 5-1011).

REFERENCES

1. R. Ramaty and R. E.Lingenfelter, Ap. J.(Letters), 213, L5 (1977).
2. W. D. Arnett, Ann. N. Y. Acad. Sci., 302, 90 (1977).
3. W. A. Mahoney, J. C. Ling, Wm. A. Wheaton, and A. S. Jacobson, Ap. J., 286, 578 (1984).
4. G. H. Share, R. L. Kinzer, J. D. Kurfess, D. J. Forrest, E. L. Chupp, and E. Rieger, Ap. J. (Letters), 292, L61 (1985).
5. T. Lee, D. A. Papanastassiou, and G. E.Wasserburg, Ap. J. (Letters), 211, L107 (1977).
6. W. A. Mahoney, J. C. Higdon, J. C. Ling, W. A. Wheaton, and A. S. Jacobson, 19th Int. Cosmic Ray Conf., 1, 357 (1985).
7. P. von Ballmoos, R. Diehl, and V. Schonfelder, Ap. J., 318, 654 (1987).
8. W. A. Mahoney, J. C. Higdon, J. C. Ling, W. A. Wheaton, and A. S. Jacobson, BAAS, 18, 979 (1986).
9. C. J. MacCallum, A. F. Huters, P. D. Stang, and M. Leventhal, Ap. J., 317, 877 (1987).
10. H. Norgaard, Ap.J., 236, 895 (1980).
11. J. W. Truran, in "Nucleosynthesis: Challenges and New Develop-

ments", eds. W. D. Arnett and J. W. Truran (Univ. of Chicago Press, Chicago, 1985), p. 292.
12. A. G. W. Cameron, Icarus, 60, 416 (1984).
13. D. S. P. Dearborn, and J. B. Blake, Ap. J.(Letters), 288, L21 (1985).
14. N. Prantzos, and M. Casse, Ap. J., 307, 324 (1986).
15. N. Prantzos, M. Casse, and M. Arnould, 20th Int.Cosmic Ray Conf. Paper,(NAUKA, Moscow 1987), OG 2.3-4.
16. R. K. Wallace, and S. E.Woosley, Ap. J. Suppl., 45, 389 (1981).
17. W. Hillebrandt, and F. K. Thielemann, Ap. J., 255, 617 (1982).
18. D. D. Clayton, Ap. J., 280, 144 (1984).
19. J. C. Higdon, and William A. Fowler, Ap. J., 317, 710 (1987).
20. R. D. Hoffman, and S. E. Woosley, BAAS, 18, 948 (1987).
21. Icko Iben, in "Nucleosynthesis: Challenges and New Developments", eds. W. D. Arnett and J. W. Truran (Univ. of Chicago Press, Chicago, 1985), p. 272.
22. S. R. Pottasch, "Planetary Nebulae, A Study of Late Stages of Stellar Evolution", (Reidel, Dordrecht 1984), p. 22, 31, 121, 275.
23. Icko Iben, in "Advanced Stages in Stellar Evolution", eds. I. Iben, A. Renzini, and D. N. Schramm, (Geneva Observatory, Sauverney Switzerland, 1977), p. 49.
24. A. G. W. Cameron, in "Essays in Nuclear Astrophysics", eds. C A. Barnes, D. D. Clayton, and D. N. Schramm, (Cambridge Univ. Press, Cambridge 1982), p. 23.
25. G. de Vaucouleurs, and W. D. Pence, A. J., 83, 1163 (1978).
26. J. N. Bahcall, M. Schmidt, and R. M. Soneira, Ap. J. (Letters), 258, L27 (1982).
27. J. C. Higdon, and William A. Fowler, in preparation.
28. W. E. Harris, and R. Canterra, Ap. J. (Letters), 231, L19, (1979).
29. S. Starrfield, W. M. Sparks, and J. W. Truran, Ap. J. (Letters), 303, L5 (1986).
30. J. W. Truran, in "Nucleosynthesis and its Implications on Nuclear Physics and Particle Physics", eds. J. Audouze, and N. Mathieu (Reidel, Dordrecht 1986), p. 97.
31. J. W. Truran, in "Nucleosynthesis: Challenges and New Developments", eds. W. D. Arnett and J. W. Truran (Univ. of Chicago Press, Chicago, 1985), p.292.
32. G. H. Share, this proceedings.
33. M. Leventhal, this proceedings.
34. R. W. Bussard, R. Ramaty, and R. J. Drachman, Ap. J., 228, 928 (1979).
35. W. H. Zurek, Ap. J., 289, 603 (1985).
36. R. E. Lingenfelter, Invited Rapporteur Talk at 20th Int. Cosmic Ray Conf. Paper,(NAUKA, Moscow 1987).
37. R. Ramaty and R. E. Lingenfelter, Nature 278, 127, (1979).
38. S. E. Woosley, submitted to Ap. J. (Letters), (1987).
39. S. A. Colgate, Astrophys. Space Sci., 8, 457 (1970).
40 G. A. Tammann, in "Supernovae: A Survey of Current Research", eds. M. J. Rees and R. J. Stoneham (Cambridge Univ. Press, Cambridge 1982), p. 371.

Constraints on Gamma-Ray Line and Continuum Emission from the Galactic Center Region at MeV Energies

R. Diehl, P. v.Ballmoos *, V. Schönfelder

Max Planck Institut für Extraterrestrische Physik, Garching, FRG

ABSTRACT

MPE Compton Telescope observations of MeV radiation from the direction of the Galactic Center lead to constraints on the central source and on diffuse processes in the Galaxy:
The extent of 1.8 MeV line emission from ^{26}Al suggests an ^{26}Al production process with pronounced concentration towards the Galactic Center. The absence of other γ-ray lines constrains nucleosynthesis and cosmic ray excitation parameters in the Galaxy.
Upper limits on continuum γ-radiation can be interpreted as a central compact source in a low emission state in contrast to the 1979 observation with the HEAO-3 instrument. In addition, these upper limits constrain continuum γ-ray emission due to interactions of cosmic ray electrons below 100 MeV with Galactic interstellar matter.

INTRODUCTION

The central region of the Galaxy is expected to be a domain of enhanced gamma ray line luminosity from mainly 2 processes: (1) thermonuclear production of radioactive matter, and (2) energetic particle interactions with interstellar matter.
Towards the Galactic Center the line of sight traverses regions of enhanced stellar density and hence high nova rates, as well as the 5 kpc ring of enhanced supernova activity [15] ; the thermonuclear processes in those events generate radioactive material, whose decay leads to emission of gamma ray lines.
The shortlived radionuclides produced during such a nucleosynthetic event are ejected at high velocities (novae: 100 – 5000 km/sec; supernovae I: $\leq$ 15000 km/sec; supernovae II: $\leq$ 5000 km/sec) [28] and result in gamma ray lines with significant Doppler broadening. Candidate radioactive supernova products emitting gamma ray lines are ^{56}Co (0.847 MeV, 1.24 MeV, and 2.60 MeV, τ = 111 days), ^{57}Co (0.136 MeV and 0.122 MeV, τ = 389 days), and ^{22}Na (1.275 MeV, τ = 3.72 years). Time history of the line fluxes could allow a study of the expansion of the supernova shell after the explosion. The expected fluxes in these lines for a 'standard supernova' at 10 kpc distance are in the range $10 - 10^{-2}$ photons cm^{-2} sec^{-1}, several months after the supernova explosion [5]. Typical Doppler broadening predicted for SN 1987A is ~5 keV [3].

Another important gamma ray line is expected from radioactive ^{44}Ti (1.157 MeV, and 0.068 and 0.078 MeV, τ = 67.8 years), generated in explosive nucleosynthesis. Supernovae of type I are believed to be the dominant ^{44}Ti sources with typical yields of 10^{-4} M$_\odot$ [28]. At any time one or a few supernova remnants should be present in the Galaxy in a state not yet generating significant radio emission, but still not depleted in ^{44}Ti. Therefore clumpy diffuse emission at 1.157 MeV is expected with flux values up to $1.5 \cdot 10^{-4}$ photons cm^{-2} sec^{-1} for a young remnant at 10 kpc distance [6]. Doppler broadening is significant for high ejection velocities of supernovae I, resulting in a ~160 keV wide γ-ray line slighly redshifted with supernova remnant age [4].
Other lines from the decay of longlived radionuclides generated in nucleosynthetic events are narrow in general, as in this case the radioactive matter is dispersed in the interstellar medium before decay and thus is not significantly Doppler broadened. The superposition of many nucleosynthetic events builds up gamma ray emissivity distributed as the supernova or nova distribution in the Galaxy [15] , and the flux is determined from the balance between (irregular)

---

* presently at CESR Toulouse, France

© 1988 American Institute of Physics

production and radioactive decay. Candidate species for narrow lines are ^{60}Fe (1.173 MeV and 1.332 MeV, $\tau = 2.2 \cdot 10^6$ years), and ^{26}Al (1.809 MeV, $\tau = 1.04 \cdot 10^6$ years). The fluxes expected from 'steady' Galactic nucleosynthesis are quite uncertain, but generally in the range 10^{-4} to 10^{-5} photons cm^{-2} sec^{-1} rad^{-1} [6]. The expected line width is in the keV range.
The interaction of low energy cosmic rays with interstellar matter (1 - 100 MeV/nucleon) leads to excitation of the interstellar matter nuclei. Among those, the deexcitation of the relative abundant species ^{12}C and ^{16}O results in broad (~110 keV) gamma ray line emission at 4.4 and 6.1 MeV, respectively. A very narrow component of the 6.1 MeV line is expected from cosmic ray excitation of Oxygen on interstellar grains [21]. The flux values estimated recently from a model of the interstellar medium based on HI and molecular cloud measurements and on recent cosmic ray ionization measurements are less than $\sim 2 \cdot 10^{-6}$ photons cm^{-2} sec^{-1} rad^{-1} in the 4.4 MeV and 6.1 MeV line, from these narrow lines [11]. When instrumental resolution as typical for our instrument is taken into account, the superposition of these contributions to the gamma ray spectrum results in the spectra indicated in figure 4 (from [11]).
Observations have demonstrated the Galactic Center region to be a source of gamma ray line radiation. The 0.5 MeV annihilation line has been found to contain a variable point source component [23], which is interpreted as either due to variable annihilation conditions for positrons generated in beta decays [27], or due to variable positron production mechanisms in the vicinity of a black hole [20]. The 1.8 MeV line radiation detected from the Galactic Center region ([17,26]) is identified with the decay of radioactive ^{26}Al, a species which is believed to be produced during thermonuclear processing in supernovae [28], novae [28], or massive stars [18]. Also, one supermassive supernova event of $5 \cdot 10^5$ M$_\odot$ about 1-2 million years ago yields the required mass of ^{26}Al [13]. As ^{26}Al nucleosynthesis should be accompanied by other nucleosynthesis products, steady ^{26}Al production leads to expectation of other γ-ray lines, while in the case of one supernova event a million years ago all radionuclides with lifetimes shorter than this time should have decayed by now. Therefore constraints on other γ-ray lines from pointlike or extended sources were investigated from our data.
The Galactic Center MeV continuum γ-ray emission was reported[22] to be variable on the timescale of 1/2 year by about a factor 7. It has been argued that this variability is correlated to the time variable 0.5 MeV point source [20].
Deexcitation gamma ray line emission was claimed to be detected from carbon. A detection of 4.6 MeV gamma radiation from the Galactic Center region was reported [10], assigned to the first excited level of ^{12}C (line energy 4.43 MeV, see above). The line flux was determined to be $9.5(\pm 2.7) \cdot 10^{-4}$ photons cm^{-2} sec^{-1}, at the 3.5 σ confidence level.
Given these prospects for gamma ray line emission and the reported measurements from the Galactic Center region, it is worthwhile to investigate the constraints imposed on gamma ray line intensity from this region by the data of the MPE Compton telescope balloon flight.
The MPE Compton telescope was described in detail by Schönfelder et al. [25]. It was launched with a stratospheric balloon from Uberaba, Brazil, on October 31, 1982, with a flight duration of 4 hours at float altitude (3.6 g/cm^2). A detailed description of the flight is given by v.Ballmoos et al. [1], the analysis of the last 2.6 hours with respect to the spatial extent of the ^{26}Al emission and emission of other candidate γ-ray lines in the energy range 0.8 - 10 MeV from the Galactic Center region is the subject of this paper.

DATA ANALYSIS

The detector events were analyzed with respect to measured total energy deposits in the Compton telescope upper and lower detectors, and with respect to directional information in the 'angular residual' parameter. (The 'angular residual' is defined as the nearest approximation of the event circle to a source direction; the event circle is the sky projection of the measured

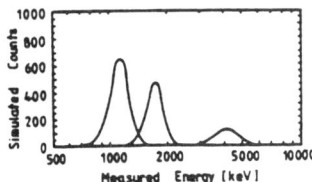

Fig. 1. Instrument response to gamma raylines at 1.16, 1.8, and 4.4 MeV (from Monte Carlo simulations)

event parameters, which indicates the photon incidence directions compatible with a particular event [2]. Energy spectral analysis was performed with event selection on angular residuals, rejecting events with parameters incompatible with an origin from the Galactic Center direction. This results in a 'software collimation' within the instrument field of view of about 1 sr. The accepted range around the Galactic Center was chosen to be 4.5°, which takes into account the instrumental resolution of about 10 degrees FWHM, the coverage of the Galactic disc during our balloon flight, and a possible extended source origin.

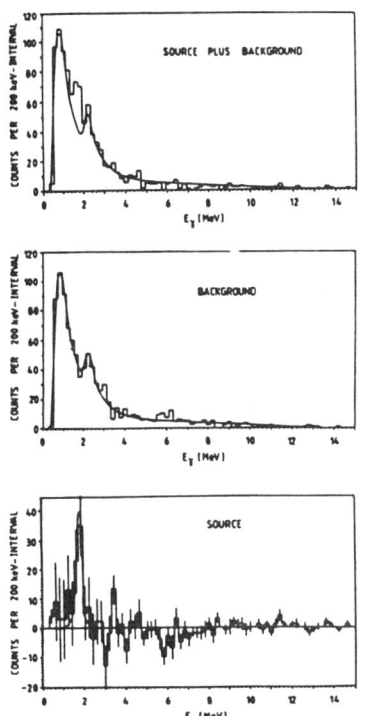

Fig.2 : Energy deposit spectra for the Galactic Center direction (source, background, and background subtracted; see also v.Ballmoos et al., 1987b)

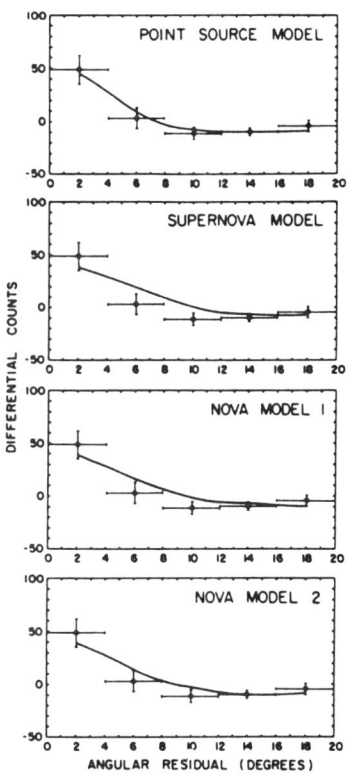

Fig.3 : Angular residual profile comparison to best fit models from point source, supernova distribution, and 2 nova models (models see Mahoney et al., 1984)

The ^{26}Al source extent was analyzed based on the angular residuals profile, which is different for concentrated and extended sources: a concentrated source yields more detector events with an event circle close to the source direction (small angular residuals), while an extended source results in a larger fraction of events with larger angular residuals.

The background was determined using the 'mirror method' [1]. This method derives an 'off source' background measurement using the same 'software collimation' for an acceptance region determined by symmetric reflection of the source acceptance region on the telescope axis ('off source'). It provides a powerful simultaneous background measurement without normalization or efficiency uncertainties involved, based on instrumental symmetry only.

The instrument response as expected from gamma ray line radiation was determined from Monte Carlo simulations of the instrument under actual balloon flight conditions, including the variable source aspect. From these simulations (including the 'mirror' background subtraction method) energy deposit spectra were determined (see figure 1). Also, count profiles with respect to the angular residuals were derived from Monte Carlo simulations (see figure 3), using the one-dimensional model distributions for novae and supernovae in the Galaxy [15].

Figure 2 shows the measured source and background spectrum for the Galactic Center selection, as well as the background subtracted resultant spectrum of measured energy deposit. This spectrum is adequately fit with a model composed of a powerlaw and the response to 1.8 MeV line radiation of ^{26}Al from the Galactic Center (χ^2-value is 82.9 for 89 degrees of freedom; fit range 0.8-10 MeV); no significant improvement of the fit can be obtained with the inclusion of responses at other candidate gamma ray lines. Upper limits for other candidate lines were derived using a method which includes the response of a candidate gamma ray line into the fit model and increases the area under this line until the fit would be rejected from the χ^2 degradation with 95 % confidence, equivalent to 2 σ upper limits [8]. For our 4 parameter fit, the 95 % confidence limit corresponds to an increase in χ^2 by 9.5 with respect to the best fit. The gamma ray line energies investgated were: 0.8 MeV, 1.16 MeV, 1.3 MeV, 4.4 MeV, and 6.1 MeV, all for point source as well as extended source origin.

Note that this test was applied for the hypothesis of single lines. A superposition of lines was not tested, as the intensity in each component of a more complex hypothesis is constrained by the upper limits derived for narrow lines. Upper limits for spectral ranges were derived from the measurement using Monte Carlo simulations of a power law spectrum $\sim E^{-2.0}$.

RESULTS AND DISCUSSION

Energy (keV)	Pointsource Flux Limit (2σ) (10^{-4}Photons cm^{-2} sec^{-1})	Supernova Model Flux Limit (2σ) (10^{-4}Photons cm^{-2} sec^{-1} rad^{-1})
800	6.0	17.2
1160	3.4	12.5
1300	7.8	13.0
4400	4.8	8.9
6100	3.6	8.7
600-3000	28	72
3500-7000	9.6	17.7

Table 1: Upper limits for γ-ray lines from MPE data for 2 source distribution models

The γ-ray line at 1.8 MeV from ^{26}Al was identified in our data at a significance of 4σ. The flux in this line was determined to be 6.4 ($\pm$2.6)$\cdot 10^{-4}$ photons cm^{-2}sec^{-1}. No other significant gamma ray line intensity was found in the data. The '2 σ' upper limits (at the 95 % confidence level) determined at the five line energies 0.8, 1.16, 1.3, 4.4, 6.1 MeV are listed in table 1 for the hypothetical point source origin as well as for an origin distributed as the supernova distribution of the Galaxy [15]. The slightly higher upper limit derived for 1.3 MeV input energy is due to a small insignificant positive line flux contribution at this energy ($1.0 \cdot 10^{-4}$ photons cm^{-2} sec^{-1} at the 0.9 σ level). Continuum flux limits listed in table 1 were derived for 2 intervals assuming power law input with spectral index -2.0; the high limit in the 0.6–3 MeV band results from exclusion of the ^{26}Al line region 1.5–2.1 MeV for this analysis.

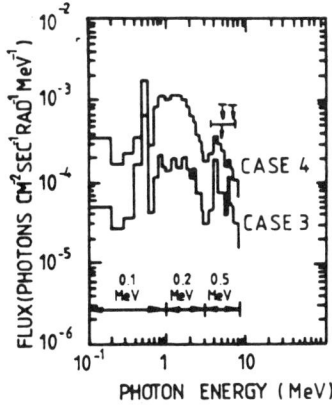

Fig.4: Gamma ray spectra for the Galactic Center region as expected from cosmic ray interactions with interstellar matter (from Ramaty, Kozlovsky, Lingenfelter, 1979)

The upper limits for broadened line emission from shortlived radionuclides are consistent with the absence of Galactic supernova events in the time period preceeding our balloon flight.

The supernova origin of ^{26}Al radiation detected in our data is questioned from directional analysis of our data: If the simulated count profiles with angular residual are fit to our measurement (see figure 3), the probabilities for the models to explain our data are satisfactory for the point source model (0.87) and the sharply peaked nova model N2 derived from M31 (0.37), while the probability value of 0.08 for the supernova model is marginally acceptable only. Supernova nucleosynthesis of ^{26}Al at the same time is expected to generate shortlived ^{44}Ti with an average decay γ-flux level about a factor 15 times the flux in the 1.8 MeV line [7]. This expectation is in clear contrast to our upper limit for the 1.157 MeV line emission, which rules out recent ($\sim$200 years) supernova type I nucleosynthesis in the Galactic Center direction.

The narrow line fluxes from superimposed supernovae in ^{60}Co gamma rays at 1.17 and 1.33 MeV were estimated by Clayton [6] in a steady state approximation. In this steady state, the interstellar medium will contain as much ^{60}Fe as produced during the last ^{60}Fe lifetime[9]:

$$N(^{60}Fe) = Y(^{60}Fe,T) \cdot \tau_{60}$$

where τ_{60} is the mean lifetime of ^{60}Fe ($2.2 \cdot 10^6$ years), and $Y(^{60}Fe,t)$ is the yield in ^{60}Fe per time interval, i.e. the supernova rate times the yield per supernova. For a supernova rate of 0.025 year^{-1} the number of contributing supernovae would be as high as $5.5 \cdot 10^4$, adequate for being approximated as a smooth distribution. Clayton[6] derives flux estimates for the 1.17 and 1.33 MeV gamma rays, which are admittedly very uncertain because of the unknown fraction of ^{60}Ni originating from decay of ^{60}Fe, estimated to be between a few and 50 % of the total ^{60}Ni yield. If corrected for a recent revision of the ^{60}Fe lifetime[14], gamma ray flux estimates are of the order of $1.2 \cdot 10^{-4} \cdot (^{60}Fe/^{60}Ni\text{-ratio})$, assuming all supernovae to be concentrated around the Galactic Center. Our 2 σ upper limit to the flux in these lines is similar to the limits derived from HEAO-3 data[16]. The line should have been detected, if the $(^{60}Fe/^{60}Ni)$-ratio were above $\sim$10%.

The derived limit for the ^{22}Na line at 1.275 MeV constrains the average mass of ^{22}Na ejected from novae. If all novae are assumed to be concentrated around the Galactic Center, the ejected mass relates to the measured flux in this line as:

$$M(^{22}Na) = F_\gamma \cdot 22 \cdot m_p \cdot 4\pi R^2 \Delta t_N$$

where Δt_N^{-1} is the average nova frequency of $\sim$1/40 years, $R = 10$ kpc, m_p the proton mass of $1.67 \cdot 10^{-24}$ g. From the observed limit we obtain $M(^{22}Na) < 2 \cdot 10^{-7} M_\odot$, if 1/4 of all novae are believed to be relevant for ^{22}Na and ^{26}Al nucleosynthesis (O-Ne-Mg white dwarf novae[12]). The ^{26}Al yields calculated for these objects[29] are $3 \cdot 10^{-7} M_\odot$. This leads to an upper limit on the ^{22}Na/^{26}Al ratio of $\sim$1.0, a value still in agreement with the very uncertain yields for the case of convectively mixed explosive hydrogen burning on O-Ne-Mg white dwarfs.

The gamma ray line fluxes predicted from low energy cosmic ray interactions with interstellar matter are still a factor of 2-4 below the upper limits for single lines derived from our data, even for the most optimistic model of Ramaty et al., where a high metallicity in the ambient matter

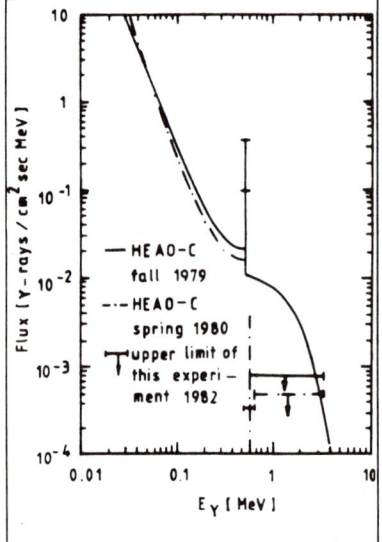

Fig.5: Upper limit on MeV continuum emission of the Galactic Center source, consistent with the low emission state reported from the HEAO-3 measurement (Riegler et al., 1985)

and in cosmic rays is assumed at 5 kpc from the Galactic Center. Thus our analysis does not allow us to derive constraints on the model parameters discussed by Ramaty et al.[21].

The 4.6 MeV line flux reported by Haymes et al.[10] of $9.5(\pm 2.7) \cdot 10^{-4}$ photons cm^{-2} sec^{-1} for an instrument opening angle of 13 degrees was demonstrated[21] to be inconsistent with the diffuse emission models based on cosmic ray interactions with interstellar matter. These authors therefore concluded that the flux detected by Haymes et al. was produced in one or more discrete sources in the Galactic Center region. Our measurement is incompatible with this interpretation, as our 2 σ upper limit for point source origin at the Galactic Center is $4.81 \cdot 10^{-4}$ photons/cm^{2}sec, a factor of 2 below the value reported by Haymes et al.[10].

From our 2σ upper limit (Figure 5) to the continuum point source flux in the MeV region we conclude that the central source was in a low intensity state at the time of our measurement, compatible with the 1980 HEAO-3 measurement.

Equivalently, this observational upper limit can be interpreted to constrain the electron component of the cosmic radiation: the predicted fluxes of diffuse γ-radiation from bremsstrahlung interaction of this component with interstellar matter are comparable to our 2σ upper limit[24]. This constrains the cosmic ray electron intensity to less than the most optimistic models discussed by Sacher and Schönfelder[24], if additional contributions to the MeV emissivity of the Galaxy are expected from the inverse Compton component and potential point sources in the direction of the Galactic Center.

ACKNOWLEDGEMENTS

We would like to thank the National Scientific Balloon Facility (USA) and the Institute de Pescquisas Espacials (Brazil) for conducting the balloon flight, and our MPE colleagues U. Graser, W. Hofmeister, N. Huber, L, Pichl, and F. Schrey for their contributions in our project.

REFERENCES

1.) v. Ballmoos, Diehl, Schönfelder (1987a), Ap.J. 312,134
2.) v. Ballmoos, Diehl, Schönfelder (1987b), Ap.J. 318, 654
3.) Chan, Lingenfelter (1987), Ap.J. (Lett.) 318, L51
4.) Chan, Lingenfelter (1987), Proc. 20. Int. Cosmic Ray Conf., 1, 164
5.) Clayton (1974), Ap.J. 188, 151
6.) Clayton (1982), Essays in Nuclear Astrophysics, edited by Barnes C.A., Clayton D.D., Schramm D.N., Cambridge University Press
7.) Clayton (1984), Ap.J. 280, 144
8.) Dunphy et al. (1981), Ap.J. 244, 1081
9.) Fowler and Hoyle (1960), Ann.Phys., 10 , 280
10.) Haymes at al. (1975), Ap.J. 201, 593
11.) Higdon, J.C. (1987), Proc. 20. Int. Cosmic Ray Conf., 1, 160
12.) Higdon and Fowler (1987), Ap. J. 201, 593
13.) Hillebrandt, Thielemann, Langer (1987), MPE preprint 1987
14.) Kutschera et al. (1984), Nucl. Instr. and Meth. B5, 430
15.) Leising and Clayton (1985), Ap.J. 294, 591
16.) Mahoney et al. (1982), Ap.J. 262, 742
17.) Mahoney et al. (1984), Ap.J. 286, 578
18.) Prantzos et al. (1987), Proc. 20. Int. Cosmic Ray Conf., 1, 152
19.) Ramaty and Lingenfelter (1980), NASA TM 82066, 1980
20.) Ramaty and Lingenfelter (1987), AIP Conf. Proc. 155, 51
21.) Ramaty, Kosslovsky, Lingenfelter (1979), Ap.J.Suppl 40, 487
22.) Riegler et al., (1985), Ap.J. (Lett.) 295, L13
23.) Riegler et al., (1981), Ap.J. (Lett.) 249, L13
24.) Sacher W., Schönfelder V. (1984), Ap.J. 279, 817
25.) Schönfelder, Graser, Diehl (1982), Astr.Astrop. 110, 138
26.) Share et al., (1985), Ap.J. (Lett.) 292, L61
27.) Webber, Schönfelder, Diehl (1986), Nature 323, 692
28.) Woosley S. (1986), Lecture Notes Saas Fee Workshop
29.) Woosley S. and Hoffman R. (1986), Bull.Am.Astr.Soc. 18, 4, 948

SPATIAL DISTRIBUTION OF INTERSTELLAR ^{26}Al GAMMA-RAYS:
PRELIMINARY RESULTS USING SMM

W. R. Purcell, M. P. Ulmer
Northwestern University, Evanston IL 60208

G. H. Share, R. L. Kinzer
Naval Research Laboratory, Washington D.C. 20375

E. L. Chupp
University of New Hampshire, Durham NH 03824

ABSTRACT

We present preliminary results of an analysis of data collected with the Solar Maximum Mission Gamma-Ray Spectrometer (GRS) in an effort to provide limits on the spatial distribution of Galactic ^{26}Al radiation. This analysis utilizes a new technique for applying the GRS data to extra-solar observations. Here we describe this technique and present data collected using four passages of the Galactic center region through the GRS field-of-view. The data are compared with three models: one representing a point source at the Galactic Center; one a distribution following that of Galactic CO; and a third which follows the distribution of novae in the Galaxy. Preliminary analysis indicates that the data are qualitatively in better agreement with the diffuse models.

INTRODUCTION

The discovery[1] of 1.809 MeV radiation from the decay of radioactive ^{26}Al in the interstellar medium by HEAO-3, and subsequent confirmation[2] by SMM, is exciting since it is the first direct observation of extra-solar radioactivity. Since ^{26}Al decays with a mean lifetime of 1.04×10^6 years, this observation indicates that significant quantities of ^{26}Al have been synthesized in the Galaxy in the last 1 - 2 million years. In addition, if the production site of the observed ^{26}Al can be determined, this will provide a unique opportunity for directly comparing observation with theories of nucleosynthesis.

The presence of a detectable gamma-ray line at 1.809 MeV from the decay of ^{26}Al in the interstellar medium was first proposed[3] nearly ten years before its detection. While modern nucleosynthesis theories predict ^{26}Al may be produced in many astrophysical sites, its presence in the interstellar medium requires a mechanism for transporting the active ^{26}Al out of the dense high temperature regions of its production. Hence, all theories predicting the presence of active ^{26}Al in the interstellar medium involve astrophysical sites which experience some form of mass loss. Examples are massive red giant[4] and Wolf-Rayet[5,6] stars with their intense stellar winds, and explosive nucleosynthesis in both novae[7,8] and supernovae[9,10].

© 1988 American Institute of Physics

The observed ^{26}Al may either have been produced in a single event, an exploding supermassive star[11] or recent nearby supernova[12], or represent the superposition of many individual events. If the observations represent the superposition of many sources, it may be possible to identify the source through a knowledge of the distribution of the ^{26}Al radiation. For this study, the current regions of star formation in the Galaxy are assumed to follow the Galactic CO distribution,[13,14] while the assumed nova distribution is based on observations of novae in M31.[13,15] In both cases, the distributions were assumed to be one-dimensional in the plane of the Galaxy.

The original satellite measurements made by HEAO-3 and SMM were unable to identify the ^{26}Al distribution due to their poor angular resolution, however the results of both instruments were consistent. Recently, two balloon experiments have independently reported the detection of ^{26}Al radiation from the direction of the Galactic Center.[16,17] The balloon observation reported by MacCallum et al. favors a diffuse source at the 90% confidence level and presents a diffuse source flux which is consistent with the HEAO-3 and SMM results. The results of the Compton telescope balloon observation reported by von Ballmoos et al., while not able to exclude the possibility of a diffuse source, appears more consistent with a point source located at the Galactic Center. The goal of this work is to place constraints on models of the distribution of ^{26}Al radiation in the Galactic plane.

INSTRUMENT AND ANALYSIS

The Solar Maximum Mission (SMM) satellite has been in nearly continuous operation since it was launched into low Earth orbit in early 1980. The Gamma-Ray Spectrometer (GRS) on the SMM satellite consists of a set of seven 3-inch diameter x 3-inch thick NaI(Tl) scintillation crystals actively shielded by CsI(Na) on the back and sides.[18] Active gain stabilization allows comparison of data collected over long periods. The GRS detector axis has remained within 10° of the Sun throughout the mission. The on-axis effective area is ~79 cm^2 and the aperture ~160° FWHM at 1.8 MeV. Recent Monte Carlo simulations[19] of the GRS have improved the knowledge of the instrument angular response at large angles.

Since the GRS detector axis remains oriented toward the Sun, the Galactic Center region passes within 6° of the center of its field-of-view in late December of each year. It was the presence of an annular modulation of the 1.809 MeV line flux which led to the SMM detection of ^{26}Al. This technique did not place significant constraints on the ^{26}Al distribution since it only used the large GRS field-of-view, and did not take full advantage of the Earth as an occulting disk. It is the combined use of Earth occultation, the large GRS field-of-view and the nearly seven years of data collected with the GRS which makes a detailed study of the distribution of the 1.809 MeV radiation possible.

The analysis technique is designed to produce an ~15° FWHM effective 'field-of-view' by using the Earth as an occulting disk in the ~160° FWHM GRS field-of-view. The basic technique consists of assigning spectra to be either source or background measurements based on whether a specific source position is occulted. If the source position is occulted, the associated spectrum is considered to be a background measurement. Otherwise, the spectrum is considered a source measurement. Additional selection criteria were applied to suppress further systematic effects caused by variations in the background. First, spectra were required to have been collected at least 10,000 seconds from the last significant passage of the satellite through the South Atlantic Anomaly. Next, spectra were selected only if the source position was within 10° of the Earth's horizon. Finally, equal numbers of source and background spectral accumulations were selected around each occultation transition, with a maximum time separation of five minutes between source and background collections. Background subtraction was performed for each occultation transition separately, and the resultant spectra accumulated. The accumulated background-subtracted spectra were then fitted with a single line fixed at 1.809 MeV with the instrument energy resolution, and a linear background term. The fit was performed over the energy range 1.6 - 2.0 MeV.

For the analysis described here, only data collected when the center of the GRS field-of-view was within 40° of the Galactic plane were used. At this time, we have only analyzed the data accumulated during 1981, 1982, 1985 and 1986. Spectra were produced for source positions at 5° intervals along the Galactic plane. As a monitor of any residual systematic background variation, the Galactic plane

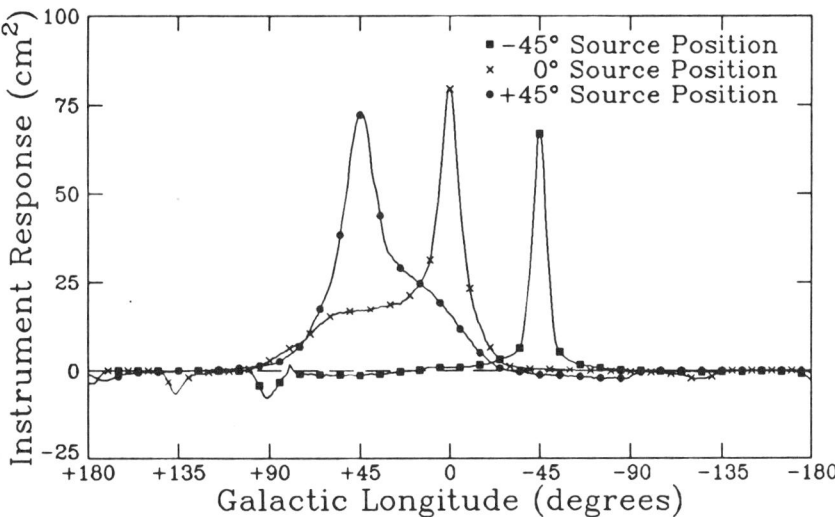

Fig. 1. - Effective 'field-of-view' of the GRS in the Galactic plane using the occultation technique described in the text for assumed source positions at the Galactic Center and at ±45° Galactic longitude.

in the anti-center direction was analyzed in a similar fashion. Since the source and background spectra accumulated for each sky position were selected from a common data set, it is possible for the source intensity determined for adjacent positions to have been derived, in part, from the same data. Therefore, these intensities are not statistically independent. There has not yet been time to perform a rigorous statistical analysis of this effect.

After the spectra had been selected based on the criteria described above, the effective 'field-of-view' in the Galactic plane (using Earth occultation) for the specific GRS observation was determined using the ephemeris information available for each one-minute collection. From the ephemeris information, the position of the Earth in the GRS field of view was determined for each source and background collection. With this, and a knowledge of the GRS field-of-view, the sky exposures for the source and background collections were determined. Since the GRS axis does not move between the source and background collections, the only difference in the sky exposures between the source and background will be due to the relative motion of the Earth in the GRS field-of-view. The difference of the respective source and background exposures was performed to generate an effective 'field-of-view' for the given observation. Figure 1 shows the effective 'field-of-view' in the Galactic plane for source positions at the Galactic Center and in the Galactic plane at ±45° Galactic longitude. The asymmetry in the effective 'field-of-view' for the 0° and +45° positions is due to data selection effects causing an asymmetric occultation of the source position. The effective angular resolution is ~15° FWHM in the Galactic plane.

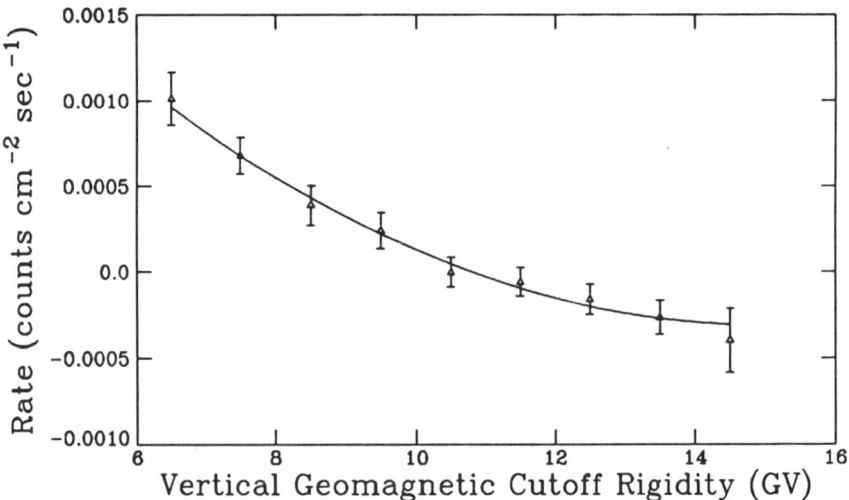

Fig. 2. - Relative intensity of the background line features near 1.8 MeV with rigidity. The solid line represents the best fit to the data.

CORRECTION OF RESIDUAL BACKGROUND EFFECTS

Residual systematic variations are observed at about the 15% level which the spectral background subtraction does not eliminate. These systematic variations are caused by neutron interactions[1] on ^{27}Al which generate background features at 1.779 and 1.809 MeV, each with half lives of less than a few minutes. Since the decay time for both of these reactions is short compared with the background subtraction time, the variability of these reactions must be determined. We have found that the intensity of these background features can be monitored by the vertical geomagnetic cutoff rigidity. The relative intensity of these features with rigidity is shown in Figure 2.

Using the best fit functional dependence of the relative background line counting rate with rigidity, the residual background contribution is estimated based on the difference in the rigidity at which the source and background spectra were accumulated. This background contribution is accumulated and subtracted after the spectral fit is performed. The maximum background contribution is estimated to be about 0.5×10^{-4} counts cm^{-2} sec^{-1}.

RESULTS AND DISCUSSION

Using the effective 'field-of-view' generated for each Galactic plane position, the expected responses are determined for three models: 1) a point source located at the Galactic Center; 2) a

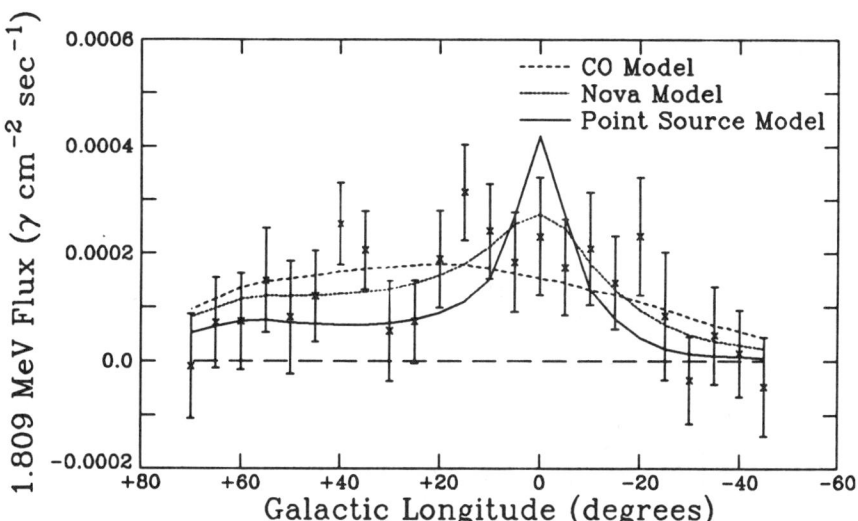

Fig. 3. - Background-subtracted 1.809 MeV line flux for positions along the Galactic plane near the Galactic Center. The data points are not statistically independent. The curves shown represent the best fits to the data.

diffuse model following the distribution of Galactic CO; and 3) a diffuse model following the distribution of Galactic novae. The models are fit to the background-subtracted data with their intensity as the only free parameter. Figure 3 shows the background-subtracted data for the Galactic Center observations and the best fit responses for these three models. The flux values determined in this analysis are consistent with previous SMM results. Additional analysis is required to accurately treat the covariant terms in order to determine the statistical significance of the model fits to the data. To determine the level of additional systematic effects, the same analysis was performed for positions in the Galactic anti-center direction. These data are found to be consistent with zero.

The current analysis of four years of data collected with SMM seems to be best fit by a diffuse distribution. A point source of ^{26}Al radiation located at the Galactic Center appears inconsistent, however a more detailed analysis, including the proper treatment of the covariant statistics, is required to determine the significance of this apparent inconsistency. Further analysis utilizing this technique, and the addition of two more years of data, should enable us to place useful limits on whether the ^{26}Al radiation originates predominantly from a point source or a diffuse distribution.

ACKNOWLEDGEMENTS

We wish to thank Steve Matz and Mark Leising for their valuable comments and suggestions throughout this analysis, and to Dan Messina for his direction of the production analysis and careful screening of the data. This work was supported by NASA contracts S-10987-C at NU and S-14513-D at NRL.

REFERENCES

1 Mahoney, W. A., et al., Ap. J., 286, 578 (1984).
2 Share, G. H., et al., Ap. J., 292, L61 (1985).
3 Ramaty, R. and Lingenfelter, R. E., Ap. J., 213 L5 (1977).
4 Norgaard, H., Ap. J., 236, 895 (1980).
5 Dearborn, D. S. P., and Blake, J. B., Ap. J., 288, L21 (1985).
6 Prantzos, N, and Casse, M., Ap. J., 307, 324 (1986).
7 Arnould, M., et. al., Ap. J., 237, 931, (1980).
8 Hillebrandt, W., and Thielemann, F.-K., Ap. J., 255, 617 (1982).
9 Truran, J. W., and Cameron, A. G. W., Ap. J., 219, 226 (1978).
10 Arnett, W. D., and Wefel, J. P., Ap. J., 224, L139 (1978).
11 Hillebrandt, W., et al., Nuclear Astrophysics, 343 (1986).
12 Clayton, D. D., Ap. J., 280, 144, (1984).
13 Leising, M. D., and Clayton, D. D., Ap. J., 294, 591 (1985).
14 Stecker, F. W., and Jones, F. C., Ap. J., 217, 843 (1977).
15 Sharov, A. S., Astr. Zh., 48, 1258 (1971).
16 MacCallum, C. J., et al., Ap. J., 317, 877 (1987).
17 von Ballmoos, P., et al, Ap. J., 318, 654 (1987).
18 Forrest, D. J., et al., Solar Physics, 65, 15 (1980).
19 Matz, S. M., and Jung, G. V., private communication (1987).

ORIGIN OF A POSSIBLE DIFFUSE GALACTIC EMISSION AT 511 keV

M. Signore *

* Ecole Normale Supérieure, 24, rue Lhomond, 75231 Paris, France et Observatoire de Paris-Meudon, 92190 Meudon, France

G. Vedrenne**

** Centre d'Etude Spatiale des Rayonnements, B.P. 4346, 31029 Toulouse Cedex, France

ABSTRACT

The latest observational results on the 511 keV annihilation radiation have suggested a possible diffuse origin for this emission - Share et al.[1] -. Recently Ramaty and Lingenfelter[2] have interpreted these observations in terms of a variable point source and a diffuse interstellar source of positrons; they have given an evaluation of ^{26}Al and ^{56}Co contribution to the diffuse 511 keV flux while that of ^{44}Ti has been given more recently by Woosley[3]. Here are discussed the nature and distribution of the various e^+ sources. It is shown that the diffuse 511 keV emission can be explained by nucleosynthesis of galactic SN and novae.

INTRODUCTION

A recent report on the SMM detection of the galactic 511 keV annihilation radiation - Share et al.[1] - compares the observed SMM flux for an assumed point source near the Galactic Center with upper limits obtained from balloon observations in 1981 and 1984 and concludes that a bulk of the radiation observed by SMM comes from a diffuse interstellar source of positrons; if the 511 keV emission is proportional to the measured galactic CO-distribution, the diffuse flux is: $F_D \sim (1.6 \pm 0.3) \times 10^{-3}$ ph.cm^{-2}s^{-1}rad^{-1} at the center, while if the total SMM flux comes from a point source at the Galactic Center: $F_p \sim (2.1 \pm 0.4) \times 10^{-3}$ ph.cm^{-2}s^{-1}. We have studied the nature of the sources of the main e^+ emitting radio isotopes - SN and novae - and their galactic distribution using:

- the observational data at 511 keV[1] and at 1809 keV[4];
- the galactic angular distributions studied by Leising and Clayton[5,6];
- the nucleosynthesis of the models of SN and novae of Woosley and his collaborators[7,8,9];
- the recent observational galactic SN rates given by Van den Bergh, Mc Clure and Evans[10].

© 1988 American Institute of Physics

THE e^+ FLUX

From Leising's thesis [6], we can write the e^+ flux, for the central radian, produced via ^{26}Al-decays and via the other decays: ^{44}Ti, ^{56}Co, ^{22}Na, respectively

$$F_{Al} : 1.33\sigma_{Al\ 0\ SN} + 2.6\sigma_{Al\ 0\ n} = 4 \times 10^{-4} \ e^+ \ cm^{-2} \ s^{-1} \ rad^{-1} \quad (1)$$

$$F_{non\ Al} : 1.33\sigma_{non\ Al\ 0\ SN} + 2.6\sigma_{non\ Al\ 0\ n} = 20.6 \times 10^{-4} \ e^+ \ cm^{-2} \ s^{-1} \ rad^{-1} \quad (2)$$

- the right hand side of equation(1) is deduced from the observed ^{26}Al-line flux of 4.8×10^{-4} ph.cm^{-2} s^{-1} rad^{-1} given by Mahoney et al [4];

- the right hand side of equation (2) corresponds to the observed [1] 511 keV line flux of 1.6×10^{-3} ph.cm^{-2} s^{-1} rad^{-1} with an assumed "Ps-fraction" f~ 0.9 supposing that the annihilation occurs in molecular hydrogen - Brown and Leventhal [11];

- $\sigma_{Al\ 0\ SN}$, $\sigma_{non\ Al\ 0\ SN}$ are the local rates, per unit area, of e^+ ejected from SN via ^{26}Al decay and via other decays respectively and similarly for $\sigma_{Al\ 0\ n}$, $\sigma_{non\ Al\ 0\ n}$.

THE σ_0's

We determine the various σ_0 using the relevant nucleosynthesis in the latest models of novae and SN of Woosley and his collaborators [7,8,9]:

- we only consider the typical SNII explosions of over 15 M_0 which eject $\sim 10^{-4} M_0$ of ^{26}Al and $0.8 \times 10^{-4} M_0$ of ^{44}Ti; they also eject a large amount of ^{56}Ni but it is covered by several solar masses of material, thus e^+ from ^{56}Co-decay are completely trapped; there are $x = x_1 \ 10^{-11}$ pc^{-2} yr^{-1} SNII explosions of over 15 M_0.

- standard SNIa - supposed to be carbon deflagrations in white dwarfs (model 2 of Woosley and Weaver [8], model W7 of Nomoto et al. [12]) - eject $\sim 0.8 \times 10^{-4} M_0$ of ^{44}Ti and 0.5 M_0 of ^{56}Ni. A very small fraction ε of e^+ from ^{56}Co decay can escape from SNI ejecta; ε depends critically on the envelope models (optical thickness ionization and trapping by magnetic fields). We introduce: $y_a = y_{1a} \ 10^{-11}$ pc^{-2} yr^{-1} SNIa and $\varepsilon = 10^{-2} \varepsilon_1$, since Arnett [13] and Colgate et al. [14] considered that ε may reach a few percents. Thus we take ε_1 as a free parameter.

- subluminous SNIb which show helium but not hydrogen lines might result from either a massive star core collapse (Wolf-Rayet model) or a white Dwarf detonation (White Dwarf Model); see Branch [15] for a detailed discussion on SNIb. If both classes of progenitors are operating in nature, the W.R class belongs to our SNII explosions of over 15 M_0 and, in the following, we call SNIb the possible W.D. class; therefore we suppose that these SNIb eject only $\sim 0.25 \ M_0$ of ^{56}Ni which could roughly explain their dim light curves; and we introduce: $y_b = y_{1b} \ 10^{-11}$ pc^{-2} yr^{-1} SNIb.

- recently, Woosley, Taam and Weaver [9] (hereafter WTW) have studied a Peculiar model of SNI: the "detonating helium dwarf" that experience explosive helium burning; these rare events can eject:

$\sim 0.8 \times 10^{-2}$ M_0 of ^{44}Ti and ~ 0.5 M_0 of ^{56}Ni (WTW, model 4). We introduce:
$y_p = y_{1p} \, 10^{-11}$ $pc^{-2} yr^{-1}$ SNIp and $y_{1p} = 10^{-2} z_1$.

- finally, we introduce: $t = t_1 \, 10^{-8}$ $pc^{-2} yr^{-1}$ novae which eject ($a_1 \, 10^{-7}$ M_0) of ^{26}Al and ($b_1 \, 10^{-7}$ M_0) of ^{22}Na; in order to explain the occurence of outbursts on O-Ne-Mg white dwarfs implied by the large amounts of Ne, Mg, Al, O found in ejecta of some novae: nova Cr A 1981, nova Aqu 1982, nova Vul. II 1984 etc. ... - see Sparks et al. [16] -, one considers the "WH$_2$-con model", which is one of the parametrized nucleosynthesis nova models of Woosley [7]: $a_1 = 1.8$, $b_1 = 1$.

With these amounts of radioactive elements ejected by SN and novae and from the expressions introduced for their local rates, the various local rates, per unit area, of e^+ are given in table 1.

Table 1: values of σ_0

$10^{-4} e^+$ $cm^{-2} s^{-1}$		^{26}Al	^{56}Ni	^{44}Ti	^{22}Na
	novae	$1.4 \, a_1 t_1$			$1.8 \, b_1 t_1$
	SNII M>15M$_0$	$1.4 \, x_1$		$0.75 \, x_1$	
	SNIa		$7.5 \, \varepsilon_1 y_{1a}$	$0.75 \, y_{1a}$	
	SNIp		$7.5 \, \varepsilon_1 y_{1p}$	$0.75 \, z_1$	
	SNIb		$\frac{7.5}{2} \varepsilon_1 y_{1b}$		

LOCAL AND TOTAL RATES OF SN AND NOVAE

Nucleosynthesis implies that x is not the local rate of all the galactic SNII but rather the local rate of SNII explosions of over 15 M_0. Moreover because the steady state situation for massive stars makes the "Initial Mass Function" equal to their present death rate, we can infer, from the latest local IMF given by Scalo [17,18]:

$$x_1 (M_* \geqslant 15 \, M_0) \sim 0.4 \qquad (3)$$

It is a factor of ~ 3 smaller, at large masses, than the IMF given by Miller and Scalo [19]; in fact there are large uncertainties in the IMF at these large masses. In order to check our value (3) for x_1, we can compare it to the more recent observational rates given by Van den Bergh, Mc Clure and Evans [10]. Their observational galactic SN frequencies, about 3 times lower than Tamman's frequencies, are [19]:

$$\nu_{Ia} \sim 0.3 \times 10^{-2} \text{ yr}^{-1}; \quad \nu_{Ib} \sim 0.4 \times 10^{-2} \text{ yr}^{-1}; \quad \nu_{II} \sim 1.1 \times 10^{-2} \text{ yr}^{-1} \quad (4)$$

On the other hand, from Leising and Clayton's study [5,6], we deduced that a galactic frequency (in yr^{-1}) can be converted into a local rate (in $\text{pc}^{-2} \text{yr}^{-1}$) dividing it by [20]:

$1.3 \times 10^9 \text{ pc}^2$ for a CO-distribution;

$3.5 \times 10^9 \text{ pc}^2$ for a nova like distribution.

In particular, if one supposes that SN have a CO-type distribution, the local rates corresponding to the Van den Bergh's frequencies are given by:

$$y_a \sim 0.2 \times 10^{-11} \text{ pc}^{-2} \text{yr}^{-1}; \quad y_b \sim 0.3 \times 10^{-11} \text{ pc}^{-2} \text{yr}^{-1}; \quad x_{II} \sim 0.8 \times 10^{-11} \text{ pc}^{-2} \text{yr}^{-1} \quad (5)$$

As the local SNII frequency given by Tamman [19]: $2.3 \times 10^{-11} \text{ pc}^{-2} \text{yr}^{-1}$ can be inferred from the IMF of Miller-Scalo [19] by invoking all stars with mass $M_* \geq 8.4 \, M_0$, we can obtain from the IMF of Scalo [17], the new local SNII frequency $x_{II} (M_* \geq 8.4 \, M_0) \sim 0.8 \times 10^{-11} \text{ pc}^{-2} \text{yr}^{-1}$ corresponding to the Van den Bergh et al. [10] galactic SNII frequency: $\nu_{II} \sim 1.1 \times 10^{-2} \text{ yr}^{-1}$.

In the following, we will adopt:

$$\left.\begin{array}{l} x(\geq 15 \, M_0) \sim 0.4 \times 10^{-11} \text{ pc}^{-2} \text{yr}^{-1} \text{ SNII of over } 15 \, M_0 \\ y_b \sim 0.3 \times 10^{-11} \text{ pc}^{-2} \text{yr}^{-1} \text{ for SNIb} \end{array}\right\} \quad (6)$$

since SNIb, which are also correlated to the star forming regions, can be described by a CO-distribution. But it is not so clear for the SNIa and SNIp distributions and therefore for their local rates.

In a recent study of the porosity in the Milky Way, Heiles [21] has found that the inner disk (R < 3 kpc) should be dominated by the old type I SN and the ring of molecular clouds (4 < R < 7 kpc) by young type II SN. Few SN should be in the outer disk. So, using this conclusion: SNII have a CO-distribution and their local rate is given by (6) while at least a part of SNI can have a distribution more peaked toward the Galactic Center - a nova like distribution. If SNIa have this distribution their local rate becomes, instead of the value given in (5):

$$y_a \cong \frac{\nu_{Ia}}{3.5 \; 10} \cong 0.1 \times 10^{-11} \text{ pc}^{-2} \text{yr}^{-1} \quad (7)$$

On the other hand, if we assume that ^{44}Ca is essentially produced by SN-nucleosynthesis via ^{44}Ti decay, its abundance X_{44} supposed to be ejected by SN, must be equal to its solar abundance $X_{44 \, 0} = 1.6 \times 10^{-6}$ and:

$$X_{44} = 0.8 \times 10^{-6} \frac{x_1 + y_{1a} + z_1}{S_1} = 1.6 \times 10^{-6} \equiv X_{44 \, 0} \quad (8)$$

where we recall that SNI and SNII of over 15 M_0 eject $\sim 0.8 \times 10^{-4} M_0$ of ^{44}Ti while SNIp of WTW eject $\sim 0.8 \times 10^{-2} M_0$ of ^{44}Ti and where the local SFR (star formation rate) is $s = s_1$ M_0 pc^{-2} Gyr^{-1}. If we adopt the value $s_1 = 3$ given by Scalo[18], one must have: $z_1 \cong 5.4$ or 5.5 if y_{1a} is given respectively by (5) or (6);

that is to say: $y \cong 0.05 \times 10^{-11}$ pc^{-2} yr^{-1} SNIp   (9)

Moreover SNIp which are nothing but helium dwarf cataclysmics, are supposed to have a nova like distribution.

RESULTS

We consider 2 possible solutions:

H_{01} "Without Peculiar SNI": the ^{44}Ti excess produced by SNIp is not considered

H_{02} "With Peculiar SNI" : this excess is taken into account;

For each solution, equation (1) leads to:

$t = 0.5 \times 10^{-8}$ pc^{-2} yr^{-1} or $\mathcal{T} = 17$ yr^{-1} novae

or rather the rate of outbursts on O-Ne-Mg white dwarf described by the "WH$_2$-con model" of Woosley. We note that, even if the expected ratio $\frac{O.Ne.Mg-WD}{C.N.O-WD}$ is low, the expected outburst ratio is 1/4.

Adopting a nova like distribution for SNIp and a CO-distribution for SNIa, equation (2) leads to:

$$4.4 \, \varepsilon_1 + \begin{cases} 0.6 \\ 11.1 \end{cases} + 2.3 = 20.6$$

where the left members give the contributions of ^{56}Co, ^{44}Ti, ^{22}Na, respectively and where the upper and lower values correspond respectively to H_{01} and H_{02}:

H_{01} implies $\varepsilon_1 \sim 4$ while H_{02} implies $\varepsilon_1 \sim 1.6$.

Rough estimates of the above contributions but for an aperture of $\sim 35°$- HEAO-3 - can also be given:

$$2.6 \, \varepsilon_1 + \begin{cases} 0.3 \\ 8.4 \end{cases} + 1.8 \sim \begin{cases} 12.5 \text{ for } H_{01} \\ 14.3 \text{ for } H_{02} \end{cases}$$

with a Al-contribution of ~ 2.9 for both solutions. These estimates lead to $\sim 10^{-3}$ $ph.cm^{-2}$ s^{-1} for H_{01} and $\sim 1.1 \times 10^{-3}$ $ph.cm^{-2}$ s^{-1} for H_{02}. Both solutions marginally explain the low value found by HEAO-3.

With a novalike distribution for SNIa and SNIp, a very similar conclusion to the above one is obtained because the type of distribution for the SNIp is the most important and not the SNIa ones.

More generally, for $0 < f < 1$, equation (2) leads to:

$$4.4 \, \varepsilon_1 + \begin{cases} 0.6 \\ 11.1 \end{cases} + 2.3 = \frac{16}{(2 - 1.5f)} - 4$$

For H_{01} solutions we have: $4.4 \, \varepsilon_1 \cong \frac{16}{(2 - 1.5f)} - 7$ and ε_1 is an increasing function of f; in particular:

$$0.66 < f < 0.9, \; 2 < \varepsilon_1 < 4$$

A rough estimate for the diffuse flux of HEAO-3 is $F_{35} \sim (2 - 1.5f)(2.6 \, \varepsilon_1 + 5)$.

For H_{02} solution we have: $4.4 \, \varepsilon_1 \cong \frac{16}{(2 - 1.5f)} - 17.4$ and ε_1 is also an increasing function of f with:

$$0.72 < f < 0.9, \; 0 < \varepsilon_1 < 1.6 \text{ and } F_{35} \cong (2 - 1.5f)(2.6 \, \varepsilon_1 + 13)$$

Tables 2 gives the percentages of e^+ from the various decays and their distribution for both H_{01} and H_{02} with $f \sim 0.66$ and $f \sim 0.9$.

H_{01}		Al	Na	Ti	Co	$\Sigma \, e^+$
$f = 0.66$	CO-type	5	0	4	44	53
$\varepsilon_1 = 2$	nova type	20	14	0	13	47
		25	14	4	57	100
$f = 0.9$	CO-type	3	0	2.5	57	62.5
$\varepsilon_1 = 4$	nova type	13	9	0	15.5	37.5
		16	9	2.5	72.5	100
H_{02}		Al	Na	Ti	Co	$\Sigma \, e^+$
$f = 0.66$	CO-type	5	0	4	0	9
$\varepsilon_1 = 0$	nova type	20	14	57	0	91
		25	14	61	0	100
$f = 0.9$	CO-type	3	0	2.5	23.5	29
$\varepsilon_1 = 1.6$	nova type	13	9	42	7	71
		16	9	44.5	30.5	100

CONCLUSION

In any case, nucleosynthesis from SN and nova ejecta can explain a diffuse 511 keV emission:

- if $0 < \varepsilon < 1.6 \times 10^{-2}$: H_{02} is the solution and ^{44}Ti is the main contributor. Moreover for $f < 0.72$ and $\varepsilon_1 = 0$: e^+ from ^{56}Co are no longer needed; this solution has been already considered by Woosley [3].
- if $2 \times 10^{-2} < \varepsilon < 4 \times 10^{-2}$: H_{01} is the solution and ^{56}Co is the main contributor. This solution has been already proposed by Ramaty and Lingenfelter [2].

Now are waited the results of more rigourous calculations of the "escape fraction" ε from SNI modelers and measured and observed values of the "positronium fraction" f.

Finally, new observations with imaging systems such as SIGMA and with high sensitivity instruments for low energy gamma-rays such as OSSE on GRO are urgently needed.

We want to thank S.E. Woosley and J. Scalo for their comments, M. Cassé and R. Chevalier for pointing out the new observational results of Van den Bergh et al., and finally S. Van den Bergh and R. Evans.

REFERENCES

1. G.H. Share et al. preprint to be published in Ap. J. March 15, 1988.
2. R. Ramaty and R.E. Lingenfelter, The Galactic Center (Am. Inst. Physics., N.Y., 1987), p. 155.
3. S.E. Woosley, preprint 1987.
4. W.A. Mahoney, J.C. Ling, W.A. Wheaton, A.S. Jacobson, Ap. J. 286, 578 (1984).
5. M.D. Leising and D.D. Clayton, Ap. J., 294, 591 (1985).
6. M.D. Leising, Thesis Rice University, Houston Texas (1985).
7. S.E. Woosley, Nucleosynthesis and Chemical Evolution (16th Adv. Course of the Swiss Acad. of A & A, Geneva, 1986).
8. S.E. Woosley, T.A. Weaver, Radiation Transport and Hydrodynamics (I.A.U. n° 89 held at Copenhague Reidel, Dordrecht, 1986).
9. S.E. Woosley, R.E. Taam, T.A. Weaver, Ap. J., 301, 601 (1986).
10. S. Van den Bergh, R.D. Mc Clure, R. Evans, Ap. J., 323, 44, (1987).
11. B.L. Brown and M. Leventhal, Ap. J., 319, 637 (1987).
12. K. Nomoto et al., Ap. J. 277, 791 (1984).
13. D. Arnett, Ap. J., 230, L37 (1979).
14. S. Colgate et al. Ap. J., 237, L81 (1980).
15. D. Branch, Atmospheric Diagnostics of Stellar Evolution Chemical Peculiarity, Mass Loss and Explosion. (I.A.U. n° 108 held at Tokyo, Lecture Notes Springer Verlag, 1987), in press.
16. W.M. Sparks et al., R.S. Ophiuchi 1985. The recurrent nova phenomenon (VNU Science Press Utrecht 1982), p. 39.
17. J. Scalo, Fund. Cos. Phys. 11, 1-278 (1986).
18. J. Scalo, private communication, (1987).
19. G.A. Tamman, Supernovae: A survey of current research (Reidel Dordrecht, 1982), p. 371.
20. M. Signore and G. Vedrenne, submitted to A & A (1987).
21. C. Heiles, Interstellar Processes (Reidel Dordrecht, 1987), p. 225.

γ-RAY LINE EMISSIONS AS TRACERS OF TENUOUS PHASES OF THE INTERSTELLAR MEDIUM

J. C. Higdon
Joint Science Department
The Claremont Colleges
11th Street and Dartmouth Avenue
Claremont, CA 91711

ABSTRACT

γ-ray line emissions, produced by e^+ annihilations and the decay of ^{26}Al, provide unique tools for probing on galactic scales the tenuous interstellar phase - the least understood component of the interstellar medium.

INTRODUCTION

Based primarily on soft (≤200 eV) x-ray emission and O VI absorption line data, some models[1,2] of the interstellar (IS) medium suggest that a major fraction of the IS volume consists of merged supernova remnants with typical interior temperatures of ≈ 10^6 K. Alternatively, Kulkarni and Heiles[3] have presented evidence, based primarily on analyses of 21 cm H I line emissions, that H I fills as much as 0.9 of IS space. However, at distances beyond the solar vicinity (> 100 pc) the relative contributions of the warm and hot tenuous phases are uncertain because x-ray emission, the primary indicator of hot plasmas, is highly attenuated by photoelectric absorption in intervening H I clouds. To investigate further the tenuous phases, new approaches must be developed for tracing plasmas.

γ-RAY LINE EMISSSIONS

The primary sources of ^{26}Al and e^+ are expected[4] to be, respectively, asymptotic branch stars and type I supernovae. They are much older than any IS cloud population. Consequently, ^{26}Al and e^+ sources are not correlated with the IS cloud phases, and occur randomly in IS phases with the greatest filling factors, the tenuous phases.

The annihilation sites of e^+ can be determined on galactic scales via high-resolution γ-ray spectroscopy. This can be accomplished because line profiles, produced by e^+ annihilations, are significantly different in the major[1] IS phases. In the cold (~ 100 K) phase e^+ annihilations produce a strong 3 γ continuum[5] and a line feature with a FWHM of ~ 5 KeV. In the warm (~ 10^4 K) phases e^+ annihilations produce again a strong 3 γ continuum[5] and a line feature with a narrower FWHM of ~ 1 KeV. In x-ray emitting plasmas[6] the 3 γ continuum is suppressed and

© 1988 American Institute of Physics

narrow line (FWHM < 1 KeV) emission dominates. Thus, IS filling factors could be determined from γ-ray emissions, if e^+ do not propagate very far after they are created.

It is suggested that e^+ propagate only small distances from where they were created because e^+ are scattered efficiently by collisionless interactions with magnetic fluctuations. I estimate a e^+ mean free path, λ, to be typical[7] of cosmic rays, ≈ 0.04 $(T/GeV)^{0.35}$ pc, where T is the e^+ kinetic energy. Quantitatively a 0.5 MeV e^+ diffuses a distance $\approx (\lambda c\tau)^{0.5} \sim 30$ pc in $\tau \approx 10^6$ yr; c is the speed of light. Such a distance is < 90 pc, the mean separation[1] between IS clouds. Since $\lambda(T)$ decreases as a e^+ decelerates, a large fraction of e^+ injected into the tenuous phases remains entrained in these phases.

Although e+ are likely to be entrained in tenuous plasmas, these e^+ annihilate very inefficiently. The annihilation ifetime[6] of thermal e^+ in dust grains of the typical[1] hot intercloud medium ($\sim 10^{5.6}$ K, $\sim 10^{-2.45}$ cm^{-3}) is $\sim 7 \times 10^6$ yr, significantly greater than $\sim 8 \times 10^5$ yr, the radiation lifetime[1] of the expanding supernova remnants that constitute the tenuous intercloud gas. Thus the annihilation radiation of these e^+ occurs primarily after the plasmas have cooled. Thus such 0.511 MeV emissions, when resolved, will tell us the ultimate fate of IS plasmas, whether such plasmas are ejected from the Galaxy via a wind.[7], form a galactic fountain[8], or recombine to H I in radiative[1] supernova remnants in the disk itself.

Similarly the 1.809 MeV emissions, when resolved by high-resolution γ-ray spectroscopy, will provide important constraints on the nature of tenuous phases. This results because ^{26}Al, ejected randomly into the tenuous plasma phase, becomes entrained in the hot expanding gas. The resultant 1.809 MeV line profiles are dominated by these high-velocity plasma motions. High-resolution spectroscopy measurements will constrain models of such expansion.

This work suggests that high-resolution γ-ray spectroscopy will provide unique information about the nature, spatial extent, and ultimate fate of tenuous IS phases.

REFERENCES

1. C. F. McKee and J. P. Ostriker, Ap. J., 218, 148 (1977).
2. D. P. Cox, Ap. J., 234, 863 (1979).
3. S. R. Kulkarni and C. Heiles in "Galactic and Extragalactic Radio Astronomy", eds. K. I. Kellerman and G. L. Verschuur (New York, Springer-Verlag 1987), in press.
4. J. C. Higdon, this proceedings.
5. R. W. Bussard, R. Ramaty, and R. J. Drachman, Ap. J., 228, 928 (1979).
6. W. H. Zurek, Ap. J., 289, 603 (1985).
7. W. I. Axford, 17th Int. Cosmic Ray Conf., 12, 155 (1981).
8. P. R. Shapiro, and G. B. Field, Ap. J., 205, 762 (1976).

A Possible Galactic Center Positron Annihilation Medium: Neutral Hydrogen

B. L. Brown[*] and M. Leventhal
AT&T Bell Laboratories
Murray Hill, New Jersey 07974

ABSTRACT

A series of laboratory experiments in neutral hydrogen has yielded several quantities that may be relevant to the 511 keV positron annihilation line coming from the direction of the Galactic center. The annihilation linewidth was measured for Ps, formed by positrons slowing down from high energies in H_2 and He. The fraction of positrons which survive below the threshold for Ps formation was measured with a pulsed beam technique. The linewidth for positrons directly annihilating with bound electrons in H_2 was measured with another pulsing technique. The linewidth measurements agreed with theoretical predictions while the survival fraction did not agree. The combined measurements in H_2 are used to construct the annihilation spectrum expected for galactic positrons in neutral molecular hydrogen regions. The resulting lineshape was fit directly to the HEAO-3 satellite data. The fits indicate that neutral hydrogen is a possible annihilation medium, contrary to previous reports.

Recent papers discussing positron annihilation data from the Galactic center have used several incompatible definitions of "the positronium fraction". A single definition applied to the data shows that the observational data are consistent with a triplet Ps annihilation spectrum included at a level of nearly 100% Ps formation. This is also consistent with a neutral hydrogen medium and with other models such as an ionized or cool dusty ionized region.

INTRODUCTION

In a series of three laboratory experiments we have simulated the annihilation spectrum for energetic positrons slowing down and decaying in a cold neutral galactic H_2 environment.[1,2,3] The general picture that emerges is as follows. Nearly all of the positrons survive down to energies of order 100 eV, losing energy by ionization and excitation of the neutrals. At this point, in-flight positronium (Ps) annihilation by charge exchange with the neutrals becomes significant, increasing in importance as the energy approaches the Ps formation threshold (8.6 eV in H_2). In-flight Ps annihilation gives rise to a Doppler-broadened 511 keV line of width (6.4 ± 0.1) keV FWHM and a Doppler-rounded triplet (3-

---

[*] Current Address: Department of Physics, Harvard University, Cambridge, MA 02138 U.S.A

photon) continuum. Only a small fraction (10.3% ± 0.3%) of the incident positrons survive below the Ps formation threshold to thermalize and eventually annihilate directly (2-photon) with bound electrons. Hence the positronium fraction f is 0.897 ± 0.003 in H_2. Here the quantity f is defined simply as the fraction of incident positrons that decay via Ps formation. The direct annihilation line is sharp with a measured width of 1.56 ± 0.09 keV FWHM and to a great extent determines the width of the composite line (2.2 keV FWHM). A similar result is expected for atomic H. The observed Galactic center 511 keV linewidth of $1.6^{+0.9}_{-1.6}$ keV is consistent with the composite linewidth for either H_2 or H and fitting the data directly to the composite lineshape for H_2 and H gives acceptable agreement.[4] The experiments leading to this conclusion will be reviewed below and the composite spectrum presented. In addition a simple expression relating f to the astronomical observations will be derived and the measured f values are also seen consistent with a neutral annihilation medium.[5]

EXPERIMENTAL APPARATUS

The magnetically guided beam is shown in Fig. 1. A 100-mCi ^{58}Co source was used in conjunction with a tungsten (111) moderator followed by an $\vec{E}\times\vec{B}$ filter, to produce low-energy positrons. The bias on the moderator (and source) was changed to produce different incident beam energies. The positron beam passes through several stages of differential pumping and through a cylindrical grid in region 3 before being injected into a target region with a typical pressure of 1 mTorr. After undergoing an inelastic collision, (typically ionization of H_2), positrons are trapped longitudinally in the target region by the biased cylinder in region 3 and the biased plate in region 1 and radially by a large ($\sim 10^3$G) magnetic field. The annihilation process is studied with various detectors which view the target region through a lead collimator.

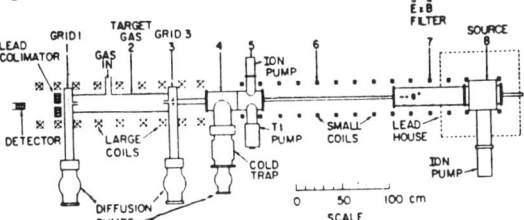

Fig. 1. Schematic diagram of the apparatus.

IN-FLIGHT ANNIHILATION

Positrons were injected into the trap with energies between 10 and 1300 eV. Ps is formed (both singlet and triplet) in the gas chamber in-flight via the charge-exchange interaction. The Ps formed in-flight has a lifetime much less than the mean collision time of Ps and a gas atom or molecule, and the experiment thus simulates the interstellar vacuum as far as the Ps 511 keV annihilation line is concerned. The Doppler broadened lineshape was studied with a Ge(Li) detector. In examining the 511 keV annihilation spectra we found a wide component from Doppler-broadened singlet Ps annihilation and a narrow component (≤ 3 keV) due to positron annihilation on Cu pumping baffles in regions 1 and 3. A good least-squares fit to the data ($\chi^2/\nu \approx 1$) was obtained with the sum of two Gaussian functions centered at 511 keV energy, a constant function below 511 keV to simulate the triplet Ps continuum, and a constant background all folded with the detector resolution of 1.5 keV (see Fig. 2). Unacceptable fits were obtained with either a single Gaussian or single Lorentzian replacing the double Gaussian. Over 100 data runs of several hours duration were taken varying the gas pressure, magnetic field and positron energy to look for systematic effects. Only the incoming beam energy affected the width of the broad singlet component. The results for several different incoming energies were extrapolated to infinite energy (simulating the case of high energy positrons in the Galaxy) using a diffusion model peculiar to our apparatus.[1] The extrapolation yielded 6.4 ± 0.1 keV FWHM in H_2 and 7.7 ± 0.2 keV FWHM in He for the singlet Ps annihilation widths. This is in excellent agreement with the theoretical calculation of 6.5 keV for H and H_2 by Bussard, Ramaty and Drachman.[6]

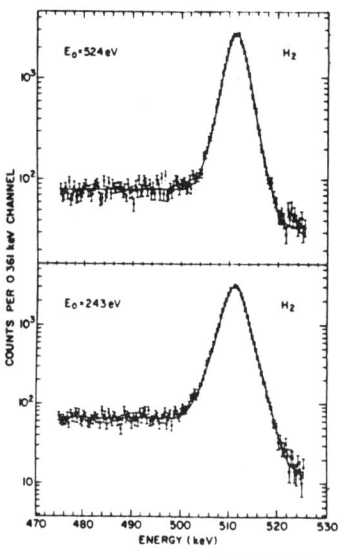

Fig. 2. Typical spectra obtained with positrons in H_2. The solid line represents a least-squares fit as discussed in the text. Typically the ratio of wide to narrow Gaussian intensities was 1 to 1.

POSITRONIUM FRACTION: SIMULATION

A pulsed beam technique was used for the measurement of the fraction of positrons that survive below the Ps formation threshold (the complement of f). The beam was gated on for 1 µs and off for 500 µs with the $\vec{E} \times \vec{B}$ region biased to a reflective potential when the beam is off. This confines the positrons to regions 1-6 until an inelastic collision is made in gas region 2. The round trip travel time from 1 to 7 is typically 1 µs. Successive collisions between the trapped positrons in region 2 and the gas result in most of the positrons forming Ps with a small fraction falling below the Ps threshold. After delay time t_d an annihilation plate in region 1 (replacing the grid) is quickly biased to an attractive potential. The surviving positrons are detected via the 511 keV annihilation gamma ray. The time t_d is much greater than the average Ps formation time in the gas and the detected positrons are thus below the Ps threshold (see Fig. 3). The cycle is repeated as t_d is varied up to $\sim 10^3$ μs. For $t_d \geq 10^2$ μs annihilation on the baffles becomes appreciable. No energy or pressure dependence was seen in the H_2 result for energies ≥ 50 eV and pressures between 0.8 and 5.0 mTorr. The surviving fraction was measured to be 10.3 ± 0.3% in poor agreement with the theoretical value[6] of 6.5%. It is believed that the disagreement arises from inaccurate values of ionization and excitation cross sections used in the Monte Carlo calculations.

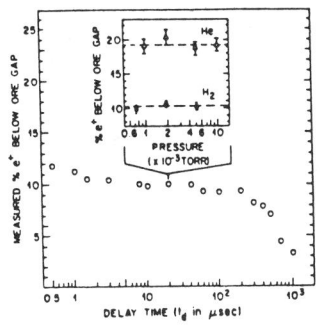

Fig. 3. Fraction of incident positrons that survive below the Ps formation threshold is shown for various delay times.

DIRECT ANNIHILATION

The direct annihilation line was studied by injection of positrons into the trap at energies below the Ps formation threshold. The basic inelastic process which resulted in trapping was vibrational excitation of the H_2 molecule (minimum energy loss 0.55 eV) although rotational excitation and elastic scattering from H_2 play an important role in the eventual thermalization of the positrons before annihilation. Once in the trap, only direct annihilation with H_2 or annihilation with the walls is possible. Because direct annihilation is a much less efficient process than Ps formation, some simple modification of the experiment was necessary. The gas baffles between regions 1-2 and 2-3 were removed to increase the containment time and the gas pressure was raised (5 to 200 mTorr) to increase the direct annihilation rate. At a pressure of 10 mTorr the confinement time and direct annihilation time were about equal (~13 ms). Data were accumulated in a pulsed mode with the beam pulsed on for 1 ms and off for 1 ms. A high purity Ge detector (1.08 keV FWHM resolution at 511 keV) was enabled only during the off cycle, after an initial delay of 100 μs. The pulsing eliminated the background from the prompt annihilation of the injected positrons on gas baffles between region 3 and 7, while the delay allowed time for the positrons to nearly thermalize before the detector was enabled.

A typical example of the data at 15 mTorr is shown in Fig. 4. As in the case of in-flight annihilation, good least squares fits ($\chi^2/\nu \approx 1$) were obtained with the sum of two Gaussian functions centered at 511 keV (one for direct annihilation and one for wall annihilations), a constant function below 511 keV for triplet Ps that could be formed on the chamber walls and be present in the instrumental background and a constant room background measured with the beam off all folded with the detector resolution. The fitting yielded a value of 1.56 ± 0.09 keV FWHM for the direct annihilation linewidth and was insensitive to pressure variation below 200 mTorr and energy variation below the Ps formation threshold. This result is in good agreement with a calculation made by Darewych[7] using a one-state approximation. His work, using a simple positron wave function, implies a rather Gaussian shape with a linewidth of 1.6 keV FWHM.

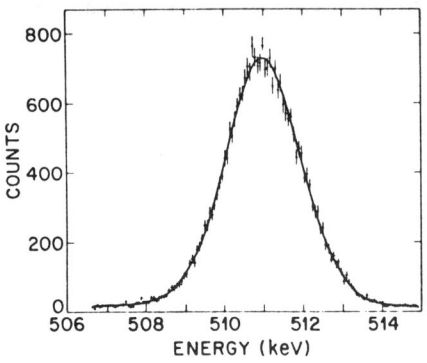

Fig. 4. Direct annihilation data obtained with a high purity Ge detector. The solid line shows a fit to the data with the model described in the text.

COMPOSITE SPECTRUM

The experimentally verified composite annihilation spectrum for positrons in a neutral H_2 medium is shown in Fig. 5 in correct proportion. Curve A is the direct annihilation line, B is the singlet Ps in-flight annihilation line and C is the triplet Ps in-flight annihilation continuum Doppler rounded with a 6.4 keV Gaussian. (Doppler profiles of singlet and triplet Ps should be identical). The composite line, for H_2, (A+B+C) has a FWHM of 2.2 keV. The corresponding composite spectrum for neutral H is expected to be almost identical. This composite spectrum has been successfully fit[4] to the highest quality Galactic center annihilation line date indicating that a cold neutral stopping medium should be considered.

POSITRONIUM FRACTION: OBSERVATIONS

Previous papers discussing positron annihilation data from the Galactic center have used several incompatible definitions of f. With f defined as the fraction of incident positrons that decay via Ps formation a simple expression can be derived relating f to the observed quantities. Ps can be formed by charge exchange with a gas atom or molecule or by radiative recombination with free electrons. In either case it will be formed in the singlet (para-Ps) state 1/4 of the time and in the triplet (ortho-Ps) state 3/4 of the time because of the quantum degeneracy. The decay from the singlet state produces two gamma rays of 511 keV energy, whereas the triplet state decay produces a "three gamma" continuum[8] which extends from 0 to 511 keV. If I_{e^+} is the number of incident positrons, then the number of annihilation-line photons $I_{2\gamma}$ is

$$I_{2\gamma}=[\frac{1}{4}(2f)+2(1-f)]I_{e^+}. \qquad (1)$$

The number of photons in the triplet-state continuum $I_{3\gamma}$ is

$$I_{3\gamma}=(\frac{3}{4})3fI_{e^+}. \qquad (2)$$

We can write f in terms of observable quantities:

$$f=\frac{4I_{3\gamma}}{4.5I_{2\gamma}+3I_{3\gamma}} \qquad (3)$$

where $I_{3\gamma}$ and $I_{2\gamma}$ are determined from the gamma-ray spectrum. If triplet/singlet conversion is present, equation (3) yields an "apparent" Ps fraction.[5]

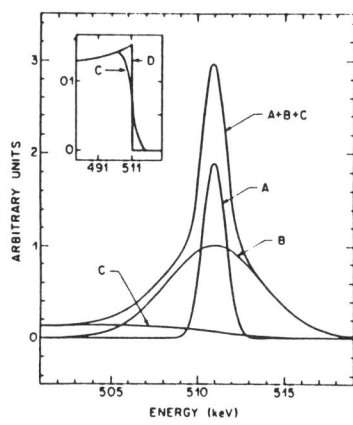

Fig. 5. Composite spectrum for positrons annihilating in a neutral H_2 medium based completely on the results of simulation experiments. The insert shows the ortho-Ps continuum without Doppler broadening.

The results of four measurements of a Ps signal from the Galactic center region are shown in Fig. 6 for the Ps fraction definition f in equation (3) and appear to call for f values consistent with a cold neutral medium and other models such as an ionized medium or one where dust plays a major role. However it is important to point out that the triplet intensities, and therefore the Ps fractions, derived from the observations depend strongly on the assumed shape of the underlying continuum spectrum. The Bell/Sandia and UNH values were based on a three-component model: the 511 keV line, the triplet continuum, and an underlying power-law continuum. The 1979 and 1980 JPL values (plotted with solid lines in Fig. 6) were derived with the use of a four-component model which includes the components of the three-component model plus a high-energy Comptonized thermal emission (hot plasma) spectrum with a peak near 700 keV.[9] Only the 1979 data, however, clearly show evidence for a high-energy component above 511 keV. The JPL 1980 data were also fitted with the three-component model used by Bell/Sandia and UNH since the data points above the 511 keV peak did not call for a high-energy component. The results of this fit are given in Fig. 6 with a dashed line. Others have suggested more sophisticated models that fit the data with very little or no positronium component at all.[10,11] It seems rather suggestive to us that the simplest and most straightforward interpretation of the data yields f values consistent with expectations.

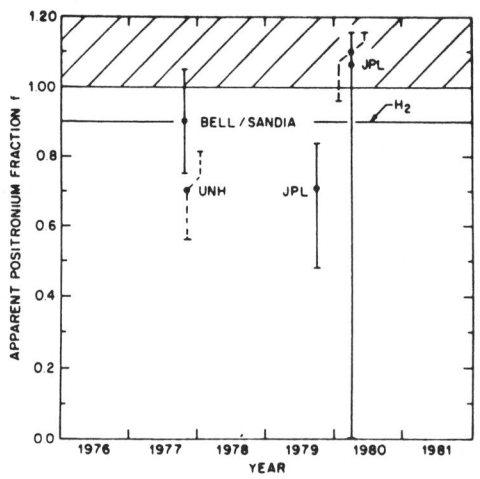

Fig. 6. The apparent Ps fraction f is shown for four different observations of the Galactic center. Values greater than f=1 are nonphysical.

REFERENCES

1. B. L. Brown, M. Leventhal, A. P. Mills, Jr., and D. W. Gidley, Phys. Rev. Lett. 53, 2347 (1984).

2. B. L. Brown, M. Leventhal, and A. P. Mills, Jr., Phys. Rev. A 33, 2281 (1986).

3. B. L. Brown, and M. Leventhal, Phys. Rev. Lett. 57, 1651 (1986).

4. B. L. Brown, Astrophys. J. 292, L67 (1985).

5. B. L. Brown and M. Leventhal, Astrophys. J. 319, 637 (1987).

6. R. W. Bussard, R. Ramaty and R. J. Brachman, Astrophys. J. 228, 928 (1979).

7. W. Darewych, Can. J. Phys. 57, 1027 (1979).

8. A. Ore and J. A. Powell, Phys. Rev. 75, 1696 (1949).

9. G. R. Riegler, J. C. Ling. W. A. Mahoney, W. A. Wheaton, and A. S. Jacobson, Astrophys. J. 294, L13 (1985).

10. R. E. Lingentfelter and R. Ramaty, in "Positron-Electron Pairs in Astrophysics," edited by M. L. Burns, A. K. Harding and R. Ramaty, p. 267 (A.I.P., New York, 1983).

11. L. Bildsten and W. H. Zureck, submitted to the Astrophys. J. (1987).

GAMMA-RAY AND NEUTRON SPECTROSCOPY OF PLANETARY SURFACES AND ATMOSPHERES*

Robert C. Reedy
Los Alamos National Laboratory, Los Alamos, NM 87545

ABSTRACT

The neutrons and gamma rays escaping from a planet can be used to map the concentrations of various elements in its surface. In a planet, the high-energy particles in the galactic cosmic rays induce a cascade of particles that includes many neutrons. The γ rays are made by the decay of the naturally-occurring radioelements and by nuclear excitations induced by cosmic-ray particles and their secondaries (especially neutron-capture or inelastic-scattering reactions). After a short history of planetary γ-ray and neutron spectroscopy, the γ-ray spectrometer and active neutron detection system planned for the Mars Observer Mission are presented. The results of laboratory experiments that simulate the cosmic-ray bombardments of planetary surfaces and the status of the theoretical calculations for the processes that make and transport neutrons and γ rays will be reviewed. Studies of Mars, including its atmosphere, are emphasized, as are new ideas, concepts, and problems that have arisen over the last decade, such as Doppler broadening and peaks from neutron scattering with germanium nuclei in a modern high-resolution γ-ray spectrometer.

INTRODUCTION

The energies and intensities of neutrons and γ rays escaping from a planet with very little or no atmosphere can be used to map the concentrations of various elements in the top few tens of centimeters of the surface. In the planet, the high-energy particles in the galactic cosmic rays (GCR) induce a cascade of particles that includes many neutrons.[1] The γ rays used for chemical mapping are made by the decay of the naturally-occurring radioactive elements (K, U, and Th) and by nuclear excitations induced by cosmic-ray particles (mainly by neutron-capture or inelastic-scattering reactions).[2,3] Certain elements, such as hydrogen, carbon, samarium, and gadolinium, can strongly affect the spectrum of neutrons in a planet, and thus can be sensed indirectly and their concentration-versus-depth profiles determined from neutron spectra[4,5] or γ rays[6,7] from other elements. The Earth's atmosphere is so thick (≈1000 g cm^{-2}) that few cosmic-ray particles reach the surface, and it also prevents γ rays made in the surface from traveling very far. Thus γ-ray spectroscopy on the Earth has been limited to low-flying searches for uranium and studies of the cosmic-ray-produced γ rays near the top of the atmosphere. Planetary γ-ray and neutron spectroscopy as considered here refer to objects with no or thin atmospheres, such as the Moon, Mars, asteroids, and comets.

Planetary γ-ray and neutron spectroscopies were both proposed around 1960 by Arnold,[8] Lingenfelter,[4] and others. However, planetary missions with

---

* This work was supported by NASA and done under the auspices of the U.S. Department of Energy.

such instruments have been very rare. On the Apollo 15 and 16 missions in 1971 and 1972, respectively, NaI(Tl) γ-ray spectrometers were flown, and spectra were accumulated over about 20% of the Moon's surface. Maps of iron, titanium, magnesium, and natural radioactivity were produced from the Apollo γ-ray data.[9,10] Most existing papers on planetary γ-ray spectroscopy date back to the Apollo era. Future missions will use advanced technologies for both γ-ray and neutron spectroscopy.

A high-purity-germanium γ-ray spectrometer is scheduled to be launched in the 1990s on the Mars Observer, which will be in a polar orbit, and others will probably be on lunar orbiters and Soviet martian orbiters. The Mars Observer Gamma-Ray Spectrometer (GRS) also will include instrumentation that can detect thermal and epithermal neutrons. A γ-ray spectrometer[11] is part of the penetrator that has been tentatively accepted for the proposed Comet Rendezvous Asteroid Flyby mission. The greatly improved detection capabilities (such as the high resolution for γ rays) and new targets (e.g., Mars with its thin atmosphere) have been changing our ideas for planetary spectroscopy considerably since the Apollo days. The Mars Observer GRS with its neutron-detection instrumentation is described below. Also discussed are the results of some simulation experiments and preliminary results for γ-ray and neutron calculations for Mars. The new instruments will produce significantly improved measurements, but they also require additional studies and calculations to anticipate possible complications arising from their greater sensitivities.

GAMMA-RAY AND NEUTRON DETECTION ON THE MARS OBSERVER MISSION

The proposed Mars Observer γ-ray detector will be a high-purity n-type germanium (hpGe) coaxial diode with a 56-mm diameter and a 56-mm length. It will be cooled to $\lesssim 100$ K by a passive radiator. The hpGe will be surrounded by a plastic scintillator, and the GRS's electronics will reject signals in the hpGe detector that are in coincidence with a signal in the plastic scintillator, eliminating background signals from the passage of charged cosmic rays through the hpGe detector. Signals from the hpGe for energies from ~0.2 to ~10 MeV will be processed in a pulse height analyzer. Below and above ≈ 2.4 MeV, the spectra will have ≈ 0.6 and 1.2 keV per channel, respectively. An entire γ-ray spectrum ($\approx 10,000$ channels) will be transmitted every ~20 seconds.

Thermal ($\lesssim 0.1$ eV) and epithermal ($\simeq 1$-1000 eV) neutrons will be detected using a boron-loaded plastic scintillator for the anti-coincidence shield. The $^{10}B(n,\alpha)^{7}Li$ reactions induced by these neutrons in the borated plastic will produce a unique signal in the scintillator's output.[12] Because the spacecraft moves at a velocity slightly faster (3.4 km/s) than that of a thermal neutron, the neutron count rates in each of the four faces of the anti-coincidence shield (which is pyramid shaped and fixed relative to the spacecraft's velocity) can be used as a Doppler filter to determine the fluxes and spectral shapes of thermal and epithermal neutrons.[13]

SIMULATION EXPERIMENTS

Several experiments have been done recently at accelerators to simulate the processes that produce γ rays in a planet's surface, and more are planned by the Mars Observer GRS team. In one series of irradiations, thick targets were bombarded with 6-GeV protons,[14] simulating the cascade of GCR particles in a large solid target. The spectra of γ rays measured in front of thick iron targets showed many narrow lines whose fluxes were in good agreement with theoretical calculations.[14]

As neutrons dominate the production of most γ rays,[2,3] another series of irradiations was done using neutrons from a 14-MeV neutron generator.[15,16] The concrete in the room around the neutron generator moderated many neutrons and produced neutrons with a continuum of energies from 14 MeV to thermal. The γ-ray spectrum from the irradiation of an aluminum target[16] is shown in Fig. 1. The relative fluxes of γ rays made by Fe(n,γ) reactions were in good agreement with calculated planetary γ-ray fluxes that only considered production by thermal neutrons.[3] Because the spectrum of neutrons in the simulation had an epithermal/thermal neutron ratio similar to that in the Moon, this agreement shows that thermal yields are good for calculating fluxes of neutron-capture-produced γ rays.[15]

Several aspects of the results for the simulations with $\leq$14-MeV neutrons were different from our experience with the low-resolution Apollo γ-ray data. As marked with diagonal lines in the top part of Fig. 1, five peaks with unusual shapes were observed. These peaks are shaped normally at their low-energy sides, and their energies correspond to those for the de-excitation of low-lying levels in germanium isotopes. The high-energy sides of these peaks extend for $\sim$50 keV, and are caused by the summing of the recoil energy from a Ge(n,n') reaction with the de-excitation γ ray.[15] Except for the peak at and above 834 keV, these sawtooth-shaped peaks made in a Ge detector should not interfere with the major γ rays expected from a planetary surface. The high-energy tail above 834 keV will be under the inelastic-scattering peaks from Al and Fe at 844 and 847 keV, respectively. Also marked in Fig. 1 is the peak at 4.438 MeV from the ^{12}C(n,nγ)^{12}C reaction (and also probably including some γ rays from the ^{16}O(n,n$\alpha\gamma$)^{12}C reaction), which has a width of 53 keV compared to the 5-keV width of an adjacent γ ray from Si. This Doppler broadening of the major carbon inelastic-scattering γ ray and the low cross section for the ^{12}C(n,γ) reaction will make the detection of carbon in a planetary surface by high-resolution γ-ray spectrometers difficult.

In my 1978 paper on the fluxes of γ rays expected from planetary surfaces,[3] I noted that cross sections for the production of γ rays by nonelastic-scattering reactions were often scarce. Usually the highest neutron energy used in measuring cross sections for the production of nonelastic-scattering γ rays is below 20 MeV, and often some energies have not been measured, such as from $\approx$6-13 or $>$15 MeV.[17] Few γ-ray-production cross sections have been measured for the proton energies of interest (hundreds of MeV to several GeV). Recently several irradiations have been done with high-energy ($\leq$78 MeV) neutrons[18] and more are planned to measure such cross sections. The lack of good cross sections for nonelastic-scattering reactions limits our ability to calculate the leakage fluxes of such γ rays, especially for those γ rays that are made by high-energy particles and interfere with inelastic-scattering γ rays. For example, the production of 1.369-MeV γ rays by the ^{28}Si(n,n$\alpha\gamma$)^{24}Mg reaction could strongly interfere with the signal from the ^{24}Mg(n,nγ)^{24}Mg reaction that is used to determine magnesium concentrations in a planetary body.[3]

Fig. 1. The γ-ray spectrum observed from aluminum irradiated with 0 to 14-MeV neutrons.[16] Most γ rays were produced in the concrete around the 14-MeV neutron generator and in the material (such as lead and borated paraffin) surrounding the Ge detector. Shaded are the five asymmetric Ge peaks from 596 to 1040 keV and the Doppler-broadened peak at 4.438 MeV from the $^{12}C(n,n\gamma)^{12}C$ reaction (and also probably the $^{16}O(n,n\alpha\gamma)^{12}C$ reaction).[15]

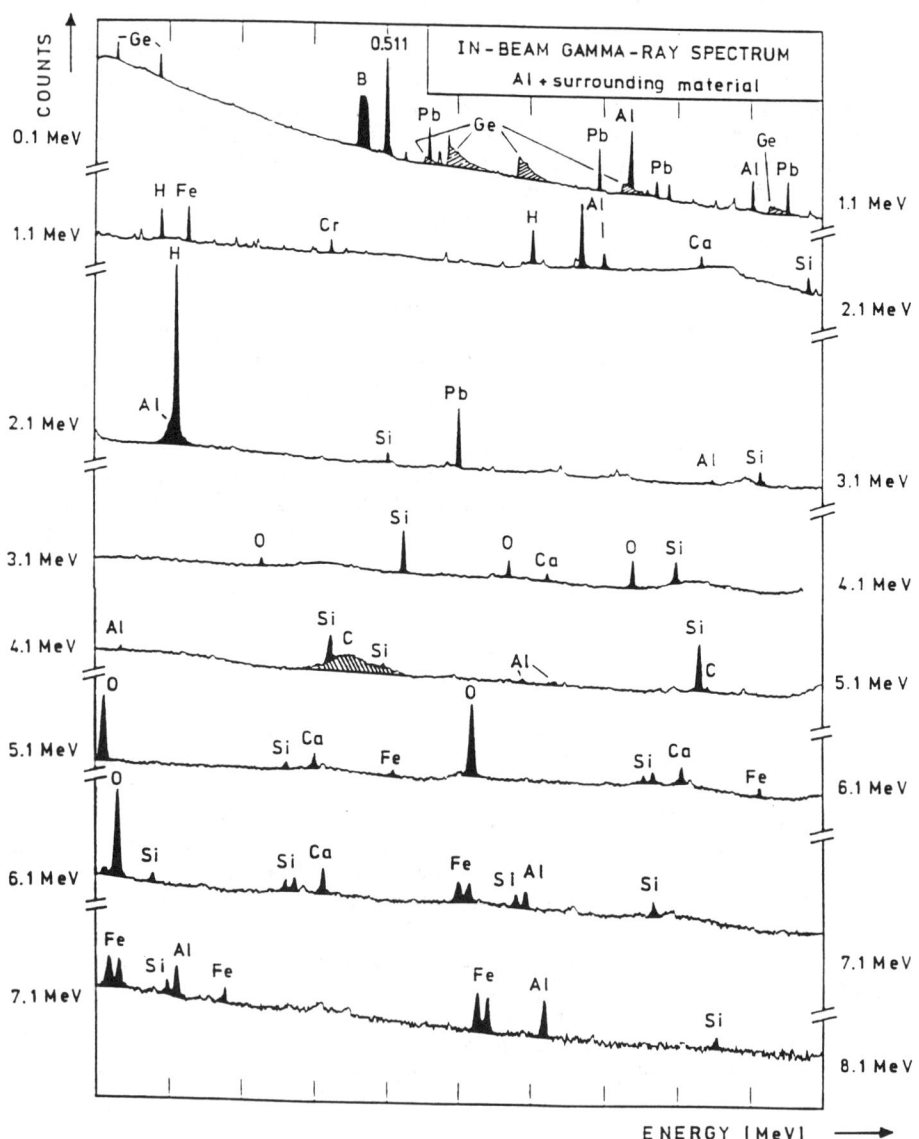

RECENT CALCULATIONS OF MARTIAN NEUTRON AND GAMMA-RAY LEAKAGE FLUXES

In planning for the Mars Observer Gamma-Ray Spectrometer experiment, calculations have recently been done for the production and transport of neutrons[5,19] and γ rays[7] in the martian surface and atmosphere. All of these calculations included a $\approx$16-g/cm^2-thick atmosphere (95.7% CO_2, 2.7% N_2, and 1.6% ^{40}Ar) and used a martian-surface composition estimated from chemical analyses by the Viking landers and of the "martian" (SNC) meteorites (the shergottites, nakhlites, and Chassigny).[5,7] Much of the emphasis in these calculations has been on the highly-variable amounts of volatiles (H_2O and CO_2) that can be present in or on the martian surface. The equilibrium distributions of neutrons in Mars were calculated using the ONEDANT[5,19] and the ANISN[7] neutron-transport codes. The ONEDANT code was modified to include the effects of gravity and the neutron's beta decay.[19] Neutrons that escape Mars with E $\leq$ 0.132 eV are gravitationally bound, although some neutrons beta decay before returning to the planet. Neutron-transport calculations done with and without the effects of gravity showed that gravity increased the flux of neutrons at the top of the martian atmosphere by $\approx$29%, but that the neutron-flux increase at the top of the soil due to gravity was only a few percent.[19]

These calculations[5,7,19] indicate that the martian atmosphere and the presence of H_2O in or CO_2 on the martian surface significantly affect the distributions of neutrons. Hydrogen rapidly thermalizes neutrons and shifts the peak of their depth distribution towards the surface. Because of its low absorption cross section, CO_2 builds a large reservoir of low-energy neutrons that can leak back into the surface.[5,19] The neutron count rates expected in the GRS's anti-coincidence shield are high enough to allow a rapid determination of the concentrations of H_2O and CO_2 in and on the surface from the observed fluxes and spectral shapes of the thermal and epithermal neutrons.[5] The depth that H_2O is below the surface can often be determined from the neutron[5] and γ-ray[7] data. Both the measured γ-ray and neutron leakage fluxes can be used together to get additional information on the concentration and stratigraphy of H_2O and CO_2 in the top meter of the martian surface.

The fluxes from the ANISN neutron-transport calculations were used to determine the production rates of γ rays by nonelastic-scattering and neutron-capture reactions.[7] The γ rays made by these reactions and by the natural decay of K, Th, U, and their daughters were transported through the martian surface and atmosphere to get fluxes at the spacecraft. The γ rays most suitable for detecting the expected major elements and the radioelements have strong enough leakage fluxes that they should be detectable with integration times of hours to several hundred hours.[20] The measurement of these elements should aid in the determination of the major rock types present and of the degree of local and global refractory enrichment. Major types of volcanic materials, especially in the southern highlands of Mars unsampled by Viking, should be identified. Readily detectable in γ-ray spectra will be S and Cl, which might be present in surface precipitates or subsurface brines. Besides elemental abundances, the γ-ray data can also be used to study the distribution of hydrogen and CO_2 in Mars by comparing γ rays made by both nonelastic-scattering and neutron-capture reactions with one element.[7]

STUDIES OF THE MARTIAN ATMOSPHERE

The martian atmosphere is very interesting in many ways. Its thickness varies over a martian year by a factor of ~2 as CO_2 frost is deposited and sublimed from the seasonal polar caps. The thickness of the atmosphere also varies with location, being the least over the huge ($\approx$27-km high) volcano Olympus Mons and the greatest above the $\approx$7-km deep canyons of Valles Marineris. These variations in the atmospheric thickness need to be considered in the neutron and γ-ray transport calculations. Nuclear interactions with the constituents of the martian atmosphere also could interfere with the γ-ray spectroscopy of carbon, nitrogen, oxygen, and potassium in the surface.[3] (The first three elements are in both locations, and potassium is assayed using the 1.461-MeV γ ray of ^{40}K, which is from the decay of the first excited level of ^{40}Ar.) Fortunately, most γ rays made by nonelastic-scattering reactions in the low density of the atmosphere should be Doppler broadened[21] and thus shouldn't interfere with the narrow γ-ray lines that are expected from most nonelastic-scattering reactions in the martian surface. A γ-ray line at 6.13 MeV with a width of $\lesssim$32 keV was seen in a high-resolution Ge(Li) spectrometer at the top of the Earth's atmosphere.[22] While much of this line could be from the decay of ^{16}N, it is also possible that it was due to the γ ray at that energy from the $^{16}O(n,n\gamma)^{16}O$ reaction being Doppler-broadened to ~30 keV.[23] Gamma rays made in the atmosphere by the decay of radionuclides (such as ^{16}N and ^{41}Ar) and by neutron-capture reactions (mainly with nitrogen and ^{40}Ar) are not expected to be broadened significantly.

The leakage fluxes of martian neutrons and γ rays also can be used to study the martian atmosphere. The martian atmospheric attenuates the intensities of γ rays from the martian surface,[24] especially for energies below ~1 MeV, and its thickness can be determined from the differences in the attenuation of several γ-ray lines with known relative intensities and very different energies, such as the 239 and 2615-keV γ rays in the thorium chain.[25,26] As the CO_2 in the martian atmosphere strongly affects the fluxes and spectra of leakage neutrons, the neutron data also can monitor atmospheric thicknesses, possibly even day-to-night variations as a function of season and location.[5] The thickness of the CO_2 deposits at the seasonal polar caps could also be determined as a function of time and location by these techniques. The γ rays made by neutron-capture reactions with ^{40}Ar (e.g., the 1294-keV line from the decay of ^{41}Ar) and nitrogen also could be used to study the martian atmosphere.

SUMMARY

Future planetary missions to the terrestrial planets and to small bodies (comets and asteroids) will have as one of their major objectives the determination of their chemical compositions. Gamma-ray and neutron spectroscopies are excellent methods for orbital or in-situ chemical studies of these objects. Such instruments are scheduled to fly on the Mars Observer Mission and have been tentatively accepted for comet penetrators. Our ideas for these missions have changed considerably since the days of the Apollo missions with their NaI(Tl) γ-ray spectrometers. New detectors (e.g., high-purity germanium) and techniques (Doppler-filter neutron spectroscopy) are available, and the new targets are different in many ways from the Moon (atmospheres and volatiles). As discussed above, much work, including simulation experiments and theoretical calculations, is being done in planning for upcoming missions.

Several problems have been identified with these future γ-ray spectroscopy experiments. The hpGe detectors can be fairly easily damaged by cosmic radiation. Experiments to understand how and when such radiation damage occurs are being done with the goal of minimizing such effects. The high resolution of hpGe γ-ray spectrometers increases our ability to measure concentrations of most elements[27,28] but means that we must be careful of effects such as Doppler broadening and interferences to major γ-ray lines from other sources. The laboratory simulations and theoretical calculations are important, especially now that we will be going to objects for which we have no "ground truth," which we had at the Moon, to normalize our measurements.[9,29]

The data obtained from both the γ-ray and neutron modes of the Mars Observer GRS will complement each other, and their use together will considerably improve the scientific return. For example, the elemental results from the γ-ray spectra are needed to help interpret the transport of the leakage neutrons. As neutrons are the major source of most γ rays, direct measurement of the neutron leakage flux can aid in interpreting the γ-ray data. The measured neutron and γ-ray fluxes also can help to infer the presence of certain elements not directly observed in the γ-ray spectra, such as relatively high amounts of neutron-absorbing gadolinium and samarium. The intensity of the leakage thermal neutrons can be used with the flux ratio for the neutron-capture γ rays from hydrogen and a major element like Si or Fe to determine the concentration and stratigraphy of H_2O in the top ~ 100 g cm^{-2} of the martian surface or the thickness of CO_2 above the martian surface. Such studies of volatiles will be very important not only for studies of Mars but also for comets and possibly for asteroids and the polar regions of the Moon.

ACKNOWLEDGMENTS

Most of the results presented here represent the efforts of the members of the Mars Observer Gamma-Ray Spectrometer team, which includes, besides myself, J. R. Arnold, J. Brückner, W. V. Boynton, D. M. Drake, P. Englert, L. G. Evans, W. C. Feldman, E. L. Haines, A. E. Metzger, S. W. Squyres, J. I. Trombka, and H. Wänke. I also wish to thank R. E. Lingenfelter for valuable discussions on leakage neutrons and Doppler-broadening effects in the martian atmosphere and for his comments on this paper.

REFERENCES

1. R. C. Reedy and J. R. Arnold, J. Geophys. Res. 77, 537 (1972).
2. R. C. Reedy, J. R. Arnold, and J. I. Trombka, J. Geophys. Res. 78, 5847 (1973).
3. R. C. Reedy, Proceedings of the 9th Lunar and Planetary Science Conference (Pergamon, N. Y., 1978), p. 2961; also Gamma Ray Spectroscopy in Astrophysics, NASA-TM-79619, p. 98 (1978).
4. R. E. Lingenfelter, E. H. Canfield, and W. N. Hess, J. Geophys. Res. 66, 2665 (1961).
5. D. M. Drake, W. C. Feldman, and B. M. Jakosky, J. Geophys. Res., submitted.

6. J. R. Lapides, Planetary Gamma-Ray Spectroscopy: The Effects of Hydrogen and the Macroscopic Thermal-Neutron Absorption Cross Section on the Gamma-Ray Spectrum, (Ph.D. Thesis, University of Maryland, College Park, 1981).
7. L. G. Evans and S. W. Squyres, J. Geophys. Res. 92, 9153 (1987).
8. M. A. Van Dilla, E. C. Anderson, A. E. Metzger, and R. L. Schuch, IRE Trans. Nucl. Sci. NS-9, 405 (1962).
9. M. J. Bielefeld, R. C. Reedy, A. E. Metzger, J. I. Trombka, and J. R. Arnold, Proceedings of the 7th Lunar Science Conference (Pergamon, N. Y., 1976), p. 2661.
10. M. I. Etchegaray-Ramirez, A. E. Metzger, E. L. Haines, and B. R. Hawke, J. Geophys. Res. 88, A529 (1983).
11. L. G. Evans, J. I. Trombka, and W. V. Boynton, J. Geophys. Res. 91, D525 (1986).
12. D. M. Drake, W. C. Feldman, and C. Hurlbut, Nucl. Instrum. & Methods A247, 576 (1986).
13. W. C. Feldman and D. M. Drake, Nucl. Instrum. & Methods A245, 182 (1986).
14. A. E. Metzger, R. H. Parker, and J. Yellin, J. Geophys. Res. 91, D495 (1986).
15. J. Brückner, H. Wänke, and R. C. Reedy, J. Geophys. Res. 92, E603 (1987).
16. J. Brückner, Neutronen-Induzierte Gamma-Spektroskopie: Beitrag zur Chemischen Fernerkundung von Planetaren Oberflächen (Doctoral Dissertation, Johannes Gutenberg-Universität, Mainz, FRG, 1984).
17. R. C. Reedy, Radiat. Eff. 94, 259 (1986).
18. P. Englert, J. Brückner, and H. Wänke, J. Radioanal. Nucl. Chem., Articles 112, 11 (1987).
19. W. C. Feldman, D. M. Drake, R. D. O'Dell, F. W. Brinkley, Jr., and R. C. Anderson, J. Geophys. Res., submitted.
20. L. G. Evans, priv. comm. (1987).
21. R. Ramaty, B. Kozlovsky, and R. E. Lingenfelter, Astrophys. J. Suppl. Ser. 40, 487 (1979).
22. J. B. Willett, J. C. Ling, W. A. Mahoney, and A. S. Jacobson, Astrophys. J. 234, 753 (1979).
23. A. E. Metzger, priv. comm. (1987).
24. A. E. Metzger and J. R. Arnold, Appl. Opt. 9, 1289 (1970).
25. A. E. Metzger, Bull. Am. Astron. Soc. 16, 678 (1984).
26. A. E. Metzger, J. R. Arnold, R. C. Reedy, J. I. Trombka, and E. L. Haines, Lunar and Planetary Science XVII (Lunar and Planetary Institute, Houston), p. 549 (1986).
27. A. E. Metzger, R. H. Parker, J. R. Arnold, R. C. Reedy, and J. I. Trombka, Proceedings of the 6th Lunar Science Conference (Pergamon, N. Y., 1975), p. 2769.
28. E. L. Haines, J. R. Arnold, and A. E. Metzger, IEEE Trans. Geosci. Electron. GE-14, 141 (1976).
29. A. E. Metzger, E. L. Haines, R. E. Parker, and R. G. Radocinski, Proceedings of the 8th Lunar Science Conference (Pergamon, N. Y., 1977), p. 949.

OBSERVATIONS of NUCLEAR EMISSIONS in SOLAR FLARES

David J. Forrest
Space Science Center and Physics Department
University of New Hampshire, Durham, N. H. 03824

Abstract

The observed energy loss spectra obtained with a gamma-ray detector during a solar flare can consist of a admixture of counts from primary electron bremsstrahlung, broad and narrow nuclear lines, gamma-rays from nuclear pion decay as well as energetic neutrons which survive to 1 AU. The details of each of these spectral components depend in a different way on the properties of both the accelerated particle species and the interaction or target region. In order to extract this information the observed energy loss spectra must be separated into its several components. We present samples of solar flare data from the Gamma Ray Spectrometer on SMM and describe the initial experimental procedures used to separate these components. These experimental unfolding and interpretive procedures along with theoretical modeling of these results have allowed the GRS data to be used to show that both electrons and ions are impulsively accelerated early in solar flares and that this acceleration process can produce electrons >100 MeV and ions >1000 MeV on time scales of 10 seconds.

Introduction

Gamma-rays and neutrons interact in a detector via a number of physical mechanisms. The resultant observed energy loss spectrum is not unique signature of the spectrum of the incident radiation in the sense that an observed count at a energy E_l, can be be produced by a broad range of incident photon or neutron energies. This property, common to all gamma-ray and neutron detectors, requires spectral testing or unfolding of the observed energy loss data before details of the incident physical flux can be determined. This testing or unfolding can be accomplished from two different perspectives; 1. testing of an assumed spectral form with the data for consistency or 2. a direct unfolding or inversion of the observed data to produce the incident physical flux. Up to the present time only the first procedure has been used to interpret the GRS flare data. The basic approach is to first test the simplest physical model, that is one with a minimum number of adjustable parameters. More complex models having several components are introduced only if required.

There are two separate sensors on the Gamma Ray Spectrometer [1] which jointly cover the energy loss band 0.3 to >100 MeV. The Main Channel Spectrometer (MCS) records the energy loss spectra 0.3 - 9 MeV with moderate energy resolution and relatively high sensitivity while the High Energy Monitor (HEM) covers the energy loss band 10 - >100 MeV with high sensitivity but poorer energy resolution. In the eight years since its launch the GRS has detected 156 solar flares above its threshold of 0.3 MeV and out of these 90 are large enough to allow a spectral determination.

The HEM on SMM has also detected more than 12 flares >10 MeV and at least two have clear evidence for nuclear pion and/or neutrons, requiring energetic ions >500 MeV.

In this paper we show examples of these observations and present some of the unfolding procedures and interpretations used as evidence for more than one emission mechanism in solar flares. This procedure is not spectroscopy of narrow gamma-ray lines but is spectroscopy of broader energy emissions which requires a different type of knowledge of the instrument's response properties.

Instrument Properties

Spectroscopy requires knowledge of many instrument parameters. Probably the most important of these is the instrument's "effective area", which gives the recorded response per unit incident flux. Figure 1 shows the "effective area" of the two GRS sensors for gamma-rays in the energy band 0.2 - 200 MeV. The effective area for the main channel spectrometer is displayed as two curves, S_t which can be interpreted as the total recorded counts per unit flux and S_{pp} which gives the counts in the photo-peak portion of the total effective area. S_{pp} goes to zero near the upper threshold of the recorded energy loss spectrum, however S_t remains finite for higher energy gamma-rays because of the continuum portion of the instrument response. The curves in Figure 1a are from the UNH GRS Response Ver 3.1 and are based on Monte-Carlo calculation for a 7.6x7.6 cm NaI [2] and preflight calibration data with the GRS instrument.

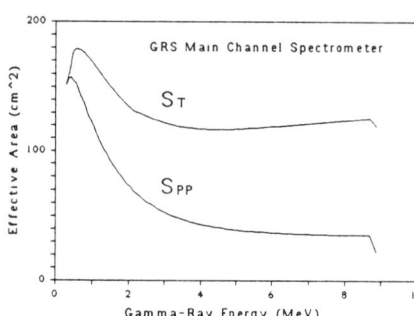
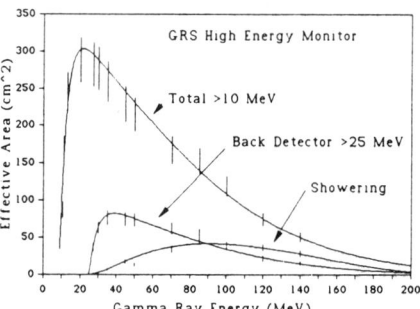

Fig. 1a. MCS Effective Area   Fig. 1b. HEM Effective Area

The recorded output of the HEM consists of an array of 19 numbers containing energy loss distribution in both the front and back portions of the detector. These elements can be combined in a number of different ways to determine both the type and spectra of the input radiation. In figure 1b three of these combinations are shown for incident gamma-rays. The top curve is the total "effective area" within the 10 - 100 MeV energy loss band for gamma-ray in the energy band 10 - 200 MeV. The two lower curves show the gamma-ray response of the large back CsI detector and the response for "showering" gamma-rays which deposit energy losses in both the front and back detectors. The vertical lines give the $\pm 1\sigma$ extent of the Monte-Carlo [3] results and the smooth lines are a functional fit to these results. Note that only gamma-rays above 50 MeV are effective at producing showering events. For reference I note that the maximum "effective area" for neutrons is 55 cm^2 at E_n = 300 MeV and that neutrons do not produce showering events [3,4].

Electron Bremsstrahlung and Nuclear Line Emissions

Before the launch of SMM it was commonly held that only rare and relatively larger flares would accelerate ions [5]. It was also felt that the yield of gamma-rays would be correlated with the number of escaping Solar Energetic Particles (SEP) and that these gamma-rays would be observed well after the impulsive phase of the flare as observed in hard X-ray bremsstrahlung. Some of the first observations by SMM in June 1980 showed this picture was not always true. The impulsive and short duration flare observed at 03:12 UT on 7 June 1980 had clear evidence for the 2.2 MeV gamma-ray line [6] and the spectrum from this flare also showed the spectral hardening at energies >1 MeV characteristic of nuclear line production [7,8].

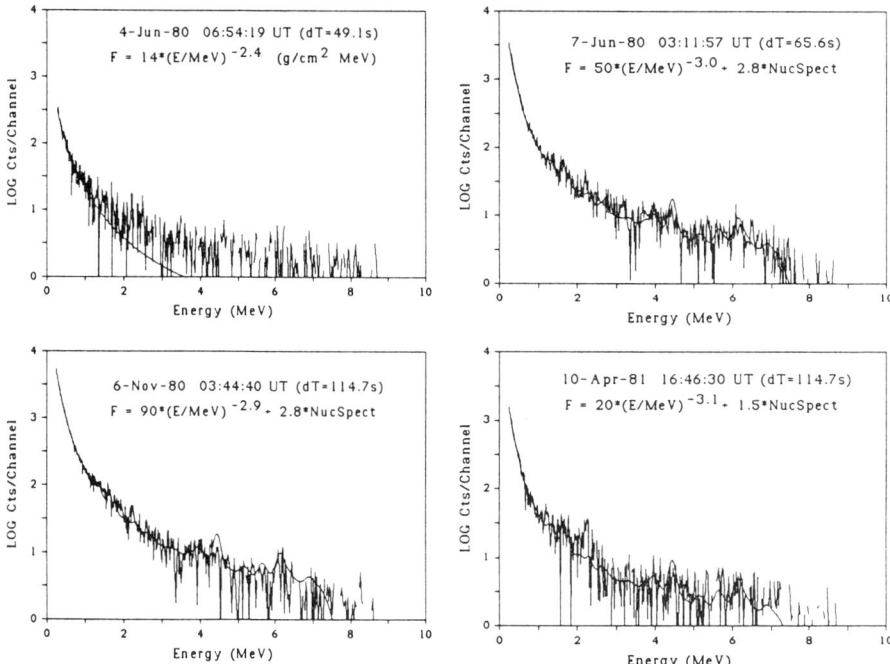

Fig. 2. The observed energy loss spectra from four sample flares. The first spectrum has been fit with just a power law model spectrum. The remaining flares have been fit with both a power law and a nuclear spectra.

Figure 2 shows the observed energy loss spectra from four sample flares. The smallest of these which also has the hardest continuum, shown in the upper left panel, has been fit with a power law continuum between 0.3 - 1 MeV. It is clear that another spectral component is needed at energies >1MeV. The remaining panels show three other flares, each fit over the full energy range with both a power law photon spectrum and a nuclear line spectrum. The nuclear spectrum used here is the spectrum determined from a detailed fit to another large GRS flare by Murphy et al. [9]. For demonstration purposes here, this spectrum has been renormalized to a unit flux in the 4.43 MeV line from ^{12}C. It is important to note that the model used to fit these three flares has only three adjustable parameters, the intensity and shape of the power law continuum and the intensity of the full nuclear spectrum. Note that spectral fits to the delayed lines at 0.511 and 2.2 MeV have not been included. Spectral test like this have been used to show that all GRS flares require a nuclear component provided the flare is large enough to be tested [10].

Nuclear Pion Gamma Rays and High Energy Neutrons

The flare observed at 01:13 UT on 21 June 1980 was the first observed with clear evidence for gamma-rays beyond 50 MeV. This flare also had a delayed signal observed at energies >10 MeV that was only consistent with the arrival at earth of solar neutrons with energies up to 500 MeV [11]. This observation was followed by an even larger flare on 3 June 1982, where the acceleration of energetic ions occurred in both the impulsive phase and a new extended phase. This extended production of both pion gamma-rays [12] and neutrons [4] resulted in a complex mixture of recorded counts from all of these emissions over a time interval >1000 seconds.

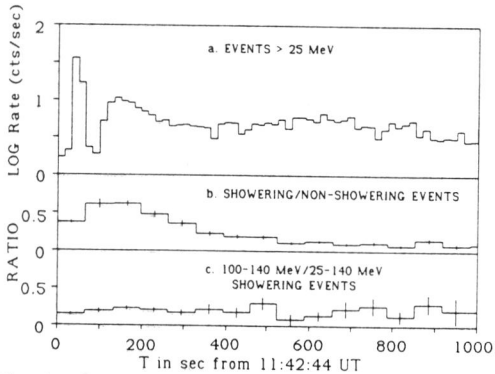

Fig. 3. Observed HEM rates and rate ratios from the flare on 3 June 1982. The top panel has a time resolution of 16 sec, the bottom two 65 sec.

In Figure 3 we show some of the information recorded in the HEM during the flare observed on 3 June 1982. Panel a. shows the total observed rate >25 MeV over a 1000 sec interval including the impulsive phase of the flare. Panel b. shows the ratio of showering events to all the events observed >25 MeV in the back CsI detector. The fact that this ratio is statistically greater than zero is clear evidence that gamma-rays >50 MeV were being produced over the entire time interval.

However the fact that this ratio is not constant in time shows that either the gamma-ray spectrum was changing or that a changing spectrum of energetic energetic neutrons was arriving at the instrument. Panel c. further investigates the high energy gamma-ray spectrum by displaying the ratio of showering events >100 MeV to those >25 MeV. This shows that the gamma-ray spectrum >50 MeV was very hard, extending above 150 MeV and that its spectral shape did not change throughout the impulsive and extended portions of the flare.

Figure 4 displays the unfolded gamma-ray spectrum from the 2nd and 3rd 65 sec intervals as displayed in Figure 3. The photon model used to unfold the observed energy loss distribution included both the direct neutral pion decay gamma-rays and bremsstrahlung from the charged pion decay positrons [13].

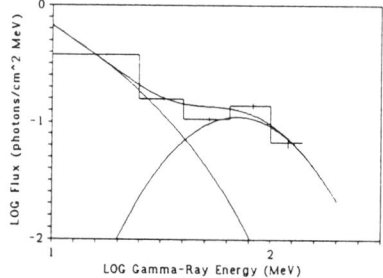

Fig. 4. The HEM gamma-ray spectrum from 11:44:55 - 47:06 UT on 3 June 1982.

This spectra and the data in Figure 3 are compelling evidence that the photon component of the signal recorded during the extended portion of this flare is due to gamma-rays from pion decay. This requires ions >500 MeV. It further tells us that energetic neutrons which are produced on the sun with these pions must be arriving at the instrument.

The total HEM rate >10 MeV together with the contributions to the total rate from the three different components is shown in Figure 5. This separation was calculated by normalizing the counting rate from the pion gamma-ray spectrum shown in Figure 4 to the observed rate >100 MeV. This contribution (shown as curve b) was subtracted from the total observed rate (curve a). During the impulsive phase the remaining events had a steep spectral distribution which is only consistent with a power law primary electron bremsstrahlung spectrum (curve c).

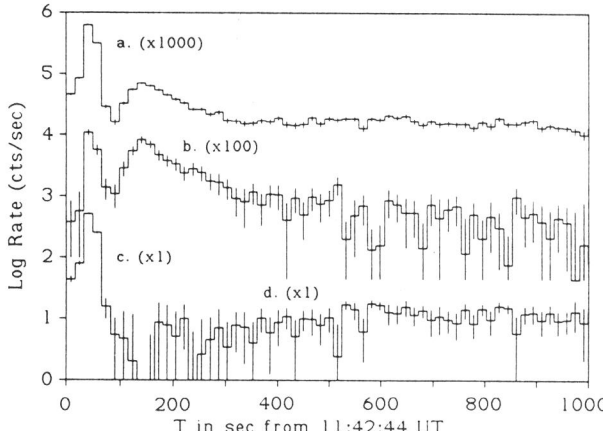

Fig. 5. The total observed HEM rate and it's components from pion gamma-rays, electron bremsstrahlung and energetic neutrons.

Energetic neutrons produced along with the pions at the sun do not arrive at the instrument until later. The gradual decrease of the ratio shown in Figure 3 panel b is caused by the increasing flux of these energetic neutrons at the instrument. It is important to note that during the extended portion of the flare the photon component of the observed rate is entirely accounted for by pion gamma-rays.

Summary

I have reviewed some of GRS response characteristics and presented sample flare observations where these properties have been used to separate the primary electron bremsstrahlung and the nuclear line emission in the MCS. I have also demonstrated how the properties of the HEM are used to separate the primary electron bremsstrahlung, pion decay gamma-rays and direct high energy neutrons at higher energies. This ability to identify and separate these components has been fundamental to a number of new discoveries in the physics of solar flares. Among these are the prompt and simultaneous acceleration of both ions and electrons in the impulsive phase of all or most flares [10,14], the heliocentric spectral dependence of the electron bremsstrahlung in flares showing that this emission is not isotropic [15], the presence of a extremely hard ion rich extended phase in some flares [4,12] and the correlation or lack thereof of trapped gamma-ray producing ions and escaping ions observed in space [16]. Beyond these broad band spectral analyses, more detailed spectral studies of these same data are being used to determine the elemental abundances in the particle interaction region [17], the meaning of the unexpected high ratio of the scattered to narrow line components of the 2.2 MeV emission from captured neutrons in the photosphere [18] and the use of this same line to determine the 3He abundances in the photosphere [19].

Spectroscopy has been fundamental to our efforts to understand the complex flare phenomena. The intense and complex gamma-ray spectra from these solar flares severely tests our spectrum "unfolding" and interpretative process. As instrumentation improves, this aspect of Gamma-Ray Spectroscopy will become more important. In particular the non-unique aspects and other limitations of our current "unfolding" procedures, mainly by testing the observations with model spectra should be supplemented with more direct and model independent procedures.

Acknowledgments

The author acknowledges contributions from his colleagues at UNH, NRL and MPI. This work was supported at UNH by NASA Contract NAS 5-28609 and grant NAG5-720.

References

1. D. J. Forrest et al., Solar Phys. 65, 15 (1980).
2. M. J. Berger and S. M. Seltzer, Nuc. Inst. and Meth. 104, 317 (1972).
3. J. F. Cooper, et al., 19th ICRC, 5, 474 (1985).
4. E. L. Chupp, et al., Ap. J. 318, 913-925 (1987).
5. J. C. Brown and D. F. Smith, Rep. Prog. Phys., 43, 125-197, (1980).
6. E. L. Chupp, et al., Ap. J. 244, L171 (1981).
7. D. J. Forrest, et al., 17th ICRC, 10, 5 (1981).
8. R. Ramaty, B. Kozlovsky and A. N. Suri, Ap. J., 214, 617 (1977).
9. R. R. Murphy, et al., 19th ICRC, 4, 253 (1985).
10. D. J. Forrest, In Positron-Electron Pairs in Astrophysics, ed: M. L. Burns, A. K. Harding and R. Ramaty, (Amer. Inst. Phys., New York, 1983), p.3.
11. E. L. Chupp, et al., Ap. J., 263, L95 (1982).
12. D. J. Forrest, et al., Adv. Space Res., 6, 115 (1986).
13. R. J. Murphy, C. D. Dermer and R. R., Ap.J. Supp. Ser., 63, 721-748 (1987).
14. D. J. Forrest and E. L. Chupp, Nature, 305, 291 (1983).
15. W. T. Vestrand, et al., Ap. J., 322, 1010 (1987).
16. E. W. Cliver, et al., 20th ICRC, 3, 61 (1987).
17. R. J. Murphy, These Proceedings, (1988).
18. X. -M. Hua and R. E. Lingenfelter, Ap. J., 319, 444 (1987).
19. R. Ramaty, X. M. Hua and R. E. Lingenfelter, These Proceedings (1988).

MODELS OF GAMMA-RAY PRODUCTION IN SOLAR FLARES

R. Ramaty
Laboratory for High Energy Astrophysics
Goddard Space Flight Center, Greenbelt, MD 20771

J. A. Miller and X.-M. Hua
Department of Physics and Astronomy
University of Maryland, College Park, MD 20742

R. E. Lingenfelter
Center for Astrophysics and Space Sciences
University of California, San Diego, CA 92093

ABSTRACT

We review gamma-ray production models in solar flares and study the transport of ions and relativistic electrons by employing a Monte-Carlo simulation that follows the individual particles throughout a flare loop. We consider energy losses, magnetic mirroring, MHD pitch-angle scattering, turbulence damping, Coulomb scattering, and drifts. For an isotropic injection of particles, MHD pitch-angle scattering in the corona could account for the rapid variability of the observed gamma-ray time profiles. Owing to damping by neutral hydrogen, such pitch-angle scattering is probably negligible in the chromosphere. Therefore, magnetic mirroring in convergent flux tubes in the chromosphere and upper photosphere is probably the dominant mechanism for producing both the anisotropies suggested by the preferential detection of >10 MeV emission from flares near the solar limb and ultrarelativistic electron bremsstrahlung observed from flares on the disc.

INTRODUCTION

The observation[1] of gamma-ray lines and continuum from many solar flares has firmly established gamma-ray astronomy as a tool for studying the active Sun[2]. Gamma-ray line emission from flares results from interactions of accelerated ions with the ambient solar atmosphere. The strongest lines are at 2.223 MeV from neutron capture on ^{1}H; 0.511 MeV from positron annihilation; and 6.129, 4.438, 1.634, 1.369, and 0.847 MeV from ^{16}O, ^{12}C, ^{20}Ne, ^{24}Mg and ^{56}Fe deexcitations, respectively. Neutrons[3-5] and gamma rays resulting from pion decay[6,7] have also been observed. Ion acceleration is invariably accompanied by the acceleration of relativistic electrons. Bremsstrahlung from these electrons is the principal source of the observed[6] gamma-ray continuum from flares. Hard X-ray bremsstrahlung from accelerated nonrelativistic electrons is also observed[8].

The mechanisms of gamma-ray production in flares have been studied in detail and are quite well understood. There are extensive studies of nuclear deexcitation lines[9-13], neutron production[14,15], pion production[14], positron production[16], neutron capture[17,18], positron annihilation[19,20], and relativistic electron bremsstrahlung[21]. There are also treatments of acceleration[22,23] and transport[24-27], but a detailed gamma-ray production model, which takes into account realistic magnetic field geometries and incorporates acceleration mechanisms in a self-consistent manner, has not yet been developed.

The models developed so far have made use of the extensive gamma-ray production studies, but have treated the transport by either ignoring the magnetic field[15] or assuming that the particles are isotropized in the interaction region by an essentially unspecified mechanism[14,28]. It is possible, nevertheless, to obtain some very important results from this approach[2]. The observed nuclear deexcitation line emissions, which are nearly isotropic for any ion angular distribution, allow an accurate determination of the total number of interacting ions of energies greater than several MeV (the typical excitation threshold). By using simple acceleration models to extend the ion spectrum to lower energies, it is then possible to estimate the total flare energy content in ions and show that this energy is at least several percent of the total flare energy[22]. Furthermore, the number of interacting ions can be compared with the number observed in interplanetary space from the same flare. This comparison shows[28] that for most flares from which gamma rays were observed, the interacting ions are much more numerous than the escaping ones. This implies that the ions which produce the gamma rays are probably accelerated and subsequently trapped in closed magnetic structures, most likely flare loops.

However, other aspects of the observations, such as angular distributions and time dependences, do require more detailed studies of particle transport. There is evidence that relativistic electrons in solar flares are anisotropic. This follows from the fact that, at energies between 0.3 and 1 MeV, gamma-ray emitting flares are observed[29] predominantly from sites near the limb of the Sun and from observations of variations of the gamma-ray continuum spectrum in this energy range with flare location on the Sun[29,30]. But we note that this evidence is still controversial, as simultaneous stereoscopic observations[31] of solar flares with PVO and ISEE-3 show no evidence for anisotropic emission in the 0.15 to 1.5 MeV range. There is perhaps better evidence for anisotropic gamma-ray emission at energies >10 MeV, where the flares are strongly concentrated near the solar limb[29,32]. It has been shown[21,33] that these observations can be accounted for by bremsstrahlung from anisotropic electron distributions in the chromosphere. We show in the present paper that electron distributions similar to those assumed in these studies could result from particle transport and interactions in a converging magnetic flux tube in the chromosphere, where the gas density increases towards the photosphere. Indeed, there is evidence that the magnetic field increases by as much as an order of magnitude from the transition region to the photosphere (see ref. 25 and references therein). As far as the angular distribution of the ions is concerned, neutron observations rule out a unidirectional downward-directed ion beam, but are consistent with the distributions which would result from transport in such a converging magnetic flux tube model[34].

The impulsive nature of the gamma-ray time profiles was observed from many flares[1]. These observations imply that the bulk of the nuclear interactions must occur below the transition region, where the density is sufficiently high to account for the rapid variability of the observed fluxes. However, if the magnetic field converges in the chromosphere, magnetic mirroring would prevent a majority of the particles from penetrating deeply into the chromosphere[25], unless the acceleration process produces a highly directed beam in the corona. We show in the present paper that pitch-angle scattering by MHD turbulence in the corona can populate the loss cone sufficiently rapidly to account for the observed time profiles, even if the initial ion angular distribution is isotropic.

Gamma-ray spectroscopy has also been used to determine elemental and isotopic compositions in the solar atmosphere. Observations of the time-dependent flux of the 2.223 MeV line from the 1982 June 3 flare were used to determine the ^{3}He/H ratio in the photosphere[18]. Observations of the deexcitation line spectrum of the 1981 April 27 flare[35] were used to show[36] that the Ne/O and Ne/C ratios in the ambient medium (probably the chromosphere) are higher by about a factor of 3 than the corresponding ratios in the local galactic environment[37]. In the present paper we investigate the implications of particle transport on this latter result.

MODELS AND PHYSICAL PROCESSES

As pointed out above, a flare loop is a likely site for particle acceleration and subsequent gamma-ray production. We consider a flux tube model consisting of a semicircular coronal portion of half-length L_c having a uniform circular cross section of radius a_c, and two straight portions parallel to a solar radius extending from the ends of the coronal portion (at the transition region), through the chromosphere and into the photosphere. We take the gas to be completely ionized in the corona and neutral below the transition region. For the density profile below the transition region we use a sunspot active region model[38] at depths -1800 km $< h < 120$ km merged with a photospheric model[39] at depths > 120 km. Zero height is the point where the optical depth for 500 nm continuum radiation is unity and $h = -1800$ km is the location of the transition region. Below the transition region, we take B proportional to a power δ of the pressure[25], $B(h) = B_c[P(h)/P_c]^\delta$, while in the corona we take the magnetic field B_c to be constant. We assume a constant coronal pressure P_c and density n_c, and calculate the pressure $P(h)$ below the transition region from hydrostatic equilibrium. The parameter δ can be calculated by specifying the photospheric magnetic field $B_p = B(h=0)$; thus $\delta = \ln(B_p/B_c)/\ln(P_p/P_c)$, where we take $P_p = P(h=0) = 1.53 \times 10^5$ dyne cm^{-2}. We assume local galactic abundances[37] for the ambient gas composition throughout the loop.

We assume that the acceleration takes place in the corona, primarily because of the requirement for pitch-angle scattering by MHD turbulence for stochastic acceleration and diffusive shock acceleration[22]. Such turbulence could be produced by the flare energy release mechanism and could exist in the ionized corona, but is expected to be quickly damped (see ref. 40 p. 27) by neutral H in the chromosphere. The energy spectrum, angular distribution, and time profile of the accelerated particles depend on the acceleration process. Energy spectra were considered for stochastic acceleration[22,41,42] and shock acceleration[23]. There are no detailed studies of the expected angular distributions; however, for stochastic acceleration the distribution is probably isotropic. Time scales were given for stochastic acceleration[22,41,42] and diffusive shock acceleration[22,23]; very short acceleration times could be achieved by nondiffusive shock acceleration[43] or acceleration in large scale field-aligned electric fields[44,45], but the spectrum of the accelerated particles has not yet been investigated for these mechanisms.

In a Monte-Carlo simulation we release energetic particles in the coronal segment with a given energy spectrum, angular distribution, time dependence, and spatial distribution. The ions subsequently lose energy through Coulomb scattering and are removed by nuclear interactions. The electrons lose energy through Coulomb scattering, bremsstrahlung, and synchrotron radiation. We employ the guiding center approximation to determine the particle's motion in the magnetic field. We note that in the presence of energy losses, $(p \times \sin\alpha)^2/B$ is not conserved, but as long as the force corresponding to these losses is antiparallel to the particle's direction of motion $\sin^2\alpha/B = $ constant[46]; here, α and p are the particle's pitch angle and momentum. For ions and relativistic electrons, the aforementioned loss processes obey this condition. In the corona, where the magnetic field is constant, the particle's pitch angle does not change, but as it penetrates the chromosphere, the pitch angle varies according to the above equation. In addition to the effect of the varying magnetic field, the pitch angle can also be changed by Coulomb scattering. However, for ions and ultrarelativistic electrons, this effect is small and can be neglected.

The pitch angle is also altered as a result of scattering by MHD turbulence. As mentioned above, such turbulence is expected to exist in the corona but not below the transition region. Alfven turbulence can scatter ions if $\gamma\beta \gg \beta_a$ and electrons if $\gamma\beta \gg 1836\beta_a$, where γ and $c\beta$ are the Lorentz factor and speed of the particle and $c\beta_a$ is the Alfven speed[47]. The first condition is easily satisfied by ions with energies greater than the gamma-ray production thresholds. The second condition, however, requires electrons of energies $\gg 10$ MeV if $B_c = 100$ G and $n_c = 10^{10}$ H cm^{-3}. Thus, even though lower energy electrons could be scattered by whistler turbulence[42], it is possible that ~ 10 MeV bremsstrahlung producing electrons will be less affected by MHD pitch-angle scattering than the ions which produce the nuclear lines. In our present calculations we take into account pitch-angle scattering for ions but ignore it for electrons.

Pitch-angle scattering is described by the diffusion coefficient $D_{\mu\mu} = (1 - \mu^2)D_{\alpha\alpha}$, where $\mu = \cos\alpha$. For protons resonating with Alfven turbulence with a Kolmogorov spectrum $D_{\alpha\alpha} \simeq 100$ sec$^{-1}[(W_a/1\text{erg cm}^{-3})/(B/100\text{G})] \times \gamma^{-1}(\mu\gamma\beta)^{2/3}$, where W_a is the total energy density of the Alfven turbulence (derived from an expression[47] with a low wave number cutoff on the spectral density equal to the resonant wave number for a 10 GeV proton in a magnetic field of 100 G). Hence, for $W_a = 1$ erg cm^{-3} and $B = 100$ G, the scattering rate (defined as $D_{\alpha\alpha}(\mu = 1)$) is approximately 50 sec^{-1} for 30 MeV protons.

The value of W_a is not known. But even this relatively low value (less than 1% of the energy density in the ambient magnetic field) would suffice to practically isotropize the particles in the corona, as a typical particle transit time is $\sim$0.5 sec. Pitch-angle scattering will affect the time profile of the interactions primarily by scattering particles in the corona from large to small pitch angles, thus enabling them to lose energy faster below the transition region. As we shall see below, even a W_a of $\sim 2 \times 10^{-4}$ erg cm^{-3} has a significant effect on the time profile of nuclear line emission. This coronal scattering, however, will not destroy anisotropic gamma-ray production in the converging magnetic field below the transition region.

Particles may be removed from the loop by drifts. If the magnetic field in the flux tube is axially symmetric, however, the drift due to a perpendicular magnetic field gradient can be ignored as this drift does not remove particles from the loop. Curvature drift in the semicircular coronal segment, on the other hand, can remove particles from the loop before they interact. If a particle begins on the loop axis, the time to escape from the loop is t_{esc} = 68sec $(m_p/m)(B_c/100G)(a_c/10^8 cm)(L_c/10^9 cm)/(\gamma \beta^2 \mu^2)$, where m_p is the proton mass, and m is the particle mass. Thus, curvature drift could be important for relativistic ions, but is probably much less important for electrons and non-relativistic ions. Curvature drift may be an important mechanism for releasing flare-accelerated particles into interplanetary space.

Our transport calculations differ from previous treatments[25-27] in that we consider both ions and relativistic electrons; we take into account a converging magnetic field below the transition region, which was ignored in a previous treatment of electron transport[26]; and we include the effects of MHD pitch-angle scattering in detail.

NUMERICAL RESULTS

Considering the model and processes discussed above, we have calculated bremsstrahlung production by ultrarelativistic electrons and nuclear line production by energetic ions in a loop with $L_c = 10^9$ cm, $a_c = 10^8$ cm, and $B_c = 100$ G. We inject the electrons and ions isotropically at the top of the coronal loop at t = 0, with an energy spectrum[2] proportional to $p^{-3.5}$ for electrons and the modified Bessel function K_2 with $\alpha T = 0.03$ for ions. As discussed, we take into account MHD pitch-angle scattering for the ions, but ignore it for electrons.

The time-integrated depth distribution of >4 MeV bremsstrahlung emission is shown in Fig. 1 for three different values of B_p. We have assumed $n_c = 5 \times 10^9$ H cm^{-3}, $P_c = 1.67$ dyne cm^{-2} (which implies a coronal temperature of 2×10^6 K), and various values of B_p. In the present calculation we have ignored synchrotron losses, which are negligible below the transition region but, for the above parameters, could lead to anisotropic distributions in the corona at late times. We see in Fig. 1 that as B_p decreases, the magnetic field converges less rapidly and particles with a given pitch angle mirror deeper in the atmosphere where the probability of producing bremsstrahlung is higher.

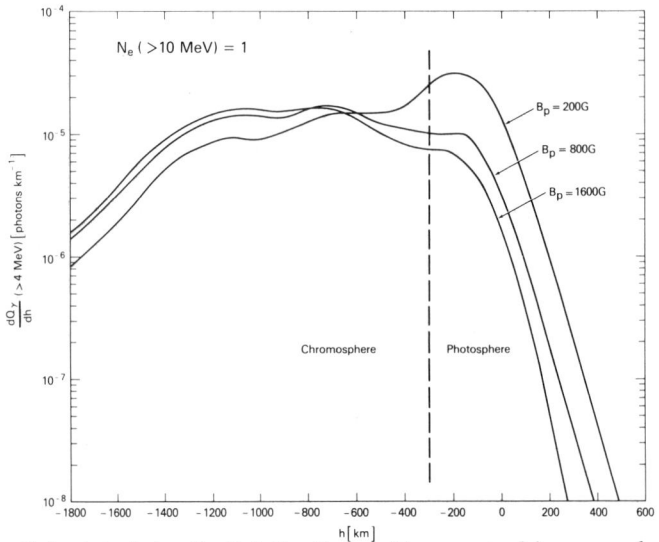

Fig. 1. Calculated depth distributions of bremsstrahlung production for a given coronal magnetic field and pressure, and three values of the photospheric magnetic field. The values of δ are 0.24, 0.18, and 0.061 for $B_p = 1600$, 800, and 200 G, respectively.

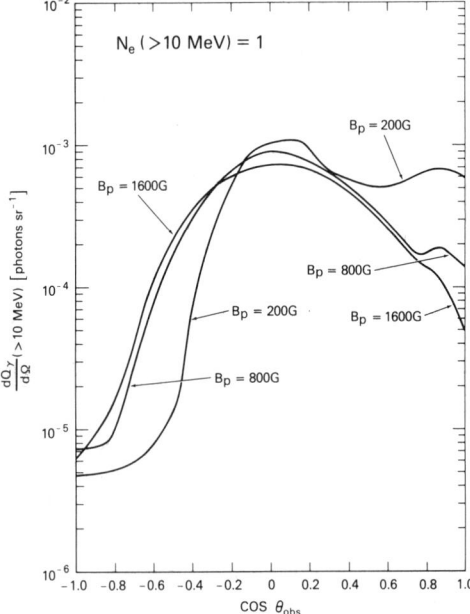

Fig. 2. Calculated angular distributions of bremsstrahlung for a given coronal magnetic field and pressure, and various photospheric magnetic fields; θ_{obs} is the angle between the direction of observation and a downward-directed solar radius.

In Fig. 2 we give the angular distribution of the height- and time-integrated >10 MeV bremsstrahlung obtained be assuming that the gamma rays are emitted along the direction of motion of the electrons. This is a good approximation for photons >10 MeV. We see that all three distributions peak at directions tangent to the photosphere. This result reflects the fact that, in the chromosphere, the bremsstrahlung yield per unit $\cos\theta_{obs}$ is maximal at $\cos\theta_{obs} = 0$, as long as $\delta < 1$ (this result can be derived analytically for exponentially varying densities and magnetic fields). The radiation at $\cos\theta_{obs}$ close to 1 is due to downward moving particles, most of which interact before they mirror. The solid angle populated by such particles is referred to as the loss cone. We see that as B_p decreases, the loss cone increases and hence more particles interact in the downward direction. Furthermore, as the number of particles in the loss cone increases, the number which can mirror and subsequently move in the backward direction decreases; this results in a steeper decrease of the emission in the backward direction for the smaller values of B_p. At $\cos\theta_{obs}$ close to -1, there is an important contribution from particles which interact in the corona. For simplicity we have integrated the emission over the azimuth of the direction of observation. However, in the curved corona—as opposed to the chromosphere—the emission does depend on this azimuth, if MHD pitch-angle scattering does not isotropize the particles.

Angular distributions such as those shown in Fig. 2 could account[21,33] for the preferential detection of gamma-ray flares from sites close to the limb. Furthermore, mirroring is probably the most effective mechanism for producing ultrarelativistic bremsstrahlung in the backward hemisphere because (i) the angular spread of such bremsstrahlung is very small; (ii) Coulomb scattering for ultrarelativistic electrons is negligible; (iii) MHD pitch-angle scattering in the chromosphere is unimportant because of damping of the turbulence; and (iv) the small amount of bremsstrahlung from the low-density corona has a very extended time dependence. Radiation in the backward hemisphere has been observed from flares on the disc (e.g., the 1982 June 3 flare[7] located at heliocentric longitude 72°).

For the nuclear line calculations, we include MHD pitch-angle scattering[48] by employing a scattering mean-free path λ, but we ignore drifts. In Fig. 3 we show the time-dependent production rate of the 4.438 MeV line for $\delta = 1/5$ and $\lambda = 1800 L_c$, as well as for $\delta = 1/5$ and $1/20$ and no pitch-angle scattering. For $B_c = 100$ G, these values of δ correspond to $B_p = 1600$ and 200 G, respectively. For a Kolmogorov spectrum and $B_c = 100$ G, $\lambda = 1800 L_c$ implies that $W_a \simeq 2 \times 10^{-4}$ erg cm^{-3}. Also shown in Fig. 3 is the observed[5] time profile of the 4–7 MeV emission of the 1982 June 3 flare; this time profile is expected to be similar to that of the 4.438 MeV time profile, provided it is dominated by nuclear deexcitation lines. We see that scattering by even such a low level of turbulence causes a very significant change in the time profile of line production by repopulating the loss cone at early times and thus decreasing the number of particles interacting at late times. Moreover, when pitch-angle scattering is taken into account, the calculated time profile provides an improved fit to the data.

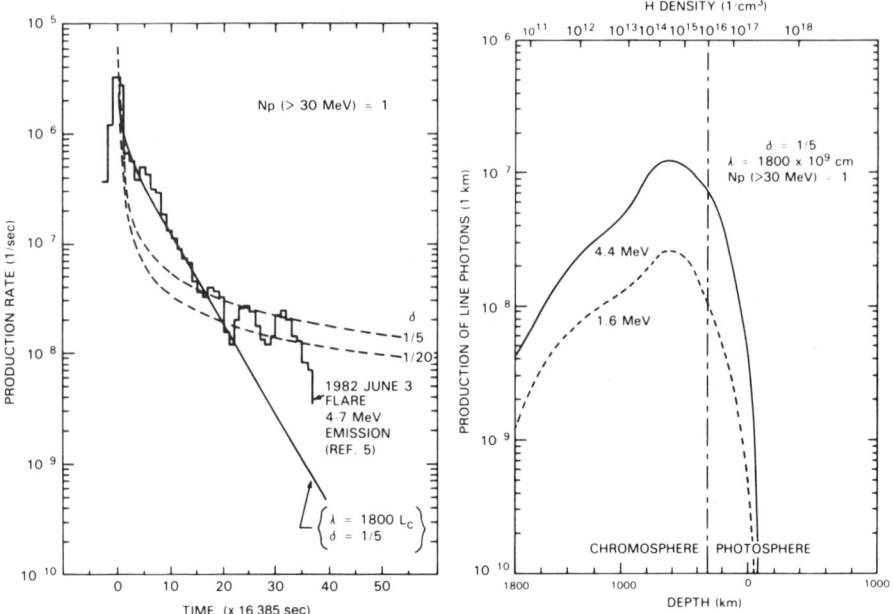

Fig. 3.(left) Calculated time dependences of 4.438 MeV line production in a magnetic field loop with $n_c = 2.5 \times 10^9$ H cm^{-3} and $P_c = 0.16$ dyne cm^{-2}; these parameters reproduce the column density and pressure at -1800 km in the sunspot active region model[38]. Solid curve: with pitch-angle scattering; dashed curves: without pitch-angle scattering.

Fig. 4.(right) Calculated depth distributions of the 4.438 and 1.634 MeV line production in a magnetic field loop with MHD pitch-angle scattering. The coronal parameters are the same as in Fig. 3. The atmospheric H density is also shown.

In Fig. 4 we show the depth distributions of the production of the 4.438 MeV ^{12}C and 1.634 MeV ^{20}Ne nuclear lines for $\delta = 1/5$ and $\lambda = 1800L_c$. These are two of the three strongest nuclear deexcitation lines observed from the 1981 April 27 flare[35]. Comparing the results of Fig. 4 with the depth distributions of bremsstrahlung from ultrarelativistic electrons in Fig. 1 (the curve with $B_p = 1600$ G corresponds to essentially the same magnetic field profile as that used in Fig. 4), we note that the bremsstrahlung production extends deeper into the photosphere than the nuclear line production. This is a consequence of the longer stopping range of ~ 20 MeV electrons compared with protons of approximately the same energy. We also note that relative to the profile of bremsstrahlung production, the profile of nuclear line production is depressed in the upper chromosphere. This is a consequence of MHD pitch-angle scattering, which was included in the ion calculations but omitted in the electron calculations.

Using the depth profiles shown in Fig. 4, we have calculated the escape probability of 4.438 and 1.634 MeV line photons from the solar atmosphere as a function of observation angle. We find that there is essentially no difference between the attenuation of these two lines. For the 1981 April 27 limb flare (θ_{obs} = 90°), these probabilities are both about 0.85. Thus, attenuation cannot explain the ratio of the 1.634-to-4.438 MeV line fluence observed from this flare[35], which is larger than the ratio expected assuming local galactic abundances[37]. As was suggested[36], this larger ratio is probably due to enhanced Ne/C and Ne/O ratios in the interaction region, which, most likely, is in the chromosphere and upper photosphere (see Fig. 4).

SUMMARY

We have briefly reviewed gamma-ray production models in solar flares, emphasizing the arguments that suggest particle acceleration and interaction within magnetic flare loops. We have argued that the particles are probably accelerated in the coronal portions of these loops because MHD turbulence, which is essential for stochastic acceleration and shock acceleration, can exist in the ionized corona, but is expected to be rapidly damped below the transition region where the gas is mostly neutral. We then discuss the physical processes that govern particle transport in such a loop model. Of particular interest is magnetic mirroring resulting from the convergence of the magnetic flux tubes in the chromosphere and upper photosphere. We find that the preferential detection of >10 MeV emission from flares located near the limb of the Sun can be explained by the anisotropies created by the mirror force. Furthermore, if the magnetic flux tubes in the chromosphere are parallel to a solar radius, then the mirror force is the most effective mechanism for producing ultrarelativistic bremsstrahlung in the upward hemisphere. Such bremsstrahlung has been observed from disc flares. We also find that, unless the initial particle distribution is highly beamed along the field lines, MHD pitch-angle scattering in the corona is the dominant mechanism for injecting the bulk of the particles into the denser regions of the chromosphere, where they react rapidly and produce impulsive gamma-ray time profiles.

Acknowledgments. We wish to thank F. C. Jones, T. G. Northrop and D. V. Reames for useful discussions. JAM was supported by NASA grant NGT-50055; XMH by NASA grant NGR-21-002-316; and REL by NSF grant ATM 8717676 and NASA grant NAG 5945. Parts of the present paper are from a dissertation to be submitted to the Graduate School, Univ. of Maryland, by J. A. Miller in partial fulfillment of the requirements for the Ph.D. degree in Physics.

REFERENCES

1. Chupp, E. L. 1987, *Physica Scripta*, **T18**, 5.
2. Ramaty, R., and Murphy, R.J. 1987, *Space Science Rev.*, **45**, 213.
3. Chupp, E.L. et al. 1982, *Ap. J.*, **263**, L95.
4. Debrunner, H. et al. 1983, *Proc. 18th Inter. Cosmic Ray Conf.*, **4**, 75.
5. Chupp, E.L. et al. 1987, *Ap. J.*, **318**, 913.
6. Forrest, D.J. et al. 1985, *Proc. 19th Inter. Cosmic Ray Conf.*, **4**, 146.
7. Forrest, D.J. et al. 1986, *Adv. Space Res.*, **6**, No. 6, 115.
8. Dennis, B.R. 1985, *Solar Phys.*, **100**, 465.
9. Ramaty, R., Kozlovsky, B., and Lingenfelter, R.E. 1979, *Ap.J. Suppl.*, **40**, 487.
10. Dyer, P. et al. 1981, *Phys. Rev.*, **C23**, 1268.
11. Dyer, P. et al. 1985, *Phys. Rev.*, **C32**, 1873.
12. Murphy, R.J. 1985, Ph.D. Thesis, University of Maryland.
13. Lang, F.L. et al. 1987, *Phys. Rev.*, **C35**, 1214.
14. Murphy, R.J., Dermer, C.D., and Ramaty, R. 1987, *Ap. J. Supp.*, **63**, 721.
15. Hua, X.-M., and Lingenfelter, R.E. 1987, *Solar Phys.*, **107**, 351.
16. Kozlovsky, B., Lingenfelter, R.E., and Ramaty, R. 1987, *Ap. J.*, **316**, 801.
17. Wang, H.T., and Ramaty, R. 1974, *Solar Phys.*, **36**, 129.
18. Hua, X.-M., and Lingenfelter, R.E. 1987, *Ap. J.*, **319**, 515.
19. Crannell, C.J. et al., *Ap. J.*, **210**, 582.
20. Bussard, R.W., Ramaty, R., and Drachman, R.J. 1979, *Ap. J.*, **228**, 928.
21. Dermer, C.D., and Ramaty, R. 1986, *Ap. J.*, **301**, 962.
22. Forman, M.A., Ramaty, R., and Zweibel, E.G. 1986, in *The Physics of the Sun*, eds. P.A. Sturrock et al., Vol. II, (Dordrecht: Reidel), p. 249.
23. Ellison, D.C., and Ramaty, R. 1985, *Ap. J.*, **298**, 400.
24. Ramaty, R., Lingenfelter, R.E., and Kozlovsky, B. 1982, in *Gamma-Ray Transients and Related Astrophysical Phenomena*, eds. R.E. Lingenfelter et al. (New York: AIP), p.211.
25. Zweibel, E.G., and Haber, D. 1983, *Ap. J.*, **264**, 648.
26. Petrosian, V. 1985, *Ap. J.*, **299**, 987.
27. Kocharov, G.E., Mandzhavidze, N.Z., and Guglenko, V.G. 1987, Acad. of Sciences of the USSR, A.F. Ioffe Physico-Technical Institute, preprint.
28. Murphy, R.J., and Ramaty, R. 1984, *Adv. Space. Res.*, **4**, No.7, p.127.
29. Vestrand, W.T. et al. 1987, *Ap. J.*, **322**, 1010.
30. Chupp, E.L. 1982, in *Gamma-Ray Transients and Related Astrophysical Phenomena*, eds. R.E. Lingenfelter et al. (New York: AIP), p. 363.
31. Kane, S.R. et al. 1988, *Ap. J.*, **326**, 1017.
32. Rieger, E. et al. 1983, *Proc. 18th Inter. Cosmic Ray Conf.*, **10**, 338.
33. Dermer, C.D. 1987, *Ap. J.*, **323**, 795.
34. Hua, X.-M. and Lingenfelter, R.E. 1987, *Ap. J.*, **323**, 779.
35. Forrest, D.J. 1983, in *Positron and Electron Pairs in Astrophysics*, ed. M.L. Burns et al. (New York: AIP), p. 3.
36. Murphy, R.J. et al. 1985, *Proc. 19th Inter. Cosmic Ray Conf.*, **4**, 249, 253.
37. Meyer, J.P. 1985, *Ap. J. Supp.*, **57**, 151.

38. Avrett, E.H. 1981, in *The Physics of Sunspots*, ed. L.E. Cram and J.H. Thomas (Sacramento Peak Obs.: AURA), p. 235.
39. Allen, C.W. 1963, *Astrophysical Quantities* (London: Athlon Press), 164.
40. Melrose, D.B. 1980, *Plasma Astrophysics* (New York: Gordon and Breach), Vol.II.
41. Miller, J.A., Ramaty, R., and Murphy, R.J. 1987, *Proc. 20th Inter. Cosmic Ray Conf.*, **3**, 33.
42. Miller, J.A., and Ramaty, R. 1987, *Solar Phys.*, **113**, 195.
43. Ohsawa, Y., and Sakai, J.I. 1987, *Ap. J.*, **313**, 440.
44. Colgate, S.A. 1978, *Ap. J.*, **221**, 1068.
45. Haerendel, G. 1987, *Proc. 21st ESLAB Symposium, ESA SP- 275*, (Bolkesjo, Norway: ESA), p. 205.
46. Northrop, T.G. 1987, private communication.
47. Melrose, D.B. 1974, *Solar Phys.*, **37**, 353.
48. Palmer, I.D., and Jokipii, J.R. 1981, *Proc. 17th Inter. Cosmic Ray Conf.*, **3**, 381.

SOLAR ABUNDANCES USING GAMMA-RAY LINE SPECTROSCOPY OF A SOLAR FLARE

R. J. Murphy*
E. O. Hulburt Center for Space Research
Naval Research Laboratory
Washington, DC 20375 USA

ABSTRACT

Elemental abundances of the ambient gas at the site of gamma-ray line production in the solar atmosphere are deduced using gamma-ray line observations of a solar flare. The resultant abundances are different from local galactic abundances which are thought to be similar to photospheric abundances.

INTRODUCTION

Knowledge of the cosmic abundances of the elements is critical for testing theories of the early universe, stellar and galactic formation and dynamics, and nucleosynthesis. The Sun has been one of the primary sources of information on these cosmic abundances because its nearness has made possible optical, UV and X-ray spectroscopic analyses of its atmospheric radiation. One of the explicit goals of the Solar Maximum Mission (SMM) was to extend these spectroscopic analyses into the gamma-ray region by using solar-flare spectra obtained with the Gamma-Ray Spectrometer (GRS). The spectrum observed[1] by SMM from the limb flare on April 27, 1981 was sufficiently detailed to allow a reliable abundance determination. This represents the first application of gamma-ray spectroscopy to an astrophysical source other than the moon. The results were published previously[2,3,4].

Gamma-ray emission from solar flares consists of lines from nuclear reactions and continuum primarily from relativistic electron bremsstrahlung[5,6]. The gamma rays are produced by thick-target interactions of energetic particles with ambient gas, most likely at densities coresponding to the chromosphere or transition region[7]. Gamma-ray lines result from nuclear deexcitation, neutron capture, and positron annihilation. Excited nuclei produce both narrow and broad lines. If an energetic proton or α-particle interacts with an ambient heavy nucleus, the resultant gamma-ray line is narrow, Doppler-broadened only by the relatively small recoil velocity of the heavy nucleus. If an energetic heavy nucleus interacts with an ambient hydrogen or helium nucleus, the line is very broad, Doppler-broadened by the velocity of the excited nucleus which has lost little of its initial kinetic energy in the interaction. The broad lines merge into a relatively featureless nuclear continuum.

*Resident Research Associate at the Naval Research Laboratory, Under the NRC Associateship Program.

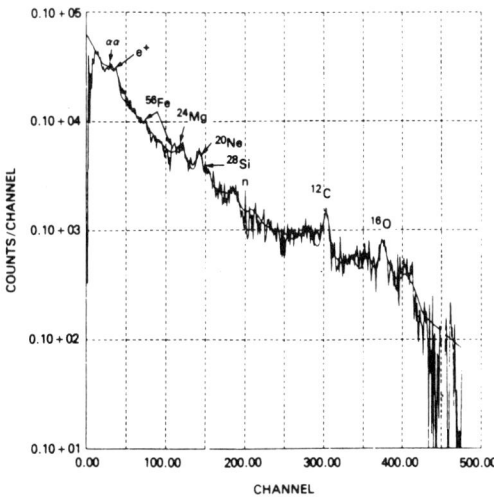

Figure 1. Pulse-height spectrum observed from the 27 April 1981 flare.

Neutrons are also produced by the interactions which may be captured on hydrogen to produce deuterium and a 2.223 MeV photon. Most of the neutrons thermalize before being captured, resulting in a very narrow line. Positrons result from the decay of radioactive nuclei and π^+-mesons that are also produced by the energetic-particle interactions. The positrons slow down and annihilate with ambient electrons either directly or via positronium. In the solar atmosphere, most of the positrons thermalize before forming positronium and the line is very narrow.

Figure 1 shows the background-subtracted pulse-height spectrum of the 27 April 1981 flare. The strongest lines are indicated along with their sources. The smooth curve is the best fit to the data as discussed below.

ANALYSIS

We have calculated nuclear deexcitation photon spectra resulting from interactions of all energetic particles with each of the 12 ambient elements in Table 1. These spectra depend on the nuclear cross sections and the energetic-particle composition, spectrum and angular distribution. Cross sections for the strongest gamma-ray lines in solar flares have recently been measured (e.g., ref. 8 and references therein) to accuracies typically better than 10-20%. The energetic-particle composition used here is given in Table 1.

The expected relative abundances in the solar atmosphere are such that interactions between particles heavier than helium can be neglected. Therefore, the spectra associated with such heavy ambient elements result primarily from energetic protons only and consist mostly of narrow lines. (The contribution of the energetic α-particles relative to that of the protons is also small due to their usually low relative abundance.) Since no gamma-ray lines are produced in p-p or p-α reactions, the ambient H spectrum results from only energetic nuclei heavier than helium and so consists of broadened lines. Similarly, the ambient He spectrum results from energetic α-particles and heavier nuclei and consists of one relatively narrow α-α fusion feature at ~0.45 MeV in addition to broadened lines. The dominant contribution to the total photon spectrum from a typical solar flare is primarily due to ambient elements heavier than He; the contributions due to ambient H and He are small since they depend predominantly on the energetic particles heavier than helium whose interactions in a thick target are suppressed by the Z^2/A dependence of the Coulomb energy loss. Also, a good fit of the calculated pulse-height spectrum to the observed spectrum is achieved primarily by fitting the narrow lines which are produced almost exclusively by energetic protons. Thus, uncertainties in the abundances of the energetic nuclei relative to the abundance of the energetic protons do not significantly affect the derived abundances of ambient C and heavier elements relative to each other as long as the energetic α/p abundance ratio is not much larger than ~0.1. However, the derived ambient H and He abundances relative to the derived ambient heavy abundances do depend on the uncertain energetic α-particle and heavy-nuclei abundances relative to the energetic proton abundance. Unless independent knowledge of the energetic-particle abundances exists for the particular flare in question, the ambient H and He abundances relative to ambient heavier element abundances cannot be reliably determined.

For the energetic-particle kinetic-energy spectrum we use a Bessel function with the spectral parameter $\alpha T=0.02$ (see, e.g., Ref. 9), a value close to the average αT determined[7] for several flares using the 2.223 to 4-7 MeV fluence ratio or the high-energy neutron arrival time profile. Neither of these methods can be used for the April 27 limb flare since the 2.223 MeV line was strongly attenuated and no neutrons were observed. Our calculations, however, indicate that variations in the energetic-particle spectrum do not significantly affect the abundance determination. The angular distribution of the energetic particles affects the shapes and central energies of the gamma-ray lines, but since these effects are smaller than the energy resolution of the SMM/GRS, we assume an isotropic distribution.

With the 12 deexcitation spectra, we include a bremsstrahlung spectrum, taken to be an unbroken power law with adjustable spectral index s_b, a neutron capture spectrum, taken to be a narrow line (<3 keV) at 2.223 MeV, and a positron annihilation spectrum, consisting of a narrow line (<2 keV) at 0.511 MeV and a positronium continuum. The fraction of positrons annihilating via positronium was taken to be 0.67 and the positronium continuum photon spectrum was taken to be proportional to photon energy up to 0.511 MeV. Each of these 15 partial photon spectra is transformed into corresponding partial pulse-height spectra using a numerical model of the SMM/GRS response. This numerical model takes into account the detector effective area, resolution, photopeak and first escape peak fractions, and the Compton continuum fraction and spectrum. The effects of uncertainties in the numerical model were investigated and were found to affect the results negligibly. The total pulse-height spectrum is the sum of the 15 partial pulse-height spectra, each multiplied by its respective intensity. To obtain the best fit to an observed spectrum, s_b and the 15 intensities are varied to minimize χ_ν^2. The resultant relative intensites of the 12 ambient

TABLE 1. Elemental Abundances

Element	Local Galactic[10]*	Corona[10]*	Energetic Particles[+]	Abundances from Gamma Rays[++]
H	2.71×10^6(1.10)	2.55×10^6(1.4)	8.66×10^5	--
He	2.60×10^5(1.25)	2.50×10^5(3.0)	5.86×10^4	--
C	1260(1.26)	600(3.0)	270	288±50
N	225(1.41)	100(1.7)	75	117±91
O	2250(1.25)	630(1.6)	600	422±62
Ne	325(1.50)	90(1.6)	85	199±27
Mg	105(1.03)	95(1.3)	144	68±25
Al	8.4(1.05)	7(1.7)	8	-15±52
Si	≡100(1.03)	≡100(1.3)	≡100	≡100±28
S	43(1.35)	22(1.7)	19	48±83
Ca	6.2(1.14)	7.5(1.5)	7	17±15
Fe	88(1.07)	100(1.5)	99	76±18

*The quantities in parantheses are multiplicative errors, f, from Ref. 10. We take m(f-1) as an estimate of a 1-σ error about the mean value m.
[+]p/α and p/O are the same as those of Ref. 11 and C through Fe relative to O or Si are similar to the mass-unbiased solar energetic particle abundances of Ref. 10.
[++]The gamma-ray abundance errors are 1-σ.

element spectra represent the derived relative abundances. The best-fit spectrum is shown by the smooth curve in Figure 1.

RESULTS

Since the derived abundances for ambient H and He have systematic uncertainties as discussed above, we have not included them in the following comparisons. The best-fit ambient abundances of C and heavier elements are given in the last column of Table 1. The statistical errors for the elements C, O, Ne, Mg, Si and Fe are sufficiently small to allow a meaningful comparison of the derived abundances with previous abundance determinations. The statistical errors for N, Al, S and Ca are too large and there is excessive interference between S and neutron capture and between He and positron annihilation. Interference arises when two sources both contribute significantly to a particular spectral feature. As a result, the two abundances cannot be determined independently. In this case, these interferences are due to the detector's inability to adequately resolve the 2.223 MeV neutron capture line from the 2.230 MeV line of S and the 0.511 MeV annihilation line from the ~0.45 MeV α-α feature.

To compare the derived ambient C and heavier-element abundances with local galactic[10] (thought to be similar to photospheric) and coronal[10] abundances (see Table 1), we have renormalized the derived abundances by minimizing the χ_ν^2 associated with the comparison. For the local galactic case, the resulting χ_ν^2 = 2.25 corresponding to a 1.8% probability that a random measurement of local galactic abundances would produce a χ_ν^2 as large or larger. This implies that the derived abundances are different from the local galactic abundances. For the coronal case, χ_ν^2 = 0.47 with a corresponding probability of 89% implying that, within errors, the derived abundances are consistent with the coronal abundances. The closed circles in Figures 2 and 3 show ratios of the renormalized derived abundances to the mean local galactic and coronal abundances, respectively. The error bars reflect 1-σ errors and the open boxes represent errors associated with the local galactic and coronal abundance measurements[10].

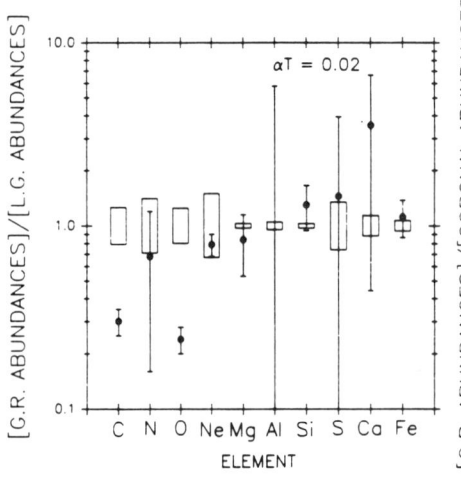
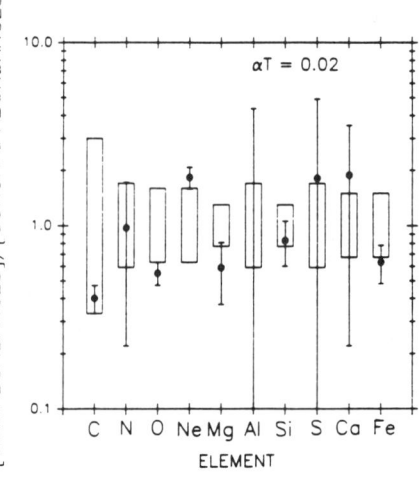

Figure 2. Ratios of renormalized gamma-ray-deduced abundances to local galactic abundances (closed circles).

Figure 3. Ratios of renormalized gamma-ray-deduced abundances to coronal abundances (closed circles).

With the normalization we have adopted, the priciple difference between the derived abundances and the local galactic abundances is the supression of the derived C and O by factors of 3 to 4. The abundances of Ne, Mg, Si and Fe are in good agreement. A similar supression of C and O has been found[10] in the coronal abundances relative to the local galactic (or photospheric) abundances. It has been pointed out (e.g., Ref. 10 and references therein) that this supression may be caused by charge-dependent mass transport from the photosphere to the corona. Since the photosphere is collisionally ionized at a relatively low temperature, such a transport could produce the first-ionization potential dependence which is observed in the coronal abundances (C, O and Ne have a high potential relative to Mg, Si and Fe). While the abundances derived from the gamma rays probably pertain to chromospheric or lower-transition-region densities in a flare plasma, similar fractionation effects could be affecting this region as well. However, if the Ne abundance in the photosphere (where it cannot be measured) is the same as that in the local galactic sample, then additional processes must be involved since the derived abundance of Ne, which has a high first-ionization potential, would be expected to be supressed similarly to C and O and it is not. Additional ionization processes, perhaps photo-ionization, could be present during flares.

A number of other solar flares with sufficiently good spectra have been observed with the SMM/GRS and new observations are expected. Planned research involves analyzing these flares individually to determine flare-to-flare abundance variations and by summing flares to obtain an average abundance with improved statistics. Future instruments with increased sensitivity will improve the statistical significance of the results and improved energy resolution will allow the determination of elements whose gamma-ray lines were not adequately resolved with the SMM detector.

ACKNOWLEDGEMENTS

I would like to acknowledge D. J. Forrest, R. Ramaty and B. Kozlovsky for essential assistance in accomplishing this research.

REFERENCES

1. Forrest, D. J., in M. L. Burns, A. K. Harding and R. Ramaty (eds.), <u>Positron-Electron Pairs in Astrophysics</u> (AIP, New York, 1983), p. 3.
2. Murphy, R. J., Ramaty, R., Forrest, D. J., Kozlovsky, B., <u>19th International Cosmic Ray Conference Papers</u>, **4**, 249 (1985).
3. Murphy, R. J., Forrest, D. J., Ramaty, R., Kozlovsky, B., <u>19th International Cosmic Ray Conference Papers</u>, **4**, 253 (1985).
4. Murphy, R. J., Ph. D. thesis, University of Maryland (1985).
5. Ramaty, R., Murphy, R. J., Kozlovsky, B., Lingenfelter, R. E., Solar Physics, **86**, 395 (1983).
6. Chupp, E. L., Ann. Rev. Astr. Ap, **22**, 359 (1984).
7. Dyer, P., et al., Phys. Rev. C, **32**, 1873 (1985).
8. Murphy, R. J., and Ramaty, R., <u>Advances in Space Research</u> (COSPAR), 4, 7, p. 127 (1985).
9. Forman, M. A., Ramaty, R., and Zweibel, E. G., in P. A. Sturrock, T. E. Holzer, D. Mihalas and R. K. Ulrich (eds.), <u>The Physics of the Sun</u> (Reidel, Dordrecht, 1986), Vol. II, Chap. 13, p. 249.
10. Meyer, J. P., Ap. J. (Suppl.), **57**, 173 (1985).
11. Cameron, A. G. W., in C. Barnes, D. D. Clayton and D. N. Schramm (eds.), <u>Essays in Nuclear Astrophysics</u> (Cambridge Univ. Press, Cambridge, 1982), p. 23.

THE NEUTRON CAPTURE LINE AND ITS COMPTON TAIL AS A FLARE DIAGNOSTIC

W. Thomas Vestrand
Space Science Center
University of New Hampshire, Durham, NH 03824

ABSTRACT

We describe the results from a series of calculations which were performed to study the Compton tail associated with the 2.223 MeV neutron capture line in solar flare spectra. We show that the relative strength of the Compton tail can be used as an independent measure of both the angular distribution and the energy spectrum of flare accelerated ions.

INTRODUCTION

One of the key observational questions for solar flare physics is: What is the number, the energy spectrum, and the angular distribution of flare accelerated ions? In a series of papers[1-3], Ramaty, Lingenfelter, and their collaborators have calculated the expected neutron fluences and γ-ray line fluences for a wide range of energetic ion distributions. These detailed calculations allow one to use measurements of direct neutrons and the strong γ-ray lines at 2.223 MeV, 4.438 MeV, and 6.129 MeV as a diagnostic for measuring ion number and spectral shape. With the exception of a few flares for which energetic secondary neutrons have been detected directly in space, comparison of the 2.223 MeV fluence with the fluence in the 4–7 MeV band has been the method used for deriving the energetic ion spectra.

The 2.223 MeV line is generated in a manner quite different from the two higher energy lines at 4.438 MeV and 6.129 MeV. Those lines are the principal nuclear deexcitation lines from ^{12}C and ^{16}O, respectively. The current thinking is that they are generated in the low chromosphere by the direct excitation or spallation of nuclei by energetic ions. In contrast, the narrow 2.223 MeV line is produced by the capture of thermal neutrons on protons in the photosphere (viz. $n + H \rightarrow D + \gamma$). These capture neutrons are thought to be generated by energetic ion interactions and thermalized by scattering in the solar atmosphere. The propagation of energetic neutrons in the solar atmosphere and the production of the narrow neutron capture line has been studied by both Wang and Ramaty[4] and Kanbach et al.[5], and it has been recently reexamined, in detail, by Hua and Lingenfelter[6].

Since the neutrons are captured primarily in the photosphere, emergent neutron capture photons must traverse the solar atmosphere. As they traverse the atmosphere, some of the photons are Compton scattered and transfer energy to the scattering electrons. Those scattered photons that eventually emerge form a low-energy "tail" on the neutron capture line. Spectral measurements of the giant 3 June 1982 event by the Gamma-Ray Spectrometer aboard the *Solar Maximum Mission* satellite clearly show such a Compton tail on the 2.223 MeV that is not detector generated.[7]

In this paper we describe a series of calculations that were performed to study the Compton tail on the 2.223 MeV neutron capture line. We show that the relative strength of the tail can be used as a measure of the 2.2 MeV line source function in the solar atmosphere. This, in turn, can independently constrain both the angular and the energy distribution of the energetic parent ions.

SIMULATION PROCEDURE

To study the formation of the Compton tail we use a Monte Carlo photon transport code to simulate the random walk of line photons in a semi-infinite plane-parallel atmosphere. The transport code contains both Compton scattering and photoelectric absorption subroutines. We can safely neglect pair production because at 2.2 MeV in a hydrogen plasma, the total pair-production cross-section is roughly three orders of magnitude smaller than the total Compton cross-section.

The influence of photon production depth was examined by running a number of simulations where all photons were generated at a given depth in the atmosphere. Runs were performed for a total of fifteen initial depths that are spaced at nearly logarithmic intervals over the range $0.1 - 30$ g cm^{-2}. For each run, two million test photons were generated. The history of each of these photons was followed until it emerged from the atmosphere or reached an energy of 30 keV.

To calculate the Compton tail for conditions likely in solar flares, we used the 2.2 MeV line source functions recently calculated by Hua and Lingenfelter[6]. Altogether, twelve source functions were digitized from the figures in that paper. Each simulation samples a model source function $S_{2.2}(d)$ in the following manner. First, the source function is renormalized so that its integral over all linear depths is unity. For each test photon, γ_i, a uniform random deviate U_i is generated and the initial linear depth of the photon D_i is derived by solving the equation

$$U_i = \int_0^{D_i} \tilde{S}_{2.2}(x) dx$$

The initial linear depth of the test photon is then transformed into an initial grammage, Z_0 g cm^{-2}). This grammage is assigned by interpolating values from the atmospheric model used by Hua and Lingenfelter. Their model is unusual in the sense that they model the density profile by joining the sunspot profile of Avrett[8] for large heights to the profile of Allen[9] for large depths. Once the initial grammage is generated, an initial photon direction is generated (isotropically) and the photon transport program is called.

The history of each photon is followed by keeping track of the photon's energy E, direction cosine μ, azimuthal angle ϕ, and column depth Z. After j collisions the photon is located at a depth Z_j, has an energy E_j, and is travelling at an angle $\theta_j = \cos^{-1}(\mu_j)$ with respect to the downward perpendicular. All photons are either absorbed or ultimately escape the atmosphere. Those photons that do escape are binned into energy and direction cosine bins so that the emergent spectra can be calculated for various viewing angles.

RESULTS

We find the generic spectrum for solar flare grammages can be broken into three morphological parts: (1) The narrow line feature; and a tail which can be further subdivided into (2) a flat featureless continuum starting at energies just below the line energy and continuing to about 1 MeV; and (3) a continuum that starts at about 1 MeV and rises at lower energies (see Figure 1.). The physical explanation for these three regions is simple. The narrow line (region 1) is composed of photons that were emitted upward and traversed the atmosphere essentially without scattering. When the source function is spread throughout the atmosphere, this component is dominated by photons generated at shallow depths. The photons in spectral region 2 have been scattered once through at most 45° or a few times through small angles. These photons were therefore initially emitted into the upward hemisphere. The photons in spectral region 3 have been scattered through large angles or have undergone a long series of scatterings. Many region 3 photons were originally emitted into the downward hemisphere and subsequently backscattered off material below the photon production region. Photons generated at grammages larger than 10 g cm^{-2}, usually emerge in region 2 or region 3.

The simulations show that the relative strength of the tail increases with the total grammage between the observer and the site of photon generation. Figure 2 shows the spectra obtained for a single view-angle bin ($\theta = 70°$) when all of the photons are released at a given initial depth. Results are shown for initial depths of 0.3, 1.0, 5.0, 10.0

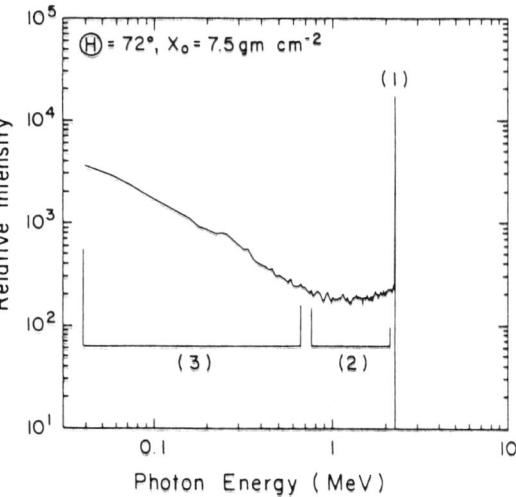

Figure 1.

Generic spectrum of neutron capture photons.

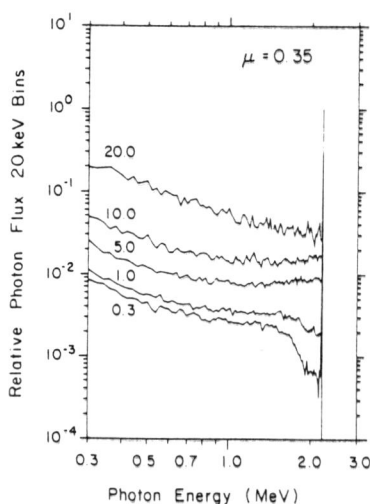

Figure 2.

The relative strength of the Compton tail for slab production models. All spectra are normalized to the same strength in the narrow line.

and 20.0 g cm^{-2}. Notice that the relative strength of the tail increases with initial photon depth. This makes sense physically, because the higher grammage will increase the probability of a photon being Compton scattered before emerging. Of course, an oblique view-angle will also increase the grammage along the line of sight. The relative strength of the tail will therefore depend on both the typical neutron capture depth and the observer's viewing angle. To quantify the relative strength of the tail for various capture depths and view angles, we use the ratio of the number of photons between 1.0 MeV and 2.2 MeV to the number of photons in the 2.223 MeV line. The ratios found for the slab models in which all photons are started at a given initial depth are shown in Figure 3. It is clear that, as expected, the ratio increases with intial depth and view-angle.

In flares, of course, the neutron capture photons are not all generated at the same initial depth. Instead, the depth at which the energetic neutrons are thermalized and captured depends on both the neutron's initial energy and angle of incidence. Wang and Ramaty[4] found that isotropically emitted monoenergetic neutrons have a most probable capture column density that is about the same as the inverse of the elastic scattering cross-section. Since 10–200 MeV neutrons have an elastic scattering cross section that decreases roughly as $1/E_n$ in hydrogen, energetic neutrons thermalize and are captured at a grammage that increases with initial neutron energy. As a consequence,

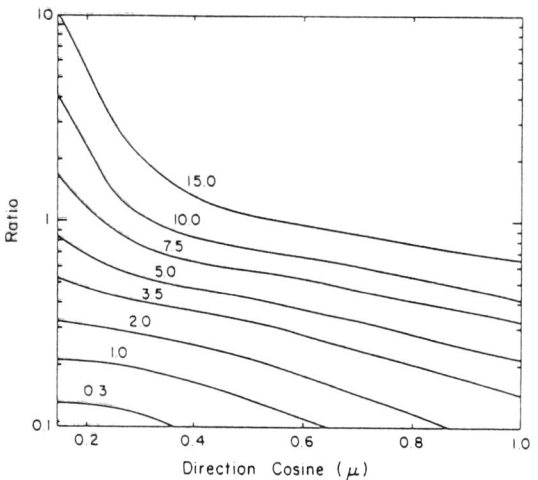

Figure 3.

Plotted is the ratio of photons in the 1.0 - 2.2 MeV band to 2.223 MeV line photons for slab models. The ratio versus viewing angle curves are labeled with the grammage for the source slab depth.

harder primary ion spectra will produce more secondary neutrons that are captured at greater depths. This reasoning is borne out by the very detailed Monte Carlo calculations of Hua and Lingenfelter[6]. Their calculations clearly show that harder ion spectra ultimately generate neutron capture line source functions whose most probable capture depths are shifted to higher grammages.

Since the fraction of photons that emerge as tail photons is a function of the source depth, one can use the tail/line ratio for neutron capture photons as a measure of the ion spectral hardness. Figure 4 shows the ratios we calculated using the source functions derived in Hua and Lingenfelter[6] for energetic ion distributions that are isotropic in the downward hemisphere and have power-law energy distributions with indices s=2, 4, and 6. Notice that the relative strength of the tail increases with spectral hardness. Our calculations using Bessel functions for the ion distributions yield similar results.

Recent studies of flares at high energies have provided evidence that the bremsstrahlung continuum from flares is anisotropic[10, 11]. To generate the observed limb brightening, the emitting electrons must have an intensity in the outward hemisphere that increases with angle from the outward normal. Two families of electron angular distributions that meet this constraint are "pancake" distributions which have peak intensitites in directions tangent to the photosphere (e.g. Dermer and Ramaty[12]) and downwardly directed beams.

If flare accelerated ions are also anisotropic then the tail/line ratio for the neutron capture line can help constrain the angular distribution. Hua and Lingenfelter[6] have calculated the capture photon source functions for the two extreme distributions of (1) a δ-function pencil beam that is downwardly directed and (2) a δ-function fan beam at 89° in the downward hemisphere (viz. "pancake" distribution). Using these source functions in our transport programs, we find that even when the primary ion spectra are identical, the downward pencil beam produces a relatively stronger tail (see Figure 5). The physical basis for these results is easy to understand. Momentum conservation forces a "pancake" distribution of accelerated ions to produce energetic secondary neutrons that preferentially have initial velocities which are nearly parallel to the photosphere. (Actually, the neutron distribution is a

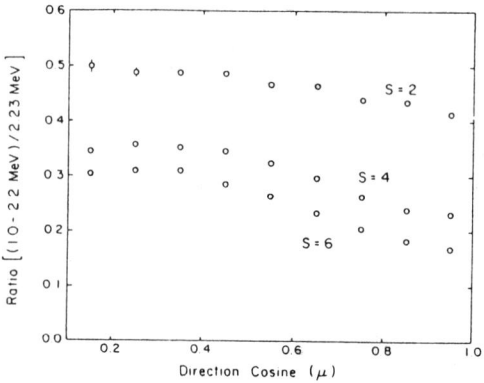

Figure 4.

The neutron capture tail/line ratio found for neutrons generated by power-law ion spectra. The parent ions have an angular distribution that is isotropic in the downward hemisphere.

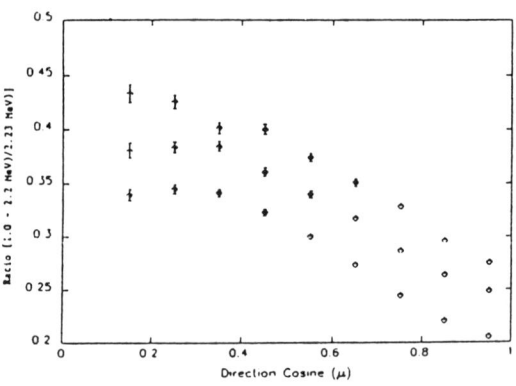

Figure 5.

The dependence of the neutron capture tail/line ratio on the angular distribution of the parent energetic ions. The curves shown are for a downwardly directed pencil beam (top), a "pancake" distribution (bottom), and a distribution that is isotropic (center). In each case the ions had an energy spectrum that is a power-law with index $s = 4$.

mixture of a nearly isotropic low-energy evaporation component and an anisotropic high-energy knock-on component.) Hence the neutrons can run out their ranges and can be captured at a relatively small depth in the atmosphere. In this case, the smaller median capture depth makes the Compton tail relatively weak. In contrast, a downwardly beamed ion distribution tends to produce downwardly beamed energetic neutrons that deeply penetrate the atmosphere. This deeper capture of neutrons results in a stronger Compton tail.

SUMMARY

We have shown that the fraction of γ-ray line photons that are Compton scattered "out" of the line and into a low-energy "tail" provide a measure of the amount of material between the source region and the detector. Specifically, we described the results from a series of calculations that examine the Compton tail associated with the neutron capture line. We showed that the relative strength of this Compton tail depends on the flare viewing angle, the spectrum of the energetic parent ions, and the angular distribution of the energetic ions. Hence for a flare with known position, the strength of the neutron capture tail can provide an important diagnostic for deriving properties of the ion population accelerated during the flare.

ACKNOWLEDGEMENTS

I am pleased to thank Dave Forrest, Gottfried Kanbach and Ed Chupp for useful discussions on neutron capture emission. I should also like to thank Mary Chupp for her efforts in "shoehorning" this manuscript into the AIP format, and thank NASA for financial support under Contract NAS 5-28609 and Grant NAG5-720.

REFERENCES

1. R.E. Lingenfelter and R. Ramaty, B.S.P. Shen (ed.), *High Energy Nuclear Reactions in Astrophysics* (W.A. Benjamin Inc., N. Y., 1967), p. 99.
2. R. Ramaty, B. Kozlovsky and R.E. Lingenfelter, Astrophys. J. Suppl. **40**, 487 (1979).
3. R.J. Murphy and R. Ramaty, Adv. Space Res. **4 (7)**, 127 (1984).
4. H.T. Wang and R. Ramaty, Solar Phys. **36**, 129 (1974).
5. G. Kanbach, K. Pinkau, C. Reppin, E. Rieger, E. Chupp, D. Forrest, J. Ryan, G. Share and R. Kinzer, 17th Internat. Cosmic-Ray Conf. Papers **10**, 9 (1981).
6. Hua X.-M. and R.E. Lingenfelter, Solar Phys. **107**, 351 (1987).
7. E. L. Chupp, Solar Phys. **86**, 383 (1983).
8. E.H. Avrett, L.E. Cram and J.H. Thomas (eds.), *The Physics of Sunspots* (AURA, Sacramento Peak Obs., 1981), p. 235.
9. C.W. Allen, *Astrophysical Quantities* (Athlon Press, London, 1963), p. 164.
10. E. Rieger, C. Reppin, G. Kanbach, D.J. Forrest, E.L. Chupp and G.H. Share, 18th. Internat. Cosmic Ray Conf. Papers **10**, 338 (1983).
11. W.T. Vestrand, D.J. Forrest, E.L. Chupp, E. Rieger and G.H. Share, Astrophys. J. **322**, 1010 (1987).
12. C.D. Dermer and R. Ramaty, Astrophys. J. **301**, 962 (1986).

NUCLEAR GAMMA-RAY LINE RATIOS AS SPECTRAL DIAGNOSTICS FOR PROTONS ACCELERATED IN SOLAR FLARES

C. J. Crannell and F. L. Lang
Code 682, Solar Physics Branch
Laboratory for Astronomy and Solar Physics
NASA/Goddard Space Flight Center
Greenbelt, Maryland 20771
and
Department of Physics
The Catholic University of America
Washington, DC 20064

ABSTRACT

New and revised gamma-ray production cross sections are employed to calculate the expected 15.10- to 4.438-MeV line ratio for proton interactions with carbon and oxygen as a function of parameters characterizing the incident proton spectrum in a solar flare. The additional cross sections for which inadequate determinations cause significant uncertainties in the results of this calculation are identified.

INTRODUCTION

A pair of gamma-ray lines from states with widely separated production thresholds can provide a measure of the spectrum of the protons accelerated in a solar flare. If the two lines result from accelerated particle interactions with the same nuclear species in the ambient solar medium, then this spectral diagnostic is unaffected by any uncertainties in relative abundances. The most promising candidate states are the isospin zero level at 4.439 MeV and the isospin one level at 15.11 MeV, both in ^{12}C. These two excited states, however, can be produced by spallation reactions with ^{16}O as well as by inelastic scattering from ^{12}C. Because ^{12}C and ^{16}O are both realtively abundant in the solar photosphere, with oxygen approximately 50% more abundant than carbon, the contributions to the corresponding gamma-ray lines by excitation of ^{16}O cannot be neglected.

The first attempts to calculate the expected fluence ratio for the 15.10- and 4.438-MeV gamma-ray lines as a function of the parameters characterizing the spectrum of the incident protons were based on the approximation that the ratio for production of the two lines from oxygen is the same as that for production from carbon.[1] The first actual measurements of the gamma-ray production cross section for the reaction $^{16}O(p,\gamma_{15.10})x^{12}C$ were obtained for proton energies ranging from 40 to 85 MeV with the University of Maryland cyclotron and reported only recently.[2] These new measurements together with a re-evaluation of previously reported measurements resulted in substantial modifications to the cross sections for both the 4.438- and the 15.10-MeV lines. The ratio of gamma-ray production cross sections for the two lines do differ for the two nuclei, so that the flux ratio of the two lines from solar flares depends on the relative isotopic abundances of carbon and oxygen as well as on the spectrum of the exciting particles. The threshold for production of 15.10-MeV gamma rays from oxygen lies below the range of optimal operating energies that were accessible with the University of Maryland cyclotron, so this crucial portion of the spectrum was not determined.

Cross sections for incident proton energies ranging from 20 to 40 MeV were measured with both ^{12}C and ^{16}O targets just one month before this conference (November 1987) using the Princeton Cyclotron Facility. These measurements span the proton energy range from just below threshold for production of 15.10-MeV gamma rays from oxygen to just past the expected peak in this cross section. In the present work, the cross sections compiled and reported in Lang et al.[2] for production of 15.10- and 4.438-MeV gamma rays are employed, together with estimates of the 15.10-MeV gamma-ray production cross section near threshold from oxygen based on preliminary analysis of the new Princeton results, to calculate the fluence ratio resulting from proton excitation of carbon and oxygen in a solar flare. The fluence ratio is determined as a function of the parameter, αT, characterizing the incident proton spectrum represented by the modified Bessel function, K, of order 2, as suggested by Ramaty.[3] A thin-target model of the gamma-ray emission process was employed in determining the fluence ratios from the incident spectra and the measured cross sections. Contributions to the fluences resulting from alpha-particle interactions have not been included, nor have the reactions involving spallation of nuclei with atomic mass other than 16. Calculations of the alpha-particle contributions, currently being investigated by Werntz and his co-workers as part of a study of spectral line shapes, will be included in a more complete analysis.

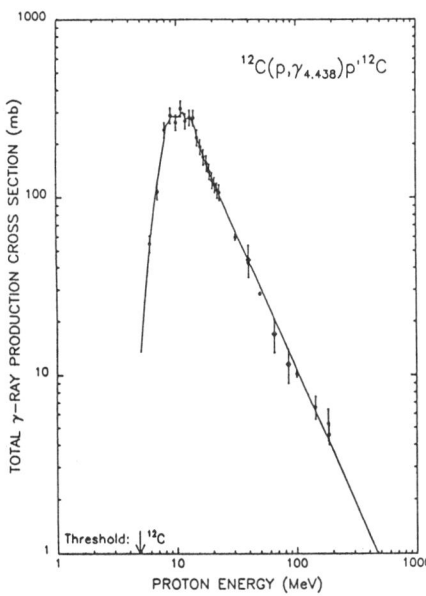

Figure 1. Total gamma-ray production cross sections for the 4.438-MeV line for proton interactions with ^{12}C as compiled and reported by Lang et al.[2]

MEASUREMENTS EMPLOYED

Values of the cross section employed in calculation of the 4.438-MeV line fluences produced by proton bombardment of ^{12}C are indicated by the solid line in Figure 1 as a function of incident proton energy. The data points are those compiled and reported as a self-consistent set.[2] Other measurements in the compilation of Lang et al., but found to be inconsistent,[4,5] are not included. The values employed for fluences of the same line from ^{16}O are indicated in Figure 2. Uncertainties in this cross section are even greater than the dearth of measurements above 25 MeV would suggest. Lang et al.[2] found that the three points that lie above the line are not only inconsistent with the new measurements they reported but also inconsistent with estimates obtained from the (p,p') cross sections and the associated branching ratios.

Values employed in calculating the 15.10-MeV line fluences produced by proton bombardment of ^{12}C and ^{16}O are indicated by the solid lines in Figure 3. The dashed line through the data points for oxygen is the same as the solid line through the carbon points except that it has been shifted by the difference in the respective threshold energies and scaled to fit the solid diamonds. The open diamonds shown for oxygen are the result of preliminary, on-line analysis of measurements recently obtained at the Princeton Cyclotron Facility. All of the other data are taken from the compilation of Reference 2.

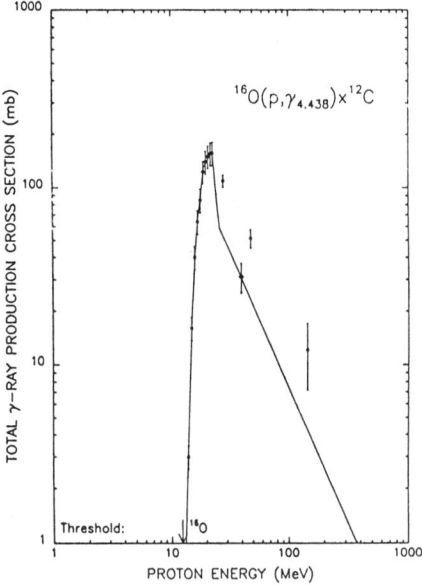

Figure 2. Total gamma-ray production cross sections for the 4.438-MeV line for proton interactions with ^{16}O as compiled and reported by Lang et al.[2]

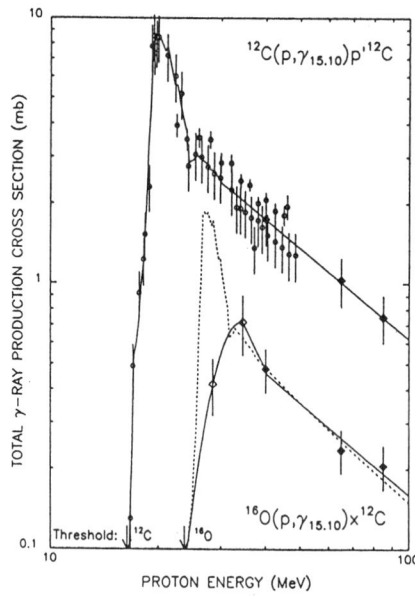

Figure 3. Total gamma-ray production cross sections for the 15.10-MeV line for proton interactions with ^{12}C and with ^{16}O as compiled and reported by Lang et al.[2] together with new estimates (open diamonds) near threshold in oxygen.

RESULTS AND CONCLUSIONS

The present calculation of the fluence ratio differs from the first attempt[1] in several significant respects. One of these is the use of newly available and improved determinations of the (p,p') cross sections. Another is the use of a different parametric form to characterize the spectrum of incident protons. Curves illustrating the shape of the spectrum of the form $K_2(\alpha T)$, for values of the parameter αT ranging from 0.020 to 0.040, are shown in Figure 4.

The fluence ratio was calculated according to the expression

$$\frac{\Phi_{15.10}}{\Phi_{4.438}} = \frac{\int dE_p \left[1.5\, \sigma^O_{15.10}(E_p) + \sigma^C_{15.10}(E_p)\right] \frac{dJ}{dE_p}}{\int dE_p \left[1.5\, \sigma^O_{4.438}(E_p) + \sigma^C_{4.438}(E_p)\right] \frac{dJ}{dE_p}}, \text{ where} \quad (1)$$

$\sigma^O_{15.10}(E_p)$ = cross section for the reaction, $^{16}O(p,\gamma_{15.10})x^{12}C$;

$\sigma^C_{15.10}(E_p)$ = cross section for the reaction, $^{12}C(p,\gamma_{15.10})p'^{12}C$;

$\sigma^O_{4.438}(E_p)$ = cross section for the reaction, $^{16}O(p,\gamma_{4.438})x^{12}C$;

$\sigma^C_{4.438}(E_p)$ = cross section for the reaction, $^{12}C(p,\gamma_{4.438})p'^{12}C$;

$\frac{dJ}{dE_p}$ = proton flux as expressed in Equation 14 of Forman, Ramaty, and Zweibel;[6] and the factor 1.5 is the abundance of oxygen relative to that of carbon, in accord with the results reported for solar flares by Murphy.[7]

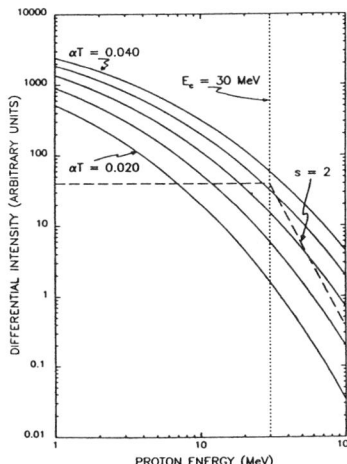

Figure 4. A family of curves (solid lines) showing the proton spectrum parameterized as the modified Bessel function K of order 2. Also shown as a dashed line is a spectrum that is a power law above 30 MeV with a slope of -2 and a constant below 30 MeV.

The resultant flux ratio is shown as a function of αT in Figure 5. This should be compared with the results obtained by Crannell, Crannell, and Ramaty[1] for a proton spectrum of the form illustrated by the dashed line in Figure 4: constant at energies below a cutoff energy, E_C, and powerlaw above E_C. The values obtained here span the same range as those obtained by Crannell, Crannell, and Ramaty for E_C less than 20 MeV. None of the values obtained for the flux ratio with the present calculation, however, is as large as those obtained by Crannell, Crannell, and Ramaty for values of E_C of 20 MeV or greater. This difference occurs because all of the proton spectra parameterized as $K_2(\alpha T)$ decrease by a factor of 6 or more between 5 and 20 MeV, the region of the spectrum in which cross sections for production of the 4.438-MeV line are maximum but below threshold for production of the 15.10-MeV line. The flux ratio is, thus, seen to be particularly sensitive to the steepness of the proton specrum below 20 MeV. Use of a thick-target model of the gamma-ray emission process, which is physically more realistic, would have the effect of hardening the spectrum of interacting protons and thereby significantly increasing the calculated ratio.

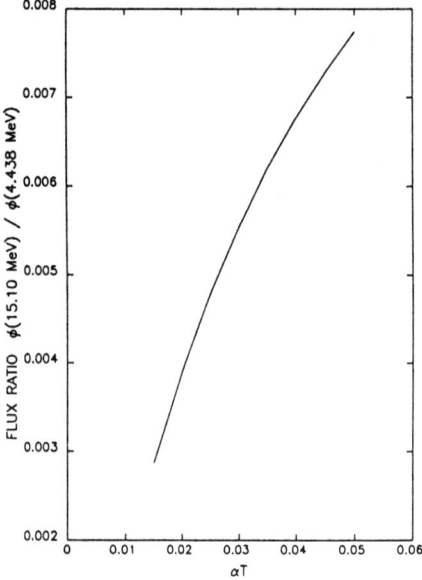 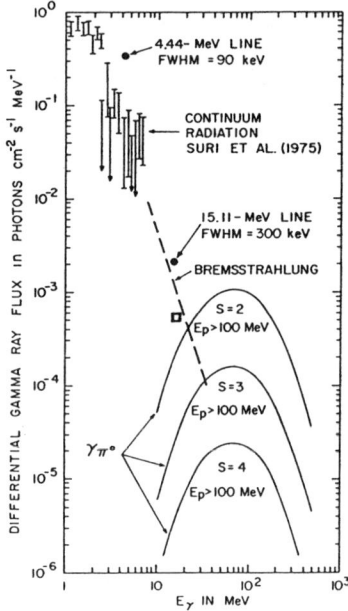

Figure 5. Ratio of the gamma-ray line fluence at 15.10 MeV relative to that at 4.438 MeV as a function of the parameter αT characterizing the spectrum of protons incident on carbon and oxygen in a solar flare.

Figure 6. Spectra of the gamma-ray line and continuum emissions taken from Figure 5 of Reference 1. The open square indicates the differential flux of 15.10-MeV gamma rays determined in the present work for protons incident on carbon and oxygen in the solar flare of 1972 August 4 with the observed 4.438-MeV line flux indicated.

If the energetic protons and heavier nuclei in a solar flare are accelerated to spectra that are similar when expressed as functions of particle energy per atomic mass unit (E/amu), then the effective threshold for gamma-ray line production by alpha-particle interactions is a factor of four less in E/amu than that for proton interactions (Cf. Figure 1 of Reference 1.). Thus, while the alpha-particle interactions contributed less than 10% to the 4.438-MeV line fluence calculated by Crannell, Crannell, and Ramaty for all but the $E_C = 0$ spectra, the alpha-particle interactions are not expected to be negligible for spectra of the $K_2(\alpha T)$ form. Furthermore, because alpha-particles are isospin zero, their interactions will contribute only to the 4.438-MeV line fluence and not to the 15.10-MeV line. Consequently, the ratios calculated here will drop by an additional amount, estimated to be approximately 30%, when contributions due to alpha-particle interactions are included.

For further comparison with previous results and for comparison of the expected 15.10-Mev line flux with other emissions actually observed in a solar flare, Figure 6 presents the value for that line emission determined in the present work superposed on spectra taken from Figure 5 of Reference 1. The calculated fluxes are normalized to the line and continuum emission observed below 8 MeV in the 1972 August 4 solar flare. The original 15.10-MeV line emission was calculated for a differential proton spectrum with $E_C = 30$ MeV and a powerlaw spectral index of 2 above that energy. The new value was calculated for a proton spectrum of the form $K_2(\alpha T)$ with the value of αT, 0.029, more recently suggested by Murphy and Ramaty using a thick-target model for that flare.[8] As noted above, this value is expected to be reduced an additional ~30% when the contributions due to alpha-particle interactions are included in the calculation. The line, nevertheless, remains detectable with the large spectrometers and their active anticoincidence shields being considered for flight during the maximum of the current solar cycle, 1991-1994.

The authors are indebted to Reuven Ramaty for informative discussions. This work was supported in part by NASA RTOP 188-38-01-04 and NASA Grant NSG 5006.

REFERENCES

1. C. J. Crannell, H. Crannell, and R. Ramaty, Astrophys. J. <u>229</u>, 762-771 (1979).
2. F. L. Lang, C. W. Werntz, C. J. Crannell, J. I. Trombka, and C. C. Chang, Phys. Rev. C <u>35</u>, 1214-1227 (1987).
3. R. Ramaty, Particle Acceleration Mechanisms in Astrophysics, J. Aron et al. (ed.), (New York, AIP, 1979), p. 135-154.
4. K. Strauch and F. Titus, Phys. Rev., <u>103</u>, 200 (1956).
5. W. Zobel, F. C. Maienschein, J. H. Todd, and G. T. Chapman, Nucl. Sci. Eng., <u>32</u>, 392 (1968).
6. M. A. Forman, R. Ramaty, and E. G. Zweibel, Physics of the Sun, Vol. <u>II</u>, P. A. Sturrock (ed.), (D. Reidel Publishing Company, 1986), p. 249-289.
7. R. Murphy, Gamma Rays and Neutrons from Solar Flares, Ph. D. Disseratation, University of Maryland, (1985).
8. R. Murphy and R. Ramaty, Advances in Space Research, <u>14</u>, No. 7, 127, (1984).

SOLAR FLARE GAMMA-RAY LINE SHAPES

C. W. Werntz* and F. L. Lang*
Code 682, Solar Physics Branch
NASA/Goddard Space Flight Center, Greenbelt Maryland 20771, and
Department of Physics, The Catholic University of America
Washington, DC 20064

ABSTRACT

Prompt gamma rays from nuclear states excited by flare-accelerated protons and alpha particles are emitted before the recoiling nuclei have slowed appreciably in the ambient solar medium. In particular, gamma-ray lines emitted after (p,p') and (α,α') excitations display characteristic shapes that are rapidly varying functions of the angle between the emitted gamma rays and the incident beam direction. This angular dependence has led to suggestions[1] that an analysis of certain strong and isolated gamma-ray lines should yield information about the directionality of the exciting ions. In the work reported here, the shapes and energy shifts of the gamma-ray lines from the $^{12}C(p,\gamma_{4.44})p'\,^{12}C$ and $^{16}O(p,\gamma_{6.13})p'\,^{16}O$ reactions have been calculated <u>ab initio</u> using standard techniques of nuclear theory. The calculated spectra are compared to laboratory spectra[2] for the carbon line, incoming proton energy 40 MeV, and for the oxygen line[3], E_p = 24 and 45 MeV.

INTRODUCTION

Nuclei at rest excited by (p,p') or (α,α') reactions recoil with sufficient velocity to induce Doppler shifts of many keV in the prompt de-excitation gamma rays. Since the excited nucleus can scatter into any angle in the forward direction the gamma-ray line from a single excited state is observed as a broad structure. This so-called coherent Doppler effect was first reported[4] when the doublet structure of the carbon 4.44-Mev line, produced by (p,p'), was observed with a high resolution gamma-ray detector. The possible utility of this complex line structure as a diagnostic in solar flare research was first suggested by Ramaty, Kozlovsky, and Lingenfelter[5].

The origin of the complex structure lies in the strongly anisotropic distribution of the gamma rays in the frame of the recoiling nucleus. This anisotropy arises because of selective excitation of magnetic substates[4] of the excited nucleus. Our goal of fitting observed line shapes in solar flares by folding in ion

---

*Supported in part by NASA Grant NSG 5066.

spectra and directional distributions requires a detailed
knowledge of line shapes in laboratory experiments. In this paper
the Doppler-broadened shapes of the carbon 4.44-MeV line and the
oxygen 6.13-MeV line are calculated and compared to laboratory
data at several energies. The success of the calculation suggests
that (p,p') and (α,α') line profiles can be calculated for any
angle over a range of projectile energies from near threshold to
150 MeV.

THEORY

In the reactions under study only the angle between the
incident beam direction and the emitted gamma, θ_γ, is fixed. It
is convenient to use the center of mass coordinate system in which
the z axis is the incoming beam direction and the gamma is
restricted to the xz plane so that $\phi_\gamma = 0$. The proton in a given
scattering event is scattered through the angles θ_p and ϕ_p and
the direction of recoil of the excited nucleus is defined by $\theta_N = \pi - \theta_p$, $\phi_N = \pi + \phi_p$.

The cross section for an inelastic scattering event
accompanied by the emission of a gamma ray can be written formally
as [5]

$$\frac{d^2\Sigma(E_p,\theta_p,\phi_p,\theta_\gamma)}{d\Omega_p d\Omega_\gamma} = \sigma(E_p,\theta_p) \, g(E_p,\theta_p,\phi_p,\theta_\gamma) \quad (1)$$

where $\sigma(E_p,\theta_p)$ is the differential inelastic scattering cross
section and $g(E_p,\theta_p,\phi_p,\theta_\gamma)$ is the gamma correlation function.

The profile of a particular gamma-ray line in the center of
mass system can be calculated by integrating over the unobserved
proton scattering angles. The Doppler shift energy E_c is defined
by $E_\gamma = E_o + E_c$, where E_o is the unshifted gamma energy in the
frame of the recoiling nucleus and E_γ is the observed gamma
energy. The cross section for the emission of a gamma into the
solid angle $d\Omega_\gamma$ within the Doppler shift interval dE_c can be
written

$$\frac{d^2\Sigma(E_p,\theta_\gamma,E_c)}{d\Omega_\gamma dE_c} = \iint d\cos\theta_p \, d\phi_p \frac{d^2\Sigma}{d\Omega_p d\Omega_\gamma} \delta(E_c - E_o \vec{\beta}_N \cdot \hat{k}_\gamma) \quad (2)$$

The delta function restricts the azimuthal angle of the recoiling
nucleus to that angle for which the gamma ray undergoes the
Doppler shift E_c. The velocity of the recoiling nucleus, in units
of c is $\vec{\beta}_N$, and $\hat{k}_\gamma$ is the unit vector in the direction of the
emitted gamma ray. It has been assumed that $\beta_N \ll 1$ so that the
gamma is assumed to have the same direction in all coordinate
systems of interest. The cross section can be calculated in the

laboratory or solar coordinate system by making the repacement $\vec{\beta}_N \rightarrow \vec{\beta}_N + \vec{\beta}_{CM}$, where $\vec{\beta}_{CM}$ is the velocity of the CM of the proton + target nucleus.

The differential cross section $\sigma(E_p, \theta_p)$ on the RHS of Eq. (1) is defined by

$$\sigma(E_p, \theta_p) = 1/2 \sum_{\nu, \nu', m} |a_{m\nu', \nu}(\theta_p)|^2 \qquad (3)$$

where the energy dependent $a_{m\nu', \nu}$ are the amplitudes for a proton in initial spin state ν to scatter through the angle θ_p into spin state ν', leaving the excited nucleus in magnetic substate m. For a $j^\pi \rightarrow 0^+$ nuclear transition the correlation function multiplying $\sigma(E_p, \theta_p)$ in Eq.(1) can be expressed[6] in terms of the nuclear density matrix $\rho_{m,m'}$ in the (2j + 1) dimensional magnetic sublevel space and the vector spherical harmonics $\vec{X}_{jm}(\theta_\gamma, 0)$ defined in Ref.7,

$$g(E_p, \theta_p, \phi_p, \theta_\gamma) = \sum_{m,m'} \rho_{m,m'}(E_p, \theta_p) \qquad (4)$$

$$\times \exp[i(m'-m)\phi_p] \vec{X}_{jm}(\theta_\gamma, 0) \cdot \vec{X}^*_{jm'}(\theta_\gamma, 0).$$

The density matrix is an incoherent sum over the unobserved proton spin states of a quadratic form in the scattering amplitudes,

$$\rho_{m,m'}(E_p, \theta_p) = \frac{1}{\sigma(\theta_p)} \frac{1}{2} \sum_{\nu, \nu'} a_{m\nu', \nu}(\theta_p) a^*_{m'\nu', \nu}(\theta_p). \qquad (5)$$

Contrary to Ref. 4, non-diagonal density matrix elements contribute to the correlation function g. The reduction of $\vec{X}_{jm}(\theta_\gamma, 0) \cdot \vec{X}^*_{jm}(\theta_\gamma, 0)$ to a sum over associated Legendre polynomials $P_L^{m'-m}(\cos\theta_\gamma)$ can be obtained from standard formulas for angular correlations[8].

The theoretical prediction of gamma line profiles reduces to a calculation of the energy and angle dependent scattering amplitudes $a_{m\nu', \nu}$. In the present work, the coupled channel code CHUCK2, supplied to us by P. D. Kunz of the University of Colorado, has been used to calculate the required amplitudes. This code numerically solves the Schroedinger equation for the proton + target nucleus in a Hilbert space of the ground state and one or more excited states of the nucleus. The complex optical potential is determined by the requirement that fits to experimental elastic and inelastic cross sections, as well as proton polarizations, be optimized. Satchler[9] has discussed the use of optical model codes to calculate the correlation function g and makes the point that combinations of density matrix elements different from the ones which determine proton spin observables are being sampled. The relatively good success of the present calculation of line profiles is gratifying both from the nuclear physics and solar physics standpoints.

COMPARISON TO DATA

The inputs to CHUCK2 are the diagonal complex potentials in each channel, the coupling potential, and the deformation parameters which characterize excited collective states. In keeping with common usage, the potentials in all channels were taken to be identical. Comfort and Karp[10] have parametrized the p + C potential at eight laboratory proton energies between 12.1 and 183 MeV. Their parameters for E_p = 40 MeV have been used to calculate the line profiles for the carbon 4.44-MeV line excited by (p,p'). The deformation parameter[11] of the excited state was chosen to be β_2 = 0.63. The comparison to 40-MeV data at θ_γ = 70° and θ_γ = 90° is shown in Fig. 1. An estimated background under the line due to detector response has been added to the theoretical curves as described in the figure captions. The theoretical curves have been normalized to fit the raw data, expressed in counts/channel, but, in general, absolute inelastic cross sections calculated with CHUCK2 fit experiments well.

Unlike the carbon 4.44 MeV line, which is broad even in solid targets, the oxygen 6.13 MeV line is broadened only when a gas target is used in an excitation experiment. Necessarily, the statistical error in the data becomes larger in such a case. Our theoretical curves for the oxygen 6.13 MeV line are compared in Figs. 2, 3 to experimental data at E_p = 23.7 and E_p = 44.6 MeV. Again a background has been added to the theoretical curves. The optical model parameters[12] were taken from a fit to experimental $^{16}O(n,n')^{16}O^*(6.13)$ data at E_p = 24 MeV. The parameters for the potential at E_p = 44.6 MeV were interpolated from the global optical model fit to $^{16}O(p,p')$ of van Oers and Cameron.[13] The deformation parameter for the excited 3⁻ state was fixed at β_3 = 0.60. An excess number of counts in the data compared to theory for $E_\gamma > E_o$ is presumably due to de-excitation gamma rays from $^{15}O^*(6.17)$ produced by the (p,pn) spallation reaction on ^{16}O.

The comparison to experimental data suggest that the shapes of gamma-ray lines arising from (p,p') excitations can be reliably calculated. One can expect that the shapes of lines following (α,α') excitations can be calculated with comparable confidence. The actual fitting of solar gamma-ray data will require that the contributions to the two lines by the spallation reactions $^{16}O(p,\gamma_{4.44})\alpha^{12}C$ and $^{16}O(p,\gamma_{6.17})pn^{15}O$ also be taken into account. A reasonably successful model of the shape of the 4.44-MeV line from spallation of ^{16}O has been presented in Ref. 2.

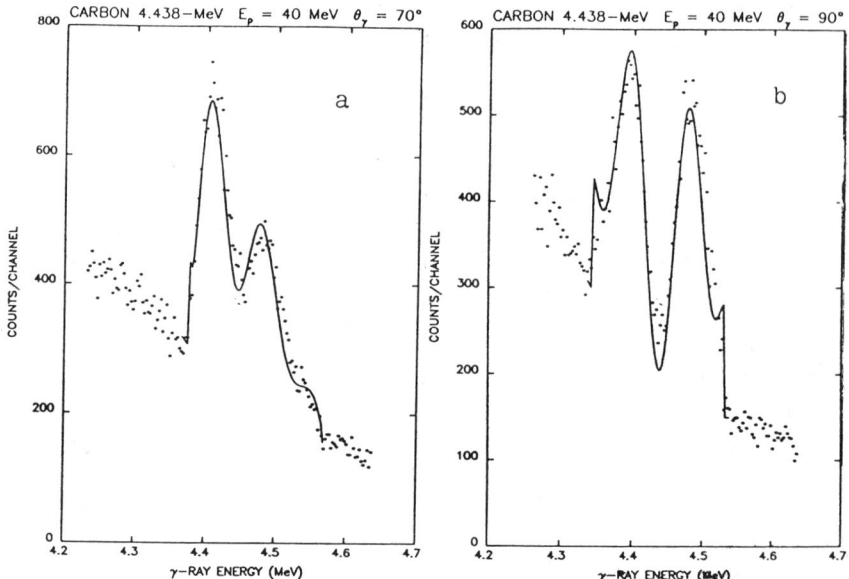

Fig. 1. Calculated line profiles compared to experimental data[2] for $^{12}C(p,\gamma_{4.44})p'^{12}C$ at E_p = 40 MeV, θ_γ = 70° (a) and 90° (b). A background of the form Nb = N1 (E- E2) / (E1- E2) + N2, where E1 and E2 are the lower and upper energy limits of the theoretical line profile, has been added. No experimental resolution has been folded in.

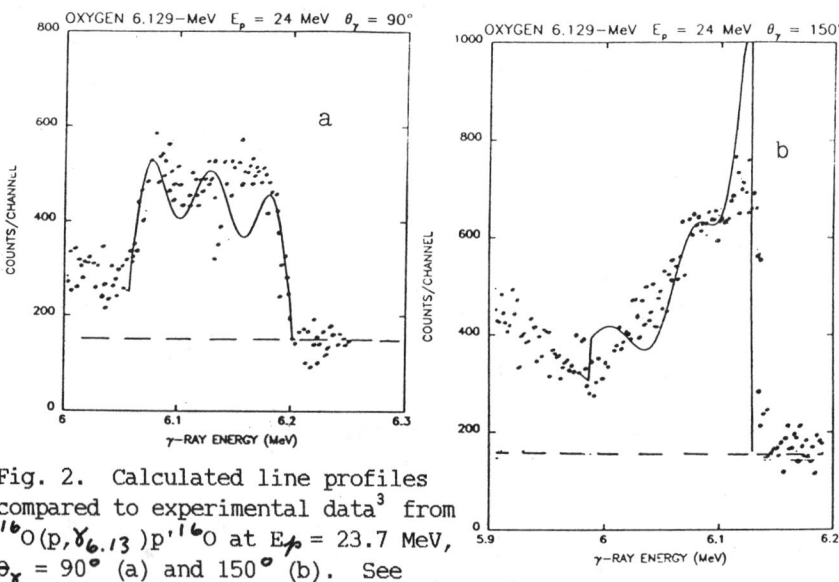

Fig. 2. Calculated line profiles compared to experimental data[3] from $^{16}O(p,\gamma_{6.13})p'^{16}O$ at E_p = 23.7 MeV, θ_γ = 90° (a) and 150° (b). See Fig. 1 for details about background and resolution.

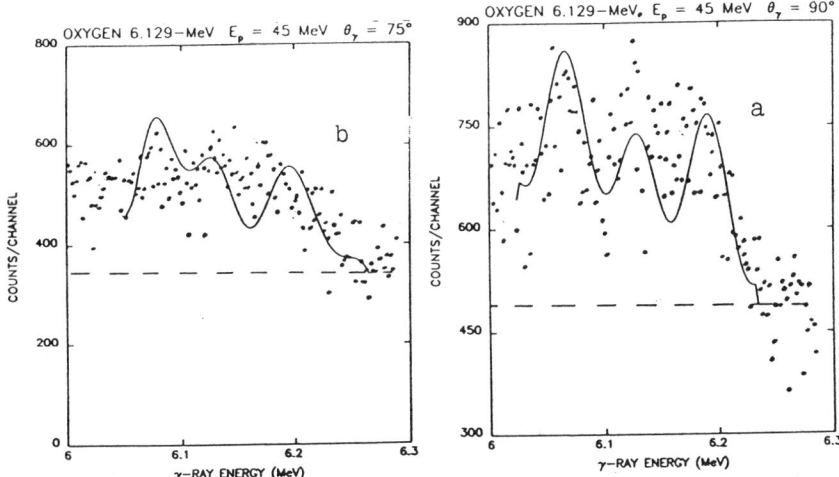

Fig. 3. Same as Fig. 2, except $E_p = 44.6$ MeV, $\theta_\gamma = 75°$ (a) and $90°$ (b). Excess of theory over data for $E_\gamma > E_0$ may be due to an unresolved $^{15}O^*(6.17)$ line.

REFERENCES

1. R. Ramaty and C. J. Crannell, Astrophys. J. 203, 766(1976).
2. F. L. Lang, C. W. Werntz, C. J. Crannell, J. I. Trombka, and C. C. Chang, Phys. Rev. C35, 1214(1987).
3. J. Narayanaswamy, P. Dyer, S. R. Faber, and S. M. Austin, Phys. Rev C24, 2727(1981) and private communication.
4. J. J. Kolata, R. Auble, and A. Galonsky, Phys. Rev. 162, 957(1967).
5. Reuven Ramaty, Benzion Kozlovsky, and Richard E. Lingenfelter, Ap. J. Suppl. Ser. 40, 487(1979).
6. F. H. Schmidt, et al., Nucl. Phys. 52, 353(1964).
7. J. D. Jackson, Classical Electrodynamics, 2nd. Ed. (Wiley, New York, 1975), p753.
8. M. Simonius, in Polarization Nuclear Physics, Lecture Notes in Physics, 30, edited by D. Fick, (Springer-Verlag, Berlin-Heidelberg-New York, 1974), p79.
9. G. R. Satchler, Nucl. Phys. 55, 1(1964).
10. J. R. Comfort and B. C. Karp, Phys. Rev. C21, 2162(1980).
11. G. R. Satchler, Nucl. Phys. A100, 497(1969).
12. P. Grabmayr, J. Rappaport, and R. W. Finlay, Nucl. Phys. A350, 167(1980).
13. W. T. H. Van Oers and J. M. Cameron, Phys. Rev. 84, 106(1969).

RELATIVE TIMING OF SOLAR PROMPT γ-RAY LINE AND X-RAY EMISSION PRODUCED BY FLARE ACCELERATED IONS AND ELECTRONS

E. Hulot, N. Vilmer and G. Trottet
UA 324 - DASOP - Observatoire de Paris
Section d'Astrophysique de Meudon, 92190, Meudon, France.

ABSTRACT

SMM and Hinotori observations show that the peak time of the γ-ray emission is sometimes delayed with respect to the one of the hard X-ray flux. Such delays may be interpreted either as an evidence of a two step acceleration process of electrons and ions or as the result of the partial trapping and/or propagation of the particles from the acceleration region to the interaction site. Here we focus on the latter hypothesis and present some preliminary calculations of the time dependent transport of energetic ions. Preliminary estimates of the 4.4 MeV line emission are used to discuss the relative timing of Hard X-ray and γ-ray emissions. One difficulty with the preliminary model discussed here is that the number of ions involved in the γ-ray line production is very large. Nevertheless, for reasonable parameters, a good agreement is found between observed and expected delays.

INTRODUCTION

SMM[1] and Hinotori[2] observations of Hard X-ray and γ-ray emissions have revealed that energetic ions and electrons are accelerated in the impulsive phase of solar flares. However in some events, the peak intensity of the higher energy channels (4-8 MeV range) is reached later (2 to a few 10s) than the peak at low energies. As the bulk of the 4-8 MeV emission is attributed to energetic ions, this led several authors to state that relativistic electrons and ions result from a second step acceleration which is delayed with respect to the acceleration of low energy electrons[3]. Alternatively these delays may be the result of partial trapping and/or propagation effects that electrons and ions suffer in their transport from the acceleration region to the hard X-ray and γ-ray emitting sources[1,4,5]. If models of the evolution of the electron population in the interaction region have been extensively developed so far[6], such models for the ions are still rare. The propagation of flare accelerated ions from the corona to the lower atmosphere taking into account partial trapping[7] as well as the relative temporal evolutions of the trapped electrons and ions[5] have been investigated independantly. However, these studies did not consider the relative timing of X-ray and γ-ray emissions taking into account both trapping and precipitation effects. The purpose of this contribution is to present preliminary calculations of the time dependant interaction of flare accelerated ions during their transport including both these effects. Preliminary results on the relative timing of the production rate of the 4.44 MeV ^{12}C deexcitation line and of the X-ray flux at different energies are discussed.

CHARACTERISTICS OF THE MODEL

The γ-ray line production is assumed to be entirely due to › 5 MeV protons. These particles are continuously injected at a rate

$$q(E,t) = q_0 E^{-\gamma} f(t) \qquad (1)$$

where E is the proton kinetic energy and f(t) the time profile of the injection. In order to discriminate between time variations due to acceleration and transport γ is kept constant during the injection. The protons are injected in the corona and propagate towards the chromosphere either directly or after escaping from a coronal trap of mean density n_0. The characteristics of the ambient medium are kept constant with time. The coronal trap is taken as a fully ionized hydrogen plasma while the underlying atmosphere is considered as a neutral gas of constant elemental abundances with an increasing density with depth (see Figure 1).

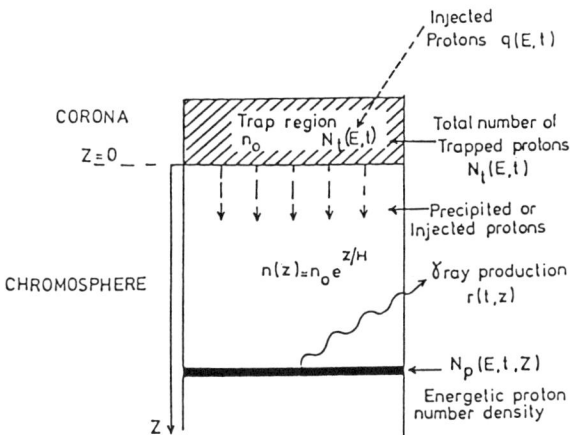

Fig. 1 : Schematic drawing of the coronal and chromospheric regions where energetic protons are injected and radiate γrays. The chromospheric density scale height is taken as H = 100 Km which is typical of a flaring atmosphere.

1. Temporal evolution of trapped and precipitated protons.

In this first approach, collective effects as well as magnetic mirroring and drifts are neglected. Energy losses are then entirely due to Coulomb collisions[8]. Scattering due to collisions, although acting on large time scales, allows some protons to escape from the trap. As a first approach, a continuity equation[9] is used to estimate the number of instantaneous protons which are trapped in the whole coronal region $N_t(E,t)$ and which precipitate $N_t(E,t)/T_f(E)$ ($T_f(E)$ is the escape time from the trap for a proton of energy E and is roughly equal to the deflection time [9,10]).

For simplicity the injection time profile is taken as :

$$f(t) = \begin{cases} t(2t_{max} - t) & \text{for } 0 < t < 2t_{max} \\ 0 & \text{elsewhere} \end{cases} \qquad (2)$$

The "injection time" t_{inj} is then $2 t_{max}$.

2. Propagation of the protons in dense layers.

The footpoint of the coronal trap is at z = 0 where the density $n_{(z)} = n_0$ (see Figure 1). The protons are supposed to have velocities $v(E)$ along z. For a neutral gas energy losses are essentially due to excitation and ionization of the ambient hydrogen atoms $(dE/dt)_I$[8]. Pitch angle scattering due to interactions with the neutral atoms occur on time scales much longer than those considered below and can be thus neglected. Following earlier calculations concerning electrons[6], the instantaneous linear density of protons $N_{p(E,t,z)}$ (Mev^{-1} km^{-1}) is related to the injected population $q'_{(E,t)}$ at z = 0 by :

$$\frac{\partial N_{p(E,t)}}{\partial t} + \frac{\partial}{\partial E}\left[N_{p(E,t,z)}\left[\frac{dE}{dt}\right]_I\right]$$
$$+ \frac{\partial}{\partial z}\left[N_{p(E,t,z)} v_{(E)}\right] = q'_{(E,t)} \delta_{(z)} \quad (3)$$

where $q'_{(E,t)}$ is either given by equations (1) and (2) or taken as $N_{t(E,t)}/T_{f(E)}$.

3. Production of prompt gamma-ray lines.

As an example, we shall only consider the 4.44 MeV line produced by the reaction $^{12}C(p,p')$ $^{12}C^{*4.439}$. The instantaneous reaction rate[4] of the protons with the ambient ^{12}C is given by

$$r_{(t,z)} = \int_0^\infty n_{c(z)} N_{p(E,t,z)} v_{(E)} \sigma_{(E)} dE \quad [Km^{-1} s^{-1}] \quad (4)$$

where $n_{c(z)}$ is the density of the ambient ^{12}C atoms. As we assumed constant abundances, the density scale height of ^{12}C is equal to H.

The energy dependent total cross section $\sigma_{(E)}$ is estimated from Ramaty et al[11] (1979) and Lang et al (1986)[12]. Assuming that γ-ray photons are produced isotropically in an optically thin interaction region, the flux line is proportional to $r_{(t,z)}$[13].

PRELIMINARY RESULTS

1. Trapped and precipitated protons.

$N_{t(E,t)}$ depends on four parameters : q_0, t_{max}, γ and n_0. For given q_0, t_{max} and γ, figure 2 shows the peak time $t_{max(E)}$ of $N_{t(E,t)}$ as a function of E for different values of n_0. For small values of n_0, energy losses are small at all energies. Protons are thus stored in the trap and reach their maximum towards $2t_{max}$ (curve 1). On the contrary for high values of n_0, particles lose energy on a time scale smaller than t_{inj} and $t_{max(E)} \approx t_{max}$ (curve 6). For intermediate values of n_0 (curves 2,3,4) there is an energy increasing delay $\Delta t_{(E)} = t_{max}(E) - t_{max}$. For a given energy $\Delta t_{(E)}$ increases with decreasing n_0. Moreover calculations show that $\Delta t_{(E)}$ increases with t_{max}. The existence of such delays for trapped protons was pointed out by Ryan (1986). At a given energy the ratio of trapped and precipitated particles varies roughly as $1/n_0$ and is about 10^{-5} for E = 20 MeV and $n = 10^{11}$ cm^{-3}.

2. Propagation in dense layers of the atmosphere.

$N_{p(E,t,z)}$ as deduced from equation (3) depends on the five parameters : q_0, t_{max}, γ, n_0 and H. As expected from the energy dependance of ionization losses the spectrum progressively hardens

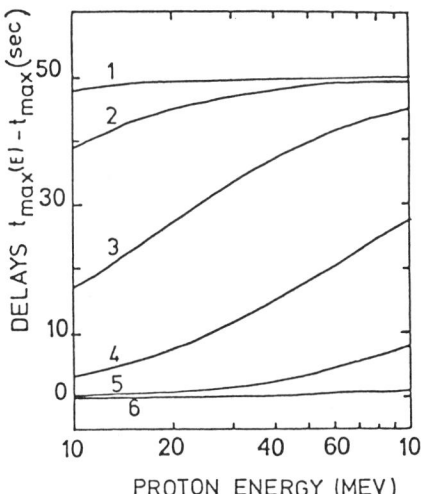

Fig. 2 : Computed delays $t_{max(E)} - t_{max}$ for $\gamma = 3.5$, $t_{max} = 50s$ as a function of the proton energy E, for different values of the trap densities. Curves 1, 2, 3, 4, 5, 6 correspond respectively to $n_o = 10^9$, 10^{10}, 10^{11}, 10^{12}, 10^{13} and 10^{14} cm^{-3}. These values for the densities are used to show the expected effects although reasonable values for coronal traps are smaller than 10^{12} cm^{-3}.

with depth. The temporal evolution of $N_{p(E,t,z)}$ is very similar to the injection profile since in the case considered in Figure 1 propagation effects occur on very short time scales as compared to the injection time. In the case where protons injected at z = 0 are precipitated from the trap, a similar hardening of the proton spectrum with depth is observed. However, in that case, the temporal evolution of $N_{p(E,t,z)}$ at a given depth reflects the delayed proton injection at z = 0. At a fixed energy E, the maximum of $N_{p(E,t,z)}$ is reached later deep in the atmosphere, because of the higher photon energy required at z = 0.

3. Time profiles of the 4.44 MeV gamma-ray line.

It is obvious from the previous section that a direct injection of protons at z = 0 leads to a 4.44 MeV time profile similar to the injection one. For the assumed parameters the line production comes essentially from layers of densities between about 3.10^{14} cm^{-3} to 6 10^{15} cm^{-3} with a maximum production at n ~ 10^{15} cm^{-3}. This reasonably agrees with densities inferred by other authors[13]. The results are then similar to those obtained from a thick target model. Figure 3 shows the temporal evolution of the 4.44 MeV production rate r(t,z) at different depths for protons first injected in the trap. As expected, the behaviour of $r_{(t,z)}$ is similar to the one of the ~ 10-20 MeV protons which essentially contributes to this line. The maximum of the production rate is delayed as compared to the injection maximum. This delay increases with depth since higher energy injected protons are required to reach higher depths. Moreover, the production rate exhibits a long tail which, for similar reasons, is longer at high energies. As in the case of a direct proton injection, the maximum of the photon production occurs at n ~ 10^{15} cm^{-3}. The total 4.44 MeV production rate obtained after integration of r(t,z) over z (figure 3) also exhibits a long tail and a delay of its maximum with respect to t_{max}. In agreement with previous results[5,7], the duration of the tail depends on the trap density, being longer for lower densities. The relative delay of the maximum of the 4.44 MeV production rate and of the injection is a decreasing function of n_o (for $n_o \geqslant 10^9$ cm^{-3}) and an increasing function of t_{max}.

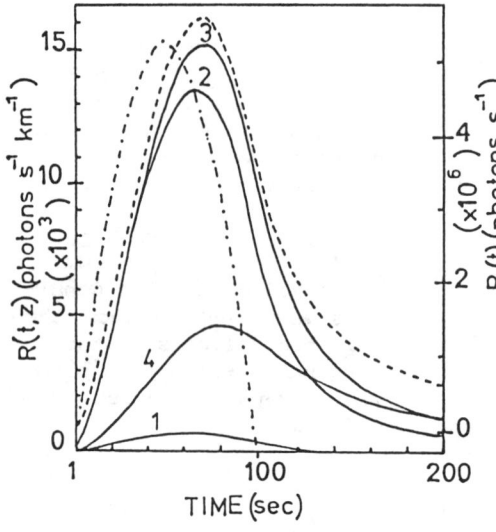

Fig. 3 : Computed values of $R(t,z) = 3\ 10^{24}\ r(t,z) / a_c$ (photons s^{-1} km^{-1}) for $\bar{\gamma} = 3.5$, $t_{max} = 50s$, $n_0 = 10^{11}\ cm^{-3}$, $H = 100$ Km, $q_0 = 1$ as a function of time for different depths z. a_c is the ^{12}C elemental abundance. Curves 1, 2, 3 and 4 correspond respectively to z = 500, 900, 1000 and 1200 Kms. The dashed curve refers to $R(t) = 3\ 10^{24}\ r(t)/a_c$ where $r(t)$ is the total 4.44 MeV production rate while the dotted-dashed one shows the injection time profile.

4. Relative timing ot the 4.44 MeV γ-ray line and of the X-ray flux at different energies.

The hard X-ray emission produced by the simultaneously flare accelerated flare electrons injected in the same trap is also estimated using a similar model as the one considered for protons. As the purpose is to evaluate the differences in the temporal evolutions of the fluxes due to the particle interactions, the electron injection function is taken to be identical to the proton one in spectral and temporal shape. This is justified because of the weak dependance of the temporal evolution of the X-ray flux on the spectral hardness of the injection[6]. Under these conditions, the timing of the 4.44 MeV γ-ray line produced by the protons compared to the maximum of the X-ray flux produced by both trapped and precipitated electrons[6] is estimated. The results are presented in Table 1. It can be noticed that with reasonable trap densities ($n_0 \simeq 10^{10} - 10^{11}\ cm^{-3}$) these delays are of the same magnitude as the ones observed in the solar flares of June 7 1980[1] and July 11 1978[14] if t_{max} is chosen to be comparable to their durations (FWHM). Moreover, these delays are an increasing function of t_{max}, thus of the burst duration as observed on several events[1].

DISCUSSION AND CONCLUSION

The preliminary calculations presented in this contribution show that the interaction of electrons and protons with the ambient medium may account for the observed relative timing between hard X-rays and γ-rays. If both electrons and protons are directly injected at the top of the chromosphere, hard X rays and γ-rays will show up similar time profiles. On the contrary, if the injection occurs in a trap of medium density (~ $10^{10} - 10^{11}\ cm^{-3}$) the γ-ray time profile will be delayed with respect to the hard X ray one, the latter peaking earlier. For reasonable parameters, the expected delays agree with the measured ones and exhibit the observed increase with the peak duration.

Table 1 : Delays between the peak times of the 4.44 MeV line and of the hard X-ray flux at different energies as a function as the trap density for two values of t_{max}.

X-Ray Energy (KeV)	Trap Density n_o (cm^{-3})		
	10^{10}	10^{11}	10^{12}
t_{max} = 5 sec			
30	2.7sec	3.2sec	1.9sec
100	1.2sec	2.4sec	1.7sec
300	0.3sec	1.2sec	1.2sec
t_{max} = 50 sec			
30	33sec	19sec	6.1sec
100	24sec	17sec	5.8sec
300	13sec	12sec	4.8sec

However, the main difficulty is that the number of protons needed to produce the observed γ ray flux is several orders of magnitude larger than that deduced from a pure thick target model. Such a difficulty may be resolved if collective effects in the trap and a broader angular distribution of the precipitated protons are taken into account.

REFERENCES

1. E.L. Chupp : Ann. Rev. Astron. Astrophys., 22, 359 (1984).
2. M. Yoshimori, K. Okudaira, Y. Hirasima, I. Kondo : Solar Phys. 86, 375 (1983).
3. T. Bai : in R.E. Lingenfelter, H.S. Hudson, D.M. Worall (eds.) Gamma Ray Transients and Related Astrophysics, AIP, 315 (1982).
4. R. Ramaty : in P.A. Sturrock (ed.), Physics of the Sun, 11, 291 (1986).
5. J.M. Ryan : Solar Phys. 105, 365 (1986).
6. N. Vilmer : Solar Phys. 111, 207 (1987).
7. E.G. Zweibel, D.A. Haber : Astrophys. J. 264, 648 (1983).
8. J.D. Jackson, Classical Electrodynamics, John Wiley and Sons, New-York (1962).
9. D.B. Melrose, J.C. Brown : Montly Notices Roy. Astron. Soc. 176, 15 (1976).
10. B.A. Trubnikov : Rev. Plasma Phys. 1, 105 (1965).
11. R. Ramaty, B. Kozlovsky, R.E. Lingenfelter : Astrophys. J. Supp. 40, 487 (1979).
12. F.L. Lang, C.W. Werntz, C.J. Crannell, J.L. Trombka, C.C. Chang : Phys. Rev. C 35, n° 4, 1214 (1987).
13. R. Ramaty, R.J. Murphy : Space Science Reviews, to be published (1987).
14. H.S. Hudson, T. Bai, D.E. Gruber, J.L. Matteson, P.L. Notan, L.E. Peterson : Astrophys. J. 236, L91 (1980).

GAMMA-RAY BURST SPECTROSCOPY

K. Hurley
Space Sciences Laboratory, University of California, Berkeley, CA 94720

ABSTRACT

The statistics and physical properties of gamma ray burst spectral features are presented, following a brief review of early theoretical predictions of, and experimental searches for, such emission. The ~100 observations of 50 keV absorption and 400 keV emission features constitute the largest data base on neutron star lines. Although the statistical significance of individual observations is often weak, and interpretation of these features as cyclotron absorption and annihilation radiation poses some theoretical problems, it is clear that future observations may have far-reaching implications.

PALEOSPECTROSCOPY

Gamma-ray bursts (hereafter GRBs: see [1] for a comprehensive review) are widely believed to be associated with galactic neutron stars in large measure because of observations of features in their energy spectra — features which may be interpreted to obtain magnetic field strengths and gravitational redshifts. It has therefore become common to associate burst spectroscopy with these observations, which have been carried out since 1978. It is less commonly realized, however, that far-sighted experimenters and theoreticians searched for and predicted the existence of such lines long before their discovery. In 1973, in one of the first observations of GRB energy spectra,[2] Cline et al. concluded that although no line structure was observed, "... great improvements in energy *and time* resolution might show fine-scale spectral variability with a variety of monochromatic lines ..." (emphasis added). It is important to recall the context of the times: GRBs had just been discovered,[3] solar flare line emission was still a novelty,[4] and apart from a tentative report of line emission from the galactic center,[5] the detection of line emission from other astrophysical sources was still several years in the future. While the remark concerning the energy resolution was obvious, that relating to the time resolution has turned out to be prophetic, as we will see below. The next mention of a search for line emission came in 1974: Metzger et al.,[6] observing an intense GRB on Apollo 16, set limits to 511 keV and 2.2 MeV line emission around 0.3% and 3%, respectively, of the total burst continuum flux. Finally, the first high time resolution search for lines with a Ge spectrometer (albeit with degraded energy resolution) was made by Imhof et al. in 1975:[7] here too the results were negative.

At the same time, theoreticians were predicting the existence of the line emissions which experimenters could not find. Stecker and Frost, in their stellar superflare model,[8] pointed out that burst spectra might have the same line emission as solar flares. Sofia and Van Horn, in their antimatter-normal star collision theory,[9] predicted 0.5 MeV line radiation. Fission and radiative capture spectra were predicted by Bisnovatyi-Kogan et al.,[10] and 511 keV emission by Prilutski and Usov.[11]

© 1988 American Institute of Physics

Few of the theories which these predictions were based on are still in vogue. Nevertheless, they played the important role of pointing out the value of GRB spectroscopy.

OBSERVATIONS

The first reports of the detection of spectral features in GRBs were due to the Ge spectrometer aboard the ISEE-3[12] spacecraft, and the Konus NaI(Tl) scintillation counters aboard Venera 11 and 12.[13] Subsequently, features have been found in the data of the HEAO A-4, SIGNE, and Venera 13 and 14 experiments. The statistical significance or confidence level of the line detection ranges from 1.4 and 3.45 σ for ISEE, to up to ~10 σ for Konus, to 99.9% confidence for HEAO. These values are obtained after spectral deconvolution; it should be noted that typical deconvolution procedures yield non-unique, model dependent, and even "obliging" results, in the sense that the corrected measurements tend to comply to assumed model fits.[14] Thus the "true" significance of a line detection may be quite difficult to estimate accurately in practice, particularly when the instrumental energy resolution is broad. (It is unfortunate in this respect that the ISEE Ge spectrometer was unable to record high resolution spectra for most of the mission due to a problem with the electronics.) Another complicating factor is that there are often many spectral accumulation time intervals which may be examined for lines of various energies. The results obtained to date are summarized in Table I. In some cases, the data reduction is continuing, and more detections can be expected. An important conclusion which can be drawn from this table is that, with roughly 100 line detections, GRBs represent a significant fraction of the line observations from all astrophysical objects, including the sun.

Table I. GRB Line Feature Detections

Spacecraft	Instrumentation	Number of events			References
		A. Total	B. With L.E. features	C. With H.E. features	
ISEE-3	Ge	$\geq$24	(not observable)	1	12
HEAO-A	NaI	21	2	1	15
Venera 11, 12	NaI (Konus)	143	~30 (~21%)	15 (10%)	13,16,17,18
	NaI (SIGNE)	39	1	1	19,20
Venera 13, 14	NaI (Konus)	130	$\geq$20 (>15%)	29 (22%)	13,21

Table II describes the characteristics of the observed features, which fall into two categories: low-energy absorption features and high-energy emission features. In both cases the features are broad, which instructs us to use caution in interpreting them as "lines." The last column in this table points out another problem in the interpretation of these features: that of confirmation. For a transient event such as a GRB, containing a transient spectral feature, "confirmation/non-confirmation" may either be non-simultaneous (i.e., features are detected/not detected by different instruments in *different* GRBs) or simultaneous (i.e., features are detected/not detected by different instruments in the same GRB: obviously a stronger confirmation; it would be best to reserve the term "confirmation" for simultaneous

Table II. Properties of GRB Line Features

Energy, keV	Type	Instantaneous line to continuum flux ratio	Line width, keV	Duration, sec	Confirmation?/ References
~27–70	Absorption	1–18%	~3–>28	≤4	Yes, but not simultaneous[15] and No[14,22]
350–500	Emission	3–30%	200–990	≲0.25–>20	Yes[12,17;20,17] and No[21,23,18]

detections). For both types of features, the picture is mixed. The HEAO A-4 experiment[15] has detected both low- and high-energy features in GRBs, but not simultaneously with another experiment. For one event, a simultaneous non-confirmation of an absorption feature has occurred.[14,22] There have been simultaneous confirmations of the high-energy emission features, although in one case the estimated "line" fluxes differ by a factor of 10,[12,17] and there has been a simultaneous non-confirmation of one as well.[21,23,18] This confusing situation has led to some scepticism concerning the reality of the results. A glance at Figure 1, however, reveals one possible explanation for the confusion: not only are the line features themselves transient on short timescales (down to 0.25 s or less[20]) but also the continuum against which they must be measured is changing rapidly not only in intensity, but also in shape. Thus in order to confirm a feature simultaneously, two experiments must have both good energy and time resolution, and spectra taken over the same time and energy intervals must be compared. This is almost never achieved in practice.

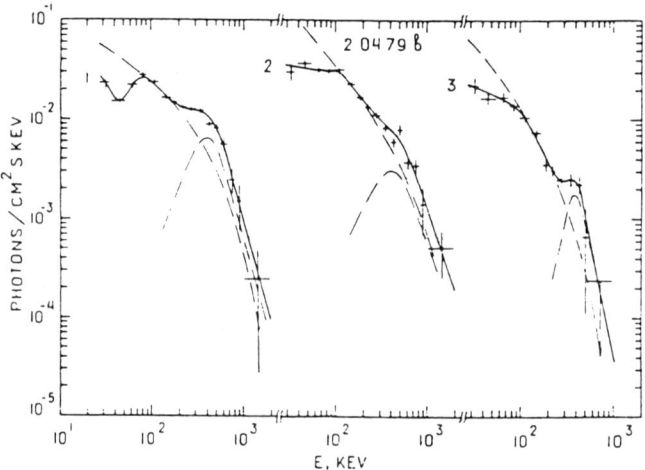

Fig. 1. Three successive energy spectra, taken over 4, 4, and 8 s, for a GRB (from Golenetskii et al.[18]). Note the strong time evolution of the low- and high-energy features, and the continuum.

INTERPRETATION

The low- and high-energy features are commonly explained as cyclotron absorption in a strong magnetic field[16] and gravitationally redshifted e^+-e^- radiation;[12,16] the values of B and z are what would be expected for a neutron star. Both explanations have their difficulties. One is that lines, even "lines" as broad as these, must be generated in cold regions, whereas the underlying continuum indicates that high temperatures (several hundred keV) are present somewhere in the source. For the absorption features, this requires that a cold region with an intense magnetic field overlie the hot emitting region.[24] Alternatively, one could attempt to explain these features by a two-component spectrum,[25] although this too requires field strengths of the same magnitude. If the high-energy emission features were generated in a high temperature region, they would be both broadened and blueshifted.[26] Golenetskii et al.[18] have suggested that their observations may in fact be consistent with e^+-e^- annihilation line production in a multi-temperature plasma. It is also possible, in some cases, that we are observing either the radiation from pairs which have cooled before annihilating,[27,19] or stimulated annihilation radiation.[29]

If we overlook these difficulties in interpretation, and accept the neutron star hypothesis, then it is clear from Table I that these observations represent the majority of the current data base on neutron star line features. The statistics of Table I, however, indicate that only about one out of every 5 bursts displays a line feature. Detector threshold effects and observing geometry may play a role in determining this number. In any case, it is interesting to compare these statistics with those of other X-ray sources. Only two definite, and one tentative, observations of cyclotron line features have been reported[30,31,32] out of about two dozen X-ray pulsators which are fairly certain to be associated with neutron stars. (Perhaps it is significant for the interpretation of gamma burst line features that these are accreting binary X-ray sources.) Only two line features in the 400 keV range have been reported, one associated with the Crab,[33] and the other with the 10 June 1974 transient.[34] Thus the frequency of "line" observations in the case of GRBs, while low, is somewhat higher than the frequency for other sources associated with neutron stars.

What is the source of the positrons needed to explain the annihilation lines? Broadly speaking, two classes of sources can be invoked: photon pair production, and nuclear positron emitters. In the first case, pairs may be produced either by single photon production in a strong magnetic field[35] or by photon-photon collisions.[36] In the second, accreting matter with cosmic abundances and energies of about 100 MeV/nucleon undergoes nuclear reactions,[37,38,39] exciting nuclei which decay by a variety of reactions including γ-ray line emission. It is not clear whether this is a significant source of positrons in GRBs,[40] since the half-lives of the nuclei must be of the order of the burst duration or less. However, the possible existence of excited nuclei raises an interesting question which is examined below.

FUTURE GOALS

It is essential for future GRB spectrometers to resolve line features with good statistics, in order to measure line shapes and energies accurately. This requires not only good energy resolution, but time resolution on the order of several hundred milliseconds or better. Several experiments now being built will contribute significantly to our understanding of GRB energy spectra. One is the Franco-Soviet Lilas instrument aboard the Soviet Phobos mission which will be launched in 1988, and the other is the spectroscopy modification to the BATSE instrument aboard GRO, to be launched in 1990. Both employ high-resolution NaI scintillators. In the future, xenon drift chambers[41,42] and Ge arrays will be needed to progress in the direction of better energy resolution and larger surface areas.

One obvious goal of such experiments is to determine the true nature of the line features and separate them from the continuum. Another is to search for nuclear line contributions in burst spectra. Figure 2 shows the high-resolution measurement of Teegarden and Cline,[12] with line contributions at 420 and 738 keV (~1.4 and 3.45 σ, respectively). If the latter is interpreted as the redshifted 847 keV ^{56}Fe line, and if matter with solar abundances is present, line contributions at higher energies are expected, as shown. Similar suggestions have been made by other authors as well.[43,44,45]

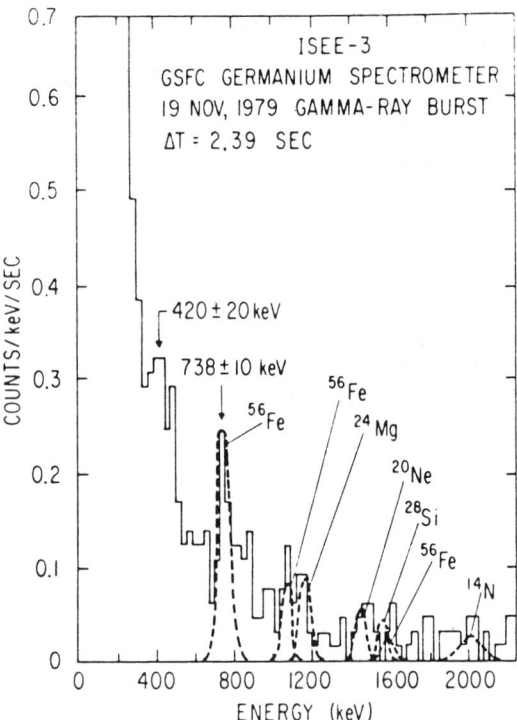

Fig. 2. High-resolution spectrum of the November 19, 1979 GRB, with line contributions at 420 and 738 keV (from Teegarden and Cline[46]). Normalizing to the latter line, possible contributions at higher energies derived from Ramaty et al.[47] are shown.

One interesting application of these measurements will be the study of the internal structure of neutron stars: gravitational redshift measurements can be used to constrain the neutron star equation of state.[48,49,50] But quite possibly the most interesting discoveries are still unpredictable. Past improvements in the sensitivities, energy ranges and time resolution of GRB detectors have consistently revealed new, unexpected phenomena, and there is every reason to believe that improvements in the energy resolution will do likewise.

ACKNOWLEDGEMENTS

This work was supported by University of California, Berkeley, Space Sciences Laboratory general funds from the State of California.

REFERENCES

1. E. P. Liang and V. Petrosian, Eds., Gamma Ray Bursts (AIP Press, N.Y., 1986).
2. T. Cline, U. Desai, R. Klebesadel, and I. Strong, Astrophys. J. Lett. **185**, L1 (1973).
3. R. Klebesadel, I. Strong, and R. Olson, Astrophys. J. Lett. **182**, L85 (1973).
4. E. Chupp, D. Forrest, P. Higbie, A. Suri, C. Tsai, and P. Dunphy, Nature **241**, 333 (1973).
5. W. Johnson, F. Harnden, and R. Haymes, Astrophys. J. Lett. **172**, L1 (1972).
6. A. Metzger, R. Parker, D. Gilman, L. Peterson, and J. Trombka, Astrophys. J. Lett. **194**, L19 (1974).
7. W. Imhof, G. Nakano, R. Johnson, J. Kilner, J. Reagan, R. Klebesadel, and I. Strong, Astrophys. J. **198**, 717 (1975).
8. F. Stecker and K. Frost, Nature Phys. Sci. **245**, 70 (1973).
9. S. Sofia and H. Van Horn, Astrophys. J. **194**, 593 (1974)
10. G. Bisnovatyi-Kogan, V. Imshennik, D. Nadyozhin, and V. Chechetkin, Astrophys. Space Sci. **35**, 23 (1975).
11. O. Prilutski and V. Usov, Astrophys. Space Sci. **34**, 395 (1975).
12. B. Teegarden and T. Cline, Astrophys. J. Lett. **236**, L67 (1980).
13. E. Mazets, S. Golenetskii, V. Il'inskii, R. Aptekar, and Yu. Guryan, Nature **282**, 587 (1979).
14. E. Fenimore, R. Klebesadel, and J. Laros, Adv. Space Res. **3** (4), 207 (1983).
15. G. Hueter, HEAO-1 Observations of Gamma-Ray Bursts, Ph.D. Thesis, University of California, San Diego, Department of Physics, 1987.
16. E. Mazets, S. Golenetskii, R. Aptekar, Yu. Guryan, and V. Il'inskii, Sov. Astron. Lett. **6** (6), 372 (1980).
17. E. Mazets, S. Golenetskii, R. Aptekar, Yu. Guryan, and V. Il'inskii, Nature **290**, 378 (1981).
18. S. Golenetskii, E. Mazets, R. Aptekar, Yu. Guryan, and V. Il'inskii, Astrophys. Space Sci. **124**, 243 (1986).
19. C. Barat, AIP Conf. Proc. No. 101 (AIP Press, N.Y., 1983), p.54.
20. C. Barat, K. Hurley, M. Niel, G. Vedrenne, I. Mitrofanov, I. Estulin, V. Zenchenko, and V. Dolidze, Astrophys. J. Lett. **286**, L11 (1984).
21. E. Mazets, S. Golenetskii, Yu. Guryan, R. Aptekar, V. Il'inskii, and V. Panov, AIP Conf. Proc. No. 101 (AIP Press, N.Y. 1983), p. 36.
22. B. Dennis, K. Frost, A. Kiplinger, L. Orwig, U. Desai, and T. Cline, AIP Conf. Proc. No. 77 (AIP Press, N.Y., 1982), p. 153.
23. P. Nolan, G. Share, E. Chupp, D. Forrest, and S. Matz, Nature **311**, 360 (1984).
24. E. Liang, Comments Astrophys. **12**, 35 (1987).

25. J. Lasota and B. Belli, Nature **304**, 139 (1983).
26. R. Ramaty and P. Meszaros, Astrophys. J. **250**, 384 (1981).
27. R. Ramaty, S. Bonazzola, T. Cline, D. Kazanas, P. Meszaros, and R. Lingenfelter, Nature **287**, 122 (1980).
28. V. Zheleznyakov, Astrophys. Space Sci. **83**, 117 (1982).
29. R. Ramaty, J. McKinley, and F. Jones, Astrophys. J. **256**, 238 (1982).
30. J. Trumper, W. Pietsch, C. Reppin, and W. Voges, Astrophys. J. Lett. **219**, L105 (1978).
31. W. Wheaton, J. Doty, F. Primini, B. Cooke, C. Dobson, A. Goldman M. Hecht, J. Hoffman, S. Howe, A. Scheepmaker, E. Tsiang, W. Lewin, J. Matteson, D. Gruber, W. Baity, R. Rothschild, F. Knight, P. Nolan, and L. Peterson, Nature **282**, 240 (1979).
32. G. Maurer, W. Johnson, J. Kurfess, and M. Strickman, Astrophys. J. **254**, 271 (1982).
33. M. Leventhal, C. MacCallum, and A. Watts, Nature **266**, 696 (1977).
34. J. Ling, W. Mahoney, J. Willett, and A. Jacobson, AIP Conf. Proc. No. 77 (AIP Press, N.Y. 1982), p. 143.
35. J. Daugherty and A. Harding, Astrophys. J. **273**, 761 (1983).
36. P. Guilbert, A. Fabian, and M. Rees, Mon. Not. R. Astr. Soc. **205**, 593 (1983).
37. V. Shvartsman, Astrofizika **6** (1), 123 (1970).
38. R. Ramaty, G. Borner, and J. Cohen, Astrophys. J. **181**, 891 (1973).
39. K. Brecher and A. Burrows, Astrophys. J. **240**, 642 (1980).
40. B. Kozlovsky, R. Lingenfelter, and R. Ramaty, Astrophys. J. **316**, 801 (1987).
41. B. Sadoulet, R. Lin, and S. Weiss, IEEE Trans. Nuc. Sci. **NS-34**, 52 (1987).
42. B. Sadoulet, S. Weiss, A. Parsons, R. Lin, and G. Smith, presented at 1987 Nuc. Sci. Symp., San Francisco, to be published in IEEE Trans. Nuc. Sci.
43. R. Ramaty and R. Lingenfelter, Phil. Trans. R. Soc. Lond. A **301**, 671 (1981).
44. R. Ramaty, R. Lingenfelter, and B. Kozlovsky, AIP Conf. Proc. No. 77 (AIP Press, N.Y., 1982), p. 211.
45. S. Matz, E. Chupp, D. Forrest, G. Share, P. Nolan, and E. Rieger, AIP Conf. Proc. No. 115 (AIP Press, N.Y. 1984), p. 403.
46. B. Teegarden and T. Cline, Astrophys. Space Sci. **75**, 181 (1981).
47. R. Ramaty, B. Kozlovsky, and R. Lingenfelter, Astrophys. J. Supp. **40**, 487 (1979).
48. K. Brecher, Astrophys. J. Lett. **215**, L17 (1977).
49. L. Lindblom, Astrophys. J. **278**, 364 (1984).
50. E. Liang, Astrophys. J. **304**, 682 (1986).

THEORIES OF GAMMA-RAY BURST SPECTRA

D. Q. Lamb

*Department of Astronomy and Astrophysics
and Enrico Fermi Institute
University of Chicago*

ABSTRACT

We review radiative processes in γ-ray burst sources and discuss the implications of spectral observations. We summarize recent work on a nonthermal Compton model for γ-ray burst spectra. We describe a new γ-ray spectral inversion method which promises to resolve the controversy surrounding the existence of line features in γ-ray bursts. Finally, we emphasize the importance of identifying quiescent counterparts to γ-ray burst sources at other wavelengths, and the need to determine new, highly accurate burst locations for this purpose.

INTRODUCTION

Twenty years after the discovery of γ-ray bursts,[1] their origin remains a profound mystery. Five years ago a consensus had developed that they originate on or near (magnetic) neutron stars, based largely on (1) the $\simeq$ 8 sec oscillations seen in the tail of the 1979 March 5 event and the rapid ($\sim$ 0.01 sec) intensity variations seen in many other events, (2) the low energy ($\simeq$ 50 keV) features reported in about 30% of events and interpreted as absorption or emission at the cyclotron fundamental, and (3) the high energy ($\approx$ 430 keV) emission features reported in about 7% of events and interpreted as a redshifted pair annihilation line (see, e.g., the reviews by Mazets and Golenetskii,[2,3] Mazets et al.,[4] Cline[5], Lamb,[6,7] Hurley,[8,9] and Katz[10]).

In the face of no established periodicities in other bursts and little confirmation of the low and high energy spectral features by other experiments, this consensus has begun to unravel. The lack of information about the γ-ray burst distance scale contained in plots of the number of events versus fluence, peak flux, or maximum count rate, has also become widely appreciated.[11-15] Indeed, several authors have recently proposed compact objects at cosmological distances as the sources of γ-ray bursts.[16-20] Nevertheless, we consider that (magnetic) neutron stars are still the most likely source of the bursts.

Here we review radiative processes in γ-ray burst sources and discuss the implications of spectral observations. These observations show that the typical γ-ray burst spectrum is roughly a broken power law, and that the break lies at approximately the rest mass-energy of the electron. We summarize recent work on a nonthermal Compton model for the continuum spectrum. Resolution of the controversy surrounding the existence of line features is crucial to progress in modelling γ-ray burst sources. We describe a γ-ray spectral inversion method which promises to do just that. Finally, it is likely that a secure understanding of the nature of γ-ray burst sources will require the identification of quiescent counterparts at other wavelengths. We discuss the problem of finding counterparts and emphasize the importance of determining new, highly accurate burst locations.

For earlier reviews of theories of γ-ray burst spectra, see Lamb,[6,7,21] Epstein,[22] and Zdziarski.[23] For comprehensive discussions of all aspects of γ-ray bursts, see the conference proceedings edited by Lingenfelter, Hudson, and Worral,[24] Woosley,[25] and Liang and Petrosian.[26]

SUMMARY OF OBSERVATIONS

Gamma-ray burst sources are typically quiescent for $\gtrsim 1$ year, emitting a γ-ray flux less than the detection limit of $\sim 10^{-7}$ erg s^{-1} cm^{-2}. They then flare for 0.1-1000 sec (usually $\sim$ 1-10 sec), reaching a flux of up to 10^{-3} erg s^{-1} cm^{-2}.[27] They often exhibit erratic intensity variations on time scales $\delta t \lesssim 0.01$ sec.[27]

Peak burst fluxes $F \approx 10^{-4}$ erg s^{-1} cm^{-2} correspond to burst luminosities

$$L \approx 10^{38} \left(\frac{F}{10^{-4}}\right) \left(\frac{d}{100 \text{ pc}}\right)^2 \text{ergs}^{-1} \approx L_{\text{Edd}} \left(\frac{F}{10^{-4}}\right) \left(\frac{d}{100 \text{ pc}}\right)^2 \left(\frac{M}{M_\odot}\right)^{-1}, \quad (1)$$

where d is the distance to the source and L_{Edd} is the Eddington luminosity. Intensity variations on a time scale $\delta t \lesssim 0.01$ sec imply an emission region radius

$$R \lesssim 10^8 \left(\frac{\delta t}{0.01 \text{ sec}}\right) \text{cm}, \quad (2)$$

provided there is no relativistic beaming.

The continuum energy spectrum of "classical" γ-ray bursts is roughly a broken power law $F_E \propto E^{-\alpha}$, with an X-ray spectral index $\alpha_X \sim 0$ and a γ-ray spectral index α_γ varying from $\sim 0.5 - 2$.[23] The transition between the two regimes occurs between 100 keV and 1 MeV, suggesting some link between the shape of γ-ray burst spectra and the rest mass-energy of the electron. This spectral form implies that most of the γ-ray burst luminosity is emitted in γ-rays.[22] The typical observed ratio of X-ray to γ-ray luminosities is $L_X(3 - 10 \text{ keV})/L_\gamma(> 100 \text{ keV}) \sim 0.02$. The Solar Maximum Mission (SMM) data, which extend to $\sim 5 - 10$ MeV or more, show no clear high energy cut-offs.[28]

Fig. 1 shows spectra from two strong γ-ray bursts.[29,30] The solid lines are least-square power-law fits to data in the range in which they are plotted.[23] The scale on the vertical axis is the received power per logarithmic photon energy interval, $\propto E^2 dN/dE$, showing directly the energy range in which most of the burst luminosity is emitted. The spectrum of GB830801b initially breaks at ≈ 0.5 MeV, with $\alpha_X \approx -0.7$ and $\alpha_\gamma \approx 1$. The break evolves downward in energy with time, while α_γ increases to ≈ 2.5.[29] The spectrum of GB840805 breaks at ≈ 0.5 MeV.[30] In general, burst spectra seem to soften with time, and with decreasing flux in individual pulses.

Mazets et al.[2,3,31,32] have reported low-energy (~ 50 keV) absorption or emission features in about 30% of burst spectra and high-energy (~ 430 keV) emission features in about 7% of burst spectra. Heuter[33,34] has reported low-energy absorption features in GB780325 and GB780608, and states that they are significant at the 99.9% and 95% confidence levels, respectively. Mazets et al.[2] reported an emission feature at ≈ 430 keV in the first 4 seconds of the burst GB781104 and a strong, broad emission feature at 400 keV in GB781119. Barat[35], with better time resolution, confirmed the Mazets et al.[2] result for

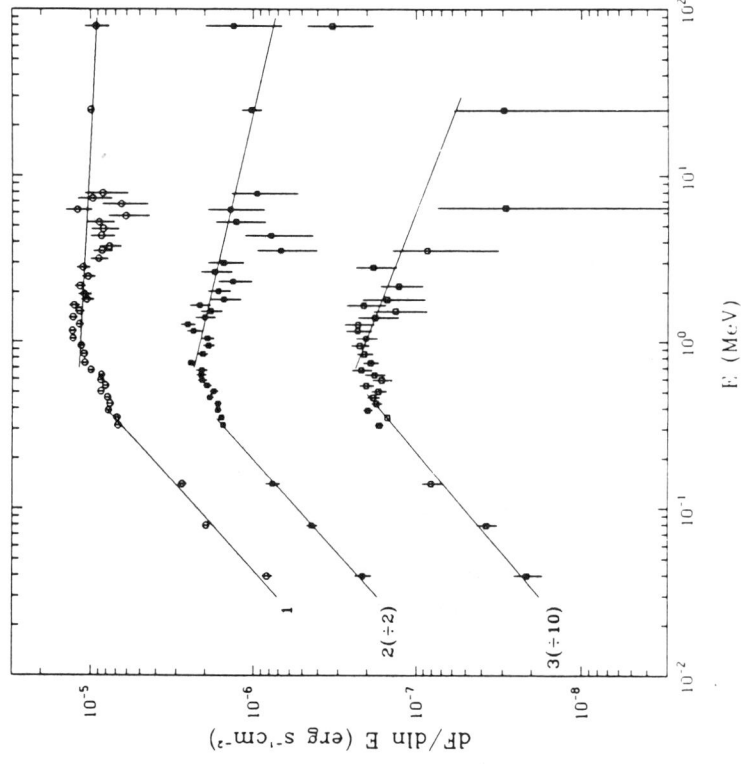
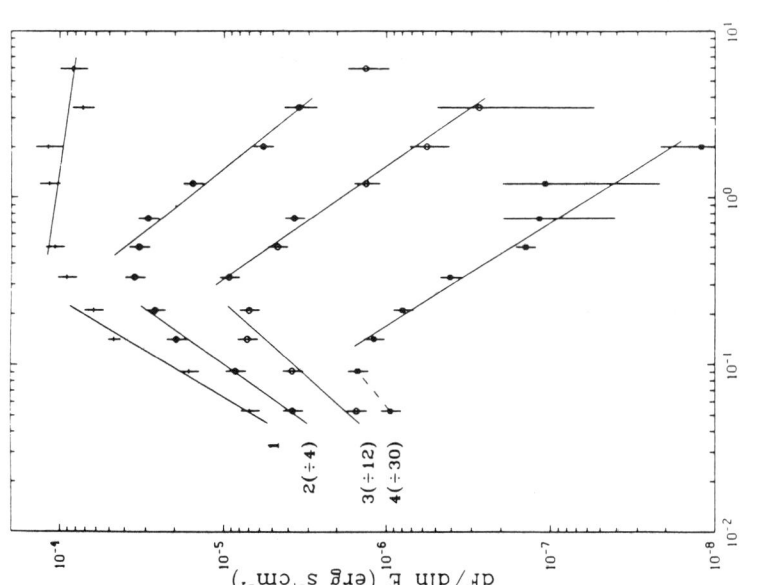

Fig. 1.—Spectra of "classical" γ-ray bursts. a) The spectra of GB840805 in three 16 s time intervals. The data are from Share et al.[30] b) The spectra of GB830801b in four 0.5 s time intervals. The data are from Kuznetsov, et al.[29] After Zdziarski.[23]

GB781104 and the presence of a strong emission feature at 400 keV in GB781119, but not its great width. Mazets et al.[2] also reported a broad emission feature at ≈ 430 keV in the spectrum of GB811231, but the feature was not confirmed by SMM.[36] No narrow (widths < 70 keV) high-energy emission features have been found in the spectra of 60 bursts observed with SMM,[36] nor have any broad features been reported.

The existence of these line features is controversial because such features can be an artifact of the spectral deconvolution, and their significance depends sensitively on the assumed form of the continuum spectrum.[37-40] Further, the horizontal and vertical error bars derived using the model fitting method usually employed have no straightforward meaning. In particular, the horizontal error bars derived in this way often significantly overestimate the actual spectral resolution (see Loredo and Epstein,[41] and the discussion below).

Mazets et al.[4,31,42] suggested that short γ-ray bursts with soft spectra form a distinct class. Additional evidence has recently emerged supporting the existence of such a class, now called "soft γ-ray repeaters" (SGRs). These include SGR 0526-66, the source of the famous 1979 March 5 event,[43,44] SGR1900+14,[42] and SGR 1806-20, the source of the 1979 January 7 event.[45-48] SGR 0526-66 has been observed to repeat 14 times on intervals ranging from days to months,[49] SGR1900+14 three times in 3 days,[42] and SGR 1806-20 more than 100 times on intervals ranging from seconds to months.[47] The bursts from these sources typically last 0.1-0.25 seconds and have color temperatures ~ 50 keV, much less than that of other γ-ray bursts. Fig. 2 shows the spectrum of GB790107 from SGR 1806-20.

The existence of repetitions is reminiscent of X-ray bursts. However, Type I X-ray bursts, which are thermonuclear outbursts on accreting neutron stars, exhibit properties which are not observed in SGRs. These include much softer spectra, sometimes regular intervals between bursts, and spectral softening in the decay phase.[50] Type II X-ray bursts from the Rapid Burster, which are thought to be due to accretion onto a magnetic neutron star, also exhibit some properties which are not observed in SGRs, including much softer spectra and a strong correlation between total burst energy and the waiting time to the next burst.[50] However, interesting similarities also exist, including the lack of spectral evolution during the burst, and the tendency of bursts to rise abruptly and decay exponentially or, when the burst duration is relatively long, to be flat-topped. The Rapid Burster is a regularly recurring transient, with intervals between active phases of about 6 months.[50] SGR 1806-20 exhibits irregularly recurring active phases,[47] while SGRs 0526-66[49] and 1900+14[42] also appear to produce bursts in clusters.

These similarities suggest episodic accretion onto a magnetic neutron star as the origin of SGR bursts.[48] Further support for this view comes from the strong emission feature at ≈ 430 keV present in the spectrum and the ≈ 8 second oscillations seen in the tail of the 1979 March 5 event from SGR 0526-66.[43,44] The disparities between the properties of SGR bursts and those from the Rapid Burster may arise from differences in the physical characteristics of the sources, e.g., the magnetic field strength. In this regard, we note that the spectra of SGR bursts strongly resemble those of bright accretion-powered pulsars, which are known to be accreting strongly magnetic neutron stars.[51]

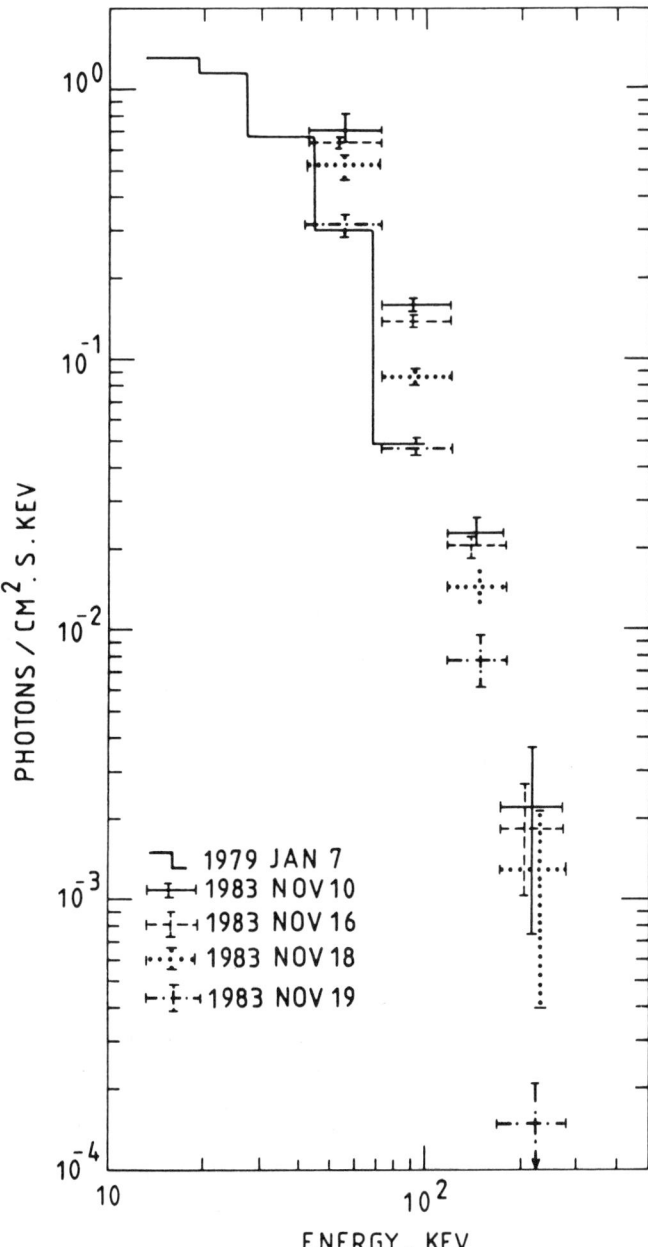

Fig. 2.—Deconvolved photon spectra of the four most intense events from SGR 1806-20 detected by *P9*. For comparison, the initial 1979 January 7 event, detected by *Prognoz* 7, is shown. The January 7 and November 10 spectra are 0.5 s integrations; all others are over 1 s. One sigma upper limits are shown for the 270-407 keV range; they are indentical for the November 16, 18, and 19 events. From Atteia *et al.*[46]

FEATURES OF THEORETICAL MODELS

We describe several basic features of theoretical models of γ-ray bursts, following Zdziarski.[23] We introduce the notation,

$$x \equiv \frac{E}{m_e c^2}, \quad \Theta \equiv \frac{kT}{m_e c^2}, \quad \ell \equiv \frac{L\sigma_T}{Rm_e c^3}, \qquad (3)$$

where E is the photon energy, L is the total luminosity, R is the source radius, and ℓ is the "compactness parameter," familiar from discussions of active galactic nuclei.

Using equation (1) and $R \lesssim 10^8$ cm, which is valid provided no relativistic beaming occurs, the sources of strong γ-ray bursts are compact ($\ell > 1$) unless $d \lesssim 20$ pc,

$$\ell \approx 30 \left(\frac{F}{10^{-4} \text{ erg cm}^{-2} \text{ s}^{-1}}\right) \left(\frac{d}{100 \text{ pc}}\right)^2 \left(\frac{R}{10^8 \text{ cm}}\right)^{-1}. \qquad (4)$$

If $\ell \gtrsim 1$, the cooling time for mildly relativistic pairs is shorter than the light crossing time for the emission region, and pairs will not escape. Furthermore, pair production due to γ-γ collisions can then affect the observed spectrum. The optical depth to pair production $\tau_{\gamma\gamma} \propto \ell$ and the Thomson depth of the pairs is roughly $\tau_{\text{pair}} \sim \min[\ell/4\pi, (\ell/4\pi)^{1/2}]$, if the plasma is in pair equilibrium. Zdziarski and Lamb[52] and Zdziarski, Lamb, and Lightman[53] find that pair absorption produces a substantial steepening of the spectrum above $x \sim 1$ or a significant step-like decrease in the spectrum at $x \sim 1$ for γ-ray burst spectra if $\ell \gtrsim 100$. This implies

$$d \lesssim 2 \left(\frac{F}{10^{-6} \text{ erg cm}^{-2} \text{ s}^{-1}}\right)^{-1/2} \left(\frac{R}{10^8 \text{ cm}}\right)^{1/2} \text{ kpc} \qquad (5)$$

for bursts with flat high-energy spectra.

If γ-ray bursts originate from neutron stars or compact objects with disks, the surface of the star or disk will intercept some fraction of the total flux. Part of the incident flux will be reflected,[54] and part of it will be absorbed and re-emitted as blackbody radiation.[55] If the γ-ray burst is not outwardly directed and the source is located close to the surface of the star or disk, reflection produces a power-law X-ray spectrum with luminosity $L_{re} \sim L$, and absorption produces a blackbody luminosity $L_{bb} \sim L$ having $kT_{bb} \approx 2$ keV. Both components are inconsistent with the observed ratio $L_X/L_\gamma \sim 0.02$. Ways around this difficulty include an emission region size $R \gtrsim 2\ R_{ns}$,[22,55] relativistic beaming, or black hole models.

Attributing the reported ≈ 50 keV features in γ-ray burst spectra to emission or absorption at the cyclotron fundamental implies a magnetic field $B \approx 4 \times 10^{12}$ G and a magnetic neutron star origin for them. However, such a strong magnetic field appears inconsistent with the overall properties of γ-ray burst spectra on several grounds. First, narrow absorption features require either that the magnetic field strength increases outward or that the absorbing layer has a thickness $h \lesssim 0.1 R_{ns}$. Both of these conditions are very difficult to achieve physically. Second, absorption at the second harmonic should be just as prominent as that at the fundamental,[56] which is not observed. Third, photons with $x > 2$ can convert into e^+e^- pairs before escaping. This produces a high energy cutoff at some $x_{\text{cut}} > 2$, whose value depends on the strength of the magnetic field. Matz et al.[28] found that, for typical γ-ray burst spectra seen by SMM, this constraint implies $B \lesssim 3 \times 10^{11}$ G. For GB840805, which has a power-law spectrum extending to ≈ 100 MeV,[30] it implies $B \lesssim 3 \times 10^{10}$ G.

The second difficulty above is ameliorated if the radiation from burst sources is optically thin synchrotron emission, since then a strong dip can appear at the fundamental without significant harmonics.[57] One way around the third difficulty is to invoke two emission regions, with the X-ray portion of the spectrum produced near the star and the γ-ray portion far away.

THERMAL MODELS

Goodman[16] and Pacsyński[17] have proposed a relativistic blackbody spectrum for γ-ray bursts, based on an expanding pair fireball model. The outflow becomes optically thin and the trapped radiation leaks out when the temperature drops to ~ 20 keV at $R \sim 30 R_{ns}$. However, the observed spectrum looks like a blackbody with $T = T_{ns}$ and $R = R_{ns}$, due to the Doppler blueshift. This spectral shape differs from that of γ-ray bursts, but a cool envelope surrounding the source might yield spectra similar to those observed. Pair absorption does not limit d, since the escaping photons have $x < 1$ in the comoving frame of the outflow.

Thermal bremsstrahlung fits to the Konus γ-ray burst spectra give $\Theta \sim 0.2 - 2$, but leave unexplained the power-law emission observed at $E \gtrsim 1$ MeV by SMM[22,23]. Furthermore, bremsstrahlung is a rather inefficient radiation process, and both Compton and synchrotron processes are faster for the physical parameters considered characteristic of γ-ray burst sources.[6,7,21]

Fenimore et al.[37] proposed a thermal Compton model for γ-ray burst spectra. They fit the model to three time intervals in GB781104, finding $\tau_T \sim 1 - 3$ and $\Theta \sim 0.3 - 0.5$. These fits also fall well below the power-law γ-ray spectra seen in other bursts by SMM. Nevertheless, it is interesting that the Comptonization in this model is nearly saturated during the first two time intervals, yielding $\alpha_X \approx 0$ and a Wien peak. This Wien peak fills in a broad excess at ~ 400 which had been interpreted as a redshifted pair annihilation line in other bursts. The amplification factor of 10-100 in the model rules out a slab geometry and requires $R \gtrsim 3 - 10 R_{ns}$, in order that the stellar surface intercept only a small fraction of the reflected radiation.[23]

Lamb[6] proposed thermal synchrotron emission as an explanation of γ-ray burst spectra, and Liang[58,61] showed that it fits the Konus data. However, these fits again fall well below the SMM power-law γ-ray spectra.[57,59] The lack of self-absorption in most burst spectra for $E > 30$ keV constrains the luminosity, and consequently, the distance of the source.[6,7,60] Taking $\Theta = 0.3$ and $B = 5 \times 10^{12}$ G as typical, one obtains $L \lesssim 10^{40}(A/10^{10}$ cm$^{-2})$ erg sec^{-1} and $d \lesssim 10(F/10^{-6}$ erg cm^{-2} s$^{-1})^{-1/2}(A/10^{10}$ cm$^2)^{1/2}$ kpc. Self-absorption at low energies appears to be seen in some Konus spectra.[2,3] The above expressions then represent actual estimates of L and d,[61] however the bursts then strongly violate the pair absorption limit [eq. (4)].[62]

All thermal synchrotron models require an extremely efficient thermalization mechanism in order to keep electrons in high Landau levels, since $\tau_{cool} \ll \tau_{Coulomb}$.[21] Hameury et al.[63] accept instead that most electrons are in the lowest Landau level and have a parallel temperature $\Theta_\parallel \sim 1$. Compton scattering off the one-dimensional electron distribution produces a power-law γ-ray spectrum. These γ-rays excite a small fraction of the electrons to higher Landau levels via inverse Compton scattering, producing secondary synchrotron emission. Hameury et al.[63] assume that the excited electrons form a thermal distribution. However, radiative excitation and deexcitation dominate[56] and will produce a nonthermal distribution.[23]

Thermal models cannot account for the power-law radiation extending to many MeV seen in classical γ-ray bursts by SMM. Thermal synchrotron emission may account for the spectra of the soft γ-ray repeaters, however. Furthermore, nonthermal models of all bursts must include a thermal component if $\ell > 1$, since nonthermal particles then cool before they escape, and this component may be partly responsible for the low energy emission.

NONTHERMAL MODELS

Single Compton scattering of soft photons by nonthermal electrons injected with a power-law distribution $\propto \gamma^{-\Gamma}$ gives rise to a photon power law with $\alpha = \Gamma/2$ (see, e.g., Svensson[68]). This process forms the basis for most models of the X- and γ-ray spectra of active galactic nuclei, but we are not aware of any γ-ray burst models of this kind. The resulting power-law photon spectrum can account for the relatively steep γ-ray portion of burst spectra, but not the flat X-ray portion.

Zdziarski and Lamb[52] proposed repeated Compton scatterings of soft photons by nonthermal electrons as an explanation of γ-ray burst spectra. They assume injection of electrons with a power-law distribution $\propto \gamma^{-\Gamma}$ and $\ell_{soft}/\ell_e \ll 1$, where ℓ_{soft} and ℓ_e are the soft photon compactness and the compactness of the injected nonthermal electrons, respectively. Solution of the kinetic equation for the electrons gives a steady-state distribution with a power-law index differing from Γ, and a broken power-law photon spectrum,

$$F_x \propto \begin{cases} x^{-\alpha_X}, & x \lesssim 1; \\ x^{-\alpha_\gamma}, & x \gtrsim 1. \end{cases} \quad (6)$$

Here α_X depends on ℓ_{soft}/ℓ_e, $\alpha_\gamma = \Gamma - 1$, and $\Gamma > \alpha_X + 1$ is assumed. The break at $x \sim 1$ is due to the relativistic cutoff in the Klein-Nishina cross section. Such broken power-law spectra are similar to the spectra of many classical γ-ray bursts (see Fig. 1).

Ramaty et al.[64] (see also Katz[65]) fit a nonthermal synchrotron emission to the low-energy exponential part of the 1979 March 5 burst from the soft γ-ray repeater SGR 0526-66, using monoenergetic pairs injected at an energy $\gamma_o = 7$ in a magnetic field $B = 10^{11}$ G. The harder emission comes from annihilation of pairs which have cooled to form a thermal distribution with $\gamma \approx 1.2$. A location for SGR 0526-66 in the Large Magellanic Cloud, as assumed by Ramaty et al.,[64] poses great difficulties because of pair absorption.[62] If this source is located in the Galaxy, however, nonthermal synchrotron emission may explain its bursts, as well as those of other SGRs.

More generally, synchrotron emission from injection of monoenergetic electrons either resembles the model of Ramaty et al.,[64] or produces harmonic structure at ~ 100 keV and hard spectra not seen in γ-ray bursts. Synchrotron emission from nonthermal power-law electron distributions can fit the power-law emission at energies $E > 1$ MeV seen in classical γ-ray bursts by SMM, but rises well above the flat ($\alpha_X \sim 0$) X-ray spectrum.[57] Brainerd and Lamb[57] have therefore proposed a two-component (thermal plus nonthermal) synchrotron model for the spectra of these bursts. Why the thermal component has a temperature $\Theta \sim 1$ is left unexplained in this model.

Sturrock[66] and Ruderman and Cheng[67] have proposed more complicated nonthermal synchrotron models as explanations of γ-ray burst spectra. In Sturrock's model, curvature γ-rays interact with the magnetic field of a neutron star and produce pairs which emit synchrotron radiation. The resulting spectrum has $\alpha = 0.5$ below the magnetic pair producton threshhold and $\alpha = 2$ above it. It resembles that of some bursts if $B \sim 3 \times 10^{12}$ G, but fails to explain the power-law emission at $E > 1$ MeV with $\alpha_\gamma \sim 0.5 - 1.5$ seen in many bursts. The model of Ruderman and Cheng was originally constructed to explain the spectrum of the Vela pulsar, which has a paucity of X-rays. In this model, a pair cascade near the light cylinder produces synchrotron radiation. The pairs escape before they cool, producing a power-law spectrum above some energy x_{escape} and little emission below it, thus satisfying the X-ray paucity constraint.

TWO-COMPONENT COMPTON MODEL

Recently, Zdziarski, Lamb, and Lightman[53] extended the earlier work of Zdziarski and Lamb[52] on the spectra produced by Compton scattering off of nonthermal electrons to include the effects of an optically thick thermal "background" of pairs. Such a component must be present when the compactness ℓ_e of the injected nonthermal electrons exceeds 1, since then the nonthermal pairs cool before they can escape from the emission region. The temperature of the thermal pairs is determined by balancing cooling due to inverse Compton scattering of soft photons, and heating due to Compton scattering of γ-rays and to Coulomb collisions between the nonthermal and thermal pairs.

Small compactnesses, $\ell_e \lesssim 10$, require source distances $\lesssim 60$ pc for the strongest bursts and $\lesssim 600$ pc for the weakest ones [see eq. (4)]. In this case, the thermal pairs have an optical depth $\ll 1$ and a temperature $\Theta > 1$. Compton heating alone would produce a relativistic temperature because the spectrum radiated by the nonthermal pairs peaks at $x \sim 1$. Coulomb heating due to collisions with nonthermal pairs further increases it. Zdziarski, Lamb, and Lightman[53] find that Coulomb collisions with the thermal pairs deplete the nonthermal pair distribution function at low energies and reduce its optical depth to well below unity.

Because the optical depth of the nonthermal pairs is small, the photon spectrum does not resemble a broken power-law. Rather, it resembles the thermal bremsstrahlung spectrum produced by a semi-relativistic plasma with only a weak power-law tail. No pair annihilation feature is visible. It is masked by a

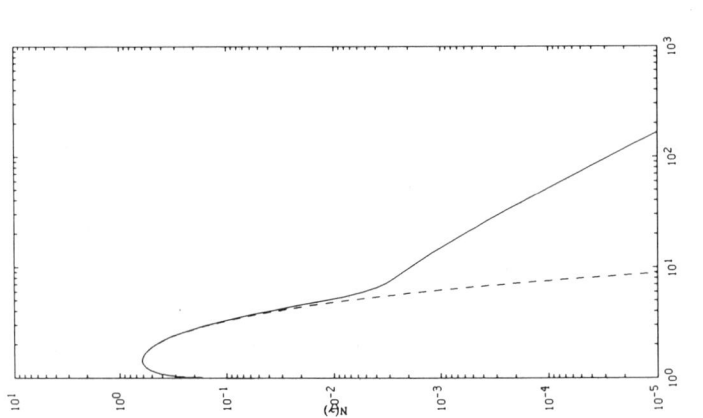
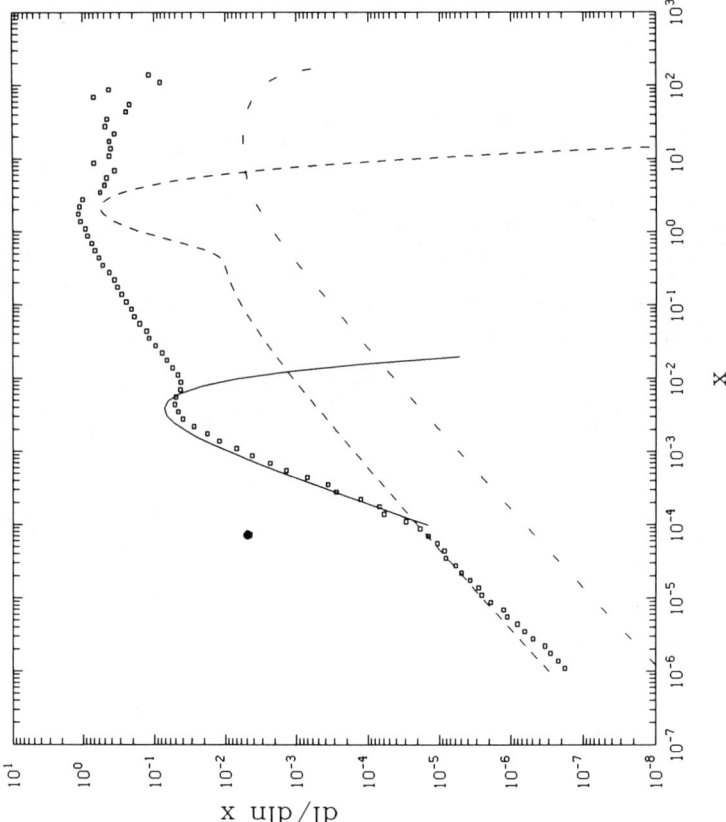

Fig. 3.—The steady-state pair distribution function (left panel) and the photon spectrum (right panel) for a two-component nonthermal Compton model with $\ell_e = 3$ and $\ell_{soft} = 0.1$. The injected pairs have a power-law index $\Gamma = 2$ and the thermal pairs have a temperature $\Theta = 0.5$. The Compton optical depth of the nonthermal pairs is $\approx 5 \times 10^{-2}$. From Zdziarski, Lamb, and Lightman.[53]

275

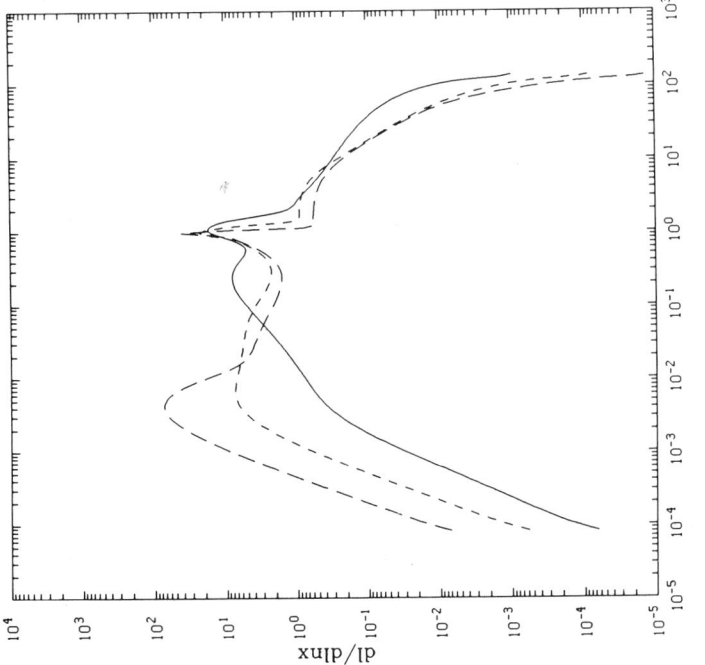
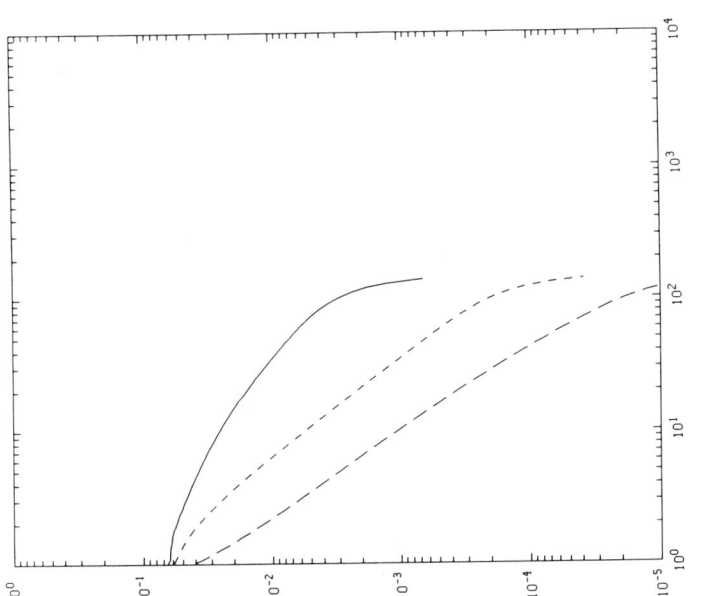

Fig. 4.—The steady-state pair distribution function (left panel) and the photon spectrum (right panel) for a two-component nonthermal Compton model with $\ell_e = 30$ and $\ell_{soft} = 1$ (solid line), 10 (short-dashed line), and 100 (long-dashed line). The injected pairs have a power-law index $\Gamma = 2.5$. The thermal pairs have a temperature $\Theta = 8 \times 10^{-2}, 2 \times 10^{-2}$, and 4×10^{-3}, and the nonthermal pairs have a Compton optical depth $5 \times 10^{-2}, 3 \times 10^{-2}$, and 1×10^{-2} for $\ell_{soft} = 1, 10$, and 100, respectively. From Zdziarski, Lamb, and Lightman.[53]

combination of Compton downscattering of the annihilation photons, and Compton upscattering of bremsstrahlung photons when $\ell_{soft} = 0$ or blackbody photons when $\ell_{soft} \neq 0$. Figure 3 shows the steady-state pair distribution function and the photon spectrum for $\ell_e = 3$ and $\ell_{soft} = 0.1$. The injected pairs have a power-law index $\Gamma = 2$ and the thermal pairs have a temperature $\Theta = 0.5$.

Large compactnesses, $\ell_e \gg 10$, apply to sources at distances $\gg 60$ pc for the strongest bursts and $\gg 600$ pc for the weakest ones. In this case, the thermal pairs have an optical depth $\gg 1$ and a temperature $\Theta \ll 1$. The nonthermal pairs have an optical depth $\ll 1$, and cool before they can escape from the emission region. All of these values are due to the fact that Compton cooling is very efficient at large compactnesses.

Because the optical depth of the nonthermal pairs is small, the photon spectrum closely resembles that produced by Compton upscattering of soft photons on a cool, optically thick thermal plasma with only a weak power-law tail. Figure 4 shows the steady-state pair distribution function and the photon spectrum for $\ell_e = 30$ and $\ell_{soft} = 1, 10, 100$. The injected pairs have a power-law index $\Gamma = 2.5$ and the thermal pairs have a temperature $\Theta = 8 \times 10^{-2}, 2 \times 10^{-2}$, and 4×10^{-4} for $\ell_{soft} = 1, 10,$ and 100, respectively. The photon spectrum exhibits a Wien peak at $x \approx 3\Theta$. A strong pair annihilation feature is visible, because the temperature of the thermal pairs is low. The effect of pair absorption is modest.

In summary, Zdziarski and Lamb[52] investigated the spectrum produced by nonthermal Compton models when $\ell_{soft}/\ell_e \ll 1$. They implicitly assumed that no thermal component was present, and found an optical depth ≈ 1 for the nonthermal pairs and a broken power-law spectrum. However, a thermal component must be present when $\ell_e > 1$, since mildly relativistic particles cool before they can escape from the emission region. Zdziarski, Lamb, and Lightman[53] find that the optical depth of the nonthermal pairs is then $\ll 1$ always. The photon spectra are therefore always approximately thermal, with only weak power-law tails. Such spectra do not resemble those of classical γ-ray bursts, but do resemble those of soft γ-ray repeaters.

GAMMA-RAY SPECTRAL INVERSION

Observations of the spectra of γ-ray bursters are usually performed with satellite-borne scintillators. The response functions of these devices are complicated and, as a result, the data they produce are related in a complicated way to the incident spectrum. Thus estimating the incident spectrum from the data and the detector response functions is nontrivial.

Gamma-ray burst spectra are usually estimated using the model fitting method. In this method, one *assumes* a model for the incident spectrum (*e.g.*, thermal bremsstrahlung, thermal synchrotron, or power-law) and uses it to calculate the expected count spectrum. The parameters of the model are then adjusted to fit the count rate data, usually by minimizing χ^2. Finally, the incident spectrum is estimated by plotting points at the energies corresponding to the midpoints of the energy loss bins of the count rate spectrum, and the photon number flux values derived by scaling the count rate data to the best-fit model spectrum. The horizontal error bars reflect the width of the energy bins of the count rate spectrum, while the vertical error bars reflect the errors in the count rate, scaled to the best-fit model spectrum. Figure 5 (a) schematically illustrates this method.

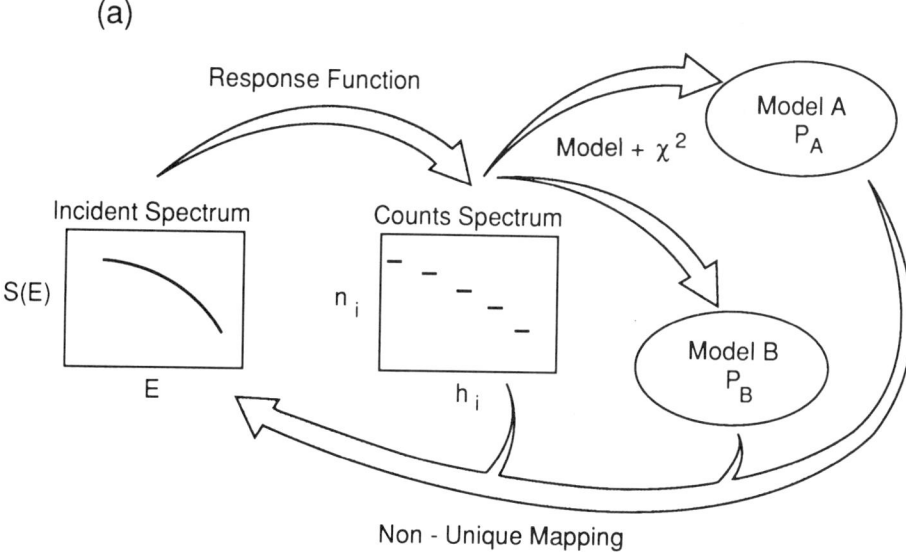

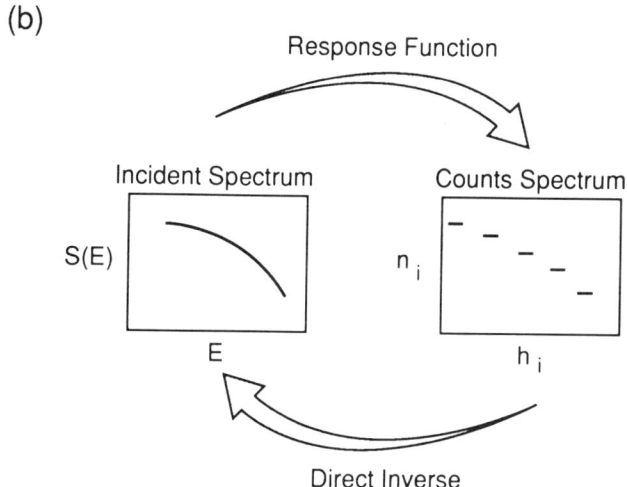

Fig. 5.—Schematic representation of γ-ray spectral inversion methods. A γ-ray spectrometer maps the photon number spectrum to a discrete counts spectrum. a) Model fitting maps the counts spectrum to sets of parameter values P_A and P_B from which a non-unique map must be constructed, using the data, back to the photon number spectrum. b) Direct inversion maps the discrete counts spectrum directly back to the photon number spectrum. From Loredo and Epstein.[41]

The model fitting method suffers from two basic problems. First, a model for the incident spectrum must be assumed. This is unjustified, given our ignorance about the origin of γ-ray bursts, yet the spectral estimate often depends sensitively on the choice of model.[38-40] Further, the spectral estimate is "obliging" in the sense that the estimate follows the model, even if the model gives a poor fit.[37] For these reasons, the spectral estimate produced using this method is also unsuitable for assessing model spectra different from that assumed.

Second, the model fitting method does not estimate the incident spectrum but, rather, the parameters of the model [see Figure 5 (a).]. Since there is no unique method for determining the spectral estimates from these parameter estimates and the data, ambiguity and arbitrariness enter. A particular consequence is that photon energy loss and incident photon energy are confused, so that the horizontal error bars produced by model fitting do not reflect the resolution of the spectral estimates but, rather, the width of the energy loss bins of the count rate spectrum, *i.e.* the photon energy loss resolution. The resolution achievable with a given set of data is often significantly worse than the error bars reflecting bin size imply.

As a result of these problems, the shape of γ-ray burst continuum spectra and, especially, the presence of line features in these spectra remain highly controversial. This is unfortunate, for confirmation of the low-energy features, interpreted as absorption or emission at the cyclotron fundamental, and of the high-energy features, interpreted as redshifted pair annihilation lines, would greatly strengthen the evidence that γ-ray bursts orginate from magnetic neutron stars. Conversely, establishing their absence would greatly weaken it. Thus, a solution to the γ-ray spectral inversion problem is crucial to progress in modelling γ-ray burst sources.

Recently, Loredo and Epstein[41] have proposed a solution. It is based on the Backus-Gilbert direct inverse method, which has a long history of success in analyzing complicated geophysical data. In this method, one constructs new response functions from a linear combination of the original functions. A linear inversion method is preferable for γ-ray spectral estimates, because it allows one to assess the effect of the new detector response functions, despite the complexity of the original ones, and to propagate the errors. Figure 5 (b) schematically illustrates this method.

The Loredo-Epstein adaptation of the Backus-Gilbert method provides accurate spectral estimates which have a well-defined resolution and which are model-independent. Hence the horizontal and vertical error bars have a well-defined meaning. Furthermore, Loredo and Epstein[41] have demonstrated that these estimates can be used in many cases for model fitting without detailed knowledge of the detector response functions. Of course, they are not as accurate for this purpose as model fitting using the raw data and the actual detector response functions; however, they are a valuable alternative, given that the actual response functions vary from burst to burst due to geometrical effects and that response functions are sometimes unavailable for political reasons.

One of the great virtues of the LE/BG method is that it provides a resolution function, which taken times the incident spectrum and integrated over incident photon energy gives the spectral estimate. This function reveals precisely how a spectral estimate relates to the unkown incident spectrum. Also, comparison of the LE/BG resolution function with the original channel response function illustrates the extent to which the LE/BG estimate is superior.

Figure 6 shows the resolution function for an inverse calculated using a system of 19 channel response functions with bin boundaries chosen to be similar to those actually used in satellite experiments. The arrow indicates the energy E_j for which the spectral estimate is being calculated. The Backus-Gilbert tradeoff parameter was set to zero, corresponding to the highest resolution inverse (see Loredo and Epstein[41] for details). Also shown is the channel response function whose peak encloses E_j, normalized to unit area. The channel response function has only about 32% of its area in its peak, whereas the LE/BG resolution function has about 83%, showing that a significant gain in resolution has been achieved.

Figure 7 illustrates the application of the LE/BG method to simulated γ-ray burst spectral data. The energy range and bin boundaries were chosen to simulate qualitatively the results obtainable with the HEAO-1 A4 experiment for γ-ray burst GB780325. The A4 data on this burst has been reported to show a low-energy absorption feature[33,34] (note that the A4 experiment employs active shielding to decrease the Compton continuum contribution to the detector response function). The accuracy of the inverse is very satisfying, and is typical of that achievable with the LE/BG method. Note, however, that the horizontal error bars overlap, showing that the rigorous measure of the spectral resolution obtained using the LE/BG method is significantly worse than the resolution implied by plotting error bars equal to the energy loss bin size of the count rate spectrum (as is done in the model fitting method).

Application of the LE/BG method to γ-ray burst spectra can resolve the controversies surrounding the shape of burst continuum spectra and, especially, the presence or absence of line features. We therefore strongly encourage experimenters to investigate this method and to apply it to their data.

COUNTERPARTS AND PRECISE LOCATIONS

As noted at the beginning of this review, the nature of γ-ray burst sources remains an enigma. The plot of log N versus log $N_{\max}$ may hold important information about the nature of the burst sources, if it can be extended down to much lower values of $N_{\max}$.[15,69] So, too, do the X- and γ-ray spectra of the bursts, particularly when the controversy surrounding the presence or absence of line features is resolved. However, the recent history of radio, infrared, and X-ray astronomy suggests that a secure understanding of these sources will require the identification of quiescent counterparts at other-than-γ-ray wavelengths.

The apparently large amplitude of the γ-ray burst phenomenon makes the task of finding quiescent counterparts particularly challenging. A number of bright, repeating optical transients have now been observed.[70-75] Yet no persistent emission from γ-ray burst sources is known down to very faint limits, $\approx 22^m$ or fainter. If these optical transients are associated with γ-ray burst sources, it implies that at optical wavelengths the burst sources brighten and fade by as much as 19^m within seconds.[70-75] Another way to put this is to say that the optical rise and decline of γ-ray bursts exceeds that of supernovae, yet occurs on a time scale which is a million times shorter.

Very precise error boxes remain the best hope for identifying quiescent counterparts at other wavelengths. Given the apparently large amplitude of the γ-ray burst phenomenon, it seems essential to carry out observations at other wavelengths simultaneously with the γ-ray bursts themselves. If counterparts can

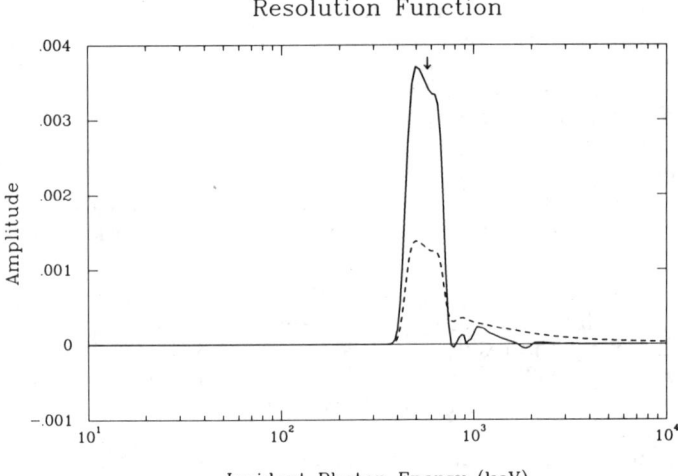

Fig. 6.–Typical resolution functions produced by the Backus-Gilbert method for estimating γ-ray spectra. A resolution function (solid curve) at energy $E_j = 570$ keV (indicated by the arrow) calculated by using the channel response functions with $E_j \geq 570$ keV and omitting those with $E_j < 570$ keV. For comparison, the dashed curve shows the channel response function with boundaries extending from 440 keV to 700 keV, normalized to have unit area. Note that the horizontal axis is logarithmic, and that the area under the dashed curve at high energies is large. From Loredo and Epstein.[41]

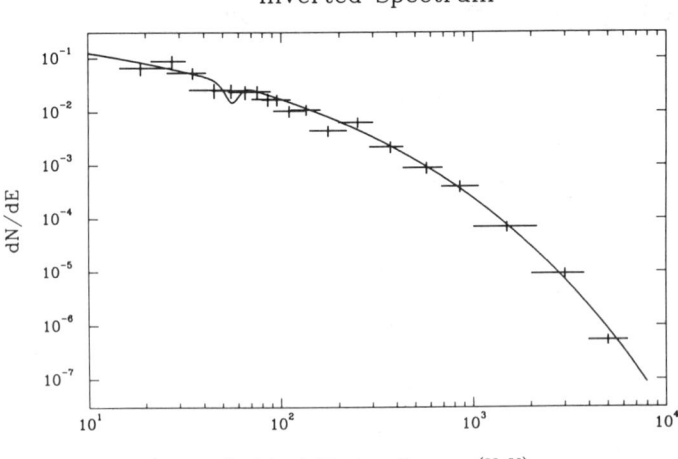

Fig. 7.–Inversion of an incident optically thin thermal bremsstrahlung spectrum with an absorption feature at 45 keV and with Poisson noise added to simulate data. The results of Backus-Gilbert inversions at 16 energies are shown. Vertical error bars show 1 σ confidence regions calculated from the propagated variance of the inverse; horizontal error bars show the energy range from which 90 % of the photons comprising each estimate come. From Loredo and Epstein.[41]

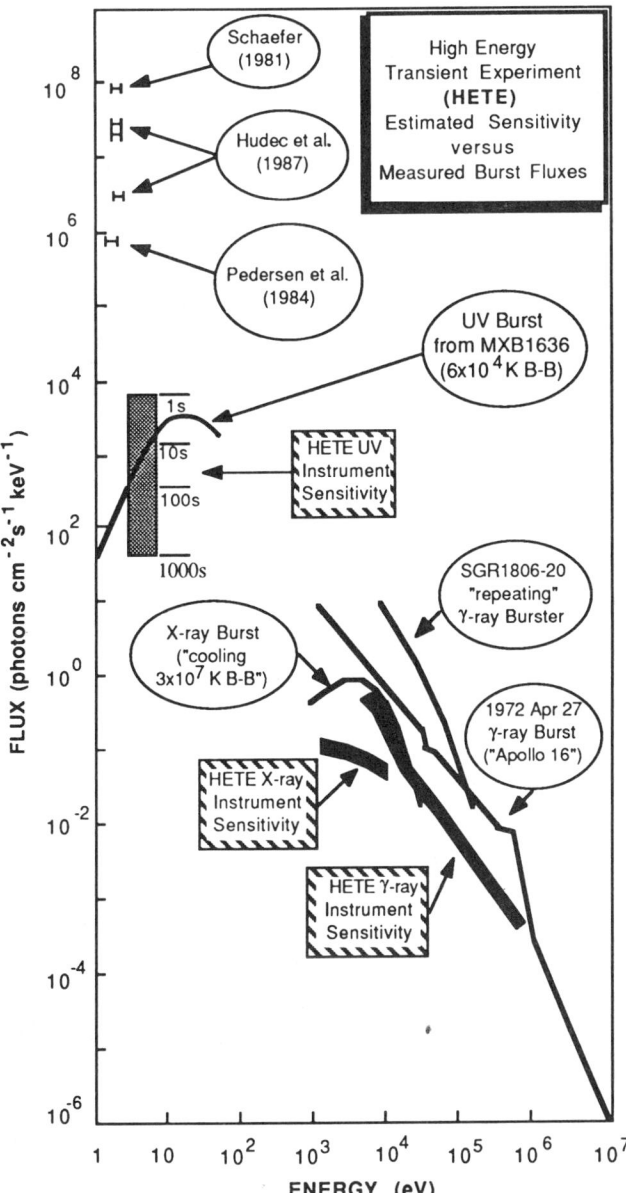

Fig. 8.–The spectral sensitivity of the High Energy Transient Experiment ultraviolet, X-ray, and γ-ray instruments relative to the spectral fluxes of typical γ-ray bursts, soft γ-ray repeaters, and X-ray bursts, and the fluxes of several optical transients thought to be associated with γ-ray bursts. After Ricker et al.[77]

be found during outburst, they will provide precise error boxes for subsequent searchs for quiescent counterparts. New, very precise error boxes also seem required, given the absence of optical candidates brighter than $\sim 22^m - 24^m$ and the abundance of candidates at fainter magnitudes in the present ≈ 2 arcmin error boxes determined with great effort by the Interplanetary Network (see, e.g. Schaefer et al.[70]). Finally, rapid availability of these very precise error boxes is important.

The High Energy Transient Experiment (HETE) achieves all three objectives. Figure 8 shows the spectral sensitivity of the HETE ultraviolet, X-ray, and γ-ray instruments compared to the spectral fluxes of typical γ-ray bursts, soft γ-ray repeaters, and X-ray bursts, and the fluxes of several optical transients which may be associated with γ-ray bursts. The three instruments provide 2π steradian sky coverage and $\approx 60°$, ± 3 arcmin, and ± 3 arcsec angular resolution at γ-ray, X-ray, and ultraviolet wavelengths, respectively. The capabilities of HETE are described in more detail elsewhere in these proceedings.[76]

We thank Tom Loredo and John Wang for comments on an earlier version of this paper. This work was supported in part by NASA Grant NAGW-830.

REFERENCES

1. R. W. Klebesadel, I. B. Strong, and R. A. Olson, *Ap. J.*, **182**, L85 (1973).
2. E. P. Mazets, S. V. Golenetski, R. L. Aptekar', Yu. A. Gur'yan, and V. N. Il'inskii, *Nature* **290**, 378 (1981).
3. E. P. Mazets, et al., *Astrophys. Space Sci.* **82**, 261 (1982).
4. E. P. Mazets and S. V. Golenetskii, in *Astrophysics and Space Physics Reviews, Soviet Scientific Reviews/Section E* **1**, ed. R. A. Sunyaev (Harwood, London, 1982), p. 205.
5. T. L. Cline, in *Proceedings Tenth Texas Symposium on Relativistic Astrophysics*, ed. R. Ramaty and F. C. Jones, *Ann. N.Y. Acad. Sci.*, **375**, 314 (1981).
6. D. Q. Lamb, in *Gamma-Ray Transients and Related Astrophysical Phenomena*, AIP Conference Proceedings No. 77, ed. R. E. Lingenfelter, H. S. Hudson, and D. M. Worrall (American Institute of Physics, New York, 1982), p. 249.
7. D. Q. Lamb, in *Proceedings Eleventh Texas Symposium on Relativistic Astrophysics*, ed. D. Evans, *Ann. N.Y. Acad. Sci.*, **422**, 237 (1983).
8. K. Hurley, in *Accreting Neutron Stars*, MPE Report No. 177, ed. W. Brinkmann and J. Trümper (Max-Planck Institut für Extraterrestrische Physik, Garching, West Germany, 1982), p. 161.
9. K. Hurley, in *Positron-Electron Pairs in Astrophysics*, AIP Conference Proceedings No. 101, ed. M. L. Burns, A. K. Harding, and R. Ramaty (American Institute of Physics, New York, 1983), p. 21.
10. J. I. Katz, in *Positron-Electron Pairs in Astrophysics*, AIP Conference Proceedings No. 101, ed. M. L. Burns, A. K. Harding, and R. Ramaty (American Institute of Physics, New York, 1983), p. 65.
11. E. P. Mazets, et al., *Ap. Space Sci.*, **80**, 3 (1981).
12. R. Klebesadel, E. Feinimore, and J. Laros, in *High Energy Transients in Astrophysics*, AIP Conference Proceedings No. 115, ed. S. Woosley (American Institute of Physics, New York, 1984), p. 429.
13. T. Yamagami and J. Nishimura, *Ap. Space Sci.*, **121**, 241 (1986).
14. J. C. Higdon and R. E. Lingenfelter, *Ap. J.*, **307**, 197 (1986).

15. B. Paczyński and K. Long, *Ap. J. (Letters)*, submitted (1988).
16. J. Goodman, *Ap. J. (Letters)*, **308**, L47 (1986).
17. B. Paczyński, *Ap. J. (Letters)*, **308**, L43 (1986).
18. B. Paczyński, *Ap. J. (Letters)*, **317**, L51 (1987).
19. A. Babul, B. Paczyński, and D. Spergel, *Ap. J. (Letters)*, **316**, L49 (1987).
20. B. McBreen and L. Metcalfe, *Nature*, **332**, 234 (1988).
21. D. Q. Lamb, in *High Energy Transients in Astrophysics*, AIP Conference Proceedings No. 115, ed. S. Woosley (American Institute of Physics, New York, 1984), p. 512.
22. R. Epstein, in *Radiation Hydrodynamics in Stars and Compact Objects*, ed. K.-H. Winkler and D. Mihalas (Heidelberg, Springer, 1986), p. 305.
23. A. A. Zdziarski, in *Proceedings Thirteenth Texas Symposium on Relativistic Astrophysics*, ed. M. Ulmer (World Scientific, Singapore, 1987), p. 563.
24. R. E. Lingenfelter, H. S. Hudson, and D. M. Worrall, ed. *Gamma-Ray Transients and Related Astrophysical Phenomena*, AIP Conference Proceedings No. 77 (American Institute of Physics, New York, 1982).
25. S. E. Woosley, ed. *High Energy Transients in Astrophysics*, AIP Conference Proceedings No. 115 (American Institute of Physics, New York, 1984).
26. E. P. Liang and V. Petrosian, ed. *Gamma-Ray Bursts*, AIP Conference Proceedings No. 141 (American Institute of Physics, New York, 1986).
27. K. Wood, U. Desai, B. Schaefer, G. Pizzichini, J. Norris, and S. Woosley, in *Gamma-Ray Bursts*, AIP Conference Proceedings No. 141, ed. E. P. Liang and V. Petrosian (American Institute of Physics, New York, 1986), p. 4.
28. S. M. Matz, *et al.*, *Ap. J. (Letters)*, **288**, L37 (1985).
29. A. V. Kuznetsov, *et al.*, *Pisma Astr. Zh.*, **12**, 755 (1986). (English translation in *Soviet Astron. Lett.*, **12**, 18).
30. G. H. Share, *et al.*, *Bull. Am. Phys. Soc.*, **30**, 745 (1985).
31. E. Mazets, S. Golenetskii, R. Apteker, Yu. Gur'yan, and V. Il'inskii, *Pisma Astr. Zh.*, **6**, 706. (English translation in *Soviet Astron. Lett.*, **6**, 38.)
32. E. P. Mazets and S. V. Golenetskii, *Astrophys. Space Sci.*, **75**, 47 (1981).
33. G. J. Hueter, in *High Energy Transients in Astrophysics*, AIP Conference Proceedings No. 115, ed. S. Woosley (American Institute of Physics, New York, 1984), p. 373.
34. G. J. Hueter, Ph.D. Thesis, University of California, San Diego (1987).
35. C. Barat, in *Positron-Electron Pairs in Astrophysics*, AIP Conference Proceedings No. 101, ed. M. L. Burns, A. K. Harding, and R. Ramaty (American Institute of Physics, New York, 1983), p. 54.
36. P. L. Nolan, *et al.*, *Nature*, **311**, 360 (1984).
37. E. E. Fenimore, J. G. Laros, R. W. Klebesadel, R. E. Stockdale, and S. R. Kane, in *Gamma-Ray Transients and Related Astrophysical Phenomena*, AIP Conference Proceedings No. 77, ed. R. E. Lingenfelter, H. S. Hudson, and D. M. Worrall (American Institute of Physics, New York, 1982), p. 201.
38. E. E. Fenimore, J. G. Laros, R. W. Klebesadel, R. E. Stockdale, and S. R. Kane, *Nature*, **297**, 665 (1982).
39. E. E. Fenimore, R. W. Klebesadel, and J. G. Laros, in *Gamma-Ray Astronomy in Perspective of Future Space Experiments* (Pergamon, Oxford, 1983), p. 243.
40. B. J. Teegarden, in *High Energy Transients in Astrophysics*, AIP Conference Proceedings No. 115, ed. S. Woosley (American Institute of Physics, New York, 1984), p. 352.
41. T. J. Loredo and R. I. Epstein, *Ap. J.*, in press (1988).

42. E. P. Mazets, S. V. Golenetskii, Yu. A. Guryan, and V. N. Il'inskii, *Ap. Space Sci.*, **84**, 173 (1982).
43. E. P. Mazets, S. V. Golenetskii, V. N. Il'inskii, R. L. Aptekar', and Yu. A. Guryan, *Nature*, **282**, 587 (1979).
44. T. L. Cline, et al., *Ap. J. (Letters)*, **237**, L1 (1980).
45. J. G. Laros, et al., *Nature*, **322**, 152 (1986).
46. J.-L. Atteia, et al., *Ap. J. (Letters)*, **320**, L105 (1987).
47. J. G. Laros, et al., *Ap. J. (Letters)*, **320**, L111 (1987).
48. C. Kouveliotou et al., *Ap. J. (Letters)*, **322**, L21 (1987).
49. S. V. Golenetskii, V. N. Il'inskii, and E. P. Mazets, *Nature*, **307**, 41 (1984).
50. W. H. G. Lewin and P. C. Joss, in Accretion-Driven Stellar X-Ray Sources, ed. W. H. G. Lewin and E. P. J. van den Heuvel (Cambridge U. Press, Cambridge, 1983), p. 41.
51. P. C. Joss and S. A. Rappaport, *Ann. Rev. Astron. Ap.*, **22**, 537 (1984).
52. A. A. Zdziarski and D. Q. Lamb, *Ap. J. (Letters)*, **309**, L79 (1986).
53. A. A. Zdziarski, D. Q. Lamb, and A. P. Lightman, *Ap. J.*, submitted (1988).
54. T. R. White, A. P. Lightman, and A. A. Zdziarski, *Ap. J.*, submitted (1988).
55. J. M. Imamura and R. I. Epstein, *Ap. J.*, **313**, 711 (1987).
56. R. Bussard and F. K. Lamb, in *Gamma-Ray Transients and Related Astrophysical Phenomena*, AIP Conference Proceedings No. 77, ed. R. E. Lingenfelter, H. S. Hudson, and D. M. Worrall (American Institute of Physics, New York, 1982), p. 189.
57. J. J. Brainerd and D. Q. Lamb, *Ap. J.*, **313**, 231 (1987).
58. E. P. Liang, *Nature*, **299**, 321 (1982).
59. G. G. Pavlov, and S. V. Golenetskii, *Ap. Space Sci.*, **128**, 341 (1986).
60. E. P. Liang, *Ap. J. (Letters)*, **308**, L17 (1986).
61. E. P. Liang, T. E. Jernigan, and R. Rodrigues, *Ap. J.*, **271**, 766 (1983).
62. A. A. Zdziarski, *Astr. Ap.*, **134**, 301 (1984).
63. J. M. Hameury, J. P. Lasota, S. Bonazzola, and J. Heyvaerts, *Ap. J.*, **293**, 56 (1985).
64. R. Ramaty, R. E. Lingenfelter, and R. W. Bussard, *Ap. Space Sci.*, **75**, 193 (1981).
65. J. I. Katz, *Ap. J.*, **260**, 371 (1982).
66. P. A. Sturrock, *Nature*, **321**, 47 (1986).
67. M. Ruderman and K. S. Cheng, *Ap. J.*, submitted (1988).
68. R. Svensson, *M.N.R.A.S.*, **227**, 403 (1987).
69. J.-L. Atteia, et al., *Ap. J. (Suppl.)*, **64**, 305 (1987).
70. B. E. Schaefer, *Nature*, **294**, 722 (1981).
71. B. E. Schaefer et al., *Ap. J. (Letters)*, **286**, L1 (1984).
72. B. E. Schaefer, *Adv. Space Res.*, **6**, No. 4, 47 (1987)
73. H. Pedersen et al., *Nature*, **312**, 46 (1984).
74. R. Hudec, et al., in *The Physics of Compact Objects: Theory versus Observation*, Proceedings of the COSPAR/IAU Symposium, Sofia, Bulgaria, 13-18 July 1987, ed. N. E. White (Pergamon, Oxford, 1988), in press.
75. E. I. Moskalenko, et al., in *The Physics of Compact Objects: Theory versus Observation*, Proceedings of the COSPAR/IAU Symposium, Sofia, Bulgaria, 13-18 July 1987, ed. N. E. White (Pergamon, Oxford, 1988), in press.
76. G. R. Ricker, these proceedings (1988).
77. G. R. Ricker, et al., "Unsolicited Proposal to NASA for a High Energy Transient Experiment to be flown as a 'Get-Away Special' " (1987).

SPECTRAL DIAGNOSTICS OF NEUTRON STAR ACTIVITY

Geoffrey J. Hueter
University of California, San Diego, CA 92093

INTRODUCTION

Because they are the most concrete evidence for a neutron star origin of gamma-ray bursts, the existence of lines in gamma-ray burst spectra has been widely debated[1] (and references therein). This controversy has overshadowed the importance of other characteristics of these observations which offer clues to the physical conditions which are present in the line producing region. This paper focusses on these line diagnostics by showing that HEAO-A4 observations of bursts on March 25, 1978 and June 8, 1978 impose strong constraints on the physical regions which produced the emission. In addition, by limiting the size, temperature, and density of the emission region, it is possible to rule out thermal bremsstrahlung as the dominant emission mechanism, as well as an extragalactic origin for bursts.

Most of the observations of absorption and emission lines have come from the Konus Experiment.[2] Although experimental defects, such as gain errors[3,4] and time-averaging over rapid spectral changes[5], have been suggested as means of spuriously producing line-like artifacts, it has not been demonstrated that these scenarios in fact can explain the entirety of the Konus data. Nonetheless, skepticism of the Konus results remains. Because of the importance of line observations to our understanding of the burst phenomena, the HEAO-A4 observations represent a crucial confirmation of the existence of lines in gamma-ray burst spectra.

The simplest and most consistent explanation for the absorption and emission lines is that they arise on the surfaces of neutron stars. The association of 10 – 100 keV lines with cyclotron absorption,

$$h\nu = \frac{eB}{m_e c} = 11.6\, B_{12}\, keV ,\qquad(1)$$

where B_{12} is the magnetic field strength in units of 10^{12} gauss, implies the terraguass magnetic fields which have been identified only with neutron stars. Similarly, the association of emission lines at 400 – 460 keV with electron - positron annihilation is consistent with the gravitational redshift,

$$z = (1 - \frac{2GM}{Rc^2})^{1/2} \approx 0.15\, \frac{M/M_\odot}{R_6} ,\qquad(2)$$

expected on the surface of a neutron star.

There have been several other explanations for the absorption lines, notably atomic transitions of magnetically distorted iron atoms[5] and the ad hoc superposition of two spectral components.[1] The former is invalidated by the predicted predominance of single photon over two photon annihilation, while it is not clear that the latter can produce sharp features with nonmagnetic spectral models. Ramaty, McKinley and Jones[6] have proposed "grasing" as a mechanism for producing a redshifted annihilation line. It is unlikely that the requisite high density ($n_e > 10^{30}$ cm^{-3}) pair plasma could be maintained in equilibrium, although it is possible that the mechanism might operate briefly following the injection of the plasma into the emission region.

HEAO-A4 LINE OBSERVATIONS

The gamma-ray burst measuring capabilities of the Hard X-Ray and Low Energy Gamma Ray Experiment on the HEAO-1 satellite, which I will refer to as HEAO-A4, have been described in detail elsewhere.[7,8] Briefly, the instrument consisted of three sets of collimated, shielded, NaI/CsI phoswich detectors which covered the energy range of .01 – 6.2 MeV. The spectrum of a burst could be measured on timescales as short as 0.64 seconds, although typically 10 seconds or more were needed to precisely determine the spectral shape. The inclusion of a burst word made possible a complete search of the 18 month mission in post flight data analysis. This search produced 21 bursts, 14 of which were previously undetected. Only two of these bursts, GB 780325 and GB 780608, were of sufficient duration and intensity that it was possible to measure spectral lines in different time intervals.

March 25, 1978

Since the observations of this burst are discussed in detail elsewhere,[7] I will only summarize them briefly here. GB 780325 was a fairly intense (1.5×10^{-5} ergs/cm^2) burst which appeared in the apertures of all three sets of HEAO-A4 detectors. Therefore, it represents the best burst measurements made by this instrument. The burst, which had a measured total duration of 87 seconds, consisted of two distinct peaks each of roughly 10 seconds duration. The spectrum of the first peak showed a cyclotron line at 76 ± 5 keV, with an equivalent width of 24 ± 7 keV. In the second peak the line centroid shifted to 49 ± 3 keV, and its equivalent width decreased to 13 ± 3 keV. Measurements of spectra within each peak, however, showed that the line centroids and equivalent widths were consistent with constant values. As discussed below, this leads to the conclusion that the lines were produced in different regions of magnetic field strength, the first time that such a measurement has been made for any x-ray source. Although the spectrum of the second peak also showed a possible emission line at 400 – 500 keV, this interpretation depends on the shape of the assumed underlying continuum and is not statistically convincing. A final feature of this burst is that the gamma ray spectrum was measured to 6.2 MeV. Because of the opacity of this emission to self pair production,[9] it must have been emitted from a much larger region than that which produced the cyclotron lines and their associated continuum.

June 8, 1978

The June 8, 1978 burst was the longest measured by the HEAO-A4 Experiment, and at 150 seconds, is one of the longest bursts measured by any instrument. The light curve of the burst is shown in Figure 1. The absence of burst activity in the shield segments (the top two panels in Figure 1) indicates that the event is in fact a gamma-ray burst and not a result of local particle activity. The transit of the event by one of the narrowly collimated Low Energy Detectors (LED) further identifies it as a gamma-ray burst and provides additional position information.

The entire burst light curve was measured in two different energy ranges, 45 – 2100 keV and 100 – 1200 keV. Comparison of these light curves shows that the burst softens with time, with the peak of the <100 keV flux lagging that of the >100 keV flux by 20 to 50 seconds. This variability is consistent with SMM results[10] for much shorter bursts.

Despite the fact that the instrument was operating in a 20.48 second spectral accumulation mode, the duration of the burst made it possible to sample the spectrum several times during the burst. Spectra for intervals A–E in Figure 1, are plotted in Figure 2 using flux units whereby a horizontal line corresponds to equal energy flux per equal log energy interval. These spectra show that during the first 82 seconds of the burst (intervals A–C), the continuum shape of the spectrum remained constant and peaked in energy flux at a few hundred keV. During the subsequent 41 seconds, the peak in continuum energy flux shifted below 100 keV and a narrow emission line formed at 400 – 450 keV. Figure 3 shows the photon spectrum averaged over the entire burst. The possible absorption feature at 59–74 keV is significant at the 2.0σ level, with an equivalent width of 5.3 ± 2.6 keV. Because of the coarse digital resolution of the pulse height analyzer during this observation, the feature cannot be further resolved in energy. The absorption line is most intense during the burst decay, Interval E, which spectrum is shown in Figure 4. The equivalent width of the line in this interval is 14.4 ± 4.3 keV (i.e., 3σ). During this time an emission line also appears, which has a flux of 0.060 ± 0.035 photons/cm^2-second and contributes 36 percent of the measured energy flux for this interval. The line has a centroid of 428 ± 52 keV and a width of 180 keV (FWHM). The spectrum was equally well fit by either a thermal bremsstrahlung or thermal synchrotron model. For thermal bremsstrahlung the fitted temperature was 333 ± 80 keV. Assuming a magnetic field strength of 5.7×10^{12} gauss, corresponding to the line centroid of 66 keV, the fitted thermal synchrotron temperature, 140 ± 28 keV, is significantly less than the thermal bremsstrahlung temperature.

INTERPRETATION

Although most of the constraints derived below are quite general, I will assume for concreteness that the site of the emission is near the surface of a neutron star. In addition to units of B_{12} for the magnetic field and R_6 for the stellar radius defined previously, it is convenient to define a dimensionless electron temperature $\Theta = kT/m_e c^2$. I will also focus the discussion on the June 8, 1978 observations presented above, although I will comment on the March 25, 1978 burst at the end of this section.

The width of an absorption line sets a limit on the homogeneity of the magnetic field. As pointed out by Fenimore et al.,[3] the variation of a dipolar field with radius,

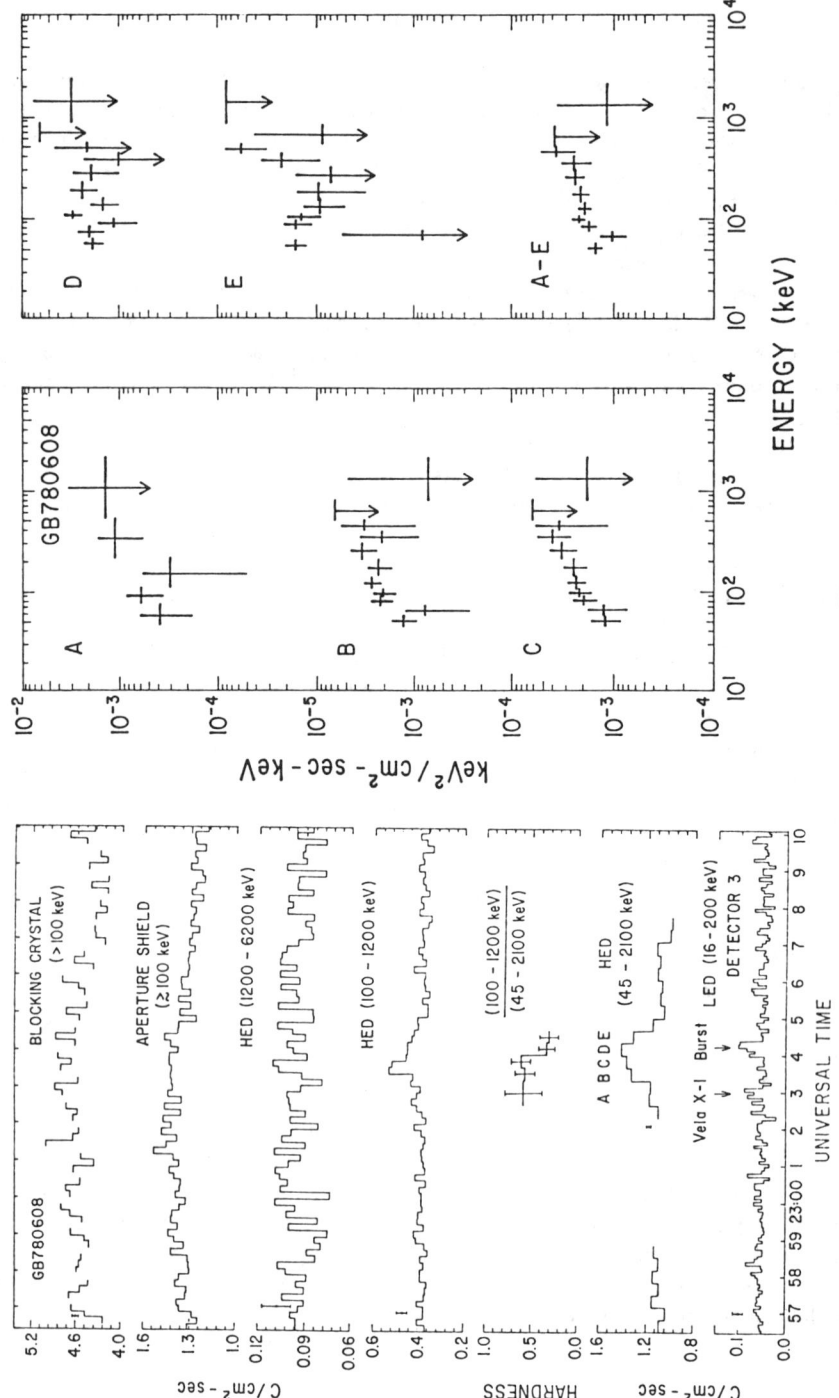

Fig. 2. Spectra for the intervals A-E in Fig. 1.

Fig. 1. GB 780608 detector and shield rates.

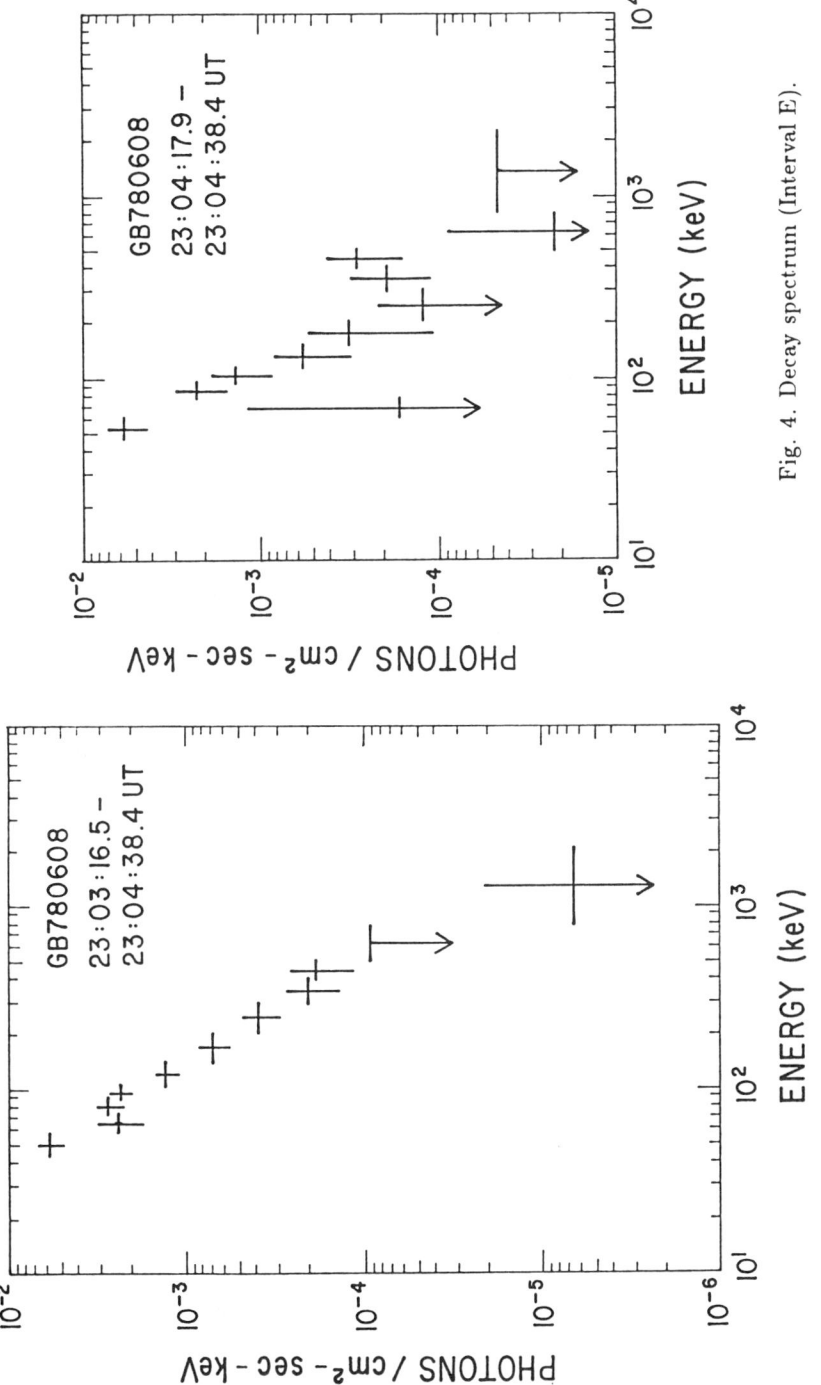

Fig. 4. Decay spectrum (Interval E).

Fig. 3. Spectrum for the entire burst (Intervals A–E).

$B \propto r^{-3}$, relates the width of the line to the size of the absorbing region according to

$$\frac{\Delta E}{E} = \frac{\Delta B}{B} > \frac{3\Delta r}{r}. \tag{3}$$

At the surface of the neutron star, the instrumental limit on the line width of <22 keV (FWHM) limits the size of the absorbing region to $r < R_6 \times 10^5$ cm, suggesting that emission comes from the polar cap. Furthermore, the similarity of the measured equivalent width and the limit on the intrinsic width of the line implies that the actual line width must be comparable to the experimental limit and that the underlying continuum cannot be produced in a region much larger than the absorbing region.

Trümper et al.[11] have shown that the line width constrains the temperature of the absorbing electrons according to

$$(\frac{\Delta E}{E})_{doppler} = 2.35\Theta^{1/2}\cos\phi, \tag{4}$$

where ϕ is the angle between the magnetic axis and the line of sight. If the absorption line in GB 780608 is observed along the line of sight, then the temperature of the absorbing electrons must be less than 6 keV. Conversely, if the absorbing electrons are much hotter than this, then the line of sight must be at a large angle to the field.

Brainerd and Lamb[12] have investigated in detail an alternative scenario in which the cyclotron structure is due to dips between emission harmonics. While the above size and temperature constraints would be qualitatively the same, the thermal synchrotron model has the simplicity of producing with a single mechanism the "lines" and "continuum" in a single population of electrons.

Although a blueshifting of an annihilation line is expected due to the influence of any magnetic field present, calculations by Daugherty and Bussard[13] show that the line shift

$$(\frac{\Delta E}{E})_{fwhm} = \frac{\log 2}{2}\frac{B}{B_{cr}}, \tag{5}$$

where $B_{cr} = m_e c^3/e\hbar = 4.414 \times 10^{13}$ gauss, in a 6×10^{12} gauss field is on the order of 20 keV or less, depending on the orientation of the field. Thus the field inferred from the absorption line would not significantly alter the observed redshift.

The observed 75 keV dispersion (180 keV FWHM) of the emission line observed in GB 780608 severely constrains the temperature of the region in which annihilation takes place. Doppler broadening of a photon of energy $E_0 = h\nu_0$ due to motion $v_s = (kT/m_e)^{1/2} = \Theta^{1/2}c$ of the annihilating pairs along the line of sight is given by

$$\Delta\nu = \nu_0\frac{v_s}{c} = \nu_0\Theta^{1/2}, \tag{6}$$

so that the plasma temperature is limited by $\Theta < (\Delta E/E)^2$. For $\Delta E = 75$ keV, this gives a limit of 13 keV to the thermal energy of the annihilating pairs. This is in rough agreement with the value of 19 keV derived for the same line width from the detailed numerical calculations[14] of electron – positron annihilation in thermal plasmas. The width of the line also limits the range of gravitational potential sampled, so that on the surface of the neutron star the height of the emission region to

$$\Delta r < R_s\frac{\Delta z}{z} = R_s\frac{\Delta E}{E} = 5 \times 10^5 R_6 \; cm. \tag{7}$$

While the size constraint for the emission line is somewhat larger than that for the cyclotron line, the temperature constraint is much more stringent. Therefore, while both emission and absorption lines must arise in localized regions near the surface of a neutron star, the possibility that these regions are spatially separated cannot be ruled out. However, since the synchrotron lifetime,[15] $\sim 10^{-16}$ seconds, is too short for the continuum emitting plasma to be thermalized, a narrow emission line naturally arises from the annihilation of cooled pairs, a scenario that has been invoked[16] to explain the GB 790305 spectrum.

Hueter and Matteson[7] have discussed in detail the additional conclusions drawn from the March 25, 1978 observations. In particular the persistence of the cyclotron lines over large changes in burst intensity requires that the magnetic field density, $B^2/8\pi$, dominate the plasma density, n_e/kT. The measured thermal synchrotron temperature for GB 780325, ~ 200 keV, implies, therefore, that $n_e < 10^{30}$ cm^{-3}. On the other hand, the thermal synchrotron luminosity,[17] $L_{syn} = 2.7 \times 10^9 n_e V \Theta^2 \kappa(\Theta) B_{12}^2$ ergs/second, exceeds the thermal bremsstrahlung luminosity,[18] $L_{ff} = 1.8 \times 10^{-22} \Theta^{1/2} n_e^2 V$ ergs/second, provided that $n_e < 10^{32}$ cm^{-3}. Therefore, bremsstrahlung is clearly ruled out as the dominant emission mechanism.

A second consequence of the noncompression of the magnetic field is that the cyclotron lines of differing energy measured in the two peaks of GB 780325 could not have come from the same spatial region. Therefore, the time between peaks, roughly 10 seconds, is an important measurement of some characteristic propagation time of the burst mechanism.

Finally, the small size of the synchrotron emitting region, $\sim 10^5$ cm, and the lack of self-absorption down to 10 keV, at a flux of .1 photons/cm^2-second, implies that the distance to the source cannot be more than 30 kpc, which strongly suggests a galactic origin for bursts.

CONCLUSIONS

Analysis of gamma-ray burst spectra suggests that bursts arise on the surfaces of galactic neutron stars. Both cyclotron and annihilation lines can be produced in the same region if there exists an electron – positron plasma which cools via synchrotron radiation before annihilating. This cooling plasma cannot simultaneously produce the extended ≥ 1 MeV emission which is measured in GB 780325 and other bursts; however, a separate spatial component is required for this emission because of the large size needed to emit this radiation without it being reabsorbed via photon – photon pair production.

The brevity of bursts, the long interval between bursts from a single source, and the dimness of the quiescent counterparts all conspire against the study of bursters at other wavelengths. Nonetheless, the gamma ray light curves and spectra provide a wealth of information on the identity of the sources, the conditions in which and the mechanisms by which the burst emission is produced, and even the distances to the bursters. The observational situation will improve significantly with the launching of GRO, since the improved timing, sensitivity, and spectral coverage of the Burst and Transient Source Experiment (BATSE) will permit the precise study of the evolution of the line and continuum, as well as the causal relationship between different spectral/spatial components.

This work was supported by NASA under contract NAS 8-36081 and under grant NAG 8-499.

REFERENCES

1. E. P. Liang and V. Petrosian, Gamma-Ray Bursts (AIP, N. Y., 1986).
2. E. P. Mazets et al., Nature **290**, 378 (1981).
3. E. E. Fenimore et al., Gamma Ray Transients and Related Astrophysical Phenomena (AIP, N. Y., 1983), p. 201.
4. E. E. Fenimore et al., Gamma Ray Astronomy in Perspective of Future Space Experiments (Pergamon, N. Y., 1983), p. 207.
5. R. W. Bussard and F. K. Lamb, Gamma Ray Transients and Related Astrophysical Phenomena (AIP, N. Y., 1983), p.189.
6. R. Ramaty, J. M. McKinley, and P. C. Jones, Astrophys. J. **256**, 238 (1982).
7. G. J. Hueter and J. L. Matteson, Astrophys. J., submitted (1988).
8. G. J. Hueter, Ph. D. Dissertation, University of California, San Diego, 1987.
9. R. E. Lingenfelter and G. J. Hueter, High Energy Transients in Astrophysics (AIP, N. Y., 1984), p. 558.
10. J. P. Norris et al., Astrophys. J. **301**, 213 (1986).
11. J. Trümper et al., Ann. N. Y. Acad. Sci. **302**, 6 (1977).
12. J. J. Brainerd and D. Q. Lamb, Astrophys. J. **313**, 231 (1987).
13. J. K. Daugherty and R. W. Bussard, Astrophys. J. **238**, 296 (1980).
14. R. Ramaty and P. Meszaros, Astrophys. J. **250**, 384 (1981).
15. E. P. T. Liang, Nature **299**, 321 (1982).
16. R. Ramaty et al., Nature **287**, 122 (1980).
17. V. Petrosian, Astrophys. J. **251**, 727 (1981).
18. W. H. Tucker, Radiation Processes in Astrophysics (MIT Press, Cambridge, 1975), p. 206.

COMPATIBILITY BETWEEN GAMMA-RAY BURST SPECTROSCOPY AND SOURCE MODELS

I.G. Mitrofanov
Space Research Institute (I.K.I.), Academy of Science of U.S.S.R.,
Moscow 117810, U.S.S.R.

C. Barat
Centre d'Etude Spatiale des Rayonnements (C.E.S.R.),
C.N.R.S./U.P.S., B.P. 4346, 31029 Toulouse Cedex, France.

ABSTRACT

Recent results of cosmic gamma-ray burst spectroscopy are compared. From these data, we discuss the main models of burst sources and the physical conditions of cosmic gamma-ray burst generation.

SPECTRAL DATA

The <u>continuum shape</u> of cosmic gamma-ray burst (GRB) spectra is a controversial topic. In the intermediate energy range (30 keV < E < 1 MeV), the 4 sec averaged continuum shapes from the Konus experiment are fitted by optically thin thermal bremsstrahlung (kT 150 - 400 keV) [1] as well as thermal synchroton models [2]. The high energy emission (E > 1 MeV) observed by SMM is significantly harder than is predicted by thermal models and therefore favors non-thermal mechanisms [3]. At low energies (E < 30 keV), the continuum shape remains unknown. However, temporal structures from soft emission are seemingly independent of the gamma-ray time histories [4]. All these results suggest that there are several spectral components for continua.

The Signe experiment is characterized by a 0.25 - 0.5 sec time resolution. The following 3 free parameter expression,

$$dN(E)/dE = CE^{-\gamma}\exp(-E/E_0) \quad (1)$$

has been used to represent the continuum shape [5]. The best fit parameter values of γ and E_0 obtained by a minimum χ^2 technique, vary from -1.0 to 2.0 and from 100 keV up to >> 1 MeV, respectively. The confidence intervals constructed with these parameters do not overlap for consecutive intervals in GB781104 and GB820827C [6,7], clearly proving a <u>spectral variability</u> at time scales of < 0.5 sec. Spectral variability explains the apparent discrepancies observed between some continuum shapes. Thus, during the 4 sec integrations of the Konus spectral data which are fitted by bremsstrahlung models, some power law spectra have been found in 0.25 - 0.5 sec intervals of the Signe data [7].

The hardness ratio between two wide energy channels also allows to study spectral variability. Assuming a thermal bremsstrahlung model, an analysis of the Konus data has shown that the temperature is correlated with the luminosity, except for the burst onsets [1]. The maximum hardness ratio in the SMM [8] and the Signe results [7], is not found in the peak top but in the rising edges of pulses. The hardness ratio monotonically decreases across the peak and during the decay [7].

Line-like features have been found at ~ 400 keV in the spectra of many intense GRB recorded by the Konus experiment [1]. They are interpreted as redshifted e^-/e^+ <u>annihilation radiation</u>. In the Signe data, these excesses appear as flash-like emission in many events [5,9,10,11,12]. These features are present at burst onset in GB781119 [10] and GB790305 [11]. The ~ 400 keV excess from GB781119 is accompanied by a high energy tail interpreted as an annihilation continuum in a weakly relativistic e^-/e^+ plasma [10]. However, another explanation is proposed hereafter. The annihilation fluence reaches up to 30 percent of the total fluence [5]. Different time resolutions lead to inconsistent spectral data. Thus, for GB811231A, an excess at ~ 400 keV is observed in the 4 sec averaged Konus spectra [1] and in six non-consecutive 0.5 sec Signe spectra [12]. This feature is also visible in the SMM results during the first spectrum accumulated over 5 sec [13]. However, it completely disappears in spectra integrated over 16 sec [14].

<u>Absorption features</u> observed below 100 keV in many Konus spectra [1] are interpreted as electon cyclotron resonances in a high magnetic field (B > 10^{12} gauss). The existence of this feature has been confirmed through the GB780325 spectrum observed by the UCSD/MIT experiment on HEAO-1 [15]. Both turnover and absorption features exist in the Signe data. They are visible in many events with a rapid spectral variability [7,16,17]. Thus, some low energy features resolved in the 0.5 sec Signe spectra are not seen in the 4 sec Konus spectra. Many burst spectra exhibit a non-monotonic variation of these features, but in some peaks, a monotonic evolution is observed from the rising edge to the decay of pulses.

A narrow line at ~ 740 keV was observed in the spectrum of GB781119 recorded by the germanium spectrometer on ISEE-3 [18]. It is interpreted as a redshifted line from the first <u>nuclear</u> 847 keV deexcitation level of ^{56}Fe. The SMM results also suggest a possible nuclear origin of the high energy continuum [19]. In the Signe experiment, some excesses of unknown origin have been observed in the 500 - 600 keV energy range for some events [16,17]. A comparison between the SMM and Signe time histories of GB820530 and GB821024 shows that some short peaks seen mainly in the 500 - 600 keV energy range in both experiments are not visible at lower energies [20,21]. Thus, the spectral variability of high energy excesses also seems probable.

SOURCE MODELS

Neutron stars are the most probable - although still uncertain - candidates for GRB sources [22]. For these objects, Disk and Corona spatial distributions have been proposed with nearest source distances of about 100 pc and 10 kpc, respectively [23-25].

For both <u>accretion and thermonuclear models</u>, some external matter is accreted onto the surface of the neutron star. However, the redshifted iron nuclear line at 740 keV is only consistent with a clean surface of the neutron star. Furthermore, if the 1928 optical flash [26] is really associated with the afterburst observed from the

GB781119 source, either the iron nuclear emission or both accretion and thermonuclear models should be rejected.

The rapid variability of flux and spectra also pose serious difficulties for these models. Alfven waves are known to provide an efficient thermonuclear energy output from the surface with very short rise times (< 1 msec) [27]. However it is difficult to explain i) the flux variability and the spectral evolution due to the large thermal inertia of the burning layer, and ii) how the thermonuclear and/or the gravitational energies are directly converted to e^-/e^+ annihilation radiation without a large X-ray emission (the observed fluence in the X-ray range represents a few percent of the total fluence).

Accretion and thermonuclear models usually provide thermal radiation from an optically thick plasma. Radiation driven winds from the surface have been proposed for a non-thermal radiation [28]. However, the relativistic factor Γ of accelerated plasma in a radiative way is restricted by the value for which the angular distribution of scattered radiation becomes quasi-isotropic in the co-moving frame [29]. Therefore, the maximum energy of scattered photons is less than 200 - 300 keV for a thermal emission with kt ~ 10 keV [29]. Hence, difficulties arise to explain i) the small X-ray fluence to the total fluence ratio and ii) the > 1 MeV emission of GRB.

Despite these difficulties, accretion and thermonuclear models are still compatible with available observational data on GRB. The physical conditions inside the source are inferred from spectral data for two spatial distributions: i) an isotropic emission with 4π angular diagram and ii) an anisotropic emission with photon beams in particular directions.

The internal source models are all associated with starquakes. In the nuclear explosion model, the energy source is related to the β-decay of super heavy nuclei [30]. The expanding cloud of heavy elements can emit photons up to 10 MeV. This emission is assumed to be isotropic. The photo-absorption inside the cloud [31] is a possible explanation of the low energy features. However, the rapid variability of GRB is difficult to explain by this model. In the magnetospheric model, neutron star vibrations create electric fields along the field lines and may induce electromagnetic cascades of e^-/e^+ pairs, nuclei and photons [32-35]. A fraction of the charged particles is trapped in the magnetosphere and may impact the surface to produce annihilation, nuclear emission and thermal X-rays. In this case, the annihilation and nuclear fluxes are smaller than the X-ray flux. The stream of electrically driven particles naturally explains the anisotropic production of high energy photons [35]. An instantaneous emission may be associated with an active region of the magnetosphere where relativistic particles moving along the field lines generate a beam of γ-photons. Due to the different orientations of the field lines, the total solid angle of this emission is about 4π. The rapid spectral variability may result from the evolution of emitting regions while the multi-peak structure may be caused by shock propagation in different regions of the magnetosphere.

PHYSICAL PROCESSES

The cyclotron interpretation of low energy absorptions at E < 100 keV needs a <u>magnetic field</u> strength B of about 5-7 10^{12} gauss. In the synchrotron models [22,36], the low energy features are related to the left edge of the first harmonic and therefore leads to a similar field estimation. On the other hand, the absence of a high energy cutoff in the SMM results limits the magnetic field strength to < 5 10^{11} gauss [37].

Similar values were obtained from the SIGNE data, particularly from GB781119. Some spectra with power low shapes were deconvolved using a transparency function with B as a free parameter in order to infer the dipole magnetic field B_* [38]. For an isotropic emission, the value of B_* does not exceed 10^{12} gauss, while the 70 keV feature [7] observed in many GB781119 spectra requires a magnetic field strength of about 10^{13} gauss for the cyclotron resonance interpretation.

For an anisotropic emission, the low energy feature E_1 is connected to the cyclotron resonance energy $E_{cyc}(B)$ through the relativistic transformation $E_1 \sim \Gamma\ E_{cyc}(B)$ where Γ is the relativistic factor of radiating particles. For B << $4.4\ 10^{13}$ gauss, the energy $E_{cyc}(B)$ is equivalent to $\sim E_{cyc}(B_*)\ B\ /\ B_*$. The relation between E_1 and E_{cyc} is compatible with the condition of magnetospheric transparency $B\ /\ \Gamma \sim B_*$ provided that $\Gamma > \Gamma_o$ and $B < B_*\ \Gamma_o$ with :

$$\Gamma_o = (E_t/E_{cyc}(B_*))^{1/2} \approx 3(B_*/10^{12} gauss)^{-1/2}(E_t/100 keV)^{1/2} \quad (2)$$

For these values of B and Γ, the anisotropic emission models remove the contradiction between the cyclotron interpretation of low energy spectral features and the observation of high energy photons and therefore, opens a way for models with strong magnetic fields.

The creation of e^-/e^+ pairs by two photons leads to γ-ray <u>self-absorption</u> if the photon density exceeds some critical value. For GRB spectra without high-energy cutoffs the upper limit for the distance d from the source may be estimated [39]. However, a self-consistent model needs to integrate both photon beaming and magnetic field effects [40]. Let us estimate this distance for the isotropic and the anisotropic emission models. A source region with dimensions of $\sim 10^6$ cm is characterized by the optical thickness :

$$\tau_{\gamma\gamma} \approx (\sigma/\sigma_T)(d/20 pc)^2 (F_\gamma/100 photons.cm^{-2}.s^{-1}) \quad (3)$$

where σ_T is the Thomson cross-section and $F\gamma$ the photon flux above the critical threshold. For the field-free case, the condition $\tau_{\gamma\gamma}$ = 1 leads to d < 20 pc. In a strong magnetic field, the cross-section depends on the geometry [40]. As compared to the field-free case, it is larger or smaller for photons propagating across or along the field lines, respectively. Numerical calculations [40] show that the estimates of d may differ significantly for different geometries. For anisotropic emission models with B << 10^{12} gauss, photon beaming both along radial directions and along field lines allows to avoid a high-energy cutoff for arbitrary galactic distributions. For $B > 10^{12}$ gauss, the photon beaming along radial directions does not remove a strong self-absorption for large scale distributions. However, photon beaming along the field lines allows an agreement between a strong magnetic field and a Corona source distribution.

<u>Synchrotron mechanisms</u> are efficient processes for $B > 10^{10}$ gauss [22]. The continuum shape is well fitted by the optically thin thermal synchrotron model in the intermediate energy range (see the spectral data section). However, these models lead to unreasonable temperatures for the power law continuum shapes observed by the SMM and Signe experiments at high energies. Moreover, the acceptance of these models requires a thermalization time shorter than the synchrotron cooling time. In a 10^{11}-10^{12} gauss field strength, particles densities as high as 10^{26} cm^{-3} are necessary if the thermalization is to be produced by Coulomb collisions [41]. Furthermore, a plasma confinement is needed, taking into account the synchrotron enhancement of radiative pressure [42] and possible surface layer instabilities [43]. At last, as this mechanism is isotropic, a large fraction of the total fluence is expected in the X-ray range.

The <u>synchrotron cooling emission</u> is another possible process for GRB [44-46]. Flares arise when electrons and positrons are injected into the magnetic field. A stationary flux could be generated if the particles are constantly excited. The flux from the stationary state is equivalent to the t_{cool} / t_{exc} ratio multiplied by the instantaneous flux, where t_{cool} and t_{exc} are the cooling and excitation time scales, respectively. However, the stationary mechanism remains uncertain.

Considering the magnetospheric model, two modes of GRB generation may be distinguished. In the <u>cold</u> mode, the neutron star surface is assumed to be cold and pulsar-like cascades of e^-/e^+ pairs occur. The main contribution to GRB is provided by annihilation with isotropic emission. When the neutron star surface becomes <u>hot</u>, electrically driven particles are excited by scattering of thermal photons ("gamma-gun" mechanism) [47]. In the presence of a magnetic field, the equilibrium energies of the e^-/e^+ pairs correspond to the relativistic factors Γ_B for which the resonant scattering of thermal photons takes place ($\Gamma_B = \hbar\omega_B / kT \sim 10\, B_{12} / T_7$). For the first synchrotron resonance, the energies of scattered photons are approximately given by $E\gamma \sim \Gamma_B^2 kT \sim B_{12} / T_7$ (in MeV), while the continuum above this energy can be attributed to higher harmonics. The "gamma gun" mechanism obviously corresponds to an anisotropic emission. Simple explanations may be proposed for the low-energy features and continuum variabilities if GRB pulses are associated with sporadic shocks of starquakes. The increasing temperature of the surface, below the active region, may lead to a decrease of Γ_B and thus to a monotonic shift of E_1. On the other hand, the radiative pressure induces an expansion of the emitting region and, therefore, a decrease of both the magnetic field and the high harmonic contribution.

This may explain the hardness ratio evolution observed during GRB peaks.

The <u>annihilation rate</u> $\dot{N}$ may be estimated as,

$$\dot{N}(511\,keV) = 10^l (d/100\,pc)^2 \left(F_{an}/10\,photons.cm^{-2}.s^{-1} \right) \quad (4)$$

where F_{an} is the observed photon flux of the annihilation process while l = 43 and 47 for Disk and Corona source distributions, respectively. This rate needs to be consistent with the positron production rate. However, the positrons may be secondary particles created by high-energy photons, or primary particles giving birth to secondary photons. In the isotropic photo-production model [48,49], the maximum rate of positron production depends on the total energy of the photons radiated above the reaction threshold. Using expression (1) for the continuum, the annihilation to the continuum fluence ratio may be estimated. The results are consistent with the Signe observations (< 30 percent) [5]. Positrons as primary particles may be produced in electromagnetic cascades initiated by powerful starquakes with a sufficient rate of e^-/e^+ pair production both for Disk and Corona distributions [35].

The annihilation process is another side of the problem. If the high-energy continuum comes from an annihilition origin [1,10], it should be emitted in an optically thin region above the surface of the neutron star.

However, for a magnetic field strength > 10^{10} gauss, relativistic e^-/e^+ pairs mainly radiate through synchrotron radiation while the contribution from annihilition is negligible [22]. Therefore, the identification of the high-energy continuum with annihilation emission of semi-relativistic e^-/e^+ pairs is not compatible with magnetospheric models.

If the annihilation emission has a line-like spectrum, electrons and positrons from the magnetosphere are cooled and annihilated below the surface of the neutron star giving rise to thermal X-rays and annihilation flashes. The maximum number of e^-/e^+ annihilations is about $N(511\,keV) = 4\pi r / \sigma_T \sim 10^{37}$ where r is the neutron star radius. A comparison between $\dot{N}$ and N leads to an estimate of the maximum life-time t (sec) of the positrons inside the surface layer of :

$$t = 10^{-6}(100\,pc/d)^2 \approx 10^{-10}(10\,kpc/d)^2 \quad (5)$$

This time corresponds to particle densities > 10^{20} cm^{-3} and > 10^{24} cm^{-3} for Disk and Corona distributions, respectively.

The rate of the possible iron nuclear emission at 740 keV [18], and the maximum number of emitting nuclei may be estimated, similarly. Since the life-time of excited nuclei is ~ 10^{-11} sec, the upper limit for the distance from GB781104 is ~ 30 kpc if the nuclear interpretation of the 740 keV feature is correct. The nuclear emission from the surface may be initiated by protons or nuclei accelerated in the magnetosphere. The excitation cross-section for ^{56}Fe have a threshold of ~ 2-3 MeV. Thus, the origin of ~ 3-5 MeV protons must be explained. Moreover, if the nuclear interpretation is correct, other observable nuclear lines should be excited [50].

CONCLUSIONS

This study leads to following conclusions concerning the physics of GRB and the main objectives of future experiments.

1) The existence of a strong magnetic field and a power law shape for the continuum at high-energies is compatible with an anisotropic emission, with photons beamed by relativistic particles moving along the field lines. Moreover, this model rules out the self-absorption restriction for the extended Disk and Corona source distributions. A fine-time spectral resolution is needed to elucidate the source geometry and the real variability time-scale.

2) An electric acceleration in the magnetosphere seems to be more appropriate than a radiative process to explain the particle energization. An important objective of future experiments is to compare the X-ray and γ-ray emissions in order to find the initial process.

3) Relativistic particles may emit γ-rays in the magnetosphere by the synchroton cooling process. The "gamma-gun" is an attractive mechanism but other processes as Alfven waves, shock acceleration, magnetic flares cannot be rejected. The polarimetry of GRB must be observed to clarify this situation.

4) The annihilition flashes are compatible with models for which the e^-/e^+ pair generation is the intrinsic process of energy output. Thus, the cross-correlation analysis of annihilation flashes with different energy ranges (high-energy γ-rays, X-rays, optical .etc..) is an interesting subject for future explorations.

5) Impacts of relativistic particles with the surface may produce nuclear emission. The utilisation of a high-energy resolution detector associated with a fine time resolution is required to detect nuclear lines and allow a better understanding of the energization and the origin of GRB.

The authors thank all the members of the Signe experiment team in Moscow and Toulouse for valuable discussions. One of the authors (I.G.M.) thanks the SMM Project for financial support.

REFERENCES

1. E.P. Mazets, S.V. Golenetskii, Y.A. Guryan, R.L. Aptekar, V.N. Ilyinskii, V.N. Panov, Positron-Electron Pairs in Astrophysics, (AIP, New York, 1983), p. 36.
2. E.P.T. Liang, Nature, 299, 321 (1982).
3. P.L. Nolan, G.H. Share, S. Matz, E.L. Chupp, D.J. Forrest, E. Rieger, High Energy Transients in Astrophysics, (AIP, New York 1984), p. 399.
4. I. Kondo et al., Spa. Sci. Instr., 5, 221 (1981).
5. C. Barat, K. Hurley, M. Niel, G. Vedrenne, I.G. Mitrofanov,

I.V. Estulin, V.M. Zenchenko, V.S. Dolidze, Ap.J.(Letters), 286, L11 (1984).
6. I.G. Mitrofanov, V.S. Dolidze, C. Barat, G. Vedrenne, M. Niel, K. Hurley, Sov. Astron. J., 61, 939 (1984).
7. I.G. Mitrofanov et al., 20 th International Cosmic Ray Conference, (OG 1.1.2., 1987), p. 19.
8. J.P. Norris, T.L. Cline, U.D. Desai, B.R. Dennis, Bull. AAS, (Baltimore, 1984).
9. C. Barat, Positron-Electron Pairs in Astrophysics, (AIP, New York 1983), p. 54.
10. C. Barat et al., Adv. Sp. Res. 6, N°4, 39 (1986).
11. G. Vedrenne, Adv. Sp. Res. 3, N°10-12, 97 (1984).
12. I.V. Estulin et al., pre-print 1063 (1986).
13. P.L. Nolan, G.H. Share, D.J. Forrest, E.L. Chupp, S. Matz, E. Rieger, Positron-Electron Pairs in Astrophysics, (AIP, New York 1983), p. 59.
14. P.L. Nolan, G.H. Share, E.L. Chupp, D.J. Forrest, S.H. Matz, Nature, 311, 360 (1984).
15. G.J. Hueter, High Energy Transients in Astrophysics, (AIP, New York, 1984), p. 373.
16. I.V. Estulin et al., pre-print 1114 (1986).
17. I.V. Estulin et al., pre-print 1218 (1987).
18. B.J. Teegarden, T.L. Cline, Ap.J. (Letters), 236, L67 (1980).
19. S.M. Matz, E.L. Chupp, D.J. Forrest, G.H. Share, P.L. Nolan, E. Rieger, High Energy Transients in Astrophysics, (AIP, New York, 1984), p. 399.
20. J.P. Norris et al., Adv. Sp. Res., 6, N°4, 19 (1986).
21. I.G. Mitrofanov et al., Subm. to Sov. Astron. J., (1987).
22. D.G. Lamb, High Energy Transients in Astrophysics, (AIP, New York, 1984), p. 512.
23. J.C. Higdon, R.E. Lingenfelter, High Energy Transients in Astrophysics, (AIP, New York, 1984), p. 568.
24. M.C. Jennings, High Energy Transients in Astrophysics, (AIP, New York, 1984), p. 412.
25. I.S. Shklovskii, I.G. Mitrofanov, Mon. Not. of R. Astr. Soc., 212, 545 (1985).
26. B. Schaefer, Nature, 294, 722 (1981).
27. I.G. Mitrofanov, V.M. Ostraykov, Astr. Sp. Sci., 77, 469 (1981).
28. S.E. Woosley, High Energy Transients in Astrophysics, (AIP, New York, 1984), p. 485.
29. I.G. Mitrofanov, A.I. Tsygan, Astrophys. Sp. Sci., 84, 35 (1982).
30. G.S. Bisnovatyi-Kogan, V.S. Imshennik, D.K. Nadexin, V.M. Ceshetkin, Astr. Sp. Sci., 35, 23 (1975).
31. G.S. Bisnovatyi-Kogan, A.F. Illarionov, Pre-print 1184 (1985).
32. A.I. Tsygan, Astron. and Astrophys., 44, 21 (1975).
33. R. Ramaty, R.E. Lingenfelter, R.W. Bussard, Astr. Sp. Sci., 75, 193 (1981).
34. E.P. Liang, High Energy Transients in Astrophysics, (AIP, New York, 1984), p. 597.
35. I.G. Mitrofanov, Astr. Sp. Sci., 105, 24 (1984).
36. E.P. Liang, Positron-Electron pairs in Astrophysics, (AIP, New York, 1983), p. 76.

37. S.M. Matz et al., Ap.J. (Letters), 288, L37 (1985).
38. I.G. Mitrofanov et al., Sov. Astron. J., 63, 1118 (1986).
39. G. Cavallo, M. Rees, Mon. Not. of R. Astr. Soc., 183, 359 (1978).
40. A.A. Kozlenkov, I.G. Mitrofanov, Sov. J. of Exp. and Theo. Phys., 91, 1978 (1986).
41. A.S. Pozanenko et al., The Physics of Compact Objects : Theory versus Observations, (Bulgaria, 1987).
42. I.G. Mitrofanov, G.G. Pavlov, Mon. Not. of R. Astr. Soc., 200, 1033 (1982).
43. I.G. Mitrofanov, Astr. Sp. Sci., 132, 155 (1987).
44. R.W. Bussard, High Energy Transients in Astrophysics, (AIP, New York, 1984), p.611.
45. J.J. Brainard, D.Q. Lamb, Ap.J., 313, 231 (1987).
46. I.G. Mitrofanov, A.S. Pozanenko, Sov. J. of Exp. and Theo. Phys., 93, 1951 (1987).
47. N.S. Kardashev, I.G. Mitrofanov, I.D. Novikov, Sov. Astr. J., 61, 1113 (1985).
48. V.V. Zheleznyakov, A.A. Litvinchuk, 20 th International Cosmic Ray Conference, (1987), p. 19.
49. V.V. Zheleznyakov, A.A. Litvinchuk, Astr. Sp. Sci., 83, 81 (1982).
50. R. Ramaty, B. Kozlovsky, R. Lingenfelter, Ap.J. Suppl. 40, 487 (1979).

SPECTRAL MODELLING OF THE SOFT GAMMA-RAY REPEATER

E. P. Liang
Lawrence Livermore National Laboratory, Livermore, Ca. 94550

ABSTRACT

This paper summarizes the results of numerical modelling of the spectra of the soft gamma ray repeater (SGR 1806-20). Parameters of the emission region can be extracted from the spectral fits under the simplistic assumption of isothermal and homogeneous emission layers. Limits on the luminosity distance of the source can also be derived if we make specific interpretations of the low frequency turnover in the observed spectrum.

INTRODUCTION

The recently discovered soft gamma ray repeater SGR 1806-20[1-3], together with the March 5, 1979[4] and March 24, 1979 soft gamma bursters, probably represent a distinct new class of high energy transients. They are distinguished by their chaotic repetition patterns and a soft exponential spectrum with an almost universal color temperature of ~45 $\pm$ 10 keV (cf. Refs.1,4). While the repetition pattern is already the subject of much speculation[5], here we concentrate on extracting the maximum amount of physical information out of the spectral data of SGR 1806-20, by making the anzatz that to first order we can approximate the emission region as a homogeneous and isothermal layer. Despite the simplistic nature of such models, these results should provide a useful framework for constructing more refined astrophysical models and confronting future observational data.

Since a summary of our results is already being published elsewhere[6], here we will concentrate on the discussion of ideas and issues which supplement and complement those in Ref. 6.

SPECTRAL MODELS

We consider four specific thermal (single-temperature) emission models which, to our judgement, provide the best fit to the published spectral data of SGR 1806-20[1,3]. These are: a. optically thin thermal bremsstrahlung (TB), b. self-Comptonized bremsstrahlung (CB), c. thermal synchrotron[7,8] (TS) and d. cooling synchrotron with thermal injection[8](CS). We have also modelled the spectra with blackbody or Wien (saturated Comptonization), but the fits are much worse than the above models so we did not pursue them further. We compute the bremsstrahlung spectra including Comptonization and self-absorption effects using a Monte Carlo code developed by Canfield[9] and the synchrotron spectra with a code generalized from that of Brainerd[8,10]. Synchrotron self-absorption is included approximately via semi-analytic techniques. In the bremsstrahlung cases we estimate the emission measure and intrinsic flux by interpreting the low frequency turnover as self-absorption. In the synchrotron cases the turnover is interpreted as either due to self-absorption or first harmonic cutoff, leading to different distance limits. Given the large error bars of the sub-10keV data, the reality and interpretation of the turnover are obviously debatable. Hence our distance estimates can be taken as conservative upper limits.

BREMSSTRAHLUNG MODEL PARAMETERS

Figure 1 illustrates the comparison of sample model bremsstralung spectra of different Thomsom depths τ and temperatures T with the published data of SGR 1806-20 (we have combined the data of GB790107 and GB831119 to maximize the energy coverage since the two spectra coincide in the overlapping range and there is no significant spectral variation among the recurrent events). The emission measure $n^2 h$ (n=electron density, h=emission layer thickness) required for the low frequency turnover comes to $\sim 10^{51} cm^{-5}$ and is relatively insensitive to the Thomson depth and temperature in the range that gives the best fit. Even though we have not yet performed a detailed chi-square analysis, just from eyeballing we conclude that in the τ-T plane a narrow ellipse covering the temperature range ~30-40 keV and Thomson depth range ~0 -1 would correspond to the best fit parameters (Fig. 1 inset). For example, for $\tau = 0$ the best fit temperature is ~ 40 keV and for $\tau = 1$ the best fit temperature is ~ 30 keV. For temperatures < 30 keV or > 40 keV the fit gets progressively worse independent of τ. Similarly, for $\tau > 1$ the fit gets progressively worse for all temperatures.

For the $\tau = 0$, T=40 keV case, the emission measure of $10^{51} cm^{-5}$ corresponds to an intrinsic flux above 30 keV of ~ $2.4 \times 10^{27} ergs/cm^2.s$. For a neutron star of 10 km radius, the luminosity distance would be ~ 3.4 kpc using the published value for the observed flux above 30 keV (cf. Ref. 1). In the opposite case of $\tau = 1$, T=30 keV the same emission measure gives rise to an intrinsic flux above 30 keV of ~ $1.9 \times 10^{27} ergs/cm^2.s$ and a luminosity distance of ~ 3.0 kpc. Since the self-absorption turnover frequency is uncertain to a factor of ~ 2 (cf. Fig.1), the emission measure and intrisic flux are uncertain to a factor of ~ 4 (since the absorption frequency $\nu_a <<$ temperature T so $\exp(\nu_a/T) \approx 1$). Hence the luminosity distance estimate is uncertain to within a factor ~ 2. In any case the distance falls short of the galactic center (~8.5 kpc) unless the stellar radius (or effective emission surface size) is significantly larger than 10 km. Since in all cases the Thomson depth is ≤ 1, the thickness of the emission layer is $\leq 2 \times 10^{-3}$ cm and the electron density $\geq 7 \times 10^{26} cm^{-3}$. Such a dense thin emission layer is hard to visualize in realistic astrophysical situations. This plus the lack of a confinement mechanism for the plasma against the superEddington flux makes the bremsstrahlung models rather implausible.

SYNCHROTRON MODEL PARAMETERS

Figure 2 illustrates the comparison of model synchrotron spectra with the spectral data. It shows that given the uncertainties of the sub-10 keV data, the turnover can be modelled either as self-absorption or first harmonic cutoff. Of course if future observations can pin down the spectrum below ~4 keV then the two interpretations can definitely be distinguished. As in all synchrotron spectral models, the spectral shape at high frequencies can only constrain the product $\nu_c = \nu_B . T^2$ (where ν_B is the cyclotron frequency) but not the field strength B or the temperature T separately. If the low frequency turnover is due to first harmonic cutoff, then the field strength and temperature are indeed obtained separately. In this case an upper limit to the intrinsic flux can be estimated from the lack of self-absorption down to the first harmonic. Note that in this case we expect to see the feature of the second harmonic as well due to the low temperature. On the other hand, if the low frequency turnover is due to self-absorption, we would need to invoke additional conditions to obtain the magnetic field strength. In the present context we propose to use the plasma confinement condition $B^2/8\pi > F/c$ (where F is total radiation flux at the source) for the field lower limit, and the condition that the first harmonic be located below 10 keV for the field upper limit. In this case a range of the intrinsic synchrotron flux can be estimated.

For the steady thermal case the formula for self-absorption can be found in Ref.7. For the cooling case (with initial thermal injection) we derive the approximate self-absorption opacity using Tucker's formula[11] by substituting ν_a/ν_B for $\gamma_\perp$ (where $\gamma_\perp$ is the transverse Lorentz factor) into the time-averaged cooling particle distribution function derived by Brainerd[8] and its derivatives. This is admittedly a crude procedure but we believe it should be adequate for the present purposes. The resultant emission parameters for both cases are listed in Ref. 6 so they will not be repeated here. We note that the resultant luminosity distance upper limit for a 10 km radius neutron star is ~5.5 kpc for both first harmonic cutoff models. The distance range for the steady thermal self-absorption model is ~3.5 – 12 kpc and the corresponding range for the cooling thermal self-absorption model is ~6–14 kpc. As expected, the cooling case gives the largest distance since the cooling particles tend to self-absorb at lower frequencies. Both self-absorption models allow a galactic center distance (~8.5 kpc).

From the microphysical point of view, a cooling particle distribution is also most plausible. Even at the low field strength of $\gtrsim 10^{10}$ G required for confinement (cf. Table I of Ref. 6), the synchrotron cooling time is $\lesssim 10^{-12}$ sec and is much shorter than the coulomb coupling time to hot protons to maintain a steady thermal distribution for densities $\lesssim 10^{24}/cm^3$. In this case however, the initial injection temperature of the electrons must exceed ~100 keV to account for the observed spectrum (cf. Ref. 6). In most x-ray pulsar models[12], the hard x-ray spectrum is produced by Comptonization with $\tau_T \gtrsim$ 1 and T ~ few – 30 keV in a super-teragauss field. Any turnover would be due to Compton saturation or bremsstrahlung self-absorption. However, for SRG 1806-20, the location of the turnover, hardness of the spectrum and lack of any harmonic feature from 8-200 keV distinguish it from most x-ray pulsars and are hard to reconcile with the strong field Compton models. Hence we opt for a weak field ($<10^{12}$ G), optically thin, high temperature scenario, such as may be produced in a collisionless accretion shock above the stellar surface. Note that to maintain a synchrotron radiation flux of ~10^{28} erg/cm^2.s (cf. Ref.6) by proton accretions the proton density in the accretion flow must be ~$5.10^{21}/cm^3$ assuming ~$0.1mc^2$ energy conversion. In a $>10^{10}$ G field the proton gyroradius and hence the minimum shock thickness $\lesssim$ 0.002 cm. This is small compared with the cooling distance of the heated electrons (~ 0.03 cm). If the shock is not pair-dominated (which seems likely) so that lepton density ~ proton density, the total Thomson depth of the shock plus post-shock cooling layer is ~10^{-4}. Interestingly this is consistent with the Thomson depth needed for the synchrotron self-absorption turnover (cf. Ref.6). A conception of this picture is depicted in Fig. 3.

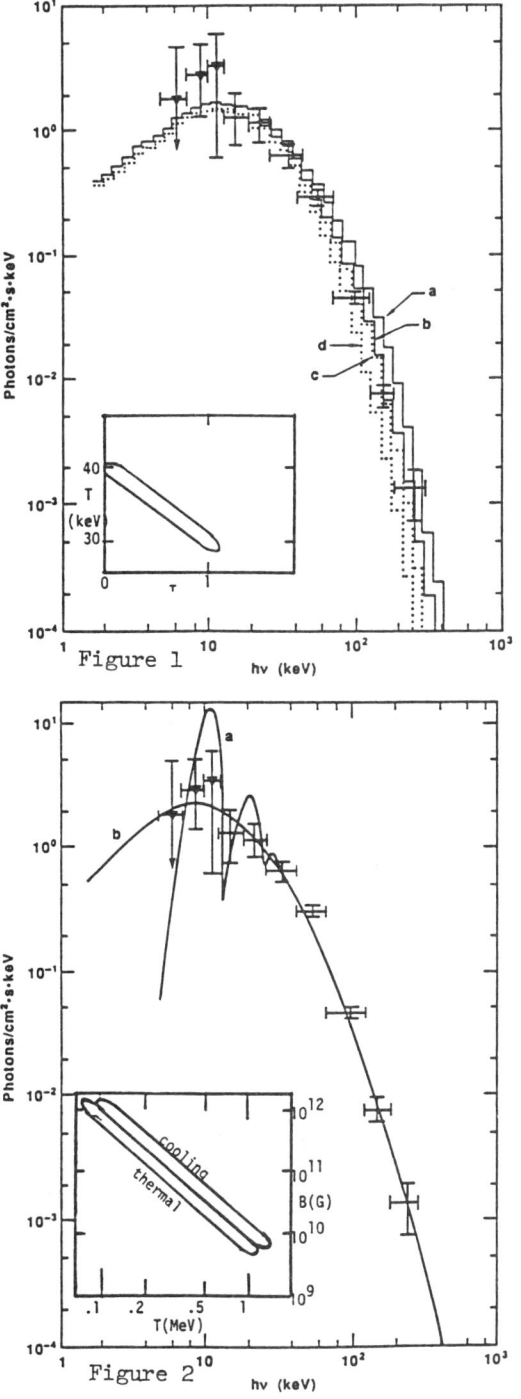

Fig. 1. Comparison of self–Comptonized bremsstrahlung spectra with data of SGR 1806–20. Data are taken from published results of Prognoz 7, Prognoz 9 and ICE (Refs.1,3). a. T=40 keV, τ =1; b. T=40keV, τ = 0.1; c. T=30 keV, τ = 1; d. T=30 keV, τ =0.01. In all cases $n^2 h \sim 10^{51}$ cm^{-5}. Inset: Schematic ellipse illustrating the best fit parameters in the τ–T plane. No confidence level can be given until chi-square analysis is performed.

Fig. 2. Comparison of model synchrotron spectra with data: a. cooling synchrotron with B=1.2×10^{12}G, injection temperature=102keV, no self–absorption; b. steady thermal synchrotron with B=6×10^{10}G, T=330 keV, τ =0.02. Inset: Schematic ellipse illustrating the best fit parameters in the B–T plane.

CONCLUSIONS

We show that rough emission parameters and luminosity distance limits for SGR 1806-20 can be obtained by modelling the spectrum with simple single temperature models. The results favor the synchrotron model over the bremsstrahlung-Compton models. However, only self-absorbed synchrotron emissions from a cooling distribution of injected hot electrons (as opposed to a true thermal distribution) appears likely if the particle density is $< 10^{24}/cm^3$ and collective plasma processes do not operate. A prediction of this model would be that the x-ray spectrum below ~ 4 keV should be very close to a quasi-Rayleigh-Jeans spectrum. Future x-ray observation of this source should be able to verify or dismiss this model.

REFERENCES

1. J. L. Atteia et al., Astrophys. J. Lett. 320, L105 (1987).
2. J. G. Laros et al., Astrophys. J. Lett. 320, L111 (1987).
3. J. G. Laros et al., Nature 322, 152 (1986).
4. E. P. Mazets et al., Astrophys. Sp. Sci. 84, 173 (1982).
5. A. Babul et al., Astrophys. J. Lett. 316, L49 (1987).
6. E. P. Liang, Nature, to appear (1988).
7. E. P. Liang et al., Astrophys. J. 271, 766 (1983).
8. J. Brainerd and D. Q. Lamb, Astrophys. J. 313, 231 (1987).
9. E. Canfield et al., Astrophys. J. 323, 000 (1987).
10. J. Brainerd, Harvard University Ph.D. Thesis (1985).
11. W. Tucker, Radiative Processes in Astrophysics (MIT, Cambridge, 1975).
12. See, e.g. P. Meszaros, AIP Conf. Proc. No. 115, p.165, S. Woosley ed (AIP, N.Y., 1984); Aron, J. et al. Ap. J. 312, 666 (1987).

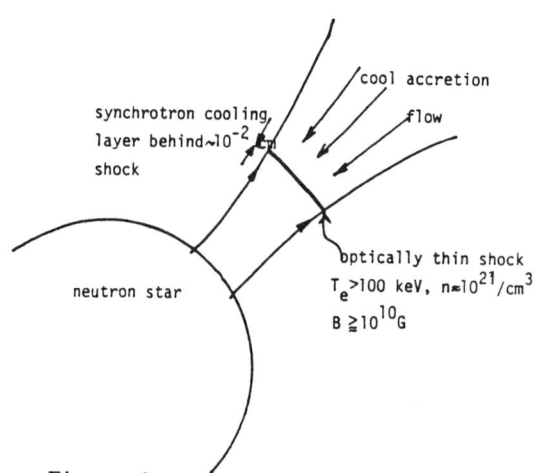

Fig. 3. Artist conception of an optically thin synchrotron cooling collisionless shock model for accretion onto a magnetized neutron star that may be relevant to the soft gamma repeater.

Figure 3

GAMMA-RAY BURSTS FROM NEUTRON STAR DETONATION

F. Curtis Michel
Rice University, Houston, Texas 77251

ABSTRACT

A phenomenological gamma-ray burst model postulating the injection of 10^{51}-10^{53} ergs into a neutron star by Paczyński[1] and by Goodman[2] is closely reminiscent of what a relativistic version of a type I supernova might look like. Burning to tightly bound quark complexes (i.e., particles) could go explosively, just as the final burning of helium to iron disrupts a white dwarf in some models of type I supernovae. Particle theorists have for some years entertained the possibility that compressing nuclear matter may lead irreversibly to an exothermic formation of "quark" matter. From stellar evolution, we know that nuclear burning does not proceed directly to the known endpoint (iron) but is halted at various intermediate stages (e.g., at helium). Thus when the central densities of a neutron star begin to rise to that at which a transition to quark matter would be expected (10-20 times nuclear) owing to accretion or fusion with a companion neutron star, explosive burning may take place. Active galactic nuclei could be powered by such events and are therefore possible source regions. The particles themselves would be massive tightly bound bosons having zero spin, zero charge, and zero magnetic moment, hence could possible evidence themselves in the form of "dark" [= nonluminous] matter or "WIMPs" (weakly interacting massive particles). The relativistic shocks from such events should accelerate cosmic rays more efficiently than the usual supernova shocks.

INTRODUCTION

Recently, Paczyński[1] and Goodman[2] have pointed out that injection of a large amount of energy (source unspecified) into a neutron star) would lead to a gamma-ray burst model despite the high opacity in such cases. The result would be a relativistically expanding cloud of ultra-hot particles. Paczyński comments, "It suggests that gamma-ray bursts may be related to some violent events on neutron stars which are far away." And Goodman notes, " ... the energy may be supplied by some still more exotic and unimagined process ... the possibility that burst sources could be optically thick and distant should not be dismissed simply because we cannot yet find a likely mechanism for providing the energy."

The present work[3] suggests both such a source and the mechanism. It came up in an entirely different context concerning what happens to a neutron star that is pushed over the Oppenheimer-Volkoff limit[4] by accretion or fusion with a companion neutron star. Under such circumstances, the central density grows beyond that of nuclear matter and as a consequence it becomes possible to form an entirely new class of particles (essentially hypernuclei) that could have much higher binding than ordinary nuclei. If so,

© 1988 American Institute of Physics

neutron star collapse could trigger a burning process which, if sufficient energy was released, would simply blow the neutron star up instead of letting it continue to collapse. (A rather similar scenario has been suggested for Type I supernova[5] wherein a white dwarf is pushed over the Chandrasekhar mass limit but burns its remaining nuclear fuel so rapidly that it explodes rather than collapsing to a neutron star.) It was not clear whether or not such events would look like gamma-ray bursts until the above papers [1,2] were recalled.

BURNING TO PARTICLES OF HIGH STRANGENESS

As we will see, the likely candidate particles have high symmetries because tight binding is typically associated with high symmetry. In nuclear physics, the alpha particle is a prototype, with zero spin and zero magnetic moment. The candidate particles additionally have zero electric charge, and therefore no electromagnetic interaction except possibly with very energetic gamma rays. They are otherwise heavy invisible particles that may still have "strong" interactions, but very small cross-sections. As such, they would thermalize with the remaining unburnt neutron star matter but thereafter be quite penetrating and probably exit the host galaxy.

The rates at which strangeness is changed in baryons when energetically favored is typically of the order of 10^{-10} sec which is quite fast compared to the free-fall time of the neutron star itself (order 10^{-4} sec) and therefore burning should shock and compress the surroundings which itself should trigger further burning since the process is expected to largely be density dependent. A sizeable fraction of the star could therefore be consumed, depending on details not yet known.

Given the central role of neutron stars as intermediaries to gravitational collapse, black hole formation could well be hindered by such explosions.

Burning to "quark" matter (an amorphous plasma of degenerate quarks) has already been suggested,[6] but the assumed energy release was not sufficient to unbind the star and the consequences were mainly to convert it from a neutron star to a "quark" star.

THE OCTET PARTICLE

The SU(3) symmetric analog of the alpha particle (spin-paired neutrons and protons all sitting in tightly bound s-wave states) would simply be a particle composed of pairs of all eight baryons instead of just two of them. We concentrate on pairs because nuclear systematics show that unpaired nucleons are only weakly bound. Being distinct, these baryons can all sit in s-states, their net attractive interactions can grow as N, and the net binding per particles can grow essentially as N^2, giving a total binding energy scaling as N^3 (or $(N-1)^3$ to be more careful). Given the already tight binding of the alpha particle (28 MeV) and the large possible enhancement in going from 2 to 8 pairs (7^3 =

343), such a particle could easily be bound well beyond that necessary to supply the roughly 1700 MeV of mass/energy required to convert nucleons into the more massive strange baryons. Because all the numbers (and uncertainties) involved are of the order of 200 MeV/nucleon, it is quite possible that the net energy release be of the order of 20%, which is adequate to unbind a neutron star with energy to spare.

Note that in atoms, increasing the attractive charge by Z increases the binding by Z^2 and the number of electrons by Z, which would similarly give a Z^3 dependence on total (atomic) binding were it not for the exclusion principle forcing electrons into more weakly bound states.

It is hard to imagine how in the laboratory one might assemble these 16 baryons (12 of which decay in 10^{-10} sec and half of which carry no electric charge with which to control them). But in the center of a collapsing neutron star, when the central density increases to 10-20 times nuclear, this "overlap" problem is solved by directly packing 16 nuclei within the space normally occupied by one. Moreover, the increasing Fermi energy converts nucleons into Λ's and Σ's to provide seed particles carrying strangeness. There must, of course, be such an activation threshold or ordinary matter would spontaneously convert into such particles.[3] The same density estimates have been made for the transition to quark matter.[7,8]

Collapsing neutron stars provide the unique setting in which such particles can form in the universe at present.

THE QUALPHA PARTICLE

The modern view is that the baryons consist of quarks, so an alternative is to consider highly symmetric configurations at the quarks themselves. One has already been predicted, the dibaryon[9] state thought to be a colorless set of spin-paired up, down, and strange quarks. The nuclear analog would be the deuteron, while the alpha particle analog would be the colorless set of spin-paired <u>triplets</u> of the up, down, and strange quarks (= the "qualpha") which exhausts the possible s-states. Indeed, this would be the hypernucleus made by absorbing two Ω^- particles on an alpha particle. All of these particles have zero spin, zero change, and zero magnetic moment. They are massive invisible bosons.

One can play with other possibilities, add charmed quarks, etc. Insofar as our interests go, we mainly care whether or not <u>some</u> stable, tightly-bound complex exists. Note that bag model calculations are not designed to reveal clustering of quarks into tight bunches. Instead the interactions are assumed to wash out and the quarks to be essentially free. What we have done is take the attractive quark interaction (which is ~1/r on moderately short scales) and postulated a near "gravitational" collapse when too many of these particles are together. Thus the particles could collapse to tight clusters before the neutron star itself!

IMPLICATIONS

We have already noted the possible relevance to gamma ray bursts, but where would these bursts take place? As Paczynski[1] and Goodman[2] point out, these events would be detectible at cosmological distances. If they took place among the general field stars, objects like PSR 1913+16 would be possible candidates. The central regions of galaxies are also another interesting possibility. A popular suggestion has been that super-massive black holes power the centers of active galactic nuclei (e.g., Quasars). Observations suggestive of a massive black hole (highly concentrated mass in the central regions of a galactic nucleus and conversion of available rest mass energy into radiation with apparently higher-than-nuclear efficiency) would also be consistent with burning to those complexes. For illustration, a dense cluster of neutron stars imbedded in an optically thick gas cloud could have a high rate of neutron star explosions, with the energy from the resultant neutron star explosions being absorbed by the cloud (rather than being seen as gamma ray bursts) and then reemitted at optical and UV frequencies. The transfer of energy out of the cluster center would, furthermore, stabilize it.

Although these events would be frequent inside the core, it is always possible that a binary neutron star pair might be ejected from the core prior to fusion, in which case it might drift free of the absorbing central regions prior to detonation and be detected as a gamma ray burst.

Is there any evidence for the existence of such particles? Octets should scatter elastically through virtual meson exchange. As such they should thermalize with ordinary matter and sink to the centers of planets and stars. In the case of the Sun, they could have observable effects on the neutrino production.[10] However, the geometrical cross-section (of about 10^{-29} cm^2) would still be too large for them to be weakly-interacting-massive-particles (WIMPs), unless the octet particle is (or decays to) an intrinsically new sort of particle that no longer exhibits strong interactions. Accumulations of octets would be invisible owing to the lack of significant electromagnetic interaction (i.e., a possible source of "dark" [= nonluminous] matter).

Acceleration of cosmic rays by shocks has come to be taken seriously, but known shocks from supernovae are too slow and weak to produce gamma rays beyond about 10^{15} eV (and even that is optimistic). The faster, more intense shocks produced by neutron star detonations might bridge the gap to the 10^{20} eV cosmic rays observed.

This work was conceived while on sabbatical leave at the Kellogg Radiation Laboratory at CalTech under the hospitality of William A. Fowler and C. A. Barnes (under NSF grants PHY-8505682 and 8604197) and personally under NSF grant AST-8511709 and NASA grant NAGW-379.

REFERENCES

1. B. Pachzynski, Astrophys. J. Lett., 308, L43 (1986).
2. J. Goodman, Astrophys. J. Lett., 308, L47 (1986).
3. F. C. Michel, Astrophys. J. Lett. to be published.
4. J. R. Oppenheimer and G. M. Volkoff, Phys. Rev., $\underline{55}$, 374 (1939).
5. V. Trimble, Rev. Mod. Phys., 54, 1183 (1982).
6. C. Alcock, E. Farhi, and A. Olinto, Astrophys. J., 310, 261 (1986).
7. G. Baym and S. A. Chin, Phys. Lett., 62B, 2451 (1976).
8. G. Chapline and M. Nauenberg, Phys. Rev. D., 16, 450 (1977).
9. R. L. Jaffe, Phys. Rev. Lett., 38, 195 (1977).
10. R. L. Gilliland, J. Faulkner, W. H. Press, and D. N. Spergel, Astrophys. J., 306, 703 (1986).

THE HIGH-RESOLUTION SPECTROSCOPY OF GAMMA-RAY TRANSIENTS

Thomas L. Cline
Laboratory for High Energy Astrophysics
NASA/Goddard Space Flight Center, Greenbelt, MD 20771 USA

ABSTRACT

The first high-resolution spectrometer flown to observe gamma-ray bursts was launched on the ISEE-3 spacecraft over nine years ago. It recorded two events before instrument failure, giving results that were suggestive but marginal. Other studies, with coarser energy resolution, also show evidence for spectral features as well as for spectral evolution on short time scales. Absolute source strength calibration will be possible only with source identification, but understanding of the burst emission processes will surely come only from the measurements having the best spectral and temporal precision. The only high- resolution gamma-ray spectrometer now planned, here or abroad, for space flight is an instrument sequel to the ISEE-3 spectrometer, to be flown on the interplanetary 'GGS Wind' mission. Much larger and higher-sensitivity, high- resolution instruments may have their optimum opportunities in conjunction with studies of solar flares in the time frame of the solar maximum of 2002.

INTRODUCTION

Nearly one decade has elapsed since our initial conference on gamma-ray spectroscopy in astrophysics[1]. It goes without saying that considerable advances have been made in many respects and that the results displayed here indicate that hope regarding the future of the discipline is justifiable. Our compliments go to Drs. Share and Gehrels for their organization of this conference; they also came up with a bright idea we didn't have: Supernova 1987a! We can expect to see some excellent results very soon, following the balloon studies of the evolutions of the SN87a continuum and gamma-ray line intensities. However, outside of the fortunate circumstance of SN87a, the original promises of gamma-ray spectroscopy in astrophysics have not yet been truly fulfilled. This slow rate of progress is due to the great difficulties in perfecting high-resolution detectors with the background rejection capabilities in the large-volume arrays needed for sensitivity in this astronomical window.
   In this note, I wish to call attention only to the area of the high-resolution spectroscopy of gamma-ray bursts. The fact that this specialty is even less mature than the traditional spectroscopy of known astronomical gamma-ray sources - and the long-standing mystery of the gamma-ray burst sources - imply that its rewards, although ill-defined, may be more tantalizing. The appropriate instrumentation for this specialty would require additional complexity (over a germanium gamma-ray spectrometer designed for point- source astronomy or for geophysics, such as on the Mars Observer) for the maintenance of performance stability and for the rapid data collection during the intense counting-rate excursions in bursts. The numbers of gamma rays collected during each changing feature within a burst may (or may not) be adequate to permit accurate, time-resolved spectroscopic studies to be made with reasonably sized

detectors, depending on how rapidly these spectra can evolve. Given the state in which the study of gamma-ray bursts still finds itself, with fresh developments and new discoveries, such as the galactic repeater, still being being uncovered, one could expect that time-resolved spectroscopy will someday provide clues towards the solutions of its puzzles.

DISCUSSION

Kevin Hurley has provided in this conference[2] a review of gamma-ray burst studies with regard to spectroscopic content. A brief summary of those results would include mention of the lower-energy features, generally attributed to cyclotron resonance phenomena, and the higher-energy features, assumed to be red-shifted annihilation lines. Detectors with characteristics entirely differing from the scintillators typically used have not yet observed the low-energy resonance features; germanium gamma-ray spectrometers, for example, have previously had the disadvantage of thresholds above the relevant 65-keV region. Thus, genuine independent confirmation of the effect, let alone an augmentation of its study with improved resolution, awaits the application of planar germanium units in this regard.

The observations of gamma ray bursts from an experiment on Solar Maximum Mission have significantly extended the energy range detected through several tens of MeV; however, neither confirmations of red-shifted annihilation lines nor indications of higher-energy nuclear lines were found. It might therefore be contended that the present shortage of evidence for, or confirmation of, narrow line features in the spectra of gamma-ray transients is an argument for their nonexistence. However, until the proper instruments (capable of high spectral resolution, with adequate sensitivity to obtain statistically significant spectra on very fine time scales) are used, this viewpoint remains negative in outlook and, actually, experimentally unfounded. The spectral evolution of bursts, as observed with several independent experiments[3,4,5], appears to be as rapid as instrumentation can indicate. This effect not only is adequate to explain the discrepancies between the results of these searches for narrow lines but may even indicate the presence of more complex behavior. It implies that much greater temporal and spectral capabilities will be required in order to properly decode and decipher the information carried in these emissions.

Finally, no events of the soft, repeating class of transients have yet been observed with any instruments employing a high-resolution spectral capability. These events appear to be an intermediate manifestation of burst behavior between the x-ray and the truly gamma-ray classes; they may be viewed as a second-order astrophysical curiosity or as indicators of the complexity of nature. Either way, there is a matrix of ignorance here that surely indicates the need for detailed experimental investigations before theoreticians can be confident of the relevance of their calculations.

What can be expected from experiments planned in the next few years? Planar germanium disks would be adequate only for the lowest-energy studies in the cyclotron resonance region; an array of large, coaxial units will be needed for the higher-energy annihilation and nuclear line studies. Certain contemplated balloon experiments, even of the 'long-duration' variety, will clearly not have the required exposure time to detect very many, if any, events. A lucky hit with one intense burst, of course, would surely provide some form of payoff. The Gamma Ray Observatory, like the various European and Soviet missions planned to be launched over the next few years, will carry no germanium gamma-ray spectrometer. The Wind-GGE mission is planned to carry a sequel instrument to the radiatively cooled germanium ISEE-3 burst spectrometer; that will be a detector that can be described as second-generation in design sophistication, but it will not be large enough to yield an order of magnitude improvement in sensitivity. The Nuclear Astrophysics Explorer is a mission proposed to provide the optimum study in the traditional discipline of gamma ray spectroscopy in astrophysics; it was not designed to perform gamma-ray transient spectroscopy, however.

The Explosive Transient Camera and its associated Rapidly Moving Telescope will soon be operational with high-precision studies of the optical radiations that may accompany gamma ray bursts. A possible sequel instrument in space is the High Energy Transient Experiment, H E T E. These may indeed make possible the optical identifications of burst source objects. Also, scrutiny with the Hubble space telescope of burst source fields triangulated with the next interplanetary networks of gamma ray burst detectors may yield identifications of candidate source objects. Given that such studies may be accomplished, one would expect that high time-resolution gamma-ray spectroscopy should contribute an extremely fruitful avenue of research, especially if one tends to associate gamma-ray emission with high-energy activity in astrophysics. The truly next generation of instruments, with greatly increased sensitivity and with spectral and temporal resolution over at least the 20-keV to 20-MeV energy range, may not be likely in the next few years. If no modifications are made to our present set of measurement priorities, then, the advantages of a combined solar-maximum/cosmic-transient mission may, instead, represent the ultimate or optimum scenario for the years centered on 2002.

REFERENCES

1. T. L. Cline and R. Ramaty, eds., Gamma Ray Spectroscopy in Astrophysics, NASA Tech. Mem. 79619, (NASA GSFC, 1978).

2. K. Hurley, this volume (1988)

3. C. Barat, K. Hurley, M. Niel, G. Vedrenne, I. G. Mitrofanov, I. V. Estulin, V. M. Zenchenko, and V. Sh. Dolidze, Astrophys. J. (Lett.) 286, L11 (1984).

4. C. Barat, AIP Conf. Proc. No. 101, AIP Press, N. Y., p. 54 (1983).

5. J. P. Norris, G. H. Share, D. C. Messina, B. R. Dennis, U. D. Desai, T. L. Cline, S. M. Matz, and E. L. Chupp, Astrophys. J. 301, 213 (1986).

GAMMA-RAY EMISSION OF CYGNUS X-1

James C. Ling
Jet Propulsion Laboratory, California Institute of Technology
Pasadena, California 91109

ABSTRACT

Observations of gamma-ray emission in the 0.4-6 MeV region from Cygnus X-1 have been reported by several groups during the last decade. The 1979 HEAO 3 observation of a broad variable feature at ~ 1 MeV has been interpreted as strong evidence for Wien and annihilation radiation produced in a hot (400 keV), pair dominated spherical cloud (r ~ 250 km) in the innermost region of the accretion disk surrounding a black hole. The temporal and spectral behavior of the 0.05-10 MeV intensity observed by HEAO 3 during its ~ 170 days exposure to the source suggested a more complex emission mechanism than that envisaged in the earlier bimodal accretion model. In this paper, I review the HEAO 3 results and summarize other gamma-ray observations in the 0.05 -10 MeV range.

INTRODUCTION

Cygnus X-1 is an accreting binary system (5.6-day period) consisting of a blue supergiant (HDE226868) and a compact object considered to be the leading candidate for a black hole. The source is a well-known x-ray (1-200 keV) emitter with time variations ranging from milliseconds to years. The spectrum, characterized by a power law below 100 keV and an exponential cutoff at ~ 200 keV, has been interpreted as unsaturated Comptonization of soft photons by hot ($<10^9$K) plasma in the inner region of the disk[1]. However, when the temperature of the plasma rises significantly above 10^9K, pairs may be formed by photon-photon interactions[2]. The positrons and electrons will then annihilate and produce gamma radiation with energy ⩾511 keV. For a 5 x 10^9K plasma, the annihilation feature is thermally blue-shifted to ~ 1 MeV with a width of ~ 1 MeV [3-4]. Pozdnyakov et al.[5] showed that if the medium is optically thick ($\tau >1$), a broad "Wien" feature around 1 MeV can also be formed due to the Comptonization of copious soft photons by the relativistic electrons and positrons.

Observations of gamma-ray emission above 0.4 MeV from Cygnus X-1 have been reported by several groups during the last two decades[6-9]. The recent results of HEAO-3[10], based on ~ 170 days of exposure to the source in 1979-1980, provided the clearest evidence for a superheated, pair-dominated plasma in the innermost region of the disk. New phenomena observed above 0.05 MeV include the long-term variability of the hard x-ray emission, a strong variable gamma-ray feature near 1 MeV, and complex spectral changes. A brief description of the HEAO 3 results, and a summary of other observations, both continuum and line emissions in the 0.05 - 10 MeV range, are presented, followed by a discussion of the implications of these results, and future observational requirements.

1979-1980 HEAO 3 OBSERVATIONS

Temporal Variability

Prior to HEAO 3, the long-term temporal behavior of Cygnus X-1 has been characterized primarily by the 1-10 keV emission. Priedhorsky et al.[11] showed that following the landmark 1971 change in behavior[12], the average x-ray flux has stayed roughly constant at the normal "low state" (LS) level except for periodic (294 day) dips and two epochs of major increases, in 1975[13] and 1980[14], to what has been known as the "high state" (HS). The former has been interpreted as an effect of the precession of the accretion disk. During a LS to HS transition, the soft x-ray (1-10 keV) flux increased by a factor of 3-5 while the hard x-ray (E>10 keV) flux decreased. This two-state behavior has been attributed to either Compton cooling of hot electrons by copious soft photons[1], or a

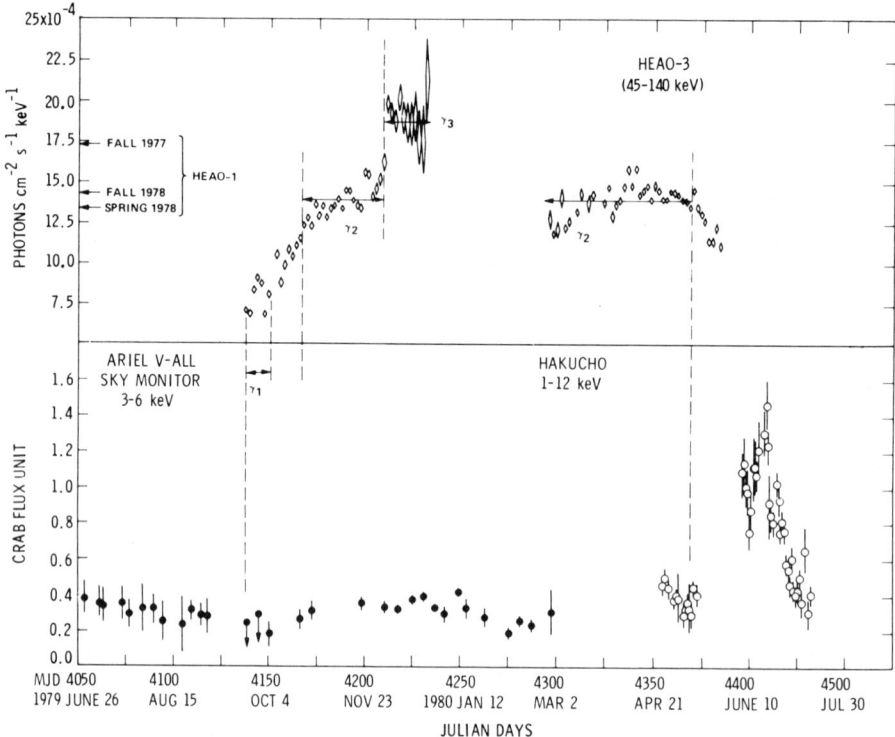

Figure 1: Two contrasting transitions were observed in 1979/1980: (1) the familiar bimodal x-ray transition in May/June 1980, and (2) a new multi-level gamma-ray transition in the fall of 1979. Time intervals γ_1, γ_2, γ_3 correspond to spectra in Figure 2.

change of the "compactness" parameter[15] in the inner region of the disk, both related to changes in the accretion rate.

A new temporal behavior was observed by HEAO 3 in 1979/1980 (Figure 1) when the 100 keV flux showed a slow (~ 70 days), steady increase in 1979, by a factor of 3, while the soft x-ray flux remained essentially constant at the normal LS level. After reaching the peak (γ_3 in Figure 1) the flux stayed fairly constant for at least 20 days until the end of 1979. When next observed in March 1980, the flux was back to the commonly observed γ_2 level where it remained for ~ 75 days until the May/June 1980 x-ray transition. Fluxes and spectra during the γ_2 and γ_3 periods were similar to those observed by HEAO-1[16] in 1978 and 1977, respectively. The absence of simultaneous changes between the soft and hard x-ray flux during an 7-month variation of the latter suggests that the emission mechanism is different and more complex than that envisaged earlier in the simple bimodal accretion model.

Spectral Variability

A comparison of the spectra during the γ_1, γ_2 and γ_3 periods is shown in Figure 2. The γ_1 spectrum (2a) has the most complex shape. It consists of a hard continuum below 400 keV and strong emission from 400 keV to 1.5 MeV. The luminosity of the latter accounts for approximately half of the total observed. The solid line shown in 2a is a fit to a Compton component[17] with an electron temperature kT = 73 keV and optical depth τ = 1.7, plus a broad (1.2 MeV FWHM) gaussian feature centered at 1 MeV with a strong flux of 1.6×10^{-2} photons/cm^2-s. A two-temperature Compton model, with kT_1 = 67 keV and kT_2 = 395 keV, and optical depth τ_1 = 1.8, and τ_2 = 5.2, also fits the data well. As the source evolved from γ_1 to γ_2 to γ_3, the MeV emission disappeared while the overall spectrum softened. Solid lines shown in 2b and 2c are fits to a Compton model[10]. The three spectra intersect at around 400 keV (2a), displaying a new type of anticorrelation between fluxes above and below 400 keV. This behavior is in contrast to the anticorrelation of fluxes above and below 10 keV observed in the x-ray transition. Observations made by the European MISO balloon instrument on 1 October 1979 and 18 May 1980, during the γ_1 and γ_2 periods, respectively, showed consistent fluxes and spectra[18-19]. Both γ_2 and γ_3 spectra are similar to those observed by HEAO 1 in 1978 and 1977, respectively. However, Nolan and Matteson[8] showed a broad feature (400 keV FWHM) centering at 500 keV in the combined spectrum of the three periods measured by HEAO 1 in 1977/1978 with a flux of 5×10^{-3} photons/cm^2-s. Such a feature was not evident in the γ_2 and γ_3 spectra measured by HEAO 3, which set a 3σ upper limit of 4×10^{-3} photons/cm^2-s. Aharonian and Vardanian[20] suggested that the feature observed by HEAO 1 is associated with the relativistic electromagnetic cascade initiated by accelerated particles in the accretion plasma. Nonthermal processes producing MeV gamma rays have also been suggested by Eilek and Kafatos[21].

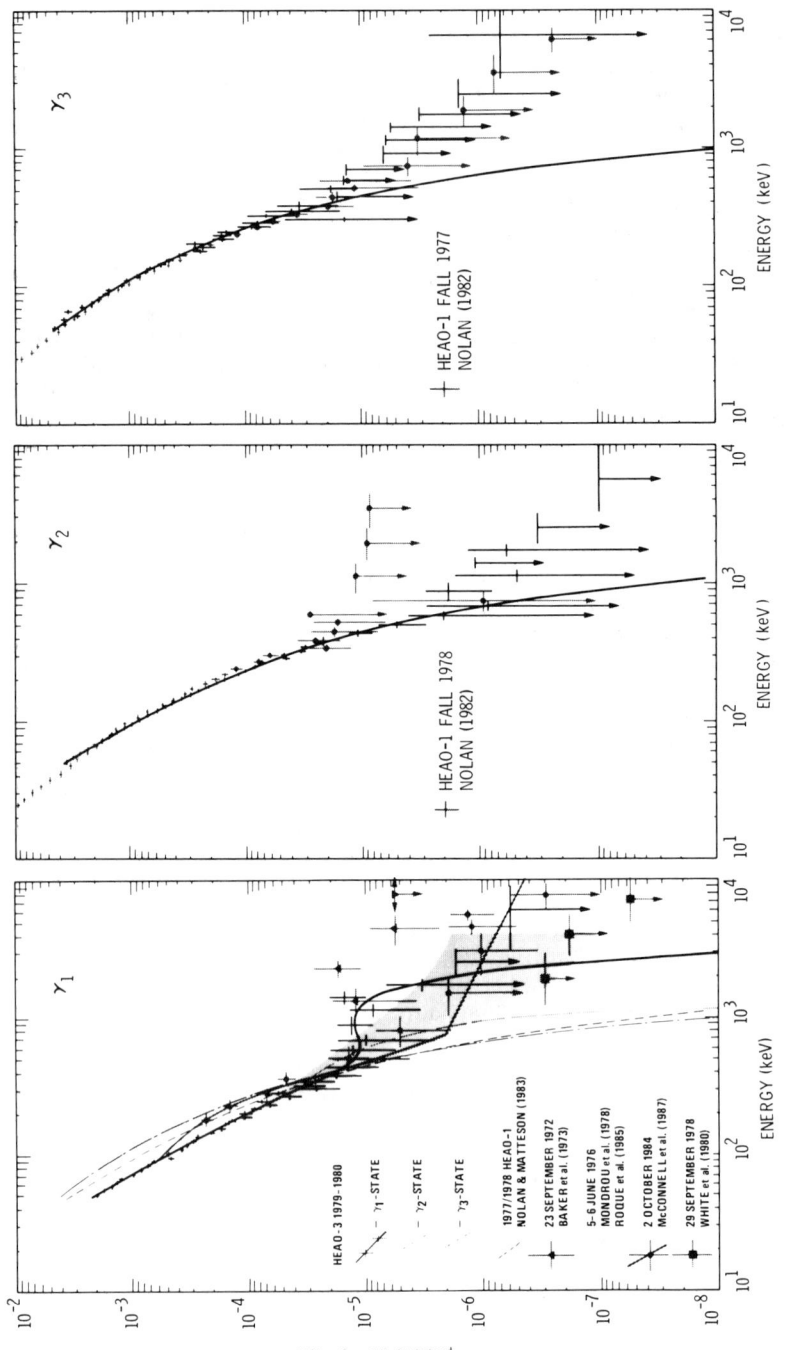

Figure 2: The γ_1 spectrum observed by HEAO 3 consists of a hard low-intensity component <400 keV and a broad feature centered at 1 MeV. As the source evolved from γ_1 to γ_2 to γ_3 in 1979/1980, the MeV feature disappeared and the spectrum softened, pivoting at ~400 keV. Other gamma-ray measurements are shown in 2a for comparison.

A Gamma-Ray Model

Based on the simple model-fits, the strong MeV feature shown in the HEAO 3 γ_1 spectrum may be interpreted as: (1) blue-shifted annihilation radiation produced in 500 keV pair-dominated plasma[3-4] or (2) a Wien feature resulting from Compton scattering of copious soft photons by relativistic electrons with a characteristic temperature of ~ 400 keV[17]. In either case, the low-energy component (E < 0.2 MeV) may be interpreted as unsaturated Comptonization of soft photons from a cooler region with temperature of the order of few tens of keV.

These are only two simple interpretations of a very complex phenomenon. More comprehensive modeling of the HEAO 3 data was recently performed by Liang and Dermer[22-23]. Using iterative relativistic Monte Carlo techniques, they computed a set of self consistent emission spectra, for various temperatures, of a pair or ion-dominated plasma. Fitting these computed spectra to the γ_1 data, critical properties (eg. temperature, radius, mean pair

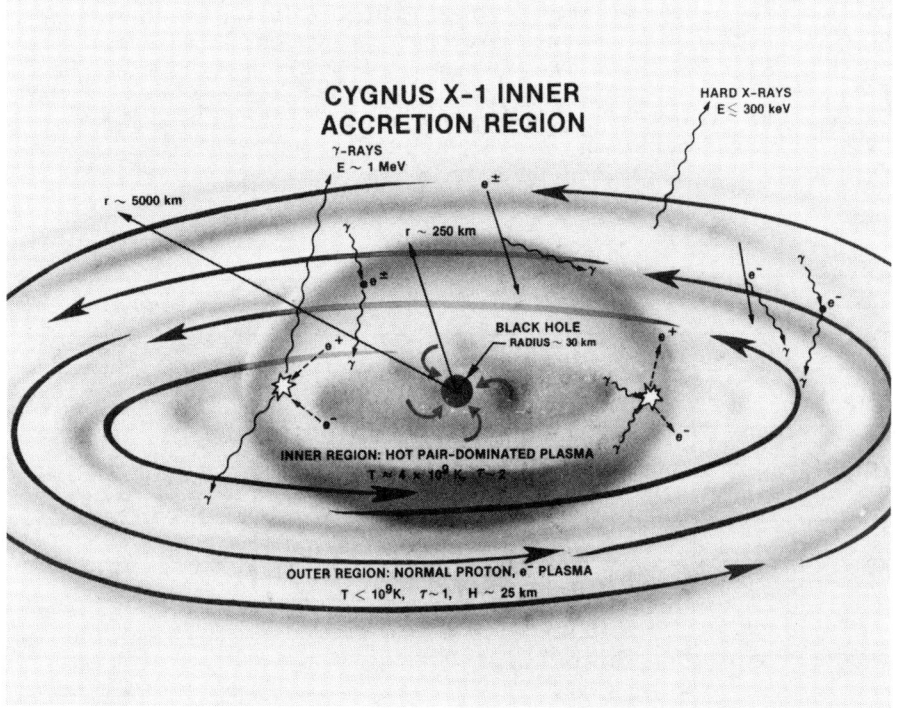

Figure 3: According to the model of Liang and Dermer[22], the gamma rays are produced in a 4×10^9 K, pair-dominated spherical cloud with a radius of 250 km in the inner region of the disk, while the x-rays are produced in the outer disk where the temperature is less than 10^9 K.

density, Thomson depth, $\gamma - \gamma$ pair production depth, etc) of the emission regions can be derived. These results, as illustrated in Figure 3, suggested that the gamma-rays (0.4-1.5 MeV) are primarily produced in a hot (400 keV), pair-dominated ($n_+ \sim 6 \times 10^{16}$ cm^{-3}), optically thick ($\tau \sim 2$) spherical cloud (r~ 250 km) in the innermost region surrounding the black hole, while the hard x-rays (E < 0.4 MeV), for both γ_1 and γ_2 spectra, are produced in a cooler (kT ~90 keV), optically thinner ($\tau \sim 0.8$), ion-dominated plasma in the outer disk (r > 250 km). Liang and Dermer[22] showed that the two regions must be weakly coupled, or the spectrum would be grossly distorted from that observed.

SUMMARY OF CYGNUS X-1 OBSERVATIONS (1965 - 1987)

Gamma-Ray Emission

The MeV feature observed by HEAO 3, as interpreted by the model of Liang and Dermer, provides us with some new insights of the condition of the innermost emission region of the disk. Other measurements of the MeV emission would further enhance our knowledge of the source. Since the discovery of Cygnus X-1 in the 1960's, there have been a total of 11 published spectra with sensitivity above 1 MeV (Table 1 and Figure 2a). Flux above 1 MeV has also been observed by three balloon experiments in 1972[6], 1976[7], and 1984[9,26]. The γ_1 spectrum is similar to the spectrum measured by Mondrou et al[7]. It

Table 1. Observations of Gamma-Ray >0.5 MeV

Experiment	Energy Range MeV	Date	Reference	X-ray State	Hard X-ray Level	Detection of MeV Emission ?
Balloon	.9-11	9/23/72	Baker et al. (1973)[6]	low	-	hard spectral component 0.9-6 MeV
Balloon	1.5-10	2/27/73	Schonfelder and Lichti (1974)[24]	low	-	no
Balloon	.1-3	6/5/76	Mondrou et al. (1978)[7]	low	γ_1	hard tail extending to 3 MeV
HEAO 1	.01-1	10/23-11/20/77 4/15-5/23/78 10/13-11/27/78	Nolan and Matteson (1983)[8]	low low low	γ_3 γ_2 γ_2	a broad 500 keV feature in the combined spectrum of the three periods
Balloon	1-20	9/29/78	White et al. (1980)[25]	low	-	no
HEAO 3	.05-10	9/27-10/10/79	Ling et al. (1987a)[10]	low	γ_1	broad 1 MeV feature
		10/27-12/8/79 3/4-5/16/80		low	γ_2	no
		12/9-12/31/79		low	γ_3	no
Balloon	.05-15	10/1/79	Bassani (1987)[19]	low	γ_1	positive flux in the 0.4-6 MeV region
		5/18/80		low	γ_2	no
Balloon	.2-9	10/2/84	McConnell et al. (1987)[4]	-	γ_2	hard spectral component extending to 6 MeV

is also consistent with the flux measured by Baker et al.[6] in the 0.9-2 MeV region only. The strong flux in the 2 -6 MeV region measured by Baker et al. was not observed by HEAO 3. Recently, McConnell et al.[9] reported the detection of a hard spectral component in the 2-6 MeV region with a flux-level at least a factor of five lower than that observed by Baker et al. They suggested that the temperature of the gamma-ray emission region could be as hot as several MeV. The hard x-ray flux measured by McConnell et al. seemed to be at the γ_2 level. The 2σ upper limits measured by HEAO 3 in the 2-6 MeV region of the γ_2-spectrum are consistent with McConnell's results. Included in Table 1 also are unpublished data[19] of the European MISO observations made in 1979 during the HEAO-3 γ_1 period, and in 1980 during the γ_2 period. They showed also a positive detection of the MeV flux in 1979, but not in 1980. These results are consistent with those observed by HEAO 3. We may conclude, based on all these data, that there is probably a variable MeV component in the Cygnus X-1 spectrum which may be associated with the hard x-ray emission state. Further observations are needed to confirm this hypothesis.

TEMPORAL VARIABILITY

Different hard x-ray flux levels and their possible association with the gamma-ray emission, as observed by HEAO 3, raises new observational issues which need to be addressed. They are: (1) the evidence for any regularity of the pattern, and (2) evidence for actual periodic behavior. Past observations provide some clues to these issues. Between 1965 and 1987, there have been at least 48 hard x-ray measurements of Cygnus X-1. A 22-year light curve of the 45-140 keV flux based on these observations[27] is shown in Figure 4. The dashed lines correspond to the three flux-levels (γ_1, γ_2, γ_3) observed by HEAO 3. Periods for either the x-ray LS to HS transition or the 294 day flux minimum are also identified. The data reveal several interesting facts: (1) while the intensities appear to be highly variable, about 90% falls within the γ_1 and γ_3 levels, (2) intensities observed as low as, or lower than γ_1, but not related to the x-ray transition, have been observed in 1967(#6), 1969(#14), 1970(#15), 1976(#33), 1979(#43), 1980(#45, #46), 1982(#47) and 1985(#48), suggesting that an excursion to γ_1 must be quite common, and (3) Cygnus X-1 seems to have made a transition between two levels similar (within 10%) to γ_2 and γ_3 in the fall of 1977. The evidence for (3) is provided by the combination of five separate measurements made in October and November of 1977: two from balloons (#36, #37), two by OSO 8 (#38, #39) and one by HEAO 1 (#40). Unfortunately, there was no information on whether the source was in the γ_1-level earlier. Nevertheless, the data suggest a repetition of the pattern with a time scale of ~ 2 years. In order to confirm this, long-term monitoring of the source, in the x-ray and gamma-ray regions, spanning several years will be required. The Hard X-Ray Burst Spectrometer (HXRBS) and the Gamma-

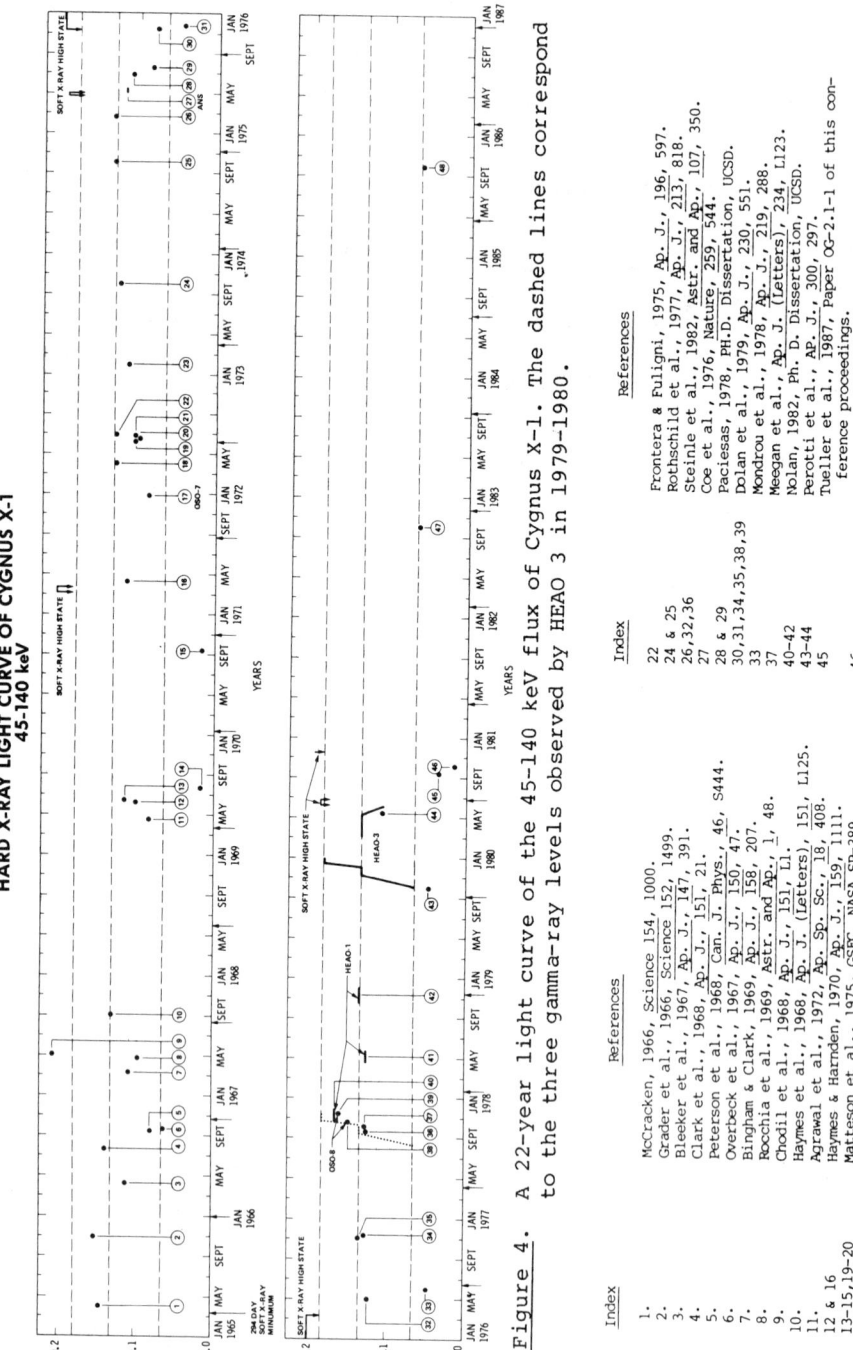

Figure 4. A 22-year light curve of the 45-140 keV flux of Cygnus X-1. The dashed lines correspond to the three gamma-ray levels observed by HEAO 3 in 1979-1980.

Index	References
1.	McCracken, 1966, Science 154, 1000.
2.	Grader et al., 1966, Science 152, 1499.
3.	Bleeker et al., 1967, Ap. J., 147, 391.
4.	Clark et al., 1968, Ap. J., 151, 21.
5.	Peterson et al., 1968, Can. J. Phys., 46, S444.
6.	Overbeck et al., 1967, Ap. J., 150, 47.
7.	Bingham & Clark, 1969, Ap. J., 158, 207.
8.	Rocchia et al., 1969, Astr. and Ap., 1, 48.
9.	Chodil et al., 1968, Ap. J., 151, L1.
10.	Haymes et al., 1968, Ap. J. (Letters) 151, L125.
11.	Agrawal et al., 1972, Ap. Sp. Sc., 18, 408.
12 & 16	Haymes & Harnden, 1970, Ap. J., 159, 1111.
13-15,19-20	Matteson et al., 1975, GSFC, NASA SP-389.
21 & 23	Ulmer, 1975, Ap. J., 196, 827.

Index	References
22	Frontera & Fuligni, 1975, Ap. J., 196, 597.
24 & 25	Rothschild et al., 1977, Ap. J., 213, 818.
26,32,36	Steinle et al., 1982, Astr. and Ap., 107, 350.
27	Coe et al., 1976, Nature, 259, 544.
28 & 29	Paciesas, 1978, PH.D. Dissertation, UCSD.
30,31,34,35,38,39	Dolan et al., 1979, Ap. J., 230, 551.
33	Mondrou et al., 1978, Ap. J., 219, 288.
37	Meegan et al., 1978, Ap. J. (Letters) 234, L123.
40-42	Nolan, 1982, Ph. D. Dissertation, UCSD.
43-44	Perotti et al., Ap. J., 300, 297.
45	Tueller et al., 1987, Paper OG-2.1-1 of this conference proceedings.
46	Frontera et al., 1985, Adv. Space Res., 5, 125.
47	Watanabe, 1985, Astro. and Space Sc., 111, 157.
48	Ma et al., 1986, preprint.

Ray Spectrometer (GRS) onboard the Solar Maximum Mission (SMM) have conducted a special series of approximately 30 pointed observations of Cygnus X-1 between mid December 1986 and first week of April 1987. The All Sky Monitor (ASM) and the Gamma-Ray Burst Detector (GBD) experiments onboard the Japanese Ginga satellite, launched in early 1987, can also provide some temporal and spectral information of the source in the 1-400 keV range. These instruments may shed more light on the issues raised above.

Gamma-Ray Line

Only one narrow line feature was observed and reported in the literature. Using a balloon-borne germanium spectrometer, Watanabe[28], in 1982, detected a line at 145 keV, with a width of 14 keV FWHM and flux of 1.3×10^{-2} photons/cm^2-s. Such a line was not seen in 1980[29] who placed a 3σ upper limit of 6×10^{-3} photons/cm^2-s. Figure 5 shows the high resolution γ_2-spectrum of Cygnus X-1 measured by HEAO 3[30]. The channels are 2 keV wide from 46 to 300 keV, 4 keV from 46 to 300 keV, 4 keV from 300 to 600 keV (top panel), and 8 keV from 600 keV to 1 MeV (bottom panel). The solid line is the best-fit Compton model. The spectrum shows no statistically significant features, particularly in the 145 keV region. The 3σ upper limit for a 14 keV wide feature at 145 keV is 3×10^{-4} photons/cm^2-s. The 3σ upper limit for a narrow 511 keV feature in the γ_2 spectrum is also 3×10^{-4} photons/cm^2-s.

CONCLUSION

The MeV emission observed by HEAO 3 has been interpreted as clear evidence for the existence of a relativistic, pair-dominated plasma in the disk surrounding the black hole as predicted by theory. Because neutron-star binaries typically emit x-rays with energy less than ~100 keV, one may consider using the gamma-ray signals as a new observational tool for distinguishing between accreting black-hole and neutron-star systems. The fact that gamma-ray emissions have also been observed from both the galactic center and NGC-4151, suggests that positron-electron pairs may also be produced in large quantity in super-massive as well as stellar-mass black-holes such as Cygnus X-1. If this is the case, Cygnus X-1 may be used as a yardstick for studying objects such as QSO's and active galaxies, as well as our own galactic center.

ACKNOWLEDGEMENTS

I wish to thank A. S. Jacobson for his leadership in the implementation of the highly successful HEAO 3 experiment, my colleagues A. Dunklee, W. A. Mahoney, R. Radocinski, Wm. A. Wheaton and A. S. Jacobson for participating in all aspects of the HEAO 3 Cygnus X-1 analysis. The research described in this paper was carried out by the Jet Propulsion Laboratory, California Institute of Technology, under contract with the National Aeronautics and Space Administration.

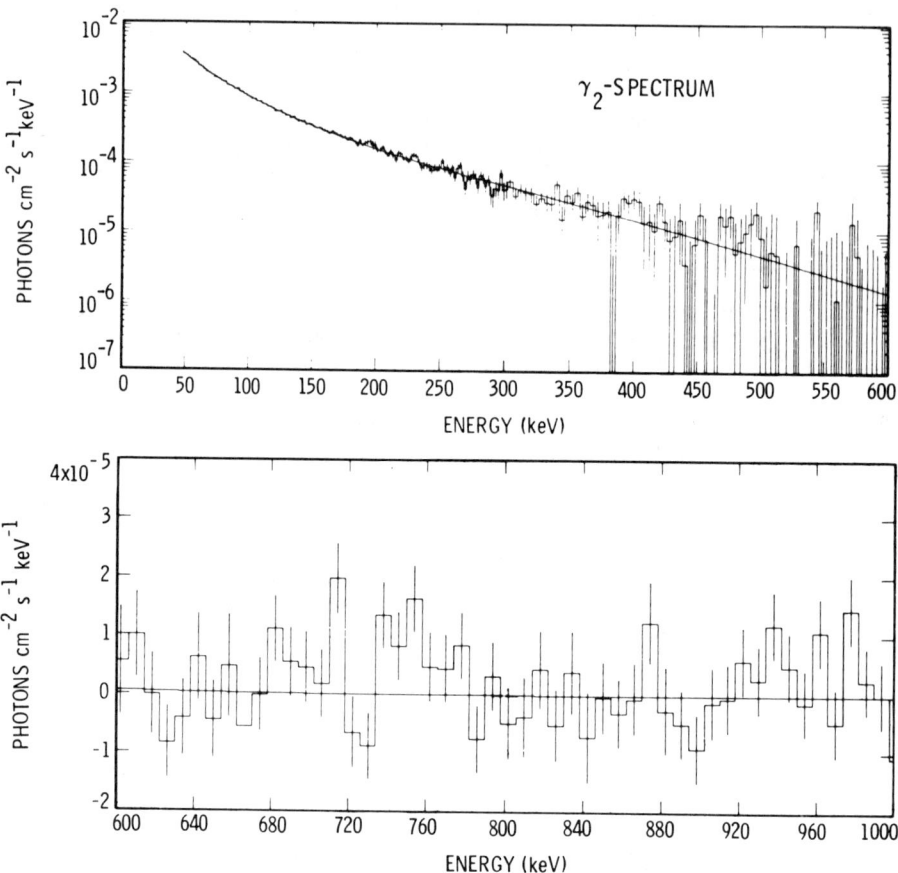

Figure 5: No statistical significant narrow line features are evident in the γ_2 spectrum.

REFERENCES

1. Shapiro et al., 1976, Ap. J., 204, 187.
2. Liang, 1979, Ap. J., 234, 1105.
3. Zdziarski, 1980, Acta Astronomica, 30, 371.
4. Ramaty and Meszaros, 1981, Ap. J., 250, 384.
5. Pozdnyakov et al., 1983, Soviet Scientific Reviews/Section E - Astro-physics and Space Physics Reviews, 2, 189.
6. Baker et al., 1973, Nature, 245, 18.
7. Mondrou et al., 1978, Ap. J., 219. 288.
8. Nolan & Matteson, 1983, Ap. J., 265, 389.

References (con't)

9. McConnell et al, 1987, 20th ICRC Proceedings, Moscow, OG-2.1-6.
10. Ling et al., 1987a, Ap. J. (Letters), 321, L117.
11. Pridhorsky et al., 1983, Ap. J., 270, 233.
12. Tananbaum et al., 1972, Ap. J. (Letters), 177. L5.
13. Holt et al., 1976 Ap. J. (Letters), 203, L63.
14. Ogawara et al., 1982, Nature, 295, 6756.
15. Kazanas, 1986, Astro. and Ap., 166, L19.
16. Nolan, 1982, UCSD Ph.D. thesis.
17. Sunyaev and Titarchuk, 1980, Astr. and Ap., 86, 121.
18. Perotti et al., 1986, Ap. J., 300, 297.
19. Bassani, 1987, private communication at this workshop.
20. Aharonian and Vardanian, 1985, Astro. and Ap., 115, 31.
21. Eilek and Kafatos, 1983, Ap. J., 271, 804.
22. Liang and Dermer, 1988, to be published in Ap. J. (Letters).
23. Dermer and Liang, 1988, this Proceedings.
24. Schönfelder and Lichti, 1974, Ap. J. (Letters), 192, L1.
25. White et al., 1980, Nature, 284, 608.
26. Owens et al., 1988, this Proceedings.
27. Ling et al., 1987b, 20th ICRC Proceedings, Moscow, OG-2.1-2.
28. Watanabe, 1985, Astro. & Space Sc., 111, 157.
29. Tueller et al., 1987, 20th ICRC Proceedings, Moscow, OG-2.1-1.
30. Ling et al., 1985, 19th ICRC Proceedings, La Jolla.

GAMMA RAYS FROM CYGNUS X-1: MODELING AND NONTHERMAL PAIR PRODUCTION

Charles D. Dermer and Edison P. Liang
Physics Department, Lawrence Livermore National Laboratory
P.O. Box 808, L-297, Livermore, CA 94550

ABSTRACT

The gamma-ray bump observed between 0.5 and 2 MeV in the spectrum of Cygnus X-1 (Ling et al. 1987) can be interpreted as the thermal emissions from a hot (kT~400 keV) pair-dominated cloud. We argue that the X-rays and gamma rays are produced in separate emission regions, and calculate the photon-photon pair production rate from X-ray and gamma-ray interactions in the vicinity of Cyg X-1 by employing a simplified geometry for the two emitting regions.

INTRODUCTION

During a period of about two weeks in the fall of 1979, Cygnus X-1 was observed with HEAO-3 to emit roughly equal fractions of its luminosity in 50-400 keV hard X-rays and 400 keV-2 MeV gamma rays (Ling et al. 1987). The X-ray spectrum of Cyg X-1 in this state (called the γ_1 state) was similar in shape to the normal (γ_2) state spectrum, and the total >50 keV luminosity in the γ_1 state differed by less than ~40% from the total >50 keV luminosity of Cyg X-1 when it is in the γ_2 state.

The gamma-ray spectrum of Cyg X-1 in the γ_1 state is very hard, with photon spectral index ~0 at 1 MeV. This gamma-ray "bump" is harder than an optically thin thermal bremsstrahlung spectrum and is suggestive of either a Wien peak or a pair annihilation feature. In other work (Liang and Dermer 1988), we showed that this bump can be self-consistently modeled in terms of the emission from a mildly relativistic pair-dominated plasma. We also argued that the X-rays and gamma rays are produced in different emission regions. Here we summarize these arguments, and calculate the rate of pair production expected from the interaction of X-ray and gamma-ray photons. We find that this mechanism produces a maximum nonthermal pair luminosity of $\sim 3 \times 10^{41}$ e$^+$/sec, which could lead to a narrow observable annihilation line if the produced positrons leave the interaction region, and rapidly slow down and annihilate in a cold surrounding medium.

© 1988 American Institute of Physics

SPECTRAL MODELING AND EMISSION REGION PARAMETERS

The allowed states of steady relativistic thermal plasmas with no pair escape are governed by a balance between the rates of pair production and annihilation. For a given proton optical depth, two solutions are possible: a low-z ($=n_+/n_p$, the ratio of positron and proton densities), low-compactness state and a high-z, high-compactness state (e.g., Svensson 1984; Zdziarski 1984). Because most of the energy in the gamma-ray bump of Cyg X-1 is emitted at photon energies above several hundred keV, we require plasma temperatures $kT \sim m_e c^2$ to fit the γ_1 spectrum. At these temperatures the $z \ll 1$ solution produces too soft a spectrum, so we focus on the pair-dominated ($z \gg 1$) solutions.

Using the known luminosity L of the gamma-ray bump and estimating kT from spectral fitting, the dimensionless compactness $l \equiv L\sigma_T/Rm_e c^3$, Thomson depth τ_T, n_+, and radius R of the hot pair cloud are uniquely determined from results of studies (Svensson 1984; Zdziarski 1984) of steady thermal plasmas in pair balance. When $kT = 400$ keV, for example, we find that $l \cong 12$, $\tau_T \cong 2$, $n_+ \cong 6\times 10^{16}$ cm^{-3}, and $R \cong 2.5\times 10^7$ cm. The spectra calculated through Monte Carlo techniques (Liang and Dermer 1988) at $350 < kT < 450$ keV provide a good fit to the γ_1 data at photon energies > 400 keV. When the Monte Carlo results are subtracted from the γ_1 data, the residual spectrum closely resembles the shape of the γ_2 spectrum, suggesting that the gamma rays are produced in the inner region of the accretion flow at the expense of the normal power-law hard (< 400 keV) X-rays (cf. Fig. 1 of Liang and Dermer 1988).

We also considered whether the emission regions can share photons or are spatially distinct. We found that if there is significant Compton reprocessing of the hard X-rays by the pair cloud, then we no longer obtain a good fit to the hard X-ray spectrum. Likewise, if the gamma rays are Comptonized by the plasma responsible for producing the hard X-rays, the gamma-ray spectrum is grossly distorted. We therefore suggested that the hard-X rays are produced by an accretion disk ($kT\sim 90$ keV, $\tau_T \approx 1$) which is physically distinct from the quasi-spherical pair cloud responsible for the production of the gamma rays. But we did not consider nonthermal pair production associated with X-ray and gamma-ray interactions outside the region occupied by the disk and thermal pair cloud. Because of the large X-ray luminosity of the X-ray disk, significant distortions of the gamma-ray spectrum from the pair cloud could occur. Moreover, a nonnegligible amount of cool pairs might be produced through photon-photon pair production and contribute to an outflowing pair wind and possibly to an observable line from annihilation of the positrons in the circumstellar medium. In the next section, we calculate the rate of pair production via this mechanism, and the differential angle and energy spectra of the positrons produced in photon-photon interactions outside the pair cloud.

PAIR PRODUCTION FROM X-RAY AND GAMMA-RAY INTERACTIONS

We approximate the X-ray and gamma-ray emitting regions by a ring of radius r and a point source located at the center of the ring, respectively, as shown in Fig. 1. The ring, representing the X-ray emitting accretion disk, is assumed to radiate isotropically at each point and to have a spectrum, when integrated over the circumference of the ring, equal to that of the residual γ_1 spectrum after subtraction of the spectrum of the gamma-ray bump. For the spectrum of the central gamma-ray source, we use the kT = 350 and kT = 450 keV spectra discussed in the previous section. Note that the results of this calculation give the maximum possible pair production rate from X- and gamma-ray interactions, as it concentrates all of the hard X-ray flux in a region closest to the gamma-ray source, and includes pair production from a region that is actually within the finite volume of the gamma-ray source. The calculation neglects, however, pair production from interactions of gamma-rays with each other outside the pair cloud. This contribution may be particularly significant if the pair cloud is quasi-spherical or toroidal in shape.

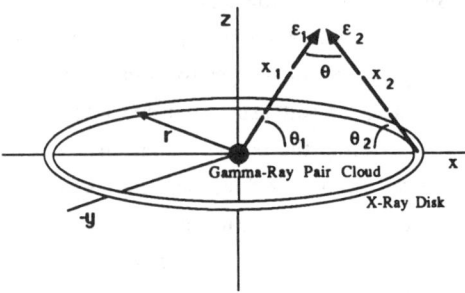

Fig. 1. Geometry of X-ray and gamma-ray photon-photon interactions used in calculations.

Let the spectral emissivity of the gamma-ray point source be denoted by $Q_1(\varepsilon_1)$ [photons/(ε_1-s^{-1})], where ε_1 is the photon energy in units of $m_e c^2$. The density of photons at a distance x_1 from the central point source is given by $n_1(x_1,\varepsilon_1) = Q_1(\varepsilon_1)/4\pi x_1^2 c$. The spectral emissivity per unit radian from the ring source is given by $dQ_2(\varepsilon_2)/d\omega$, and the photon density at a distance x_2 from an element $\delta\omega$ of the ring is given by $n_2(x_2,\varepsilon_2) = \delta\omega(4\pi x_2^2 c)^{-1}(dQ_2(\varepsilon_2)/d\omega)$. The differential pair production rate from photon-photon interactions in the volume element δV is therefore given by

$$\delta \dot{N}_+ = \delta\varepsilon_1 \delta\varepsilon_2 \delta V \, n_1(x_1,\varepsilon_1) \, n_2(x_2,\varepsilon_2) \, c \, \sigma_{\gamma\gamma}(s)(1-\cos\theta), \quad (1)$$

where the invariant quantity $s = 1/2 \, \varepsilon_1 \varepsilon_2 (1-\cos\theta)$ ($s > 1$ for pair production to occur; see Gould and Schreder 1967), θ is the angle between the directions of the interacting photons with energies ε_1 and ε_2, and $\sigma_{\gamma\gamma}$ is the total photon-photon pair production cross section. In the following derivation, we use the cross section $d^2\sigma^*_{\gamma\gamma}/d\mu^* d\phi^*$ differential in the direction cosine $\mu^* = \cos^{-1}\theta^*$ and angle ϕ^* of an outgoing positron in the center-of-momentum system of the colliding photons (Jauch and Rohrlich 1976).

Let θ_1 and θ_2 represent the angles between the x axis and the directions of two interacting photons with energies ε_1 and ε_2, respectively (Fig. 1). The differential volume element $dV = d\phi d\cos\theta_1 x_1^2 dx_1$, where ϕ is the angle between the x-z plane and the plane defined by the directions of the colliding photons. Taking into account the threshold condition, the total pair production rate is given by

$$\dot{N}_+ = c \int_0^\infty d\varepsilon_1 \int_{1/\varepsilon_1}^\infty d\varepsilon_2 \int_{-q}^1 d\cos\theta_1 \int_0^{x_1^{max}} dx_1 \, x_1^2 \int_0^{2\pi} d\phi \, n_1(\varepsilon_1, x_1) \int_0^{2\pi} d\omega \frac{dn_2(\varepsilon_2, x_2)}{d\omega}$$

$$\times (1-\cos\theta) \int_{-1}^1 d\mu^* \int_0^{2\pi} d\phi^* \frac{d^2\sigma^*_{\gamma\gamma}(s;\mu^*,\phi^*)}{d\mu^* d\phi^*} , \qquad (2)$$

where $q = (\varepsilon_1\varepsilon_2-2)/\varepsilon_1\varepsilon_2$, $x_1^{max} = r\cos\theta_1 + rq\sin\theta_1/\sqrt{(1-q^2)}$, and $x_2 = \sqrt{(r^2+x_1^2-2rx_1\cos\theta_1)}$.

We integrate equation (2) using the Monte Carlo method (see Ramaty and Mészáros 1981; Dermer and Ramaty 1986). Select a set of 8 numbers $r_1,...,r_8$ chosen randomly between 0 and 1. Because of the exponential cutoff in the gamma-ray spectrum, we sample ε_1 over only a limited energy range by taking $\varepsilon_1 = 1 + 9r_1$ (i.e., maximum gamma-ray photon energy ≈ 5 MeV). For the spectrum of the hot pair cloud, we fit the numerical results of Liang and Dermer (1988) at $kT = 350$ and 450 keV. We approximate the X-ray spectrum of Cyg X-1 in the γ_1 state by the expression $dN/dAdtdh\nu = 126 [h\nu (keV)]^{-2.75}$ below 511 keV. This expression is analytically invertible. Choosing a random number r_2, the energy ε_2 is determined from the relation $\varepsilon_2 = [1-r_2(1-\varepsilon_1^{7/4})]^{-4/7}$. The remaining variables are determined through the relations $\omega = 2\pi r_3$, $\cos\theta_1 = -q+(1+q)r_4$, $x_1 = x_1^{max}r_5$, $\phi = 2\pi r_6$, $\mu^* = 2r_7-1$, and $\phi^* = 2\pi r_8$.

The pair production rate, equation (2), becomes

$$\dot{N}_+ = \frac{(4\pi d^2)^2}{8\pi c} \frac{4f_2}{7N_{tot}} \sum_{i=1}^{N_{tot}} 9\phi_1(\varepsilon_1) (\varepsilon_1^{7/4} - 1) (1+q) \frac{x_1^{max}(1-\cos\theta)}{x_2^2} \cdot 2 \cdot \frac{d\sigma^*_{\gamma\gamma}}{d\mu^*} , \qquad (3)$$

where d is the distance to Cygnus X-1, $\phi_1(\varepsilon_1)$ is the fit to the gamma-ray flux of the hot pair cloud as observed at Earth, N_{tot} is the total number of Monte Carlo trials, $f_2 = 126 \times 511^{-1.75}$, and $\cos\theta = (x_1-r\cos\theta_1)/x_2$. The differential energy and angle spectrum of the produced positrons can be determined from a 4-fold transformation (3 rotations + boost) from the laboratory system reference frame (shown in Fig. 1) to the center-of-momentum system of the colliding photons. In this exercise, we adopted the conventions of Dermer

(1985). The transformation equations for the 4-vector $\pi = (p_x, p_y, p_z, E)$ of the produced positron are

$$\begin{cases} p_x = \bar{p}_x \cos u - \bar{p}_z \sin u \\ p_y = \bar{p}_y \cos\phi - (\bar{p}_x \sin u + \bar{p}_z \cos u) \sin\phi \\ p_z = \bar{p}_y \sin\phi + (\bar{p}_x \sin u + \bar{p}_z \cos u) \cos\phi \\ E = \bar{E}, \end{cases} \quad (4a)$$

where

$$\begin{cases} \bar{p}_x = \gamma_c [(p_x^* \cos\omega - p_z^* \sin\omega) + \beta_c E^*] \\ \bar{p}_y = p_y^* \\ \bar{p}_z = p_x^* \sin\omega + p_z^* \cos\omega \\ \bar{E} = \gamma_c [E^* + \beta_c (p_x^* \cos\omega - p_z^* \sin\omega)], \end{cases} \quad (4b)$$

and $\cos\omega = (\varepsilon_2 - \varepsilon_1)/\eta$, $u = \theta_1 + \theta - \xi$, $\cos\xi = (\varepsilon_2 + \varepsilon_1 \cos\theta)/\eta$, and $\eta = \sqrt{\varepsilon_1^2 + \varepsilon_2^2 + 2\varepsilon_1\varepsilon_1\cos\theta}$. The results were binned in kinetic energy E_k and direction cosine $\mu = p_z/p$.

RESULTS AND DISCUSSION

In Fig. 2 we show Monte Carlo calculations of the differential energy spectra of positrons produced in collisions of X- and gamma-ray photons, binned in intervals of μ. The temperature of the hot pair cloud is 450 keV, $N_{tot} = 10^5$, and $r = R = 3 \times 10^7$ cm. As can be seen, the positron spectra are flat below $E_k \cong 1$ MeV and have an exponential turnover above this energy. The rate of positron production is greatest in the range $0 \le \mu < 0.25$, that is, for those traveling parallel to the plane of the disk. These positrons also have a slightly harder energy spectrum than those produced in other solid angle intervals. This behavior

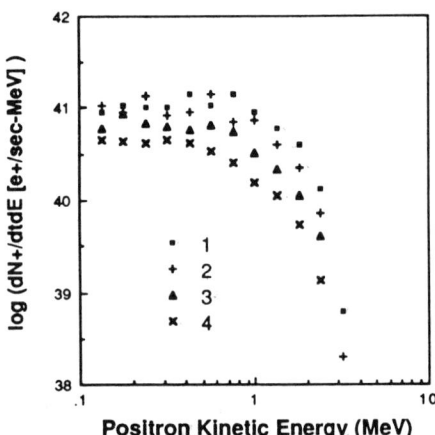

Fig. 2. Monte Carlo calculations of positron differential energy spectra produced in X-ray and gamma-ray photon-photon interactions. Labels 1, 2, 3, and 4 stand for positrons produced in the intervals $0 \le \mu < 0.25$, $0.25 \le \mu < 0.5$, $0.5 \le \mu < 0.75$, and $0.75 \le \mu < 1.0$, respectively, where μ is the cosine of the angle that the outgoing positron makes with respect to the z-axis.

reflects the fact that the secondary electrons and positrons roughly maintain the directions of the colliding photons, and that the most energetic collisions are the head-on photon-photon collisions which take place with photons emitted at small angles with respect to the plane of the disk.

We have also repeated the above calculation with parameters appropriate to a 350 keV pair cloud, but now with $r = R = 2.1 \times 10^7$ cm. We find pair production rates of 1.8×10^{41} and 2.6×10^{41} e^+/sec at kT = 350 and 450 keV, respectively. If these pairs rapidly slow down and annihilate in a cool medium, the maximum annihilation line luminosity that could be expected is $\sim 2 \times 10^{35}$ erg/s. This represents $\sim 1\%$ of the 400 keV - 1.5 MeV luminosity in the γ_1 state, and would yield a specific 511 keV photon flux at Earth $< 3 \times 10^{-4}$ photons/cm^2-s-keV, using a line width of ~ 1 keV. This flux is above the limit of the threshold of the HEAO-3 Gamma-Ray Spectrometer, but reflects the overestimates of the pair production rate from X-ray and gamma-ray interactions discussed earlier. The narrow annihilation line luminosity in the direction of Cygnus X-1 during the the γ_1 state period has not yet been reported. During the γ_2 state, the reported 3σ upper limit is 3×10^{-4} photons/cm^2-s (J. Ling; these proceedings). This flux cannot, however, be compared to the results of this calculation. We leave to further work to determine whether this mechanism could provide the $\sim 4 \times 10^{43}$ e^+/sec (Lingenfelter and Ramaty 1983) needed to explain the HEAO-3 observations of the narrow galactic center annihilation line in the fall of 1979, if one models the galactic center gamma radiation in a manner similar to our model for Cyg X-1. Should the mechanisms suggested here provide insufficient positron production compared to the observed annihilation line luminosity of the Galactic Center, one would be required to invoke additional escape from the central hot pair cloud.

REFERENCES

Dermer, C.D. 1985, *Ap. J.*, **295**, 28.
Dermer, C.D. and Ramaty, R. 1986, in *Accretion Processes in Astrophysics*, eds. J. Audouze and J. Tran Thanh Van (Editions Frontieres), p. 85.
Gould, R.J. and Schreder, G.P. 1967, *Phys. Rev.*, **155**, 1404.
Jauch, J.M. and Rohrlich, R. 1976, *The Theory of Photons and Electrons*, (New York: Springer-Verlag).
Liang, E.P. and Dermer, C.D. 1988, *Ap. J. (Letters)*, **325**, in press.
Ling, J.C., Mahoney, W.A., Wheaton, Wm.A., and Jacobson, A.S. 1987, *Ap. J. (Letters)*, **321**, L117.
Lingenfelter, R.E. and Ramaty, R. 1983, in *Positron-Electron Pairs in Astrophysics*, eds. M.L. Burns, A.K. Harding and R. Ramaty (New York: AIP), p. 267.
Ramaty, R. and Mészáros, P. 1981, *Ap. J.*, **250**, 384.
Svensson, R. 1984, *MNRAS*, **209**, 175.
Zdziarski, A. A. 1984, *Ap. J.*, **289**, 514.

PROPERTIES OF HYDROGEN/HELIUM ACCRETION PLASMAS

Nidhal Guessoum
Department of Physics, University of California, San Diego
La Jolla, California 92093

Charles D. Dermer
Physics Department, Lawrence Livermore National Laboratory
P.O. Box 808, L-297, Livermore, California 94550

ABSTRACT

We study the properties of impulsively-heated plasmas initially composed of hydrogen and helium. We follow the time-dependent behavior of the ion and electron temperatures, the pair density, and the densities of hydrogen, helium, and nuclei formed in fusion and breakup reactions. We also consider neutron production and escape, and calculate the 0.431 and 0.478 MeV line luminosities from $\alpha-\alpha$ fusion reactions, and the 2.22 MeV line luminosity from neutron capture on protons.

INTRODUCTION

High-temperature ion plasmas may be produced in the accretion flow near a compact object, or through shock heating. If the ion temperature T_i exceeds ~5 MeV, as is commonly assumed (e.g., Eilek 1980; Hogan and Applegate 1987), nucleosynthetic fusion and breakup reactions will occur. The breakup reactions limit the nuclear deexcitation line luminosity from elements such as C, N and O (Aharonian and Sunyaev 1984; Gould 1986; Guessoum, these proceedings), and provide a source of neutrons which, being electrically neutral, can escape from the plasma. If the only process available for exchanging energy between the ions and electrons is elastic Coulomb scattering, the electron temperature $T_e \lesssim$ few MeV $< T_i$ (Shapiro, Lightman and Eardley 1976; Higdon and Lingenfelter 1977).

In this paper, we investigate the processes responsible for the formation and destruction of the light isotopes d, ^{3}H and ^{3}He in a plasma originally composed solely of hydrogen and helium. We treat neutron production, capture, and escape, and evaluate the line emissivities from α-α fusion reactions and from neutron capture on protons. For the parameters considered here, the abundances of the light isotopes are maximized when the initial ion temperature is ~15 MeV. The luminosity of the 0.431 and 0.478 MeV lines is found to be $\lesssim$ 10^{-5} of the continuum luminosity radiated by the electrons. Using a source size characteristic of a 10 M_o galactic black hole, we find that most neutrons formed in the breakup of

© 1988 American Institute of Physics

α-particles leave the hot ion plasma. Consequently, the 2.2 MeV line luminosity in the ion plasma is small, but a narrow 2.2 MeV line could be produced if a significant fraction of the escaping neutrons are captured by protons in the atmosphere of a companion star.

ANALYSIS

We model the system by a sphere of hot plasma with radius $R = 10^7$ cm. Protons and α-particles with density n_p and n_α, respectively, are impulsively heated to temperature $T_{i,o}$ at initial time. Assuming equality of the proton and ion temperatures, we calculate the time-dependent behavior of T_e, T_i, the isotopic compositions, and the emitted spectrum. We employ roughly solar composition for the initial α-particle density $n_{\alpha,o}$, letting $n_{\alpha,o} = 0.1 n_{p,o}$. The magnetic field is set equal to zero in the present work.

For the ion-electron Coulomb energy exchange rate, we use equation (11) of Dermer (1986), valid for nonrelativistic protons or ions and electrons at arbitrary temperatures. The Coulomb logarithm Λ is set equal to 20. The electrons lose energy by emitting bremsstrahlung and by Comptonizing soft photons. We follow the treatment of Dermer (1988) for calculating the pair density and spectra resulting from the thermal Comptonization of bremsstrahlung and annihilation radiation (see also Guilbert and Stepney 1985). However, we now employ accurate fits to the exact thermal bremsstrahlung and annihilation source spectra. The low density of multi-MeV photons renders photonuclear breakup reactions unimportant compared to nuclear breakup reactions in this system, and are therefore neglected (see Guessoum, 1988).

In order to achieve an accuracy in our calculations of ~10%, we need to take into account reactions involving protons, neutrons and α-particles with each other and with the light isotopes d, ^{3}H and ^{3}He. Reactions of the light isotopes with themselves can be neglected as they affect results at the 1% level or less. We also treat the production and breakup of ^{6}Li, ^{7}Li and ^{7}Be, which are formed in α–α fusion reactions, and are accompanied by the emission of gamma-ray lines at 0.431 and 0.478 MeV. We have derived the thermal-averaged reaction rates for these reactions, and will present the detailed results in subsequent work (Guessoum and Dermer 1988). A typical rate has the form

$$r_{12} = n_1 n_2 A_{12} \theta^s \exp(-a/\theta) g(\theta), \qquad (1)$$

where A_{12} and s are constants, n_j is the density of particle j, $\theta = T_i/m_e c^2$, and $g(\theta)$ is a polynomial function. The constant a is proportional to the threshold energy of the reaction. This form follows from the basic expression for the reaction rate at nonrelativistic temperatures (see Guessoum, these proceedings). For example, the important breakup reaction $p + \alpha \to {}^3\text{He} + d$ has a rate given by the expression

$$r = n_p n_\alpha \, 2.0 \times 10^{-15} \, \theta^{-1/2} \exp(-35.9/\theta) \, (1 + 0.025\theta) \tag{2}$$

The form of equation (1) also holds for breakup reactions involving incident neutrons. But because neutrons are electrically neutral, they will not be confined by electromagnetic fields and may leave the system. The fraction of neutrons that freely escape from a spherical volume in time δt is $\sim \delta t \, v(T_i)/R$, where the average neutron velocity $v(T_i) \sim (3T_i/m_n)^{1/2}$. The escape of neutrons is impeded by neutron-proton and, to a lesser extent, neutron-ion collisions. We generalize the fractional neutron escape rate by letting $R \to R(1+\tau_{pn}/3)$, where $\tau_{pn} = n_p \sigma_{pn}(E_n) R$ and $E_n \approx T_i$. For the neutron-proton cross section σ_{pn}, we use the fit of Gammel (1963). Neutron decay is accounted for by a loss term employing the neutron decay lifetime. Time-dilation effects to the neutron lifetime are small and can be neglected. We also neglect gravitational effects to neutron escape (cf. Aharonian and Sunyaev 1984).

RESULTS

In Fig. 1 we show the time-variation of the electron and ion temperatures following the impulsive heating of protons and α-particles to temperature $T_{i,o} = 10$ MeV. In this simulation, the density of the injected protons corresponds to unit optical depth of the system, that is, $n_p \sigma_T R = \tau_p = 1$, implying that the initial proton density $n_p = 1.5 \times 10^{17}$ cm^{-3}. The pair density $\ll n_p$ throughout the simulation. As can be seen, T_e hovers at ~ 0.5 MeV during the 100 or so Thomson times that it takes the ions to transfer their energy to the electrons. The average luminosity radiated by the electrons during this period is $\sim 3 \times 10^{35}$ erg/s. (In general, the average bolometric luminosity $\propto T_{i,o} \tau_p^2 R$ for nonmagnetic systems with $T_{i,o} \lesssim 50$ MeV; pair effects increase the average luminosity at higher initial ion

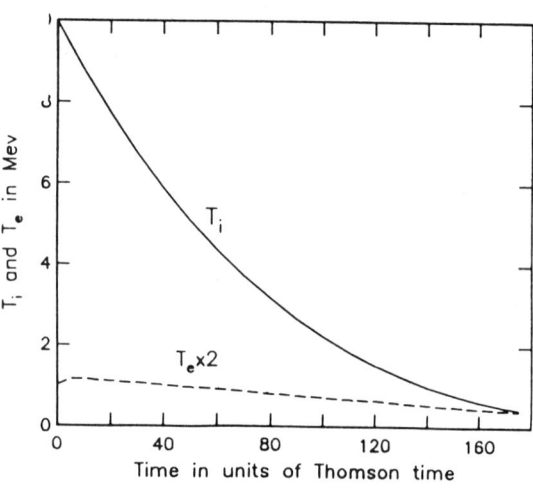

Fig. 1. Time-variation of the electron and ion temperatures as a function of time in units of Thomson time $t_T = (n_{p,o} \sigma_T c)^{-1} = 3.3 \times 10^{-4}$ sec. The initial ion temperature is 10 MeV.

temperatures [Guilbert and Stepney 1985].) In Fig. 2 we show the time-variation of the abundances of the various isotopes. Due to nuclear breakup reactions, there is a depletion of the α-particle abundance that appears in the form of n, d, ^{3}H and ^{3}He. After reaching a peak value, the abundances of the light isotopes declines somewhat due to the increasing importance of light element breakup compared to helium breakup.

The breakup of α-particles is dominated by the reaction p + α → ^{3}He + d. The abundance of ^{3}He at late time is much greater than that of ^{3}H, which is primarily formed in the less important reaction p + α → ^{3}H + p + p. The ^{3}He abundance exceeds that of d, partly because d has a lower nuclear binding energy and is therefore more rapidly broken down at low (~1 MeV) ion temperatures than ^{3}He. The final abundances of the various isotopes as a function of initial ion temperature are shown in Fig. 3 for the case τ_p = 1. At higher values of $T_{i,o}$, a decreasing fraction of α-particles survives to low temperature. Indeed, matter that has been processed through a high (>15 MeV) ion-temperature state may show ^{3}He and d abundances

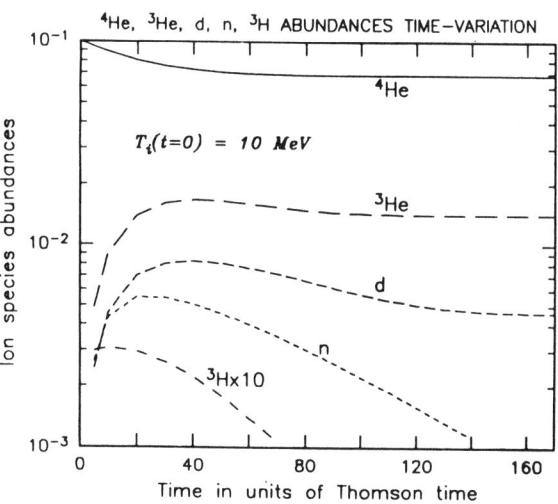

Fig. 2. Time-variation of ratios of the neutron, α-particle, d, ^{3}H and ^{3}He densities to the proton density as a function of time in units of Thomson time. The initial ion temperature is 10 MeV.

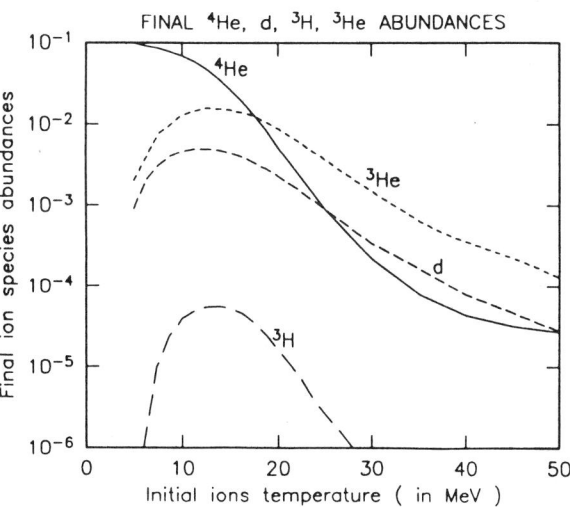

Fig. 3. Final α-particle, d, ^{3}H and ^{3}He abundances of the plasma following the impulsive heating of H and He to various initial ion temperatures.

which are comparable to or exceed the final α-particle abundance.

Time-averaged spectra emitted by this system are typical Comptonized thermal bremsstrahlung spectra, characterized by the form $E^{-q}\exp(-E/\langle T_e \rangle)$, where the photon spectral index $q \lesssim 1$ and $\langle T_e \rangle$ is an effective electron temperature. Deexcitation of ^{7}Li* and ^{7}Be* which are produced in the reactions $\alpha + \alpha \to {^7}$Li* $+ p$ and $\alpha + \alpha \to {^7}$Be* $+ n$ yields gamma-ray lines at 0.478 and 0.431 MeV, respectively. The other important line is at 2.224 MeV from the neutron capture reaction $n + p \to d + \gamma$. A measure of the line strength is determined by the ratio of the line luminosity to the continuum luminosity, the latter being equal to the energy transferred from the ions to the electrons through Coulomb coupling. In Table 1 we show the relative line and continuum luminosities as a function of $T_{i,o}$. The relative line luminosity initially increases with $T_{i,o}$ as the thermal reaction barrier is overcome. At higher $T_{i,o}$, the reaction rate approaches a constant value, and the relative line luminosity decreases as the ion thermal energy available for the formation of the continuum increases. The ratio of line luminosity to continuum luminosity for the α-α fusion lines is $\lesssim 10^{-5}$, and would be hidden by the continuum unless $\langle T_e \rangle \ll 0.4$ MeV or $n_\alpha \approx n_p$. The relative 2.224 MeV line luminosity in the source is likewise small since most neutrons escape rather than form deuterium through neutron capture on protons. This is apparent from Fig. 2, where one sees that although the composition of the isotopes stabilizes at late times, the neutron abundance continues to decline due to escape. In addition, the maximum ratio of the neutron to proton density is only ~0.0055, whereas a value of 0.037 would be expected if the neutrons remain in the source (Table 1).

TABLE 1

RELATIVE LINE LUMINOSITIES FROM AND NEUTRON PRODUCTION IN HOT ION PLASMA

$T_{i,o}$ (MeV)	$\dfrac{L_{0.431+0.478}}{\varepsilon_{p \to e}}$	$\dfrac{L_{2.224}}{\varepsilon_{p \to e}}$	$\dfrac{N_n}{N_p}$	$\dfrac{\phi_{2.224}}{f_{2.224}}$ [a]
5.0	2.0×10^{-6}	2.7×10^{-7}	1.7×10^{-3}	2.8×10^{-6}
10.0	1.0×10^{-5}	1.2×10^{-6}	3.7×10^{-2}	3.1×10^{-5}
20.0	5.3×10^{-6}	8.6×10^{-7}	0.17	7.2×10^{-5}
30.0	1.7×10^{-6}	2.9×10^{-7}	0.197	5.5×10^{-5}
40.0	7.8×10^{-7}	1.4×10^{-7}	0.198	4.2×10^{-5}
50.0	4.8×10^{-7}	9.1×10^{-8}	0.199	3.4×10^{-5}

[a] Estimated 2.224 MeV line flux from neutron capture on protons in the companion of Cygnus X-1.

A line signature at 2.224 MeV could be formed, however, if the neutrons leaving the hot ion plasma are captured on a companion star. In the case of Cyg X-1, Coulomb coupling by hot ions at $T_i \sim 50$ MeV could provide the heating source of the electrons and pairs which produce the continuum luminosity (Liang and Dermer 1988). Scaling to values appropriate to Cyg X-1, we have estimated the 2.224 MeV line flux expected if some fraction of the neutrons are captured by protons in the atmosphere of the companion star HDE226868. The results are listed in Table 1, using 2.5 kpc for the distance to Cyg X-1. The factor $f_{2.224}$ represents the ratio of 2.224 MeV photons escaping from the companion star to the number of neutrons leaving the hot ion plasma. Observation of a narrow 2.224 MeV line at the level of $\sim 10^{-6}$-10^{-5} photons cm^{-2}-s^{-1} from Cyg X-1 would provide strong support for a two-temperature plasma model, but may reqire the next generation of gamma-ray telescopes.

REFERENCES

Aharonian, F.A, and Sunyaev, R.A. 1984, *MNRAS*, **210**, 257.

Dermer, C.D. 1986, *Ap. J.*, **307**, 47.

Dermer, C.D. 1988, in *Georgia State University Conference on Active Galactic Nuclei*, eds. H.R. Miller and P.J. Wiita (Springer-Verlag), in press.

Eilek, J. A. 1980, *Ap. J.*, **236**, 664.

Gammel, J. 1963, *Fast Neutron Physics*, part II, eds. J.B. Marion and J.L. Fowler (Interscience Publishers: New York), p. 2209.

Gould, R.J. 1986, *Nuclear Phys.*, **B266**, 737.

Guessoum, N. 1988, in preparation.

Guessoum, N. and Dermer, C.D. 1988, in preparation.

Guilbert, P.W. and Stepney, S. 1985, *MNRAS*, **212**, 187.

Higdon, J.C. and Lingenfelter, R.E. 1977, *Ap. J.*, **215**, L53.

Hogan, C. J. and Applegate, J. H. 1987, *Nature*, **330**, 236.

Liang, E.P. and Dermer, C.D. 1988, *Ap. J. (Letters)*, **325**, in press.

Shapiro, S.L., Lightman, A.P., and Eardley, D.M. 1976, *Ap.J.*, **204**, 187.

NUCLEAR BREAKUP REACTIONS AND GAMMA-RAY LINES IN HIGH-TEMPERATURE PLASMAS

Nidhal Guessoum
Department of Physics, B-019
University of California, San Diego
La Jolla, CA 92093

ABSTRACT

In plasmas at temperatures above about 1 MeV, nuclei get broken up, either through particle- or photon-induced reactions. These breakup reactions alter the abundances of light elements in such plasmas, and affect the production of gamma-ray lines from deexcitations or radiative captures by the nuclei, reducing their emissivities at high temperatures. Results are presented here for the characteristic temperatures at which various light nuclei break up and for the effect of these processes on the production of certain characteristic gamma-ray lines.

INTRODUCTION

Thermonuclear breakup reactions, in which thermal protons, alphas, or photons destroy a nucleus, have received very little attention as to their importance in astrophysical conditions, campared to the large amount of effort that has been devoted to other types of nuclear reactions such as spallation reactions, radiative capture, and deexcitation processes. However, there has recently developed a great deal of interest in relativistic plasmas, particularly in relation to gamma-ray bursters, central energy sources of active galactic nuclei and quasi-stellar objects, and in accretion disks around compact objects. In such environments, the plasmas may attain a temperature of a fraction of the rest masses of nuclei, enabling the particles to induce thermonuclear breakup of the (light) nuclei constituting the plasma. The nuclei can also be broken up in photodisintegration processes induced by internally-generated photons. This latter problem is investigated elsewhere (Guessoum[1]). The present work deals with proton- and alpha-induced reactions and their effect on plasma composition and gamma-ray line production. The importance of such processes in astrophysics was first pointed out by Gould[2], it was followed by a formal treatment[3] of the breakup of Helium-4. The present work is a continuation of that treatment to all astrophysically important light species; it also extends to some of the important applications in gamma-ray astrophysics.

We will thus first present calculated reaction rates for the proton- and alpha-induced breakup processes, and then from them compute the temperatures needed for the destruction of every species, at typical densities and timescales.

The second part will deal with the astrophysical applications of such processes. We will thus briefly consider their relevance with regard to the conditions prevailing in accretion disks existing around compact objects, and then recalculate certain gamma-ray line emissivities as a function of the ion temperature, showing the effect of breakup reactions on those emissivities.

General applications of this work will be briefly discussed in the last section.

BREAK-UP REACTIONS

We consider thermonuclear reactions in a plasma where all interacting species are assumed to have the same temperature kT of the order of 1 to 10 MeV. The nuclei are thus non-relativistic, and consequently, the reaction rate (number of reactions per unit volume per unit time) for a binary collision between two species i and j is given by

$$r_{ij} = \frac{dN_{ij}}{dV\,dt} = \frac{1}{1+\delta_{ij}} \int\!\!\int \sigma_{ij}(v_{ij})\, v_{ij}\, dn_i\, dn_j \qquad (1)$$

where σ_{ij} is the cross for the process, v_{ij} is the relative velocity of the interacting particles, $dn_i = n_i f(v_i) d^3 v_i$, $f(v_i)$ being the Maxmellian distribution of velocities for species i.

Integrating over the motion of the center-of-mass and expressing σ in terms of the c.m. available energy, eq. (1) becomes

$$r_{ij} = \frac{n_i n_j}{1+\sigma_{ij}} \left(\frac{8kT}{\pi \mu_{ij}}\right)^{\frac{1}{2}} \frac{1}{(kT)^2} \int_{E_{th}}^{\infty} E\sigma(E) e^{-E/kT} dE \qquad (2)$$

where $\mu_{ij} = \frac{m_i m_j}{m_i + m_j}$, E_{th} is the threshold energy for the reaction in the c.m. frame.

It is obvious that the breakup channels that will be favored will be those which necessitate the least threshold energy. It then becomes clear that, whenever possible, a nucleus will be broken up by emission of an α-particle. For example, the main channel for breakup of ^{12}C will be: $p+^{12}$C $\rightarrow$ ^{9}B+α which requires only 7.56 MeV instead of 16.5 MeV for the channel $p+^{12}$C$\rightarrow^{11}$C+d. (Binding energies used here are taken from Gibson[4], and deWet[5]).

We thus proceed to calculate these reaction rates as functions of the ions temperature kT, using experimentally measured cross sections.

Proton-induced breakup reactions are strongly favored to those induced by helium nuclei essentially due to the fact that the alpha particle is tightly bound, and also because hydrogen is ten times more abundant than helium in a natural plasma. However, the breakup of Helium requires comparable energies when carried out by protons or alphas:

$$E_{th}(p + \alpha \rightarrow\ ^3He + d) = 18.354 MeV \quad (3a)$$

$$E_{th}(\alpha + \alpha \rightarrow\ ^7Li + p) = 17.347 MeV \quad (3b)$$

$$E_{th}(\alpha + \alpha \rightarrow\ ^7Be + n) = 18.991 MeV. \quad (3c)$$

Other effects (Coulomb barriers, for example) will also affect the reaction rates differently, but we can see that in this case at least, both p- and α-induced breakup reactions are to be considered.

Figure 1 shows the reaction rates (in sec^{-1}) for the breakup of the (light) species, as functions of the ion temperature.

The temperature at which a species is broken up is obtained by solving equation (2) for kT, setting $r_{px} = \frac{n_x}{\tau}$ where τ is the available time for the reaction. We then see that $(kT)_b$ depends on the product τn_p, which can take a wide range of values for plasmas of interest. A typically quoted (average) value for the proton density[5,7] is $\approx 10^{11}$cm-3. The timescale for the reaction can be much shorter than the free-fall time in the accretion disk (due to Coulomb-exchange energy loss from the ions to the electrons), which itself varies over many orders of magnitude. We will thus calculate $(kT)_{breakup}$ for two values of τn_p (10^{11} yr/cm^3 and 10^{16} yr/cm^3) for definiteness. Results are as follows:

	$\tau n_p = 10^{11}$ yr/cm^3	$\tau n_p = 10^{16}$ yr/cm^3
$p + d \rightarrow p + p + n$	$kT_b = 0.43$ MeV	$kT_b = 0.17$ MeV
$p + ^3He \rightarrow d + p + p$	$kT_b = 0.8$ MeV	$kT_b = 0.31$ MeV
$p + ^4He \rightarrow ^3He + d$	$kT_b = 2.25$ MeV	$kT_b = 0.91$ MeV
$p + ^{12}C \rightarrow ^9B + \alpha$	$kT_b = 1.2$ MeV	$kT_b = 0.44$ MeV
$p + ^{14}N \rightarrow ^{11}C + \alpha$	$kT_b = 0.65$ MeV	$kT_b = 0.27$ MeV
$p + ^{16}O \rightarrow ^{13}N + \alpha$	$kT_b = 1.36$ MeV	$kT_b = 0.46$ MeV
$p + ^{20}Ne \rightarrow ^{17}F + \alpha$	$kT_b = 0.67$ MeV	$kT_b = 0.27$ MeV
$p + ^{24}Mg \rightarrow ^{21}Na + \alpha$	$kT_b = 1.29$ MeV	$kT_b = 0.44$ MeV

And thus we see that breakup reactions must be considered in any plasma with ion temperature higher than (depending on τn_p) about 0.2 MeV. It is also important to note that ^{4}He nuclei are broken last on the temperature scale. Consequently, if the plasma ion temperature happens to fall between that of the breakup of the CNO elements and that of ^{4}He (for example, between 1.36 MeV and 2.25 MeV for $\tau n_p = 10^{11} yr/cm^3$), the plasma will be reduced to pure hydrogen and helium. If kT is larger than $kT_b(^4$He$)$, the plasma will be reduced to pure hydrogen. For more complete results and discussions, see Guessoum[1].

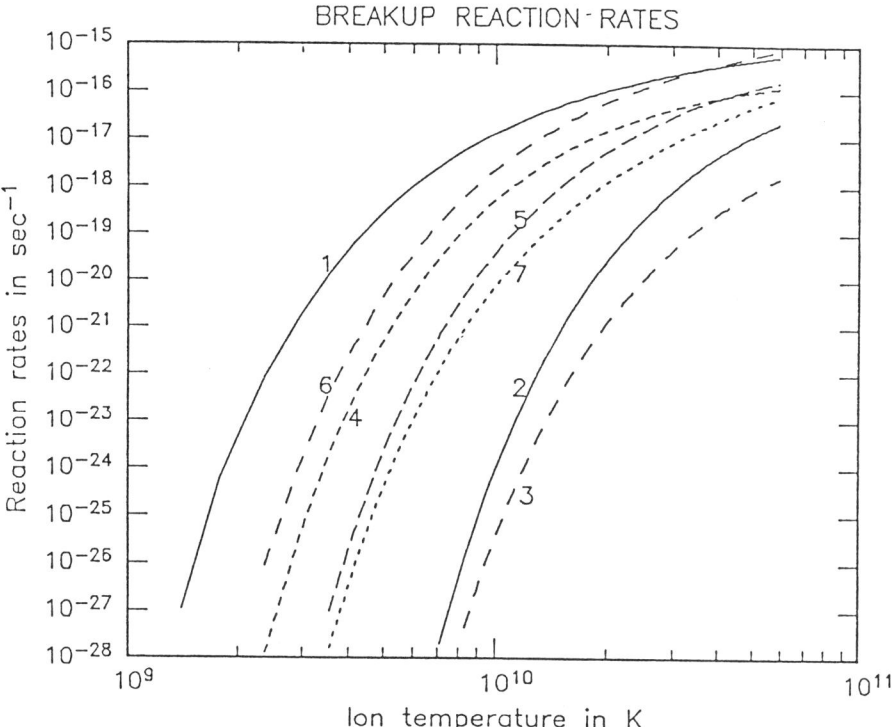

Figure 1. Light Nuclei Breakup Reaction Rates for: d(p, np)p (1), ^{4}He(p,d)^{3}He (2), ^{4}He(^{4}He, b)y (3), ^{3}He(p, pp)d (4), ^{12}C(p, ^{4}He)^{9}B (5), ^{14}N(p, ^{4}He)^{11}C (6), ^{16}O(p, ^{4}He)^{13}N (7).

The most probable place where such processes can be found is in accretion disks around black holes or neutron stars. There, a significant fraction of the gravitational potential energy can be converted into heat and, depending on several physical parameters in the plasma (mass of the star, magnetic field, accretion rate, etc.), the temperature can attain values as high as 100 MeV. The available time for the reactions depends, in such conditions, on viscosity parameters, accretion rates, and other not-very-well-known factors; the density is related to the mass of the object and to the accretion rate. Preliminary investigations[1] show that it is very likely that these processes will occur in such astrophysical conditions.

GAMMA-RAY LINE EMISSIVITIES

One specific consequence of these processes is their effect on the emissivities of gamma-ray lines produced in the deexcitation or proton-capture by nuclei. For example, ^{12}C can be excited by collisions with protons or alphas, then deexcite by emitting the characteristic 4.438 MeV photons, and similarly for ^{16}O (6.129 MeV). Lines can also be produced in proton-capture by nuclei: $p + d \rightarrow\,^3He + \gamma_{5.494+}$, and $p+^{12}C \rightarrow\,^{13}N + \gamma_{2.37+}$. Their emissivities will be similarly affected. Here, we will give explicit results for the 4.438 MeV line only. Previous treatments [8,9,10], calculated such emissivities as functions of the temperature, but did not include the breakup effects. One then writes

$$Q = qn^2 = \sum_{i,j} 4n^2 a_i a_j (2\pi\mu_{ij})^{-1/2}(kT)^{-3/2} \int_0^\infty dE_{ij} E_{ij} \sigma_{ij}(E_{ij}) e^{-\frac{E_{ij}}{kT}} \quad (4)$$

which defines q as the number of photons produced per cm^3 per sec, i and j are the interacting particles in the reactions [11]

$$p +^{12}C \rightarrow p' +^{12}C^*$$
$$\alpha +^{12}C \rightarrow \alpha' +^{12}C^*$$
$$p +^{16}O \rightarrow X +^{12}C^*$$
$$\alpha +^{16}O \rightarrow X +^{12}C^*$$
$$p +^{14}N \rightarrow X +^{12}C^*$$
$$\alpha +^{14}N \rightarrow X +^{12}C^*$$

a_i is the abundance of species i.

However, species i and/or j get broken up and thus a_i and a_j must be replaced by the following relations in eq. (4)

$$a_i(t,T) = a_i^{(0)} exp(-t/\tau_i(T)) \quad (6)$$

where $\tau_i(T)$ is the breakup timescale (or inverse of breakup rate) of species i, as calculated before. We therefore obtain a corrected emissivity $q(T, t, n_p)$

given by

$$q(T, t, n_p) = \sum_{i,j} 4a_i^{(0)} a_j^{(0)} e^{-t/\tau_i} e^{-t/\tau_j} (2\pi\mu_{ij})^{\frac{1}{2}} (kT)^{-3/2} I(kT) \quad (7)$$

where $I(kT)$ is the integral in eq. (4).

Emissivity curves for different values of τn_p are plotted in fig 2. against the uncorrected emissivity as a function of T.

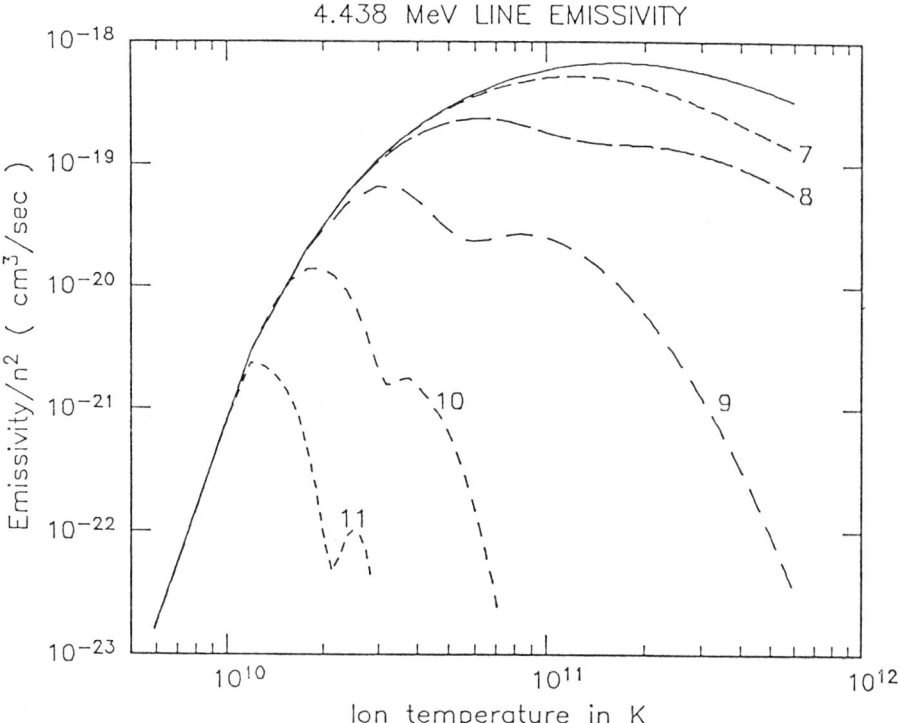

Figure 2. Emissivity of the 4.438 MeV line produced in the deexcitation of $^{12}C^*$. The solid curve is the uncorrected emissivity. The dashed curves labeled $n = 7-11$ refer to the value of τn_p as 10^n yr/cm^3. (For example curve 7 is the emissivity with breakup included for $\tau n_p = 10^7$ yr/cm^3.

CONCLUSIONS

We have presented here a simple investigation of the importance of breakup reactions and of their application and relevance in astrophysical conditions. It is clear that these processes must be taken into account in all plasmas where the temperature reaches values of about 0.5 MeV. And as the second part of this work has shown, gamma-ray line production is the first field to be affected by these reactions.

Another such application was presented at this conference and treated the line spectral features of a hydrogen/helium plasma at high temperatures.[12] In addition, one can think of many other potential applications of this investigation. For example, matter processed in accretion disks and ejected in jets around x-ray binary sources will probably show such effects. Also, studies of gamma-ray lines from Active Galactic Nuclei[13] must take these processes into account in explaining spectral features. Hopefully, the study of such effects will bring new information on the conditions and the physical processes taking place in such places as accretion disks, AGNs, QSOs, and x-ray binary sources.

I wish to thank Dr. Robert J. Gould for his continuous support and advice and Dr. Richard E. Lingenfelter for useful discussions and suggestions.

REFERENCES

1. N. Guessoum, Ph.D. Thesis, University of California, San Diego.
2. R.J. Gould, Astrophys. J. **263**, 879 (1982).
3. R.J. Gould, Nuclear Physics **B266**, 737 (1986).
4. W.M. Gibson, The Physics of Nuclear Reactions (Pergamon Press, Oxford, 1980) p. 278.
5. J.A. de Wet, Foundations of Physics **12**, no. 3, 285 (1982).
6. R.Z. Yahel, Proceedings of the 5th Göttingen-Jerusalem Symposium, ed. K.J. Fricke and J. Shaham, p. 141.
7. S.T. Shapiro and S.A. Teukolsky, Black Holes, White Dwarfs, and Neutron Stars (Wiley & Sons, N.Y., 1983) p. 450.
8. R. Ramaty, R.E. Lingenfelter and B. Kozlovsky, Gamma-Ray Transients and Related Astrophysical Phenomena, ed. R.E. Lingenfelter, et al. (AIP, 1982) p. 211.
9. J.C. Higdon and R.E. Lingenfelter, Astrophys. J. **215**, L53 (1977).
10. R.E. Lingenfelter, J.C. Higdon and R. Ramaty, Gamma-Ray Spectroscopy in Astrophysics, ed. T.L. Cline and R. Ramaty (NASA, 1978) p. 252.
11. R. Ramaty, B. Kozlovsky and R.E. Lingenfelter, Astrophys. J. Supp. **40**, 487 (1979).
12. N. Guessoum and C.D. Dermer, presented at this conference.
13. A.S. Zentsova, Sov. Ast **27**, n. 6, p. 627 (1985).

COMPTON REFLECTION OF GAMMA-RAYS BY COLD ELECTRONS

Timothy R. White
Department of Physics, Harvard University

Alan P. Lightman
Harvard-Smithsonian Center for Astrophysics

and

Andrzej A. Zdziarski
Space Telescope Science Institute

I. INTRODUCTION

We have derived an approximation to the Green's function for the Compton reflection of relativistic, $h\nu \gtrsim m_e c^2$, photons by cold electrons, extending previous treatments of this problem for nonrelativistic photons (Lightman et al. 1981). In Compton reflection, an external flux of photons impinges upon a semi-infinite slab of electrons, gives a portion of its energy to the electrons (or receives energy if the electrons are hotter), and re-emerges with an altered spectrum. Our Green's function may be used to easily calculate the emergent spectrum and albedo for any incident spectrum. Such reprocessing of γ-rays may be astrophysically important when a source of γ-rays coexists with a cool, nonrelativistic plasma, such as for nonthermal coronae above accretion disks, high energy emission near the surface of neutron stars, or a chaotic, two phase medium near the centers of active galaxies.

We assume that the incident radiation is isotropic, an assumption that should be widely applicable. When the scattering medium subtends a large solid angle as seen by the source of radiation, the incident radiation can be considered isotropic unless it is highly beamed. On the other hand, when the scattering medium subtends a small solid angle as seen by the source, the plane-parallel approximation will typically breakdown for general shapes and orientations of the scattering medium. In this case, however, each local region of the medium sees radiation from a different direction and is approximately plane-parallel on a small scale, so that the assumption of a plane-parallel geometry and an isotropic source again becomes approximately valid for describing the spatially integrated reflected spectrum.

Using a Monte Carlo method, we calculated the Green's functions, G, for 6 values of x_0: 0.1, 0.316, 1.0, 3.16, 10., and 31.6. Here $x_0 \equiv h\nu_0/m_e c^2$. (For the nonrelativistic regime, $x_0 \ll 1$, G has already been calculated semi-analytically.) We then found an approximation to G by considering a simple analytic functional form whose free parameters could be determined by fitting to the 6 Monte Carlo runs. We refer the reader to our work submitted elsewhere (White et al. 1988) for more detailed discussion of the results.

II. RESULTS

a) Green's Functions

The results of the Monte Carlo calculations are shown in Figures 1a-c. The solid curves through the Monte Carlo data are our analytic approximation to the Green's function. That approximation is best expressed using a dimensionless wavelength variable y instead of the dimensionless frequency variable x, according to $y \equiv 1/x = m_e c^2/h\nu$. The input wavelength is y_0 and the output wavelength y can be expressed in terms of the wavelength shift $\Delta y \equiv y - y_0$. In terms of these variables, our fitted Green's function is

$$G(\Delta y, y_0) = \begin{cases} B\left[(y_0+2)/(y_0+\Delta y)\right]^\beta , & \Delta y < 2 \\ A(\Delta y)^{-\frac{1}{2}}(\Delta y_c/\Delta y)^\alpha , & 2 < \Delta y < \Delta y_c \\ A(\Delta y)^{-\frac{1}{2}} , & \Delta y_c < \Delta y \end{cases} \quad (1)$$

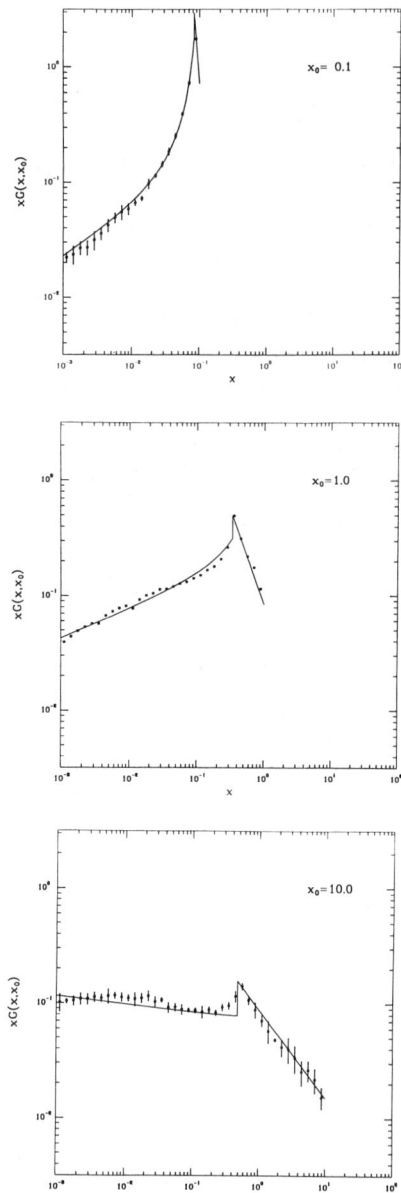

Fig. 1a-c – Compton reflected energy spectra for various values of the frequency $x_0 \equiv h\nu_0/m_e c^2$ of the monochromatic incident spectrum. Here and in all figures the incident spectrum is isotropic and the scattering medium is a semi-infinite, plane-parallel slab at zero-temperature. Results of Monte Carlo calculations, with 20,000 photons for each value of x_0, are shown by the small boxes with 3 σ error bars. The solid lines follow from our analytic approximation to the Green's function, given in the text.

Here $\Delta y_c = 10^3 - y_0$, and the coefficients A, α and β are

$$A = 0.56 + 1.12y_0^{-0.785} - 0.34y_0^{-1.04}$$
$$\alpha = -0.30y_0^{-0.51} + 0.06y_0^{-0.824} \quad (2a).$$
$$\beta = 0.37 - 1.0y_0^{0.85}$$

The coefficient B is given by

$$B(A,\alpha,\beta) = \frac{1 - A\left[2 + \left[(\Delta y_c/2)^{(\frac{1}{2}+\alpha)} - 1\right]/(\frac{1}{2}+\alpha)\right]/(\Delta y_c)^{\frac{1}{2}}}{y_0^{(1-\beta)}(y_0+2)^{\beta}[(1+2/y_0)^{(1-\beta)} - 1]/(1-\beta)}, \quad (2b)$$

for $\alpha \neq -1/2$, and

$$B(A,\alpha,\beta) = \frac{1 - A\left[2 + \ln(\Delta y_c/2)\right]/(\Delta y_c)^{\frac{1}{2}}}{y_0^{(1-\beta)}(y_0+2)^{\beta}[(1+2/y_0)^{(1-\beta)} - 1]/(1-\beta)} \quad (2c)$$

for $\alpha = -1/2$. The expression for B guarantees that G is properly normalized:

$$1 = \int_0^\infty G(\Delta y) d\Delta y. \quad (3)$$

One can use the simple relation $G(x,x_0) = x^{-2}G(\Delta y, y_0)$ to go from wavelength space to frequency space.

The form of eq. (1) was chosen according to the criteria that it be simple, that it be analytically integrable to insure exact normalization, and that it reduce to the nonrelativistic Green's function for $y_0 \gg 1$. The nonrelativistic Green's function calculated by Lightman et al. (1981), for isotropic photon input, may be approximated by

$$G_{nr}(\Delta y) \approx \begin{cases} C, & \Delta y < 2 \\ D(\Delta y)^{-\frac{3}{2}}, & \Delta y > 2 \end{cases}, \quad (4)$$

where $C = 0.10$ and $D = 0.56$. It can be verified that eq. (1) reduces to eq. (4) in the limit $y_0 \gg 1$. The coefficients $A(y_0)$, $\alpha(y_0)$, and $\beta(y_0)$ were chosen to provide the best overall fits to the Monte Carlo data for the 6 values of y_0. The approximate Green's function given by eqs. (1) and (2) never deviates from the Monte Carlo data by more than 17%, and fits much better than this at most frequencies (see Figure 1). Figure 2 gives an example of reprocessing of a power law input spectrum.

b) The Albedo

The albedo, a, is defined as the energy in the reflected spectrum divided by the energy in the incident spectrum. For a monochromatic incident spectrum at frequency x_0 the monochromatic albedo $a(x_0)$ is given by

$$a(x_0) = \int_0^\infty \left(\frac{x}{x_0}\right) G(x,x_0) dx. \quad (5)$$

Figure 3 gives the monochromatic albedo based on our approximation to the Green's function for isotropic incidence of photons, eq. (1). The small boxes indicate the more precise values of $a(x_0)$ obtained from several Monte Carlo runs at various values of x_0.

For the nonrelativistic Green's function, eq. (4), eq. (5) can be integrated to yield an analytic expression for the nonrelativistic ($x_0 \ll 1$) albedo:

$$a_{nr}(x_0) = \frac{C}{x_0}\ln(1+2x_0) + 2Dx_0^{\frac{1}{2}}\left\{\tan^{-1}\left[(2x_0)^{\frac{1}{2}}\right] + \frac{1}{(2x_0)^{\frac{1}{2}}} - \frac{\pi}{2}\right\}. \quad (6)$$

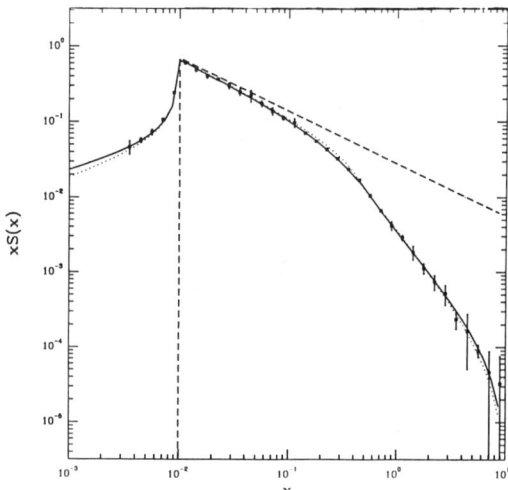

Fig. 2 – Compton reflected spectrum for an incident power-law spectrum with a low-energy cutoff, shown as the dashed line. The boxes and solid line follow from Monte Carlo results and our analytic approximation to the Green's function, respectively. The dotted line follows from an analytic approximation to the nonrelativistic Green's funtion.

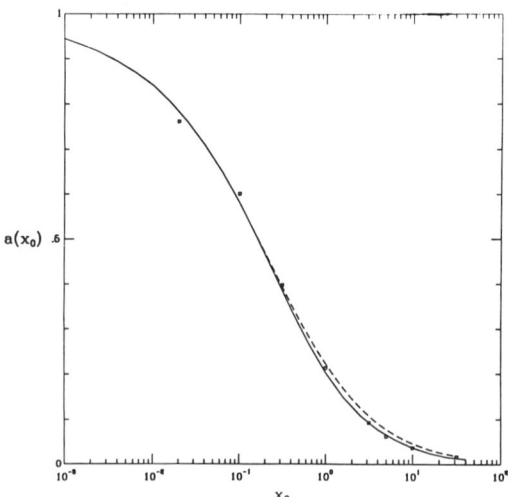

Fig. 3 – The monochromatic albedo, $a(x_0)$, for cold Compton reflection as a function of the frequency of the incident, monochromatic incident spectrum. The albedo is the ratio of total energies in the reflected and incident spectra, respectively. Results of Monte Carlo calculations are indicated by boxes. The solid line follows from our analytic approximation. The dashed line follows from an analytic approximation to the nonrelativistic Green's function.

The dashed line in Figure 3 follows from this expression, which deviates from the exact result by about 15% in the relativistic region. Considering the approximation, the nonrelativistic albedo does very well at estimating the exact albedo for all x_0 and has a relatively simple closed form expression.

REFERENCES

Lightman, A.P., Lamb, D.Q. and Rybicki, G.B. 1981, *Ap. J.*, **248**, 738.
White, T. R., Lightman, A. P., and Zdziarski, A. A. 1988, *Ap. J.*, submitted.

MODEL OF GAMMA-RAY EMISSION FROM LARGE-SCALE JETS

M. W. Anyakoha, P. N. Okeke and S. E. Okoye
Department of Physics and Astronomy
University of Nigeria, Nsukka

ABSTRACT

A model for gamma-ray emission from large-scale jets of quasars and active galactic nuclei is proposed. It is shown that if ~ 10^7-10^{10} $M_\odot$ of ambient matter is entrained by the jets in their interaction with the interstellar medium, the π° decay arising from proton collisions can constitute a observable source of gamma ray photons from the jet regions. A relativistic proton spectrum from the black hole, with spectral index between 2 and 2.66 can lead to gamma ray flux densities of the right order of magnitude expected from large scale jets.

INTRODUCTION

The quasar 3C273 is the only extragalactic radio source that has been clearly identified as a source of high energy ($\geq$ 100 Mev) gamma rays[1].

Some authors[2,3] have proposed a model of gamma ray emission from this source, ascribing it to the decay of nuetral pions and to the bremsstrahlung process in ambient gas. Morrision, Roberts and Sadun[3] showed that the gamma rays originate from the jet ascribing their origin to a relativistic proton beam which produced gamma rays on collision with clumped 'warm' entrainments in the outer jet. Although their model requires a high beam power and an unexpected degree of clumping of jet material, it provides a natural link between the gamma rays and the radio optical frequency spectra. They suggest that the diffuse γ-ray background might be the result of many unresolved transient jet events of this sort among active galaxies.

Encouraged by the above, we seek in this paper to investigate the possibility of producing gamma rays from the jets of active galactic nuclei in general, especially from those of high luminosity sources like M87, Cygnus A and the nearby radio source Centaurus A. We show that gamma-ray flux of the right order of magnitude can be produced from these jets through the decay of neutral pions resulting from proton-proton (p-p) collisions involving relativistic protons from the active centre and thermal protons entrained by the jets in their interaction with interstellar medium.

THE MODEL FOR γ-RAY GENERATION

We assume that high energy gamma-rays ($\geq$ 100 Mev) are produced in large scale jets through the decay of neutral pions generated by proton-proton collision occurring all along the jet. More specifically the interaction is:

© 1988 American Institute of Physics

$$p + p \rightarrow p + p + \ell\ (\pi^+ + \pi^-) + t\pi^0 \quad (1)$$

where ℓ and t, are zero or positive integers. Each π^0 produced decays as:

$$\pi^0 \rightarrow \gamma + \gamma \quad (2)$$

For this process to be viable enough as to produce sufficient γ-ray flux that can account for the expected flux from the 3C273 jet and possibly other large scale jets, we have relied on De Young's suggestion[4] that about 10^7-10^9 $M_\odot$ of interstellar materials can readily be entrained by large scale jets in their interaction with their environment.

We assume a differential proton flux interacting with warm entrained matter. Let the proton flux be given by:

$$dJ_p = K_p E_p^{-\Gamma} dE_p \quad (3)$$

where J_p is the number of incident protons per cm² per second and having a Lorentz factor $\gamma_p (= E_p/m_p c^2)$. K_p and Γ are constants.

The rate of production of γ-ray photons per target proton is given by:

$$q_\gamma(e_\gamma)dE_\gamma = \int_b^a dE_p\ J_p(E_p) \int_{E_{\pi min}}^{E_{\pi max}} dE_{\pi^0} M^0 \sigma_{\pi^0}(E_{\pi^0}, E_p) f_\gamma(E_\gamma, E_{\pi^0}) \quad (4)$$

where $f_\gamma(E_\gamma, E_{\pi^0})$ is the γ-ray distribution, $a=E_{pmax}$, $b=E_{pmin}$ are maximum and minimum proton energy, σ_{π^0} is the cross-section for π^0 production which is equal to $8.4 \times 10^{-27}\ \gamma_p^{0.75}$ (Eilek and Kafatos[5]). We have deduced the $\pi^\pm$ multiplicity as

$$M = 3\gamma_p^{1/3} \quad (5)$$

from a plot of data obtained by Lee et al.[6] The neutral pion multiplicity is obtained from the relation[7]

$$\frac{\pi^0\ \text{multiplicity}}{\pi^\pm\ \text{multiplicity}} = 0.5, \quad (6)$$

The required multiplicity for π^0 is therefore given by

$$M^0 = 1.5 E_p^{1/3} \quad (7)$$

The γ-ray distribution function is given by[8]

$$f_\gamma(E_\gamma, E_{\pi^0}) = (E_{\pi^0}^2 - M_{\pi^0}^2 c^4)^{-1/2}$$

for $\quad \frac{E_{\pi^0}}{2}(1-\beta_\pi) \le E_\pi \le \frac{E_{\pi^0}}{2}(1+\beta_{\pi^0}); \quad \beta = \frac{V_{\pi^0}}{c}$ \hfill (8)

and $\quad f_\gamma(E_\gamma, E_{\pi^0}) = 0$ otherwise

Using the kinematics for the π^0 decay,[8] it is easy to show that the maximum and minimum values of E_π are given by:

$$E_{\pi min} = E_\gamma + \frac{M_\pi^2 c^4}{4\pi E_\gamma} \tag{9}$$

$$E_{\pi max} = \infty$$

Our equation (4) becomes

$$q_\gamma(E_\gamma) = 2 \times 1.5 K_p n \sigma \int_{E_p} E_p^{-(\Gamma-1/3)} \int_{E_{\pi min}}^{E_{\pi max}} \delta\left[\frac{E_\pi - (0.14 E_p^{3/4})}{E_\pi}\right] dE_p dE_\pi \tag{10}$$

photons $cm^{-3} s^{-1} \Delta E_\gamma^{-1}$

The factor 2 arises from the fact that each π^0 decays into two photons. The π^0 distribution function has been approximated by a delta function[9] and $f_\gamma(E_\gamma, E_\pi) = (E^2 - M^2 c^4)^{-1/2}$ approximates to E^{-1} for $E \gg M_\pi c^2$. Carrying out the integration after making the necessary substitution for σ_{π^0} we obtain:

$$q_\gamma(E_\gamma) = 5.6 \times 10^{-26} K_p n (0.14)^{4/3(\Gamma-4/3)} \frac{E_\gamma^{-S}}{S} \tag{11}$$

where $\quad S = 4/3 (\Gamma - 7/12)$ \hfill (12)

The thermal particle density $\underline{n}$ is inferred through the relation

$$\underline{n} = \frac{M/m_p}{V} \tag{13}$$

where M_E is the total mass of interstellar matter entrained by the jet of volume V. The proton mass is represented by m_p. Thus for example, the jet of Centaurus A (NGC5128) with a volume $\sim 10^{66}$ cm^3 will have an inferred thermal particle density of $\sim 10^{-2}$ cm^{-3} if

~$10^7 M_\odot$ of interstellar material is entrained[4] throughout the length of the jet.

The gamma-ray flux is given by:

$$I_\gamma(E_\gamma) = \frac{1}{R^2} \int_V q_\gamma(E_\gamma) \cdot n \cdot dv \ (cm^2 sec \ Gev)^{-1} \quad (14)$$

where R is the distance to the source of volume, V. The integral photon flux is given by

$$I_\gamma(\geq E_\gamma) = \int_{E_\gamma}^\infty I_\gamma(E_\gamma) dE_\gamma) \ (cm^2 \ sec)^{-1} \quad (15)$$

$$= \frac{5.6 K_p \times 10^{-26} (M_E/m_p)(0.14)^{4/3} \ (\Gamma-4/3)}{R^2 \times \frac{4}{3} \ (\Gamma-7/12)} \quad (16)$$

$$\int_{E_\gamma}^\infty E_\gamma^{-4/3(\Gamma-7/12)} dE_\gamma$$

DISCUSSION OF RESULT

The constants K_p and Γ of the relativistic proton flux are not really known. K_p is usually taken to lie in the range $1 < K_p < 2000$ and assumed to be 100 for secondary electrons arising from p-p collision[10]. If the proton flux from the black hole at the galactic centre is taken to be the same as universal cosmic ray spectrum, then Γ = 2.66. However, Protheroe and Kazanas[2] in their model for a class of quasars and active galactic nuclei gauged to 3C273, used the value, Γ = 2 to obtain results in good agreement with obervations of 3C273. As the value of Γ has not been clearly established, we make two sets of calculations based on both values of the constant Γ = 2, and 2.66 and also K_p = 100.

TABLE 1

Table 1 shows the model jet parameters on the basis of the above. The mass (M_E) of the entrained matter is taken as equal to that used to obtain radio flux densities comparable to observations[11]. The radio emission arises from the electrons resulting from p-p collision via $\pi^\pm$ decay, while the gamma rays arise via π^0 decay also from the same proton interactions. We expect that the same target matter density in the jet should give the right photon fluxes since both the radio and gamma-ray emission arise as decay products of proton collisions occuring in the jet.

For the quasar 3C273 and the nearby radio galaxy NGC5128 (Centaurus A) for which Houston and Wolfendale[12] had estimated gamma-ray fluxes from SAS2 data, our model estimates are about an

TABLE 1
PARAMETERS FOR THE γ-RAY EMISSION FROM THE MODEL JET

Source (Jet)	Mass Entrained by Jet, ($M_\odot$)	Γ	Gamma-ray Flux, I_γ (this model) $(cm^2 s)^{-1}$	Other Estimates of I_γ $(cm^2 s)^{-1}$	Reference
3C273	2×10^{11}	2.0	4×10^{-7}		
		2.66	5×10^{-6}	$(6^+_{-4}) \times 10^{-7}$	(12)
Cyg A	5×10^9	2.0	8×10^{-9}		
		2.66	1×10^{-6}	4×10^{-8}	(13)
M87	8×10^7	2.0	2×10^{-8}		
		2.66	3×10^{-6}	6×10^{-7}	(13)
Cen A (NGC5128)	10^7	2.0	5×10^{-8}	3.4×10^{-8}	(13)
		2.66	7×10^{-6}	1.5×10^{-6}	(12)

order of magnitude below theirs for $\Gamma = 2$, but about a factor of four above for $\Gamma = 2.66$. This would imply that the value of Γ should lie somewhere between 2 and 2.66 for $K_p = 100$.

Another source of uncertainty is the total amount of warm protons entrained by the jet. The entrained mass depends on jet velocity, ambient gas density, effective entrainment radius and the efficiency factor. All the above are not known with any high degree of certainty.

CONCLUSION

High energy gamma-ray emission from 3C273[1] believed to be located at the large-scale jet region[3] has motivated this model that attempts to estimate the gamma-ray flux expected from this jet and those of other high luminosity sources. It is suggested that the high energy gamma-ray protons ($E_\gamma > 100$ MeV) arise mainly from the decay of π° resulting from collisions of a flux of relativistic protons from the black hole at the galactic centre and target matter entrained by the jet in its interactions with interstellar medium. The model predicts a correlation between the radio spectrum and gamma-ray energies.

Future observations with more sensitive gamma-ray telescopes may lead to the correlation of the gamma ray data with those obtained at other wavelengths and could be used to check the different emission mechanisms (π° decay, bremstrahlung and inverse compton). It could lead to better estimates of the constants (K_p, Γ) of the relativistic proton flux believed to originate in the black hole at the nuclear center, and to power the extended components of quasars and active galactic nuclei. It should then be possible to know if a universal spectral profile apply to all such sources, and if the contribution from the unresolved transient jet events of the sort here described can constitute a substantial fraction of the diffuse gamma-ray background.

REFERENCES

1. G. F. Bignami et al., Astron. Astrophys. 93, 71 (1981).
2. R. Protheroe and D. Kazanas, Astrophys. J. 265, 620 (1983).
3. P. Morrison et al., Astrophys. J. 280, 483 (1984).
4. D. S. De Young, Astrophys. J. 307, 62 (1986).
5. J. A. Eilek and M. Kafatos, Astrophys. J. 271, 804 (1983).
6. C. Y. Lee et al., Nuovo Cimento 38, 189 (1983).
7. M. Basil et al., Phys. Lett. 99B, 247 (1981).
8. F. W. Stecker, Cosmic Gamma Rays, Mono Books (1971).
9. G. G. Fazio, Annu. Rev. Astron. Astrophys. 5, 481 (1967).
10. A. G. Pacholczyk, Radio Astrophys., Freeman, SF (1970).
11. M. W. Anyakoha et al., Astrophys. J., in press (1988).
12. B. Houston and A. Wolfendale, Vistas Astron. 26, 107 (1982).
13. G. Garmire and W. Kraushar, Space Sci. Rev. 4, 123 (1965).

HIGH ENERGY GAMMA RAYS FROM THE VELA PULSAR: LONG-TERM VARIABILITY AND ENERGY DISTRIBUTION

I.A. Grenier[1], W. Hermsen[2], and J. Clear[3]
[1] Service d'Astrophysique, Centre d'Etudes Nucléaires de Saclay, France.
[2] Laboratory for Space Research Leiden, Leiden, The Netherlands.
[3] Space Science Department of ESA, Noordwijk, The Netherlands.

ABSTRACT

New results on the temporal and spectral characteristics of the high energy (50 MeV to 5 GeV) gamma ray emission from the Vela pulsar are presented. A new sensitive analysis method using maximum likelihood techniques has been applied to the COS-B data. The whole pulsed flux is found to exhibit long term variability. The data show strong evidence for a spectral break at approximately 300 MeV in the energy spectrum of this total emission. Five discrete emission regions within the pulsar lightcurve have been identified, with the spectral characteristics and long term behaviour being different. These results support the idea that various production regions simultaneously exist in the pulsar magnetosphere and that the physical processes generating the gamma rays differ with location.

INTRODUCTION

The study of the high energy gamma ray emission from celestial objects provides an excellent method in helping our understanding of the physical processes which occur within such sources. Tümer et al. (1984) at 0.3 - 30 MeV, Thompson et al. (1977) above 35 MeV and Bennett et al. (1977) at 50 MeV - 5 GeV all report a similar Vela lighcurve structure which is characterised by two sharp peaks separated by 0.42 in phase and bracketing the optical pulses. The energy spectrum of the pulsar still remains uncertain. An extrapolation to X-ray energies of the COS-B results of Kanbach et al. (1980) lies several orders of magnitude above the reported upper limits, while at TeV energies the flux values reported by Bhat et al (1980) clearly require a break in the spectrum. Earlier results from COS-B data by Bennett et al. (1977) and Kanbach et al. (1980) indicate a possible break in the spectrum at a few hundred MeV, but the level of significance of these results is low. This article presents the latest results on the Vela pulsar in the 50 MeV to 5 GeV energy range. The data have been collected using the COS-B satellite which observed the Vela region during 10 observation periods between October 1975 and June 1981. The total COS-B gamma ray lightcurve together with the radio and optical ones are given in figure 1.

THE MAXIMUM LIKELIHOOD METHOD

A maximum likelihood analysis has been applied to the data to estimate the flux and spectral properties of the source using a

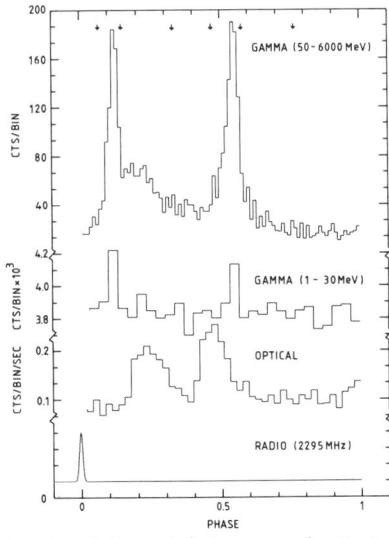

TABLE I Definition of the selected phase intervals of the Vela pulsar lightcurve.

Region	Phase interval
Peak 1	0.07 - 0.15
Interpulse-1	0.15 - 0.33
Interpulse-2	0.33 - 0.47
Peak 2	0.47 - 0.58
Trailer	0.58 - 0.77
Background	0.77 - 0.07

Fig. 1. Vela lightcurves.

testmodel, which may include up to four major components: 1. the pulsed emission from the pulsar; 2. a possible steady component from Vela; 3. a constant and isotropic instrumental background; 4. the steady diffuse galactic emission around the pulsar. The latter term results from the interaction of cosmic rays with interstellar matter. Its spatial structure can be traced by the HI and CO surveys (Dame et al., 1987; Strong et al., 1987). For further details on the method, see Grenier et al., 1988.

Table I gives the details of the phase domains which have been selected for the analysis. The background region (phase 0.77 - 0.07) has been analysed for a possible steady emission from Vela. The resultant upper limit shows that such emission accounts for less than a few percent of the pulsed luminosity, and may thus be negelected from the model. For those observations with large source aspect angles (>15°), where the efficiency and the energy resolution of the instrument are poor, this likelihood analysis is limited by the available statistics and has not been performed. For the remaining observations a detailed analysis has been made for the 5 phase intervals. Within each phase interval the analysis has been performed for 3 energy intervals (50 - 5000 MeV, 50 - 300 MeV and 300 - 5000 MeV). The division of the data into these energy intervals is based on the evidence for a break in the pulsed spectrum at approximately 300 MeV found in the analysis of Kanbach et al. (1980) using only 2 observations. Due to computing time considerations the position of the spectral break has not been used as an additional free parameter within the model, however the results of the independent modelling of the spectrum above and below 300 MeV strongly support this choice of value.

RESULTS

The likelihood analysis has been performed for the 5 selected phase intervals, for the 5 best observation periods and for 3 energy

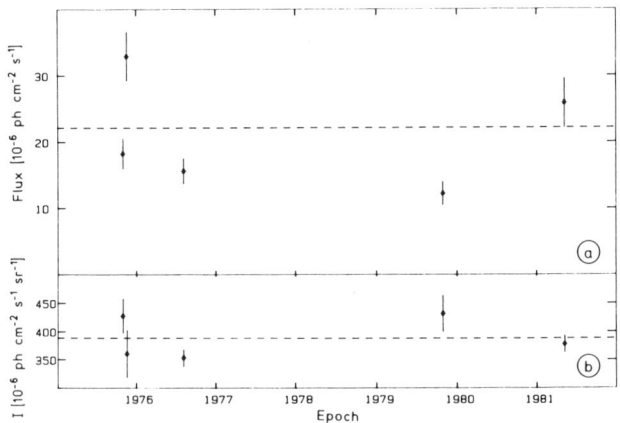

Fig. 2. Integrated flux/intensity (I) values (50-5000 MeV) for 5 observations: a) the pulsed Vela flux; b) the diffuse galactic background and instrumental emission.

ranges, giving a total of 75 spectra and flux values. Only the main results will be highlighted (for detailed results see Grenier et al). The spectrum of the whole emission derived independently at energies above and below 300 MeV and using all observations is:

$$F(50 \leq E \leq 300) = 2.74 \pm 0.21 \times 10^{-4} . E_{MeV}^{-1.72 \pm 0.07} \text{photons/cm}^2 \text{ s MeV}$$

$$F(300 \leq E \leq 5000) = 2.71 \pm 0.04 \times 10^{-3} . E_{MeV}^{-2.12 \pm 0.07} \text{photons/cm}^2 \text{ s MeV}$$

The significant difference between the high and low energy spectral indices clearly shows that the time averaged Vela pulsed spectrum from 50 to 5000 MeV cannot be described by a single power law. At least two are required and in this representation the good connection between the two spectra at 300 MeV strongly supports the choice of 300 MeV as the energy of the break. The flux of the pulsed emission from Vela has been determined by integrating the two-power-law phase averaged spectrum for each observation. The result, displayed in figure 2a, indicates the striking long-term variability of Vela. In addition, the likelihood analysis provides simultaneously the background intensity for each observation, and its stable behaviour from one observation to the next, displayed in figure 2b, confirms the true source origin of the pulsed variability phenomenon. To study the energy dependence of this variability, figure 3a shows the evolution of the pulsed flux for the energy ranges 50 - 300 MeV and 300 - 5000 MeV. The main contribution to the flux variability is clearly due to the low energy emission. The different evolutions of the flux from the two energy ranges denote a distortion of the Vela spectrum from one epoch to the next. The observed fluctuation in the spectral ratios (the ratio of low to high energy flux) has a probability of 10^{-10} of being due to a random effect. Figures 3b and 3c show the spectral indices as a function of epoch for the 3 energy ranges. As expected, the spectral index for the entire energy range is seen to vary with time with a probability of $2 \; 10^{-3}$ that this effect is due to chance, whereas simultaneous analysis of the background spectra shows no indication for variability.

Figure 4a shows the time averaged spectral index as a function of phase. The probability that the emission may be best represented by a homogeneous spectrum over all phases ($E^{-1.84}$) is 10^{-4}. A further understanding of the phase dependency of the pulsed emission may be obtained from figure 4b, which shows the spectral characteristics of the high and low energy emissions as a function of phase. Analysis of the spectral characteristics of the high energy emission shows that there is no significant variation in its spectral index with phase, giving a phase averaged value of 2.12 ± 0.07. Thus, the spectral index of the GeV radiation from Vela displays both a stable and homogeneous behaviour. In contrast, the spectral index of the low energy emission shows evidence for phase dependence (chance probability $5 \cdot 10^{-3}$). It is evident from the change in the spectral index at low energies (from the first peak to interpulse-2, 4.5σ, and interpulse-2 to the second peak, 3.6σ and from the softness of the first peak and trailer over the entire energy range, that 5 separate phase components exist in the Vela gamma ray lightcurve.

The evolution of the flux from each of the phase components for each observation and for both high and low energy ranges has been studied. The long-term variability is obvious and the relative amplitude of the variability is significantly larger for the two

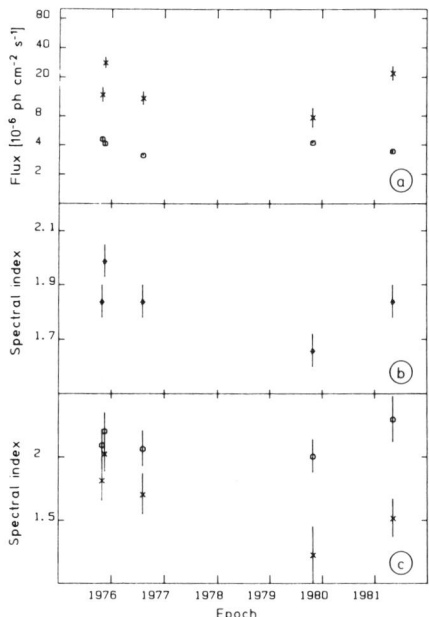

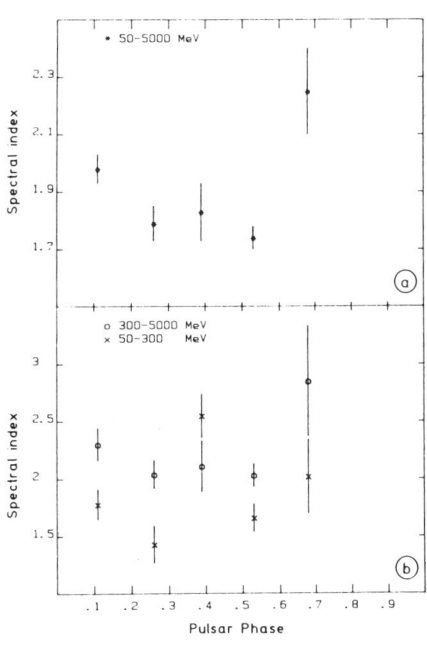

Fig. 3. a) Vela pulsed flux: 50-300 MeV(X), 300-5000 MeV(O); b) phase averaged pulsed spectral index (50-5000 MeV); c) as b), for 50-300 MeV(X) and 300-5000 MeV(O).

Fig. 4. Time averaged spectral index as a function of phase; one power-law fit over a) total energy range, b) high and low energy ranges.

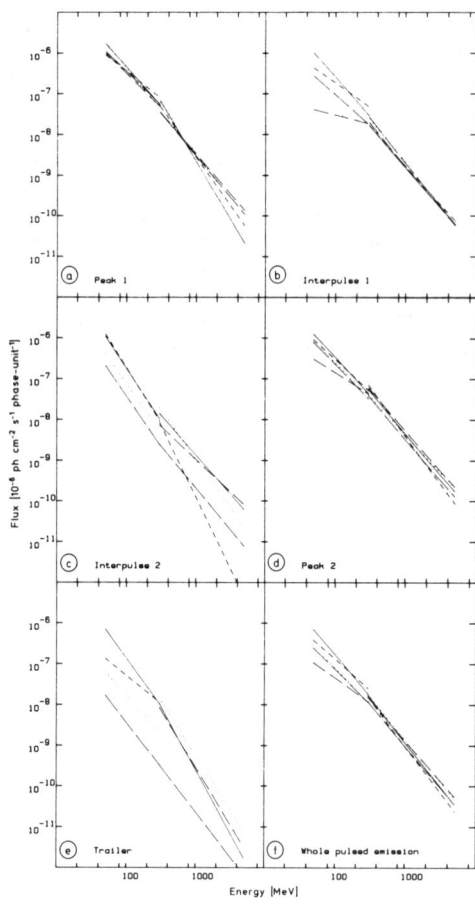

Fig. 5. Differential pulsed gamma-ray spectra from Vela for 5 phase intervals and the total phase averaged emission for each observation period. The power law spectra giving the max. likelihood fits for the 50-300 MeV and 300-5000 MeV energy ranges are shown.
Period 2 ...
Period 3 ——
Period 12 _._
Period 45 _ _
Period 59 ----

interpulse emissions.

Furthermore, the flux variations in the separate energy intervals are different resulting in spectral distortions (probability levels are given in Grenier et al. 1988). To illustrate their long term evolution, the energy spectra giving the maximum likelihood fits to the data for each phase component and for each observation are displayed in figure 5. Good agreement is observed between the low and high energy fits at approximately 300 MeV for most phase components. The two power law representation is therefore a good approximation to the real spectra (for the trailer, the poor statistics make interpretation of the results rather difficult). The actual position of the break or bend in the real spectrum will vary somewhat in energy following the apparent long term variability.

Total PSR0833-45 Spectrum: The two-power-law fit time averaged spectrum of the whole pulsed emission has been plotted in figure 6 together with the data recorded at other wavelengths in order to show the full energy distribution of the Vela pulsed radiation over the entire electromagnetic spectrum. Over this spectrum, the pulsar reaches its maximum luminosity in the MeV-GeV domain. At very high energies, the recent results of Bhat et al. (1987) indicate a spectral slope of -3.5 at TeV energies which does not require an excessive steepening when extrapolating over three decades of energy from the COS-B gamma ray range. The observed time variability presented in this article may support the proposed time variability at TeV energies, e.g. Bhat et al. (1980), however, it should be noted that the variations recorded above a few hundred MeV are relatively small ($\lesssim 25\%$).

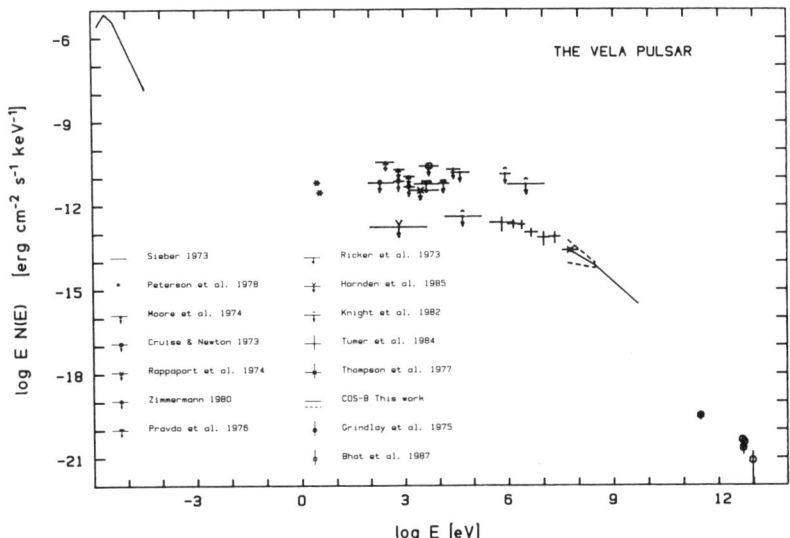

Fig. 6. Total energy spectrum from the radio up to the very high energy gamma ray domain. The upper limits (3σ) displayed in the X-ray range are calculated for an assumed duty cycle of 0.2 similar to the optical one.

CONCLUSIONS

The identification of the different components in the Vela gamma ray lightcurve supports the idea that several source regions exist in different parts of the pulsar magnetosphere. The significant contrast observed in the characteristics of the peak and interpulse emissions evidently suggests that there may be at least two types of sources in the magnetosphere according to their location. The different spectral characteristics of the components indicate that different generating processes exist in each source. The origin of the variability observed in the low energy part of the gamma ray emission on time-scales from weeks to months must still be explained. In particular it is not possible to absolutely associate this phenomenon with the occurrence of giant glitches (Grenier et al. 1988).

REFERENCES

Bennett, K. et al.: 1977, Astron. Astrophys. 61,279.
Bhat, P.N. et al.: 1980, Astron. Astrophys. 81,L3.
Bhat, P.N. et al.: 1987, Astron. Astrophys. 178,242.
Dame, T.M. et al.: 1987, Astrophys. J. 322,706.
Grenier, I.A., Hermsen, W., Clear, J.: 1988, Astron. Astrophys. (in press).
Kanbach, G. et al.: 1980, Astron. Astrophys. 90,163.
Strong, A.W. et al.: 1987, Proc. 20th ICRC, Moscow, 1,125.
Thompson, D.J. et al.: 1977, Astrophys. J. Letters 214,L17.
Tümer, O.T. et al.: 1984, Nature 310,214.

ARE THE γ-RAY FROM THE VELA PULSAR POLARIZED?

P.A. Caraveo, G.F. Bignami
Istituto di Fisica Cosmica del C.N.R., Milano, ITALY

I. Mitrofanov
Institute for Space Research, Academy of Sciences, Moscow, USSR

G. Vacanti
Iowa State University, Ames, IOWA, U.S.A.

ABSTRACT

The azimuthal distribution of planes containing e^+/e^- pairs from high-energy photon materialization is reminiscent, through a quadrupole anisotropy, of the degree and position angle of linear polarization of the incident photons. Data from the COS-B spark chamber are used in a search for such an effect in > 50 MeV photons from bright sources, such as Vela, Crab, Geminga and a reference galactic plane region in Cygnus. Only for the Vela pulsar a low-chance-probability effect is found apparently implying a high (~ 100%) degree of linear polarization for the Vela photons. This has important implications on the physics of the production mechanisms as well as of their geometry.

INTRODUCTION

A property like linear polarization would be very useful in understanding the nature of the high-energy γ-ray sources discovered by the COS-B mission. Certainly, polarization is a common property of radio pulsars (e.g. Rankin 1983a,b, and Taylor and Stinebring 1986, for a recent review). For instance, PSR0833-45 is the one showing the highest percentage (~ 100% at 4.8 GHz) of linear polarization, and PSR0531+21 too shows very significant polarization in both the radio and the optical. The detailed properties of radio pulsar linear polarization are often complex, and include such features as rotation of the polarization position angle, presence of orthogonal modes, etc. However, it is well possible that all such features be explained starting from the basic fact that the radiating particles have acceleration components parallel and perpendicular to the magnetic field and radiate their transversal energy due to the magnetobremsstrahlung mechanism (see, e.g. Bjornsson 1984; Taylor and Stinebring 1986). In this case, it is not inconceivable that, irrespective of possible different light curve morphologies, also the pulsed high-energy γ-rays be polarized.

The basic idea on how to search for polarization in high-energy γ-rays from cosmic sources has been proposed by Kozlenkov and Mitrofanov (1985, and references therein; hereafter KM), and is based on the expected effect of quadrupole modulation of the azimuthal distribution of the planes containing the e^+/e^- pairs produced in the γ-ray interactions. This anisotropy of the pair plane orientations is due to the linear polarization of the incoming photons (Maximon and Olsen 1961).

In a system with the z-axis along the wave-vector K of the incident gammas, the azimuthal distribution $\Phi(\varphi)$ of the e^+/e^- pairs has a quadrupole-like anisotropy with amplitude and direction determined, respectively, by the degree of polarization and by the polarization position angle of the incident radiation. Such a distribution can be written as:

(1) $$\Phi(\varphi) = \frac{N_t}{2\pi}\{1 + RP\cos[2(\varphi - \varphi_0)]\}$$

where φ_0 is the polarization position angle, P is the degree (between 0 and 1) of linear polarization, N_t is the total number of observed photons, and R is a numerical factor which depends on the intrinsic asymmetry ratio in the process physics and on the detection characteristics. In the ultrarelativistic case ($E_\gamma >> m_e c^2$), and allowing for screening (KM), a value

of $0.14 < R < 0.2$ is obtained, considering also the COS-B detection requirements. This is consistent with the higher values proposed by Maximon and Olsen(1961).
The quadrupole asymmetry of the effect allows for the use of a simple method (KM) for its determination: for Nt observed events, let N1 and N2 (with (N1 + N2 = Nt)), be the two numbers of observed events corresponding to two intervals of the observed azimuth angle $(\varphi, \varphi + \pi/2)$ and $(\varphi + \pi/2, \varphi + \pi)$; and let $S(\varphi)$ be the quadruple deviation function

(2) $$S(\varphi) = \frac{[N_1(\varphi) - N_2(\varphi)]^2}{N_1(\varphi) + N_2(\varphi)}$$

Because of the theoretical law (1), the deviation function should have a well-pronounced sinusoidal-like shape:

(3) $$S(\varphi) = \frac{4 N_t R^2 P^2}{\pi^2} \sin^2[2(\varphi - \varphi_0)]$$

Using the function (2) one may then check the presence of quadrupole modulation in the data: for the effect to be real, $S(\varphi)$ should take the shape (3). In this case, from the minimum, $S(\varphi) = 0$ of (3), one can derive the position angle, while from the maximum, $S(\varphi) = S^*$, one can estimate the degree of linear polarization:

(4) $$P = \frac{\pi}{2R} \sqrt{\frac{S^*}{N_t}}$$

Even for a totally polarized radiation ($P = 1$), the amplitude of the expected azimuthal modulation is rather small ($< 20\%$), and several factors conspire to "blur" the effect, the most important of which is certainly the limitation on Nt, the available statistics. Accordingly, one may estimate the Nt necessary for measuring a 100% polarization using $S(\varphi)$ to estimate the probability that the observed azimuthal modulation result from a statistical fluctuation. For a constant phase angle, the value S can be considered as the number of standard deviations of the random variable N1 (or N2) from the expectation value for the case of zero polarization, in the assumption that its dispersion equals (N1 + N2)/2. The probability of such deviation gives the probability of the zero hypothesis. Following this method, for $E_\gamma > 50$ MeV, Nt is about 2,000 events, so that, it appears worthwhile to perform a search for the polarization effect in the COS-B data, also because for the first time in the history of γ-ray astronomy, an acceptable photon number is available for a few sources.

DATA ANALYSIS AND RESULTS

During its ~ 7 year lifetime, COS-B performed 65 pointed observations of > 1 month each (Mayer-Hasselwander et al. 1985); we have used the data of periods referring to the brightest observed sources. The reconstructed arrival directions for Crab and Vela were within 10 deg of the source for $50 < E_\gamma < 150$, 7 deg for $150 < E_\gamma < 300$ and 5 deg for $E_\gamma > 300$ MeV, while for Geminga and the Cygnus Region only events with arrival direction within 5 deg from the source position were selected. From the azimuthal $\Phi(\varphi)$ distribution of the event planes one can then estimate the amplitude and phase of their quadrupole modulation using the deviation function $S(\varphi)$ for the numbers of events N1 and N2 as described earlier. For the determination of the position angle of the effect it is necessary to vary the starting azimuth value in steps of 2 deg, so that 45 steps are necessary to cover the $\pi/2$ interval of the $S(\varphi)$ function. Such steps are of course taken into account in the final evaluation of the event probability.
Fig.1 (curve a) shows $S(\varphi)$ for the 2526 events from the 2CG263-02 source, or the sum of all the COS-B events satisfying our selection from the Vela pulsar region. For comparison, curve b shows the same distribution for data from another region of the sky, that of Cygnus, probably dominated by diffuse emission and for which no polarization is expected. Specifically, 3109 photons were selected from a region of 5 deg radius centered at $l = 80$ deg, $b = 0$ deg. The two curves are very different: while for curve a the suggestion exists of a well-ordered modulation of the azimuthal dependence of the event planes, curve b does not show any effect, proving that the observed anisotropy for the Vela data cannot be due to some systematics of the spacecraft. To evaluate the statistical significance of the effect, a Montecarlo simulation was performed.

The maximum value of the deviation function ($S(\varphi)$) was computed for 300,000 random 2500-event distributions, each going through the 45 steps phase coverage cycle. The Montecarlo results thus give the chance occurence probability for a given S value for the Vela data. Such a probability is shown in Fig.1. From curve a, it is seen that azimuthal anisotropy for the Vela data has a chance occurrence probability $\sim 1.5 \times 10^{-4}$. As a further check, we have constructed the same type of distribution but with the event plane azimuths not corrected for the satellite spin, i.e. in the sky of the spark chamber rather than in the true one. No deviation from the random expectations was observed for the Vela or Cygnus data, nor any sign of sinusoidal modulation is present. Therefore, the low chance probability effect is also not due to some systematics of the spacecraft.

Another COS-B source investigated was 2CG185-5, the Crab. As in the case of Vela, we consider all class 22 events with the usual source distance selection, yielding 1533 photons over 6 observing period. Curve (a) of Fig.2 shows the result for this "total" Crab emission: no significant evidence is apparent for a modulation. On the same Figure, curve (b) shows the distribution for the Geminga (2CG195+5) data, referring to 1204 photons with the usual selections and coming from 5 observing periods, most of which were in common with the Crab. Here too no significant modulation is present.

PULSAR PHASE ANALYSIS

Considering, as mentioned earlier, the high degree of linear polarization in radio photons from pulsars in general and for the Vela pulsar in particular (up to $\sim 100\%$), it seems natural to approach the azimuthal anisotropy suggested by the data as due to linear polarization of the incoming γ-rays. In this case, according to the predictions of the effect as outlined above, eq. (4), the value of $S^* = 20$ (Fig.1, curve a) implies $RP = 0.14$, or a very high degree of polarization (bigger than 50%, and up to 100%, depending on the uncertainty on R) for the high-energy γ-rays from the Vela source. One must now, of course, remember that the majority of the COS-B γ-rays from 2CG263-02 are in fact pulsed at the period of PSR0833-45 (see,e.g., Kanbach et al. 1980), and it seems worthwhile to investigate the effect in the various region of the pulsars light curve (or phase diagram), even with necessarily reduced statistics. The 2526 photons from Vela were than divided in 4 phase intervals

(1) First pulse   : containing   584   photons
(2) Second pulse   : containing   666   photons
(3) Interpulse   : containing   791   photons
(4) Background   : containing   485   photons

Accordingly, curves (1) to (3) of Fig. 3 show the usual $S(\varphi)$ distributions for the corresponding intervals, while curve (4) shows the total "pulsed", or $(1)+(2)+(3)$ and curve (5) shows the "background" emission, as defined above.

The severe limitation introduced in statistics by such a breakdown renders any firm conclusion from Fig. 3 very difficult; it looks, however, as if the interpulse (plus may-be the II pulse) region is the one contributing most to the possible polarization. It would also appear that the S^* values of the interpulse and total pulsed radiation require RP to approach its maximum value, or a $\sim 100\%$ polarized radiation.

As to the value of the polarization P.A., bearing in mind the $\pi/2$ periodicity of the $S(\varphi)$ function, from Fig. 1 (curve a) and Fig. 3 (curve (4)) one finds $\sim 130 \pm 15$ deg. A comparison appears natural with the polarization results in the radio region, where, it should be remembered, the light curve shape is completely different, featuring only one short peak not in phase with the γ-ray peaks, but rather in a γ-ray "background" region . The classic result of Radhakrishnan and Cooke (1969) shows a significant P.A. shift (90 deg) centered at 120 deg or 154 deg, depending on the frequency, within the 4 msec peak. Our data would seem to support a picture very different from that at radio frequencies, considering also the high percentage value requested to interpret the observed anisotropy as linear polarization: the P.A. of the polarization plane either swings of a relatively small value (< 30 deg) or remains constant over a big interval of the pulsar phase (= interpulse + II pulse, or 44% of the phase).

Going back to Fig. 3, since the interpulse data are the ones that, taken singly, would seem to suggest some non-random behaviour in spite of the limited statistics, it appears interesting to look at the raw azimuth distribution of the 791 event planes. This is shown in Fig. 4, with the full π azimuth range covered in 10 deg bins. It makes sense to use the run test on such a distribution, to gauge its regularity of shape vis a vis the expectations. As apparent (and expected), the depth of modulation is shallow, but the data can indeed be divided in two separated 9-bin families, yielding a chance probability of 4×10^{-5} (Beyer 1968). Note that this is a one-shot test, with the initial azimuth fixed and no steps performed.

A similar pulsar phase analysis, albeit in the context of worse statistics, can be tried for the Crab data. The total 1533 events were divided in phase corresponding to I peak, II peak and background (see Wills et al. 1982). As could be expected from the result for the total events, no significant deviation from random expectations is observed for the Crab data subsets. One must also take into account the nebular emission (see e.g. Clear et al. 1987), concurring in this case to blur a possible pulsar signal, expecially at low energies, where a lot of the Crab events are. In any case, it would appear that much better statistics is necessary for a meaningful investigation.

CONCLUSIONS

The method outlined above has shown that, in spite of obvious difficulties, it is meaningful to search for polarization in high-energy γ-ray data in the COS-B spark chamber data. A rather significant anisotropy has been found for 2500 selected events from the Vela pulsar, and has been shown not to be present in other strong sources or in the Cygnus region of the galactic plane. With the proviso of the poor statistics and of other experimental limitations discussed, a linear polarization of the incoming photons appears as the natural explanation for this finding which has important implications on the emission mechanisms responsible for the pulsed emission (see Caraveo et al. 1988) Unfortunately, not much more can be done now to improve the significance of and confirm the experimental result of the COS-B data. This task, for both the Vela and Crab pulsars as well as for other sources, rests with the next generation of γ-ray telescopes.

REFERENCES

Beyer W.H. 1968, Handbook of Probability and Statistics CRC,p.414

Bjornsson, C-I. 1984, Ap. J., 277, 367

Caraveo, P.A., Bignami, G.F., Mitrofanov, I, Vacanti, G. 1988, Ap. J. in press (Apr.1)

Clear, J., Bennett, K., Buccheri, R., Grenier, I.A., Hermsen, W., Mayer-Hasselwander, H.A., and Sacco, B. 1987, Astron. Ap., 174, 85

Kanbach, G., Bennett, K., Bignami, G.F., Buccheri, R., Caraveo, P., D'Amico, N., Hermsen, W., Lichti, G.G., Masnou, J.L., Mayer-Hasselwander, H.A., Paul, J.A., Sacco, B., Swanenburg, B.N., and Wills, R.D. 1980, Astron. Ap., 90, 163

Kozlenkov, A.A., and Mitrofanov, I.G. 1985, Sov. Astron., 29, 159

Maximon, L.C., and Olsen, H. 1962, Phys. Rev., 126, 310

Mayer-Hasselwander, H.A., Bennett, K., Bignami, G.F., Bloemen, J.B.G.M., Buccheri, R., Caraveo, P.A., Hermsen, W., Kanbach, G., Lebrun, F., Paul, J.A., Sacco, B., and Strong, A.W. 1985, Proc. 19th Int. Cosmic Rays Conf., la Jolla, 3, 383

Radhakrishnan, V., and Cooke, D.J. 1969, Ap. Letters, 3, 225

Rankin, J.M. 1983(a), Ap. J., 274, 333

Rankin, J.M. 1983(b), Ap. J., 274, 359

Taylor, J.H., and Stinebring, D.R. 1986, An. Rev. Astron. Ap., 24, 285

Wills, R.D., Bennett, K., Bignami, G.F., Buccheri, R., Caraveo, P.A., Hermsen, W., Kanbach, G., Masnou, J.L., Mayer-Hasselwander, H.A., Paul, J.A., and Sacco, B. 1981, Nature, 296, 723

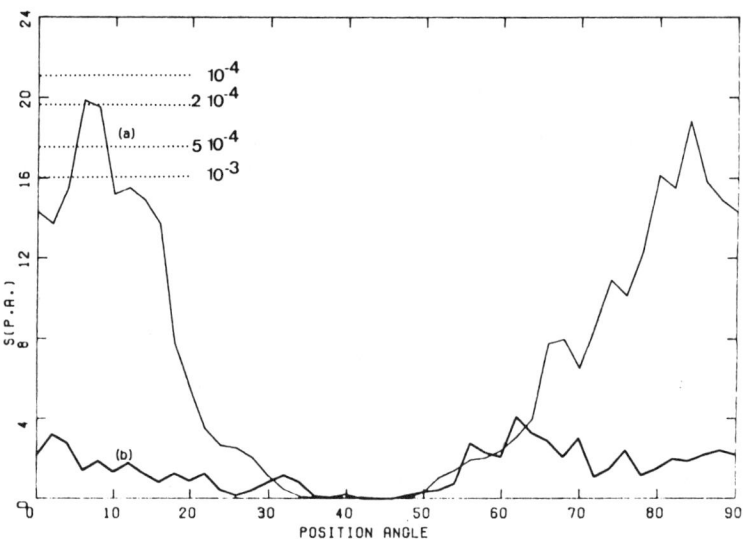

Fig.1 Curve (a) : deviation function $S(\varphi)$ for the 2526 photons from Vela. Following its definition, the numerical values of S are equivalent to a chi-square value with one degree of freedom. Overall chance occurence probability levels are also shown, according to Montecarlo simulations (see text).
Curve (b): as above, but for 3109 photons from the Cygnus Region of the galactic plane (l = 80 deg, b = 0 deg).

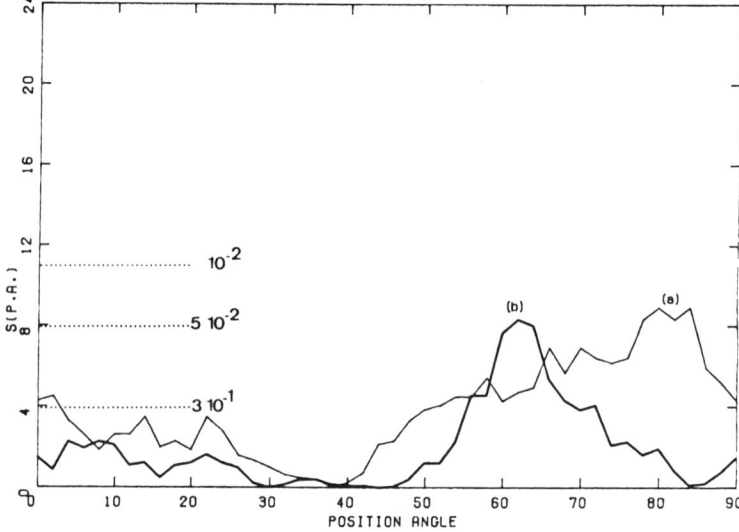

Fig.2 Curve (a) $S(\varphi)$ for 1533 photons from the Crab.
Curve (b) as above, for 1204 photons from Geminga. Units as in Fig.1. The overall chance occurrence probability levels shown are derived from Montecarlo simulations for 1500 events.

367

Fig.3 Deviation functions for different intervals of PSR0833-45 light curve (see text for exact definition). The "total pulsed" is the sum of I pulse, II pulse and interpulse distributions.

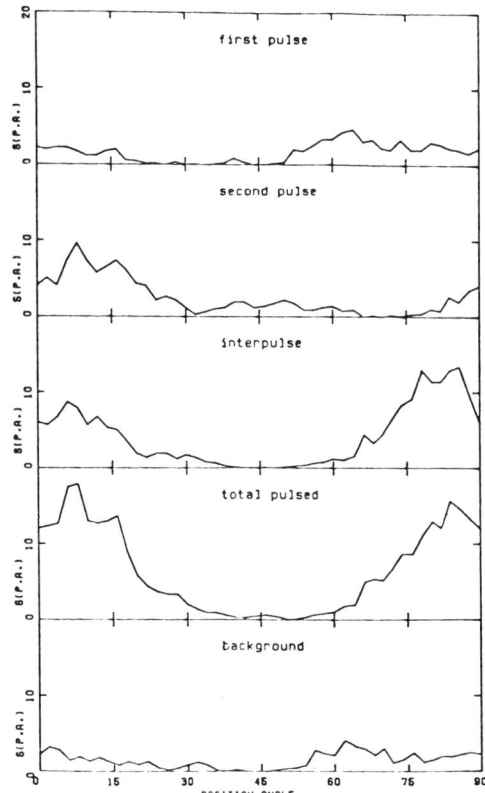

Fig.4 Raw data distribution: pair plane orientation histogram (N(P.A.)) for the PSR0833-45 interpulse data (791 photons).

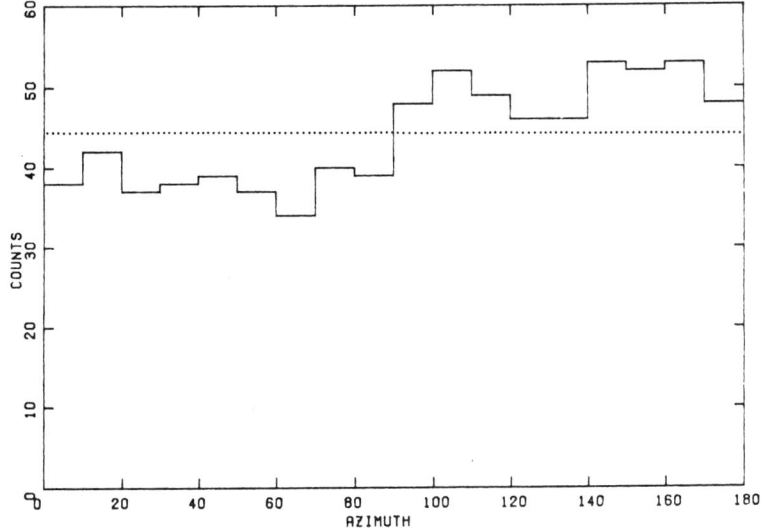

CAPABILITIES OF GRO FOR OBSERVATIONS OF SUPERNOVAE AND NOVAE

J. D. KURFESS
E. O. Hulburt Center for Space Research
Naval Research Laboratory
Washington, D. C. 20375

ABSTRACT

The Gamma Ray Observatory (GRO), scheduled for launch in March, 1990, represents NASA's next major mission in high-energy astrophysics. GRO will carry four large instruments whose main objectives are to undertake comprehensive observations of astrophysical sources throughout the 20 keV to 30 GeV energy range. This region includes those gamma-ray lines which are expected to reflect the explosive nucleosynthetic processes occurring in supernovae and novae. The capabilities of GRO to study these objects is discussed.

INTRODUCTION

Heavy element nucleosynthesis is expected in a variety of sources, during normal stellar evolution (particularly massive stars), and in supernovae and novae. Gamma-ray astronomy provides an excellent technique for testing models of nucleosynthesis in these several different sources via detection of radioactivity associated with nucleosynthetic products. A remarkable start toward these goals has been made with the observation of ^{26}Al in the interstellar medium by HEAO-3 (Mahoney et al. 1984) and SMM (Share et al. 1985) and the detection of ^{56}Co radioactivity associated with SN1987A (Matz et al. 1988, Sandie et al. 1988). The Gamma Ray Observatory will greatly extend our knowledge of nucleosynthesis with improved sensitivity and angular resolution on a long-lived mission. In this paper, we specifically discuss the capability of GRO for observations of supernovae and novae, and describe the instruments on GRO with which these observations will be made.

SUPERNOVAE

A major objective of gamma-ray astrophysics has been the detection of line emission from supernovae to test the theories of explosive nucleosynthesis. SN1987A, the nearest known supernova in several hundred years, has provided the best opportunity for such observations. Based on the idea that explosive nucleosynthesis is responsible for production of most of the heavy elements, and the similarity in the late time decay of supernova light curves with the mean life of cobalt-56 Clayton, Colgate and Fishman (1969) initially suggested the possibility of detection of Type I supernovae at distances out to the Virgo Cluster. Detection of lines associated with radioactive cobalt would provide the first direct test for the models for explosive nucleosynthesis which have been developed extensively during the past 15 years.

In addition to the confirmation of the theory of explosive nucleosynthesis, high quality observations would enable the time history of the cobalt lines to be determined with good precision. Such data could be used to indicate the degree of spherical symmetry in the supernova cloud and the degree of mixing of the core radioactive material with the outer envelope via instabilities (e. g. Rayleigh-Taylor) or due to the production of jets in the initial explosion. Line profiles can also be used to provide information on the velocity distribution of the exposed radioactive material. Observations of other radionuclides with rather short half-lives (see Table 1) will also provide information on the relative production of isotopes, and thereby provide important information on the details of the environment, such as the peak temperatures, neutron excess, and mixing, in the nucleosynthetic region.

Table 1. Supernova Gamma Ray Lines

Isotope	Mean Life	Gamma Rays
Nickel-56	8.8 days	0.163 (99%), 0.276 (31%) 0.472 (35%), 0.748 (48%) 0.812 (85%), 1.56 (14%)
Cobalt-56	112 days	0.847 (100%), 1.040 (15%) 1.238 (67%), 1.76 (15%) 2.02 (11%), 2.60 (17%) 3.26 (13%)
Cobalt-57	270 days	0.014 (9%), 0.122 (87%) 0.136 (11%)
Titanium-44	68 years	0.068 (90%), 0.078 (98%) 1.158 (100%)
Cobalt-60	7.6 years	1.173 (100%), 1.332 (100%)
Aluminum-26	1.04 million yrs	1.12 (40%), 1.809 (100%)
Sodium-22	3.75 years	1.275 (100%)

Table 1 lists the prominent lines expected from observations of supernovae. The details of the production of these isotopes have been discussed extensively by others and will not be repeated here.

The sensitivities of the OSSE and COMPTEL instruments on GRO to line gamma radiation are indicated in Figure 1. These sensitivities are calculated for a total observation time of 10^6 seconds. This is typical of the observation periods planned for the GRO instruments. The OSSE instrument covers the 0.05-10 MeV range which includes essentially all of the gamma ray lines of interest. The COMPTEL instrument covers the 1-30 MeV region with some extension into the region between 0.5 and 1.0 MeV. The combined capabilities of OSSE and COMPTEL provide sensitivities of $2\text{-}3 \times 10^{-5}$ photons/cm^2-s below 1 MeV and improving to better than 10^{-5} photons/cm^2-s above 3 MeV. This will enable Type I supernovae to be detectable to 10-20 Mpc, bringing the several supernovae which will occur in the Virgo Cluster during the GRO mission into the detectable range. It will be extremely important to have rather complete coverage for ground-based supernova patrols in order to maximize the return from the GRO observations.

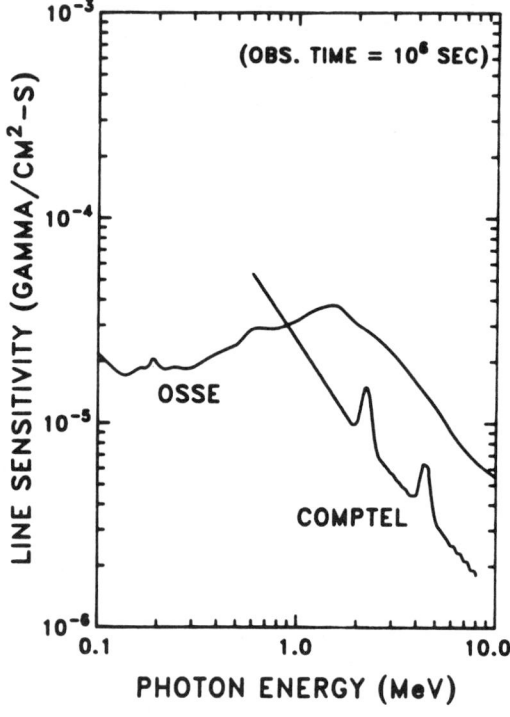

Figure 1. Sensitivity (3σ) of the OSSE and COMPTEL experiments to line gamma rays for an observing period of 10^6 seconds.

When GRO is launched in 1990, the ^{56}Co radiation from SN1987A will have decayed to somewhat below the sensitivity of the OSSE and COMPTEL instruments. However several other objectives will remain a very high priority. First, several longer-lived radioactivities which will provide specific information on the environment in SN1987A will be of interest; ^{57}Co, ^{24}Na, and ^{60}Co. The ^{57}Co line at 122 keV has been estimated by Woosley (this conference) to have an intensity approaching 10^{-4} photon/cm^2-s in this time frame and therefore should be detectable by OSSE. In addition, radiation associated with a central pulsar may become detectable during the GRO mission. All of the GRO instruments have excellent sensitivities for pulsed emission with a time resolution of 10^{-4} seconds and therefore a search throughout the 0.02 MeV to 30 GeV region will be of high priority.

Recent supernovae which have occurred during the past several hundred years in the Galaxy, but which were undetectable because of optical obscuration in the galactic plane, may be detectable in the lines of ^{44}Ti which has a 68-year meanlife. Figure 2 indicates the distance to which GRO would be sensitive to such supernovae as a function of the age. Note that supernovae which occurred at the distance of the galactic center within the last 100 years may produce an observable signal.

If the GRO mission can achieve a long lifetime, the probability for a galactic supernova or nearby novae during the mission could become significant. Extensive study of such events would, of course, become a major program during the mission.

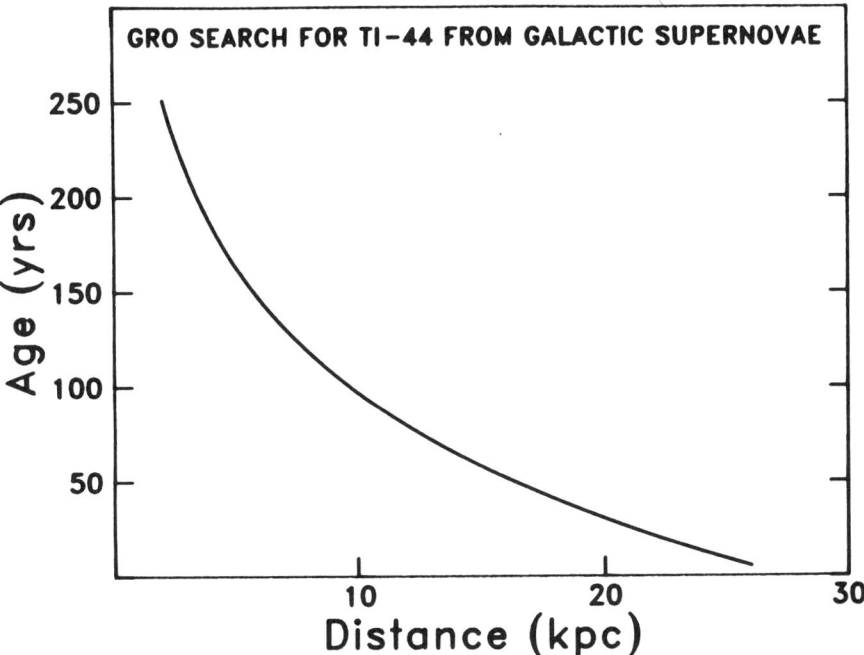

Figure 2. Distance vs. Age relationship for detectability of ^{44}Ti ($\tau \sim 68$ yr) line at 1.158 MeV from recent unobserved galactic supernovae.

NOVAE

Gamma ray observations of novae provide the opportunity to test the model of explosive hydrogen burning on the surface of an accreting white dwarf. Line gamma radiation predictions are listed in Table 2.

Several intermediate mass positron emitters are expected to be produced during novae. Leising and Clayton (1987) and Leising (1988) have calculated the intensities of these lines. The visibility of the short-lived lines depends critically on the convective mixing of the radioactivity from the hydrogen burning region to the surface of the expanding envelope. The positron annihilation lines associated with 159-min ^{18}F and 862-sec ^{13}N are the most likely candidates for observation, and may be detectable for novae which occur at distances up to about 10 kpc.

Table 2. Nova Gamma Ray Lines

Isotope	Mean Life	Gamma Rays
Sodium-22	3.75 years	1.275 (100%)
Nitrogen-13	862 sec	0.511
Oxygen-14	102 sec	0.511, 2.31 (99%)
Oxygen-15	176 sec	0.511
Fluorine-17	93 sec	0.511
Fluorine-18	159 minutes	0.511
Beryllium-7	53 days	0.478 (10.3%)

Detection of these short half-life radioactivities will be most probable for a nova which occurs in the approximately 45 degree field-of-view of COMPTEL (or, less likely, the smaller-field-of view of the OSSE instrument) or for a nearby nova which is detected by the BATSE instrument. In a special mode implemented for solar observations, it may be possible to program the OSSE instrument to be oriented toward the direction of a likely nova (e. g. toward the galactic center region) in response to a transient X-ray flash detected by BATSE if such bursts are found to be a characteristic of novae.

Longer-lived activities which have been predicted in association with novae are gamma radiation associated with ^{7}Li (Clayton 1981) and ^{22}Na (see e. g. Clayton and Hoyle, 1974). These observations can be made in response to a detection of optical novae. The intensities are likely to be rather weak and most probably will be detectable only for novae which occur within 1-2 kpc.

GAMMA RAY OBSERVATORY

The general configuration of the GRO is shown in Figure 3. Three of the instruments (OSSE, COMPTEL and EGRET) are located on the +Z side of the spacecraft. The fourth, BATSE, consists of eight modules designed to provide a full sky monitor for gamma-ray burst and transient sources. GRO is a three axis stabilized spacecraft with solar panels which can rotate $\pm$ 90 degrees about the Y-axis. This enables GRO to point to any position on the sky at any time, including the Sun. This capability will ensure that GRO can undertake well-timed observations of transient sources such as novae and supernovae in response to discoveries using ground-based optical techniques. GRO will be launched into a 400-450 km orbit with an inclination of 28.5 degrees. The baseline mission is for two years, although plans are being implemented for an extended mission. Current plans are likely to result in the space shuttle placing GRO at or near its operational orbit which will result in adequate propellent available for orbit maintenance and a resulting lifetime of 5-10 years.

Table 3 summarizes the characteristics of the GRO instruments.

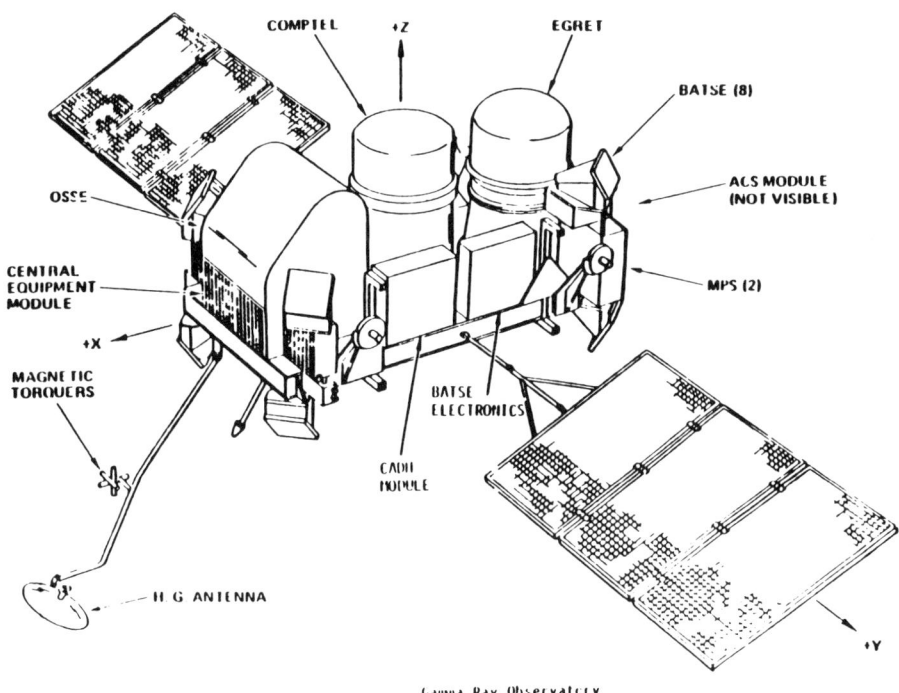

Gamma Ray Observatory

Figure 3. Illustration of the Gamma Ray Observatory and associated instruments. Three instruments (OSSE, COMPTEL, and EGRET) are mounted on the +Z side of the spacecraft. The eight BATSE modules are located on the corners of the S/C to provide full sky coverage. Other systems include the High Gain Antenna for communications with the TDRSS, Modular Power Systems (MPS), Command and Data Handling (CADH) system and the Attitude Control System (ACS). The solar panels can rotate ±90 degrees about the Y-axis. The Sun is constrained to be located in the +X hemisphere and is further constrained to be not closer than 62 degrees to the Y axis.

1. BURST AND TRANSIENT SOURCE EXPERIMENT (BATSE)

The Burst and Transient Source Experiment (BATSE) consists of eight detector modules positioned on the GRO spacecraft to provide full coverage of the sky which is unocculted by the earth. For a description of the BATSE instrument and its capabilities see Fishman et al. (1985). Each module contains two un-collimated scintillation detectors and associated electronics. A large area detector (LAD), utilizing a 20-inch diameter by 0.5-inch thick NaI crystal provides high sensitivity and directional capability in the 50-1000 keV region. The position determination capability for transient sources, with accuracies of about 1 degree, is achieved by comparing detector rates with projected areas for assumed source directions. The second unit in each module is a 5.0-in dia. by 3.0-in thick NaI crystal which provides better energy resolution and covers the energy range from 20 keV to 20 MeV. The front face of the LAD is actively shielded by a plastic scintillator charged-particle detector and the rear of the unit is passively shielded. Primary objectives of the BATSE are to investigate the spatial, temporal and spectral characteristics of cosmic gamma-ray bursts with very high sensitivity.

Table 3. GRO INSTRUMENT SUMMARY

INSTRUMENT	BATSE	OSSE	COMPTEL	EGRET
ENERGY RANGE (MeV)	0.05-1.0 (LAD) 0.02-20(spect)	0.05-10 10-200 (solar)	1-30	30-20000
ENERGY RESOLUTION	30% (LAD) 8.0% (spect)	8.0% (0.66 MeV) 3.2% (6.1 MeV)	5-8%	15%
TIME RESOLUTION				
BROAD-BAND	10 usec	125 usec	125 usec	100 usec
SPECTROSCOPY	1 sec	4 sec		
EFFECTIVE AREA (cm**2)	1800 (LAD) 125 (spect) (per module)	1950 (0.5 MeV)	30	1000
FIELD-OF-VIEW (deg)	FULL SKY	3.8 x 11.4	45 FWHM	45 x 51

Using earth occultation techniques, slower transients including observations associated with novae and supernovae under discussion here, can be obtained with BATSE, as has been demonstrated by the detection of line gamma radiation from SN1987A by the wide FOV spectrometer on SMM (Matz et al. 1988).

BATSE will have extensive capabilities for transient observations due to its large sensitive areas and versatile data system. During normal operations, BATSE will accumulate 16-channel spectra with 2-second resolution from all detector systems. Full 256-channel resolution spectra are acquired from each detector every one minute. Upon detection of a transient event, as defined by a programmable burst trigger algorithm, up to 4Mb of dedicated data can be stored for subsequent transmission to the ground. This can include, for example, 192 energy spectra from the spectroscopy detectors with time resolutions based on a time-to-spill algorithm, of as short as 64 msec.

In addition, BATSE's burst mode will provide full sky coverage for rapid transient phenomena associated with an X-ray flash which might be associated with the first observations of either the nova or supernova mechanism. Also, BATSE can use its excellent time resolution to provide sensitive measurements of pulsars which are also expected to be associated with supernovae.

2. ORIENTED SCINTILLATION SPECTROMETER EXPERIMENT (OSSE)

The Oriented Scintillation Spectrometer Experiment is designed to provide high sensitivity in the nuclear line region of the spectrum. OSSE includes four identical detector systems. Each of the detectors is mounted in a single axis pointing system which provides a rotational freedom of 192 degrees about the S/C Y-axis. A description of the OSSE instrument is provided by Kurfess et al. (1983).

The primary element of each detector system is the NaI portion of a 13-in. diameter NaI-CsI phoswich. A NaI annular shield assembly, together with the CsI portion of the phoswich, forms an active anticoincidence shield for background reduction. A tungsten alloy passive collimator provides a 3.8 degree by 11.4 degree field-of-view for the study of localized sources.

Spectra in the 0.1-10 MeV energy range are processed by two 256-channel PHA's. The energy resolution performance will be 8.0% FWHM at 0.661 MeV and 3.2% at 6.13 MeV. Individual spectra are accumulated every four seconds. Above 10 MeV, 16-channel gamma-ray and neutron energy-loss spectra are accumulated, also with typical 4-second time resolutions. An event-by-event data mode for selected energy ranges (e.g. the 4.4 MeV and/or 6.1 MeV prompt lines for ^{12}C or ^{16}O radiation) with a time resolution of 0.125 milliseconds is available for high time resolution observations. This capability has been included primarily for the purpose of observing fast pulsars.

The operation of the OSSE instrument is controlled by redundant on-board microprocessors. Varied data formats are employed to optimize the spectral acquisitions for the objectives of interest. In response to a BATSE burst signal indicating detection of a solar flare, the four detectors can automatically re-orient to the Sun if the Sun is located near the X-Y plane, (see Figure 3), which due to the constraints on the solar panels, will often be the case. Although not highly probable, this technique could be considered for use in responding to novae which might occur in a selected direction (e. g. toward the galactic center) in response to an X-ray burst detected by BATSE which met the pre-selected directional criteria.

3. IMAGING COMPTON TELESCOPE (COMPTEL)

COMPTEL is designed to observe gamma rays in the energy range 1-30 MeV in a broad (approx. 1 sr) field-of-view centered on the spacecraft Z-axis. The instrument consists of two detector arrays; and upper array, D1, comprised of seven 28-cm diameter NE-213A liquid scintillation detectors, and a lower array, D2, which consists of fourteen 28-cm diameter NaI scintillation detectors. The basic properties of the instrument are described in Schoenfelder et al. (1981). Operationally, an incoming gamma ray Compton scatters in detector array D1, and the scattered photon is detected in array D2. In those events where the scattered gamma ray is totally absorbed in D2, the energy of the incident gamma ray is the sum of the energy losses in the upper and lower detectors, and the arrival direction is confined to a cone whose axis is determined by the locations of the interactions in D1 and D2 and the Compton scattering angle derived from the scattering kinematics.

The large field of view of the COMPTEL, in conjunction with the high sensitivity (see Figure 1), will provide a high probability for serendipitous detections of transient phenomena such as novae. The line sensitivity above 1 MeV will result in a detection capability for many of the emission features associated with novae and supernovae. A more complete description of the COMPTEL experiment is given in Schoenfelder et al. (1981).

4. ENERGETIC GAMMA RAY EXPERIMENT TELESCOPE (EGRET)

EGRET covers the high-energy portion of the spectrum from 20 MeV to 30 GeV. The instrument consists of two spark chamber modules with tantalum converters which convert incoming gamma rays to positron-electron pairs. The particle trajectories are determined by using a wire readout of the tracks in the spark chambers. The arrival directions of incident gamma rays can be determined to several degrees at 100 MeV and to less than one degree at 1 GeV. EGRET's field-of-view is centered on the +Z axis and has a full-width-at-half-maximum of about 45 x 51 degrees. Beneath the spark chamber is a 30-inch x 30-inch x 8-inch thick NaI crystal assembly which serves as a total absorption shower calorimeter (TASC) and provides good energy resolution for gamma rays extending into the several GeV region. A detailed description of EGRET is given in Fichtel et al. (1983).

Although EGRET does not operate in the spectroscopy region, which is the primary thrust of this conference, observations of high energy gamma-ray emission associated with the pulsating neutron stars which are known to be associated with young supernovae will provide an exciting capability. Harding et al. (1988) have discussed expected gamma-ray emissions from SN1987A. The other important fact to remember is the synergism provided by the four instruments on GRO wherein sources will be observed over six orders of magnitude in the wavelength band thereby providing comprehensive high-energy coverage of all energetic phenomena.

5. GRO GUEST INVESTIGATOR PROGRAM

A vigorous Guest Investigator program is being planned for GRO, the details of which are currently being established. The nature of the Guest Investigator opportunities will vary with each instrument because of the different observational techniques employed. COMPTEL and EGRET both have very large fields-of-view with a significant data processing overhead before scientific analysis can be undertaken.

In general, the large data bases will be available at the

Co-I institutions, and Guest Investigators will be encouraged to make use of the expertise at these institutions and the associated data analysis capabilities. Significant guest investigator opportunities will become more available after the first year of the mission when the instrument response and data analysis techniques are well understood.

REFERENCES

1. Clayton, D. D., 1981, Astrophys. J. (Letters) 294, L97.
2. Clayton, D. D., and F. Hoyle, 1974, Astrophys. J. (Letters) 187, L101.
3. Fichtel, C. E., Bertsch, D. L., Hartman, R. C., Kniffen, D. A., Thompson, D. J., Hofstadter, R., Hughes, E. B., Campbell-Finman, L. E., Pinkau, K., Mayer-Hasselwander, H., Kanbach, G., Rothermel, H., Sommer, M., Favale, A. J., and Schneid, E. J.: 1983: 18th Int'l Cosmic Ray Conference, Vol 8, p.19
4. Fishman, G. J., Meegan, C. A., Parnell, T. A., Wilson, R. B., Paciesas, W., Matteson, J. L., Cline, T., and Teegarden, B.: 1985 Proc. 19th Int'l Cosmic Ray Conf, Vol. 3, p. 343
5. Harding, A. K., Gaisser, T. K., and Stanev, T., these proceedings (1988).
6. Kurfess, J. D., Johnson, W. N., Kinzer, R. L., Share, G. H., Strickman, M. S., Ulmer, M. P., Clayton, D. D., and Dyer, C. S.: 1983 Adv. Space Research, Vol. 3, No. 4, p. 109
7. Leising, M. D., these proceedings (1988).
8. Leising, M. D., and Clayton, D. D., 1987, Astrophys. J. 323, 159.
9. Matz, S. M., Share, G. H., Leising, M. D., Kinzer, R. L., Chupp, E. L., and Reppin, C., these proceedings (1988).
10. Sandie, W., Nakano, G., Chase, L., Fishman, G., Meegan, C., Wilson, R., and Paciesas, W., these proceedings (1988).
11. Schoenfelder, V., Diehl, R., Kanbach, G., Lichti, G. G., Deerenberg, A., Swanenburg, B. N., Visser, A., Lockwood, J., Simpson, G. A., Webber, W. R., Bennett, K., Bibbs, G., Taylor, B. G., and Wills, R. D.: 1986 Proc. 17th Int'l Cosmic Ray Conf., Vol. T1, p. 28
12. Winkler, C., Schoenfelder, V., Diehl, R., Lichti, G., Steinle, H., Swanenburg, B. N., Aarts, H., Deerenberg, A., Hermsen, W., Lockwood, J., Ryan, J., Simpson, G. Webber, W. R., Bennett, K., Dordrecht, A., and Taylor, B. G.: 1985
13. Woosley, S., these proceedings (1988).

THE GAMMA-RAY BURST CAPABILITIES OF BATSE
AND THE GAMMA RAY OBSERVATORY

G. J. Fishman
Space Science Laboratory, NASA Marshall Space Flight Center
Huntsville, AL 35812 USA

for the BATSE Collaboration and the GRO Investigators

Abstract

The Gamma Ray Observatory (GRO), scheduled for launch in 1990, will provide new and enhanced capabilities for the study of gamma-ray bursts. These include higher sensitivity, increased time resolution, broader energy coverage, rapid burst data dissemination and burst location by a single spacecraft. All four of the GRO instruments have burst capabilities, however the Burst and Transient Source Experiment (BATSE) is designed primarily for the study of gamma-ray bursts. The capabilities of BATSE and the GRO for gamma-ray burst studies are described.

INTRODUCTION

More than 15 years after their discovery, gamma-ray bursts remain one of the least understood phenomena in astrophysics. Even the distance and luminosity of the burst sources are uncertain by many orders of magnitude. One can find recently published models of bursts which place them in the solar neighborhood[1] (~50 pc) or at cosmological distances.[2]

In recent years it has become evident that gamma-ray bursts are most likely caused by a variety of emission mechanisms and/or objects. Separate classes of bursts are now known to have different spectral, temporal, and repeating characteristics. For example, there is one type of burst which is characterized by a dominant, single peak several seconds wide with no apparent rapid fluctuations. This type of burst is extremely hard initially and then softens considerably.[3,4] Other classes of bursts include the soft repeaters,[5] the March 5, 1979, burst, and the typical KONUS impulsive bursts[6] with numerous randomly spaced, narrow spikes scattered throughout their duration. The only characteristics that these separate classes of bursts may have in common is their high-energy emission and transient nature. The situation is analogous to that in optical astronomy just after the first telescopes were invented. All non-stellar objects were classified as nebulae; it wasn't until many years later that the extreme diversity of these objects became known.

Precise locations of gamma-ray bursts have not produced x-ray, radio, or optical counterparts, with the important exception of the March 5, 1979, burst (N49 in the LMC). However, these precise locations have led to the discovery of archival optical transients[7] whose relation to the gamma-ray bursts remains problematical. The statistics of gamma-ray bursts (Log N-Log S) and their implications are similarly controversial and inconclusive.[8]

The present lack of our understanding of gamma-ray burst sources is a direct result of their random, transient nature and their lack of identifiable counterparts. Future experiments cannot overcome these inherent observational limitations but will provide a much larger data set with improved time and energy resolution and burst locations.

The next generation of gamma-ray burst experiments include five experiments on the Soviet GRANAT spacecraft to be launched within the next year, the Gamma Ray Observatory (GRO) to be launched in 1990, gamma-ray burst monitors on the Soviet Phobos Mars spacecraft scheduled for launch this year, and the French/US Ulysses spacecraft to be launched in 1990. This paper presents an overview of the gamma-ray burst capabilities of the GRO.

THE GAMMA RAY OBSERVATORY (GRO)

The GRO is intended to be NASA's principal spacecraft for gamma-ray astronomy through the 1990's. The four experiments (Figure 1) have capabilities that are at least an order of magnitude

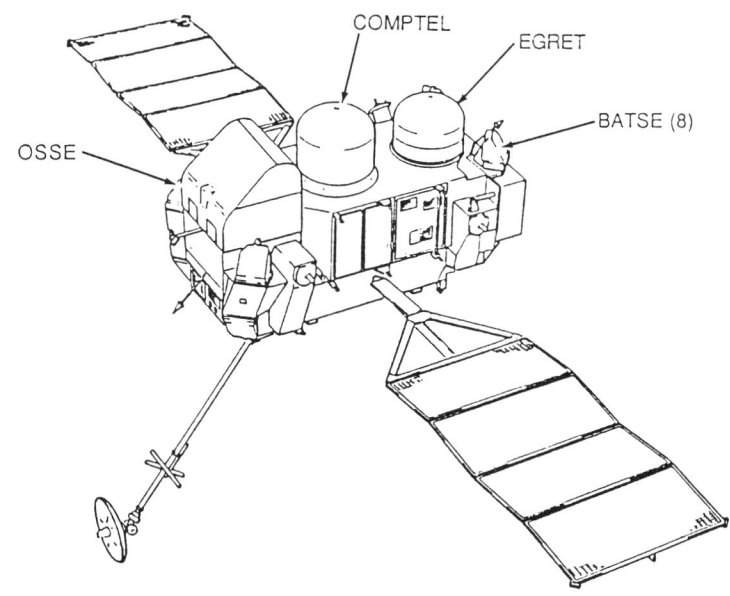

GAMMA RAY OBSERVATORY

FIGURE 1

greater than those of previous instruments. GRO is scheduled for launch in 1990, having been delayed considerably following the Challenger accident. The current launch date is, of course, dependent upon the present schedule for the resumption of Shuttle flight operations. An on-board propulsion system with refueling capability has been added to the GRO spacecraft which will allow flight extensions greatly beyond the planned 2-year mission. The GRO will be placed in a low inclination orbit at an altitude between 350 km and 450 km.

All four of the GRO instruments have gamma-ray burst capabilities but only the Burst and Transient Source Experiment (BATSE) is designed primarily for the study of bursts. BATSE will provide a burst trigger signal to the other three instruments, as described below. Figure 1 shows the configuration of the GRO spacecraft and the location of the experiments. Table 1 lists some key parameters of the four experiments.

Table 1. GRO Experiment Parameters

EGRET

Energy Range	20 MeV to 30 GeV
Energy Resolution	15% (100 MeV to 20 GeV)
Maximum Effective Area	200 cm^2 (>200 MeV)
Position Resolution	5 arc min (strong source)
Maximum Effective Geom. Factor	1000 cm^2 sr

COMPTEL

Energy Range	1.0 to 30 MeV
Energy Resolution	8% (@ 1 MeV)
Maximum Effective Area	50 cm^2
Position Resolution	7.5 arc min (strong source)
Maximum Effective Geom. Factor	30 cm^2 sr

OSSE

Energy Range	0.1 to 10 MeV (or greater)
Energy Resolution	8% (@ 0.66 MeV)
Effective Area	2310 cm^2 (@ 0.5 MeV)
Position Resolution	10 arc min (strong source)

BATSE

Energy Range - Primary Detector	30 keV to 1 MeV
Spectroscopy Detector	20 keV to 40 MeV
Energy Resolution - Primary Detector	30% (@ 0.88 keV)
Spectroscopy Detector	8% (@ 0.66 MeV)
Maximum Effective Area - Primary Detector	1800 cm^2 (ea. det.)
Spectroscopy Detector	120 cm^2 (ea. det.)
Position Resolution	1° (strong burst)

EGRET

The Energetic Gamma Ray Experiment Telescope (EGRET), shown in Figure 2, will perform a comprehensive, systematic survey of the gamma-ray sky above 20 MeV. The instrument includes a large calorimeter module to obtain accurate gamma-ray energy measurements lacking in previous instruments of this type. The incoming gamma-ray is converted to an electron-positron pair in a large spark chamber module; the track of the pair is then determined by the spark chamber. Its energy is measured by the scintillation crystal calorimeter at the lower end of the instrument. A large number of plastic scintillators provide triggering, anticoincidence, and time-of-flight signals for the detector system.

The large NaI(Tℓ) scintillation crystal (20 cm thick x 5000 cm^2) will accumulate four gamma-ray burst spectra when a burst trigger signal is received from BATSE. These spectra will cover the energy range from 650 keV to 150 MeV in 256 channels. The energy channels are derived from seven different dispersions in order to cover a large energy range. Each of the four spectra are accumulated in sequence. The integration times are pre-set by command from 0.125 to 16.0 sec.

The typical energy resolution of the crystal scintillator is 10% FWHM at 2.2 MeV and 6% at 6.3 MeV. Because of its location within the experiment and on the spacecraft, the observability of each burst will depend greatly on its location with respect to the spacecraft. Roughly one-half of the total solid angle will be relatively unobstructed.

The large plastic anticoincidence dome of EGRET can also be used for burst observations. This dome is 165 cm in diameter and responds to photons with energy depositions in the plastic greater than approximately 400 keV. The time resolution of the dome counter is 0.25 sec, which will be monitored continuously. A special microsecond burst circuit has been included which will record the arrival of a burst of gamma rays within a 2 µs period.

COMPTEL

The Compton Telescope (COMPTEL), Figure 3, is a low-background imaging telescope, optimized for the energy range 1 to 30 MeV. Two arrays of scintillation detectors are used to characterize the incoming gamma rays. The gamma ray is first Compton-scattered by one of the seven upper liquid scintillation detectors; the scattered photon is then detected by one of the fourteen lower NaI(Tℓ) crystal scintillation detectors. Both sets of detectors can localize the gamma-ray interaction to within a few centimeters. The energy losses and the interaction locations in the detectors are used to constrain the arrival direction to a circle on the celestial sphere. The intersections of numerous circles within the field-of-view are used to build a flux map or "image" of the region. A detailed description of COMPTEL is given in another paper in these proceedings.

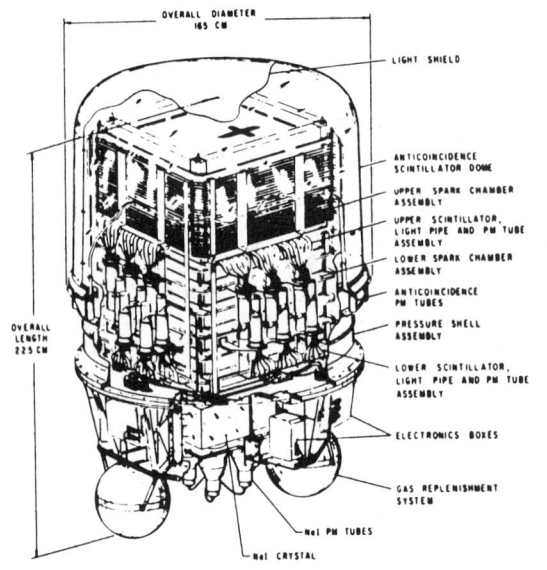

FIGURE 2

EGRET

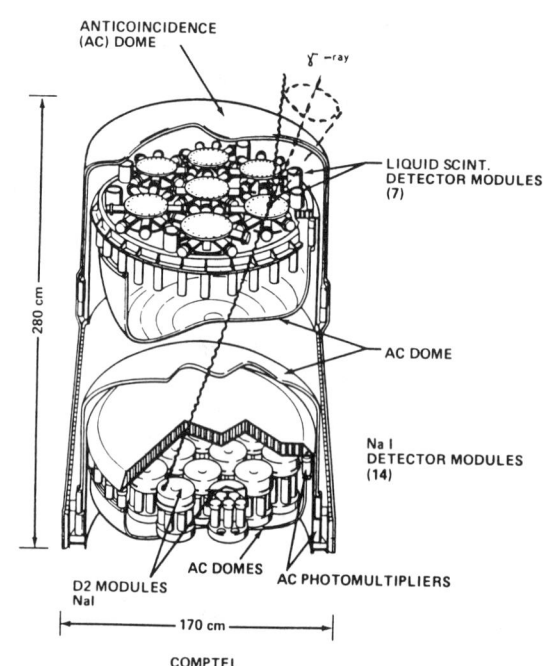

FIGURE 3

COMPTEL

Data from two of the crystal scintillation detectors will be used for gamma-ray burst studies. These detectors are each 28 cm in diameter and 7.5 cm thick. They have an energy threshold of 100 keV and an energy resolution of about 8% at 1 MeV. When a burst trigger signal is received from BATSE, six spectra of 128 channels will be accumulated from the detectors with pre-set integration times from 0.1 to 25 sec. After these six spectra are stored, up to 256 additional spectra can be obtained with integration times from 2 to 512 sec each.

Several gamma-ray bursts per year will be within the field-of-view of COMPTEL which will provide a sensitive measure of burst spectra in the energy region from 1 to 30 MeV. It should be noted that because of its excellent background rejection capability, detection of only a few photons by COMPTEL will allow the determination of a burst location to within approximately 1°.

OSSE

The Oriented Scintillation Spectrometer Experiment (OSSE), Figures 4a and 4b, consists of four large, collimated scintillation detectors. Both active and passive collimators are used to reduce background and define the field-of-view. The passive collimators have a field-of-view of 3.5° x 11°. The thick NaI(Tℓ) detectors are optimized to study the nuclear gamma-ray region, 0.1 to 10 MeV, although observations can be made at lower and considerably higher energies. Charged particle active shielding completely surrounds each central detector. Capabilities of OSSE for discrete source observations are described elsewhere in these proceedings.

The sixteen sodium iodide shield scintillators have a projected area of 1840 cm^2 each and an average thickness of 8.5 cm. The energy resolution is typically 11% at 662 keV. Housekeeping provides continuous rate information on a 0.5 sec time scale from the shield annuli.

Upon receipt of a burst trigger signal from BATSE, or from an internal burst trigger, a spectrum from the large crystal shields is accumulated over the energy range from 0.1 to 8 MeV. This spectrum consists of 256 channels from one of four shield annuli. The accumulation time is pre-set from 2 to 64 sec. In addition, the counting rate of events >100 keV will be telemetered for each of the sixteen shield segments with a minimum time resolution of 4 ms. Nominally, 4 sec prior to the burst trigger and 12 sec after the trigger will be stored into a 4096 channel memory at the 4 ms resolution. These parameters can be changed by pre-set commands.

High sensitivity observations of gamma-ray bursts can be made for the few gamma-ray bursts which happen to be in the field of the central phoswich detectors. Pointed observations at expected concentrated sources of gamma-ray bursts such as the LMC and M31 are planned.

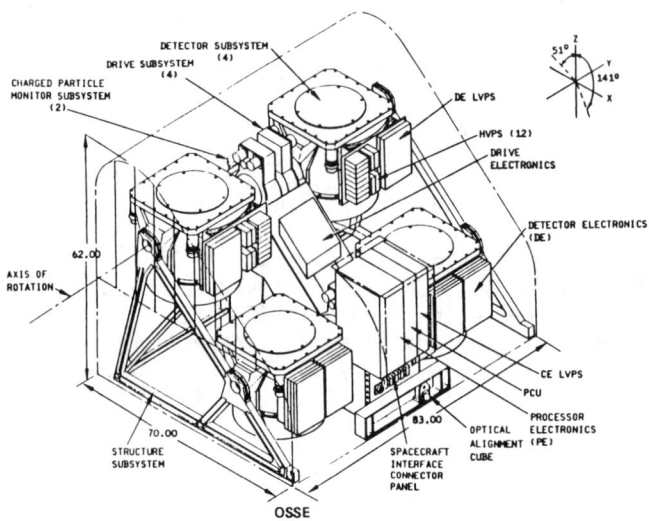

FIGURE 4a

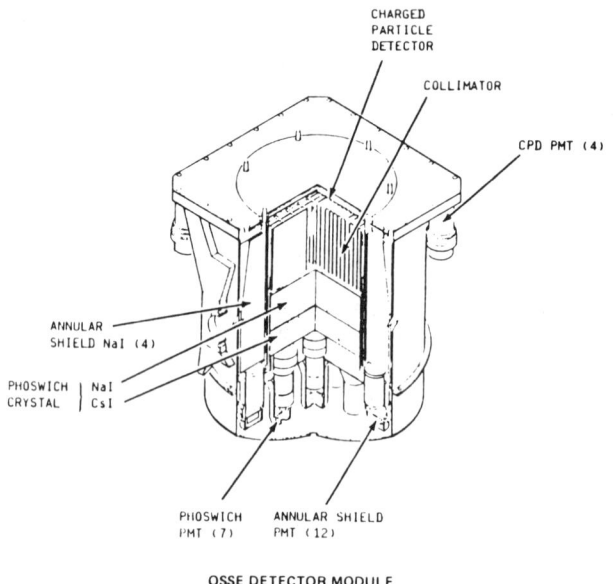

FIGURE 4b

BATSE

The Burst and Transient Source Experiment (BATSE) is an all-sky gamma-ray monitoring experiment designed primarily for the detection and detailed study of gamma-ray bursts and other transient high-energy sources in the energy range 20 keV to 10 MeV. Eight uncollimated detector modules are positioned around the spacecraft to provide an unobstructed view of sky, as shown in Figure 1.

Each detector module, shown in Figure 5, contains a large area detector, optimized for high sensitivity and directional response, and a spectroscopy detector, optimized for greater energy coverage and moderate energy resolution. Bursts are studied over the energy range from 20 keV to 40 MeV with time resolution approaching 2 µs.

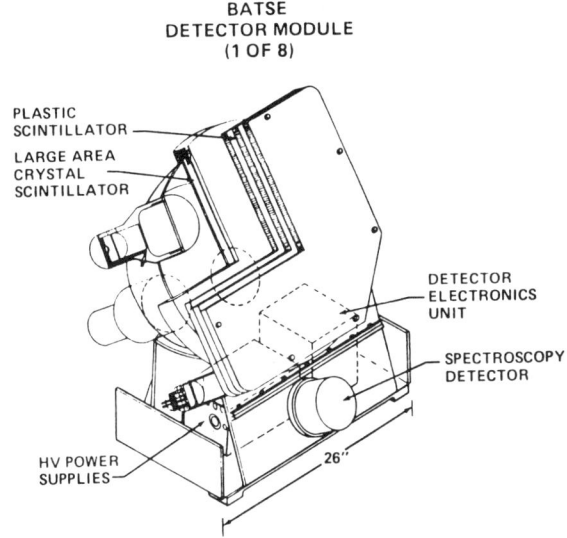

FIGURE 5

It is expected that BATSE will detect 200 to 400 bursts per year. This large number will reveal the galactic distribution of burst sources and may indicate the existence of an extragalactic burst population. Individual bursts can be studied with unprecedented sensitivity to determine temporal structure and spectral evolution. Burst locations within a few degrees can be obtained sooner than previously possible. This may allow longer-lived optical emission to be detected.

The eight BATSE detector modules have identical configurations as shown in Figure 5. The main detector is a disk of sodium iodide scintillation crystal, 20 inches in diameter and 1/2-inch thick. A light collector housing on each detector couples the scintillation light into three 5-inch-diameter photomultiplier tubes. An anticoincidence shield on the front side is used to reduce the background due to charged particles, and a thin lead and tin shield inside the light collector housing reduces the amount of background and scattered radiation entering the back side. The spectroscopy detector is an unshielded 5-inch-diameter, 3-inch-thick sodium iodide scintillation crystal. Signals from the detectors are processed in the detector electronics unit and routed to a central electronics unit for digital processing.

The central electronics unit is a microprocessor-based data system which processes and stores large amounts of data from all sixteen detectors in various formats, providing various data types with different energy and time resolution. It has the capability of detecting a gamma-ray burst onboard and then storing various types of data from the burst for later transmission through the GRO telemetry system. It takes one orbit (90 min.) to read out the data from a burst. At other times, background and calibration data are recorded, along with data used to study long-lived transients and pulsing sources. Of particular importance for obtaining calibration data are the observations of solar flares and the emission from hard x-ray pulsars such as that in the Crab Nebula. Other long-lived hard x-ray and gamma-ray sources can be observed as they are occulted by the Earth each orbit. The sensitivity of BATSE to these steady or long-lived sources has been calculated by Paciesas et al.[9].

The BATSE data system contains an onboard burst trigger system to detect bursts and accumulate rapidly 4 Mb of data during the burst. The basic trigger requirement is that at least two of the eight detectors observe the burst at a pre-specified level above background. Data from the large-area detectors will be tested for bursts on three time scales: 64 ms, 256 ms, and 1024 ms, in the energy region from 50 to 200 keV. Other trigger criteria can be included, such as limits on the rates in the charged particle detectors. These other criteria are intended to eliminate false burst triggers. An onboard determination that a triggering event is most likely a solar flare is derived by testing the ratios of selected detector count rates to see if they are consistent with that expected from a burst from the solar direction. The burst trigger signal and the solar flare trigger signal are routed to the other three GRO experiments. The OSSE investigators plan to use the solar flare trigger signal to re-orient their detectors for observation of solar gamma-ray emission soon after an intense flare. COMPTEL will use the solar flare trigger to observe energetic neutrons emitted by a solar flare.

The BATSE experiment will become an important component in the interplanetary burst timing network in the 1990's. In the timeframe of the GRO, other instruments of the network will include detectors on Ulysses and on Soviet-French interplanetary space probes. All detectors are expected to have absolute timing accuracies of 16 ms or better, providing arc-sec locations for strong bursts with fine time structure. It is expected that burst locations can be derived on a much shorter time scale than has previously been possible so that short-lived accompanying optical, x-ray, or radio emission may be detected.

The sensitivity of the BATSE detector array to typical gamma-ray bursts will be $\sim 10^{-8}\sqrt{t}$ erg cm^{-2}, where t is the burst duration in seconds. However, this sensitivity is dependent upon the burst spectrum and time profile; significantly increased sensitivity is expected for bursts with soft spectra or rapidly varying emission. The large area of the detectors will produce counting rates approaching 10^6 s^{-1} during the peaks of perhaps ~20 bursts per year. These high rates will allow statistically significant intensity variations on a time scale of tens of microseconds or spectral variations on a time scale of milliseconds to be observed. These measurements will put severe constraints on emission mechanisms and models of burst sources and on the source geometry and beaming characteristics. The time-tagged data will allow periodic emission to be detected from strong bursts on time scales as short as tens of microseconds. Such periodic emission may be expected from burst models involving oscillations of a neutron star.

The use of multiple, anisotropic detectors on a single spacecraft to determine burst locations has been successfully accomplished by the KONUS experiment on the Venera spacecraft.[8] In principle, the BATSE/GRO experiment could achieve location accuracies better than 1° for bursts with intensities greater than 10^{-6} erg cm^{-2}, if the measurements were limited only by counting statistics. In reality, it is expected that systematic uncertainties will dominate the statistical uncertainties in deriving locations for the stronger bursts. The primary sources of systematic error are: (1) unequal and/or uncertain response of different detector modules, (2) scattering of burst photons by spacecraft components and the Earth's atmosphere into detectors, and (3) uncertainties in detector alignment and aspect.

Additional details of the observational capabilities of the BATSE/GRO experiment have been described in previous publications.[9-13]

CONCLUSION

BATSE and the GRO will provide unprecedented sensitivity to gamma-ray bursts over an extended energy range, from 20 keV to above 40 MeV. One important observational capability, high resolution spectral measurements, will be lacking and must await a future space flight opportunity. The Nuclear Astrophysics Explorer, discussed at this workshop, may provide this opportunity. BATSE/GRO will also become an important element of an international burst network of the 1990's. This network will include burst experiments from the Soviets, French, Danish, and Japanese.

REFERENCES

1. F. Melia, Ap. J. Lett. 324, L21 (1988).
2. B. Paczynski, Ap. J. Lett. 308, L32 (1986).
3. A. V. Kuznetsov et al., Sov. Astron. Lett. 12, 315 (1986).

4. G. J. Fishman, W. S. Paciesas, C. A. Meegan, and R. B. Wilson, Adv. Space Res. 6, 23 (1986).
5. J. Laros et al., Ap. J. Lett. 320, L111 (1987).
6. E. P. Mazets et al., Ap. and Space Sci. 80, 3 (1981).
7. B. Schaefer, Nature 294, 722 (1981).
8. R. Epstein and K. Hurley, preprint (1988).
9. W. S. Paciesas, R. B. Wilson, G. J. Fishman, and C. Meegan, 19th ICRC, La Jolla Paper OG9.2-16 (1985).
10. G. J. Fishman, C. A. Meegan, T. A. Parnell, and R. B. Wilson, AIP Conf. Proc. No. 77, 443 (1982).
11. G. J. Fishman, C. A. Meegan, T. A. Parnell, R. B. Wilson, and W. S. Paciesas, AIP Conf. Proc. No. 115, 651 (1984).
12. G. J. Fishman, C. A. Meegan, T. A. Parnell, R. B. Wilson, W. S. Paciesas, T. Cline, and B. Teegarden, 19th ICRC, La Jolla Paper OG9.2-14 (1985).
13. J. L. Matteson, G. J. Fishman, C. A. Meegan, T. A. Parnell, R. B. Wilson, W. S. Paciesas, T. Cline, and B. Teegarden, 19th ICRC, La Jolla Paper OG9.2-15 (1985).

ON THE CAPABILITIES OF GRO TO PRODUCE SURVEYS OF THE GALACTIC PLANE IN THE LIGHT OF GAMMA RAY LINES

V. Schönfelder
Max-Planck-Institut für extraterrestrische Physik, 8046 Garching, FRG

ABSTRACT

GRO will provide the opportunity to perform high sensitivity surveys of the galactic plane in the light of gamma ray lines. The two low-energy gamma ray telescopes OSSE and COMPTEL together will cover the entire nuclear spectroscopy range from about 100 keV to 10 MeV. Both instruments have an energy resolution of a few to about 10 % FWHM, depending on energy. The collimated field-of-view of 3.8 x 11.4 degrees of OSSE together with its scanning capability, and the angular resolution of a few degrees within a 1 ster field-of-view in case of the imaging Compton telescope COMPTEL make these two intruments ideal for a first in depth- survey. Galactic gamma ray lines are expected from intersellar space as well as from discrete sources. The line-surveys will allow us to separate between these two components and to put the discussion on the origin of the line-emission on solid grounds. This will be especially true for the two so far most widely discussed gamma ray lines at 511 keV and 1.81 MeV. The present paper describes the capability and sensitivity of GRO for establishing such surveys.

1. INTRODUCTION

Surveys along the galactic plane provide a powerful means to investigate and study the distribution of certain objects or interstellar components throughout the Milky Way. Famous examples are the 21 cm-survey, the CO-survey or at gamma ray energies, the COS-B-survey.

GRO will - for the first time - provide the opportunity to perform surveys in the light of γ-ray lines. Galactic γ-ray lines are expected from interstellar space and from discrete sources. The line-surveys will allow us to separate these two components, and therefore put the discussion on the origin of the line-emission onto solid grounds. This is especially true for the two so far most widely discussed γ-ray lines at 511 keV and 1.8 MeV. However, it is expected that other γ-ray lines will be found in addition.

In the first part of this paper the GRO capabilities for performing γ-ray line surveys will be discussed. In the second part the sensitivities of the surveys will be compared with expected or predicted fluxes.

2. THE GRO CAPABILITIES FOR PERFORMING GAMMA RAY LINE SURVEYS

A schematic view of GRO with its four instruments BATSE, COMPTEL, EGRET, and OSSE is shown in Figure 1.

Fig. 1: SCHEMATIC VIEW OF GRO

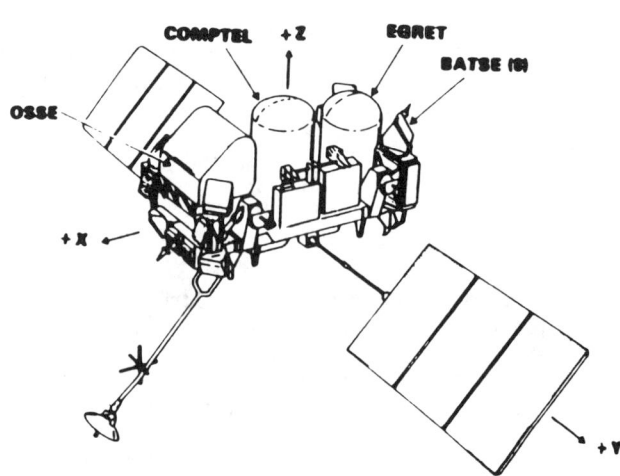

BATSE is the "Burst And Transient Source Experiment" of the Marshall Space Flight Center (Fishman et al., 1985), it consist of 8 single detectors - one at each corner of GRO, and operates above 20 keV.

COMPTEL is an imaging Compton Telescope in the MeV-range. It is built by an international collaboration between the Max-Planck-Institut für extraterrestrische Physik in Germany, the Laboratory for Space Research in Leiden /Holland, the University of New Hampshire, USA, and the Space Science Department of ESA (Schönfelder et al., 1984). Its nominal energy range is 1 to 30 MeV.

EGRET (Energetic Gamma Ray Experiment Telescope) is a spark-chamber experiment for high energy γ-ray astronomy above 20 MeV. It is built by an international collaboration between the Goddard Space Flight Center, USA, the Stanford University, USA, and the Max-Planck-Institut für extraterrestrische Physik in Germany (Fichtel et al., 1983).

OSSE is the "Oriented Scintillation Spectroscopy Experiment" of the Naval Research Laboratory (Kurfess et al., 1983). Its highest sensitivity is in the transition region between hard X- and soft γ-radiation (0.1 to 10 MeV).

The four instruments together cover more than 5 decades in photon energy. The only two instruments that are able to perform surveys in the light of γ-ray lines are OSSE, and COMPTEL. These two instruments cover the nuclear spectroscopy range and at the same time provide directional information on the infalling γ-radiation. BATSE is supposed to add some information to these surveys by using Earth occultation of sources. The following discussion, however, is restricted to OSSE and COMPTEL only.

OSSE consists of four identical shielded phoswich scintillation detectors. A NaI annular shield together with the CsI-portion of the phoswich detector form an active shield for each detector. A tungsten passive collimator within the NaI-annular shield defines the field-of-view of the detector. Each detector is mounted in a single axis orientation control system which provides offset pointing possibilities. OSSE covers the energy range 0.1 to 10 MeV. Its collimated field-of-view is 3.8 x 11.4° FWHM. The energy resolution is 8 % FWHM at 662 keV and 3.2 % at 6.1 MeV. The background is determined by a simultaneous measurement with an identical detector with offset pointing direction.

COMPTEL consists of two detector assemblies, an upper one of 7 cells of liquid scintillator NE213 and a lower one of 14 NaI-crystals. Both assemblies are entirely surrounded by anticoincidence shields of plastic scintillator. Infalling γ-rays are identified by a Compton collision in the upper detector and a subsequent interaction in the lower detector. From the locations and energy losses in both detector assemblies the arrival direction of the infalling γ-ray can be determined. The COMPTEL energy range is 1 to 30 MeV. COMPTEL is an imaging telescope with a wide field-of-view of about 1 ster. The angular resolution within the field-of-view is 2° to 4° FWHM. The photo peak energy resolution is 5 % to 8 % FWHM, depending on energy. Like in each imaging system, COMPTEL measures the background simultaneous to the source observation.

During the first year of the mission GRO is expected to perform a complete sky survey. For this task 23 different pointings of 2 weeks duration, each, are foreseen; 9 of the pointings are needed for the galactic plane survey. The sequence of the survey is constrained by various aspects (sun position, visibility of secondary source candidates of OSSE, effective observation time of COMPTEL, which is limited by the influence of the earth horizon).

3. SENSITIVITIES OF GRO TO PERFORM GAMMA RAY LINE SURVEYS

The γ-ray lines to be investigated by GRO are either produced by nucleosynthesis products or by nuclear interactions of energetic particles with matter. The most important isotopic decay chains from nucleosynthesis processes are those listed in Table 1.

TABLE I

Isotopic Decay Chains from Nucleosynthesis Processes

	Decay Chain	Mean Life (yrs)	Emission	
1.	$^{56}Ni \rightarrow {}^{56}Co \rightarrow {}^{56}Fe$	0.31	e^+	
			0.847	MeV
			1.238	MeV
2.	$^{57}Co \rightarrow {}^{57}Fe$	1.1	0.122	MeV
			0.014	MeV
3.	$^{22}Na \rightarrow {}^{22}Ne$	3.8	e^+	
			1.275	MeV
4.	$^{44}Ti \rightarrow {}^{44}Sc \rightarrow {}^{44}Ca$	68	e^+	
			1.156	MeV
			0.078	MeV
			0.068	MeV
5.	$^{60}Fe \rightarrow {}^{60}Co \rightarrow {}^{60}Ni$	$2.2 \times 10^{6*}$	1.332	MeV
			1.173	MeV
			0.059	MeV
6.	$^{26}Al \rightarrow {}^{26}Mg$	1.1×10^6	e^+	
			1.809	MeV

* (see Kutschera et al., 1984)

Gamma ray lines from nuclear reactions of energetic particles with interstellar matter have been extensively discussed by Ramaty et al., 1979. The most prominent lines are expected at 4.4 MeV and 6.1 MeV. Gamma ray lines from nuclear reactions should also be produced in discrete sources. Candidates are interstellar clouds (Silberberg et al., 1985), supernovae in clouds (Morfill and Meyer, 1981), and accreting compact objects (Brecher, 1978). The predicted line fluxes are generally low : of the order $10^{-5}/cm^2$ sec or even lower. The sensitivities of OSSE and COMPTEL for detecting γ-ray lines from point sources are shown in Figure 2. The sensitivities are identical near 1 MeV, namely $2 \cdot 10^{-5}/cm^2$ sec. Below 1 MeV OSSE is more sensitive, above 1 MeV COMPTEL has a higher sensitivity. The sensitivities of OSSE and COMPTEL for detecting diffuse γ-ray lines in interstellar space are given in Figure 3. The sensitivities strongly depend on the longitude bin-size. For OSSE the wider collimator angle of $11.4°$ was choosen. For COMPTEL any value within the field-of-view can be selected. At 1 MeV, again, the sensitivities of OSSE and COMPTEL ($10°$ longitude bin) are comparable : $10^{-4}/cm^2$ sec rad. In case of COMPTEL the sensitivity can be improved by a factor of 3 for a $40°$-longitude bin (see Figure 3).

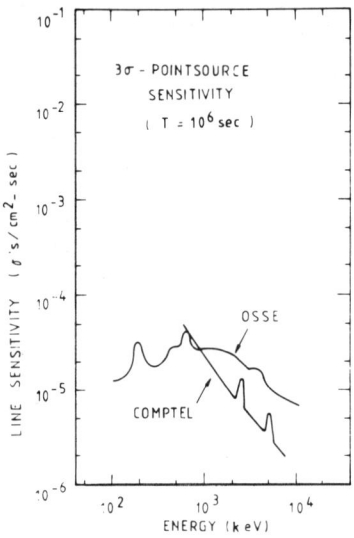

Fig. 2. Sensitivities of OSSE and COMPTEL to detect γ-ray lines from point sources.

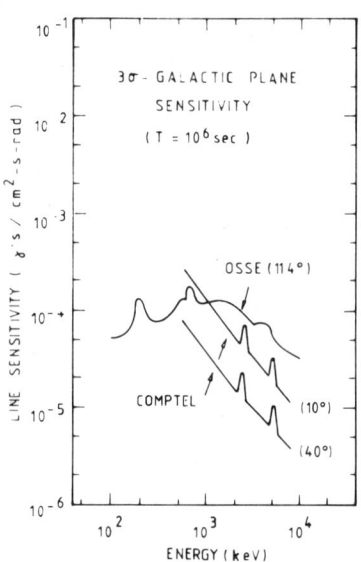

Fig. 3. Sensitivities of OSSE and COMPTEL to detect diffuse γ-ray line in interstellar space.

The sensitivities of Figure 2 and 3 are now discussed in the light of recent γ-ray measurements and calculations. Figure 4 shows possible angular dependences of the 1.8 MeV ^{26}Al-line on galactic longitude. The top figure is for a supernova origin of the line, the other two figures are for a nova origin (the N2 nova distribution is more concentrated around the galactic center than the N1 nova distribution). It is quite clear that the question of the origin of this line can only be solved by generating a map of the line-intensity along the entire galactic plane. The sensitivities of OSSE and COMPTEL for generating such a map are indicated in Figure 4 for different longitude bins. (Note that the flux normalization on the right ordinate of Figure 4 is based on the HEAO-C fluxes (Mahoney et al., 1985). If the normalization would be based on the balloon data of v. Ballmoos et al., 1987, then the sensitivities of OSSE and COMPTEL would appear at a relatively lower level).

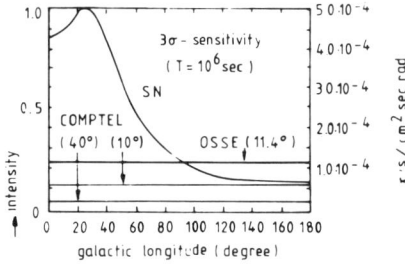

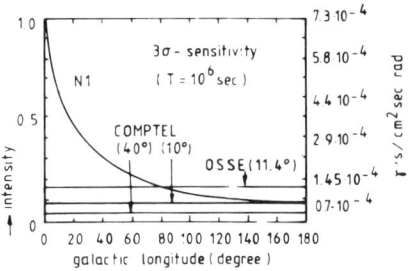

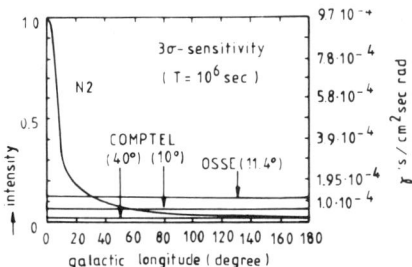

Fig. 4. Sensitivities of OSSE and COMPTEL to map the 1.8 MeV ^{26}Al-line along the galactic plane for the three different longitude distributions discussed by v. Ballmoos et al., 1987.

If any of the three models shown in Figure 4 is correct, then COMPTEL will be able to map the entire plane from the center to the anticenter region. Even in case of the sharply peaked N2 distribution the line should be still detectable in the anticenter region, if the longitude bin is made sufficiently wide. OSSE will mainly concentrate on mapping the plane around the galactic center. If all the power of the line is contained in a single point source, then no emission should be detected in any direction other than the center direction. A possible point source on top of a diffuse component should be easily detectable. The point source resolution of both instruments lies in the range 3° to 4° FWHM.

The 511 keV annihilation line at present seems to consist of two components : one is a point source component at the position of the galactic center, which is timevariable in intensity (Riegler et al., 1981). The other one shows a wide angular spread and is produced in interstellar space (Share et al., 1987). The positrons of the latter component are supposed to be mainly generated by radioactive isotopes from supernova explosions.

In Figure 5 it is therefore assumed that the 511 keV-line intensity follows the supernova-distribution. Normalization is made to the SMM-flux (Share et al., 1987). The sensitivities of OSSE and COMPTEL for mapping this line are again indicated. Here OSSE is the most sensitive instrument. The annihilation line is outside the nominal COMPTEL energy range (1 to 30 MeV). However, the COMPTEL energy threshold can be lowered by telecommand, when COMPTEL is in orbit. The estimated sensitivity of COMPTEL at 511 keV for the lowest possible energy thresholds is indicated in Figure 5 by a dashed line.

Another promising γ-ray line is the 1.156 MeV-line from the ^{44}Ti → ^{44}Sc → ^{44}Ca chain. Due to its 68 year decay time this line should be visible from the few most recent supernovas. With the GRO point source sensitivity of $2 \cdot 10^{-5}$/cm^2 sec a map of the galactic plane at 1.156 MeV therefore should show a few - say half a dozen point sources.

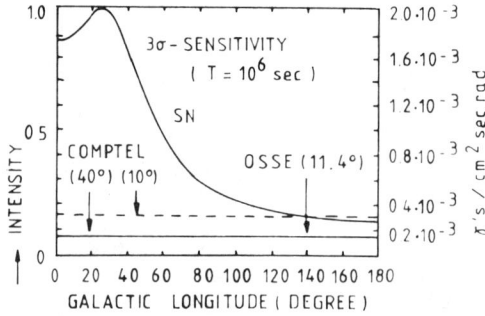

Fig. 5. Sensitivities of OSSE and COMPTEL to map the 511 keV annihilation line along the galactic plane.

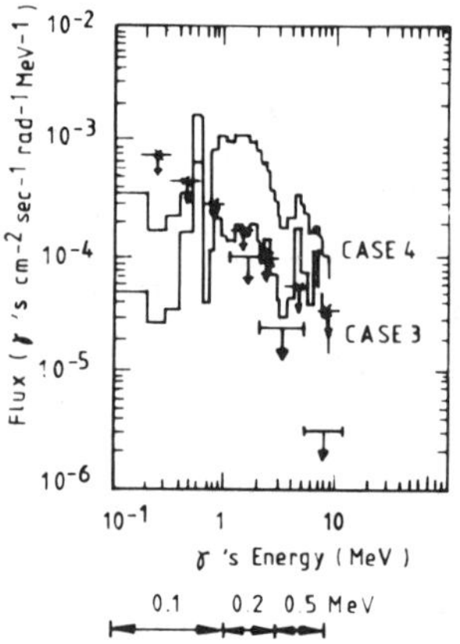

Fig. 6. Gamma ray line emission from nuclear interactions in interstellar space from the galactic center region. The γ-ray line profiles are smeared out over energy bins defined by the detector resolutions as marked below the abscissa (Ramaty et al, 1979). The two models labelled case 3 and case 4 differ in the assumed metallicity of cosmic rays. The sensitivities of OSSE and COMPTEL to detect the integral effect of the line emission are indicated at the 3 σ confidence level for 10^6 sec effective observation time. (⊥ OSSE (11.4°), ⊤ COMPTEL (20°)).

Finally, the detectability of γ-ray lines from nuclear interactions of energetic particles in space has to be addressed. Because OSSE and COMPTEL have an energy resolution of few percent only, the interstellar γ-ray line spectrum as predicted by Ramaty et al. (1979) cannot be measured with the required accuracy to reproduce the γ-ray line profiles. Instead, the spectrum will be smeared out over energy bins which are defined by the energy resolution of the detectors. The expected properly smeared out spectrum is shown in Figure 6. Discrete γ-ray lines are hardly visible anymore, however, a broad-band structure now becomes visible. Again, the OSSE and COMPTEL sensitivities are indicated. Especially above 1 MeV the instruments will have sufficient sensitivity to measure the integral effect of the line-emission for the cases considered by Ramaty et al., 1979.

4. CONCLUSION

In summary I would like to conclude that γ-ray line astronomy will certainly make a major step forward, when GRO is in orbit. From our present days knowledge we can expect most interesting measurements especially of the 511 keV and 1.8 MeV γ-ray lines. I expect that the origin of these lines will be solved by GRO.

ACKNOWLEDGEMENT

The author would like to thank Dr. W. A. Mahoney for suggesting some improvements.

LITERATURE

P. v. Ballmoos et al., 1987, Ap. J. 318, 654
K. Brecher, 1978, in "Gamma Ray Spectroscopy in Astrophysics" ed.: by T. L. Cline, and R. Ramaty, NASA TM 79619, page 275
C. E. Fichtel et al., 1983, Proc. of 18th International Cosmic Ray Conf.at Bangalore, Vol. 8, 19
G. J. Fishman et al., 1985, Proc. of 19th International Cosmic Ray Conf., La Jolla, Vol. 3, 343
W. Kutschera et al., 1984, Nucl. Instr. & Meth. B5, 430
J. D. Kurfess et al., 1983, Adv. of Space Research, Vol. 3, No 4, 109
W. A. Mahoney et al., 1985, Proc. of 19th International Cosmic Ray Conf., La Jolla, Vol. 1, 357
G. Morfill, and P. Meyer, 1981, 17th International Cosmic Ray Conf., Paris, Vol. 9, 56
G. R. Riegler, J. L. Ling, W. A. Mahoney, W. A. Wheaton, J. B. Willet, A. S. Jacobson, T. A. Prince, 1981, Ap. J. Letters, 248, L13
R. Ramaty et al., 1979, Ap. J., Suppl. 40, 487
V. Schönfelder et al., 1984, IEEE-Transactions on Nucl. Sc., NS-31, No 1, 766, and "The Imaging Compton Telescope COMPTEL", 1978, proposal submitted to NASA.
G. H. Share et al., 1987, Proc. of 20th International Cosmic Ray Conf., Moscow, Vol. 1, 156
R. Silberberg et al., 1985, 19th International Cosmic Ray Conf., La Jolla, Vol. 1, 369

GASOL : A NEW EXPERIMENT FOR THE STUDY
OF γ-RAY EMISSION FROM SOLAR FLARES

C. Barat, F. Cotin, M. Niel, R. Talon, G. Vedrenne
Centre d'Etude Spatiale des Rayonnements (CESR)
CNRS/UPS, B.P. 4346, 31029 Toulouse Cedex, France.

M. Pick, G. Trottet, N. Vilmer
UA 324 - DASOP - Observatoire de Paris
Section d'Astrophysique de Meudon, 92190, Meudon, France.

K. Hurley
University of California, Space Sciences Laboratory
Berkeley, CA 94720, USA.

A. Kuznetzov, R. Sunyaev, O. Terekhov
Institute for Space Research
Academy of Science 117810, Moscow, USSR.

ABSTRACT

Since 1980, the observations of the γ-ray spectrometers aboard the SMM or Hinotori satellites have considerably improved our knowledge of the γ-ray line emissions produced during solar flares by accelerated ions. However, both experiments did not detect the predicted lines above 10 MeV and had too low sensitivities to address problems such as the precise relative timing of individual γ-ray lines and of the bremsstrahlung component produced by the flare accelerated electrons above a few hundred keV. In an attempt to solve such problems, the GASOL experiment has been designed for providing solar γ-ray data with a higher sensitivity than previous experiments. The main characteristics of the experiment are described and some aspects of its scientific impact are discussed.

INTRODUCTION

Although the first evidence for solar γ-ray lines was obtained in 1972, most of the results came after 1980 with the launch of the γ-ray spectrometers aboard SMM[1] and later on Hinotori[2]. Since then, ~ 200 events above 300 keV have been detected. The study of these events has led in particular to new insights into the characteristics of the solar flare accelerated particles (spectral form, timing of the interaction of different species) as deduced from the production of prompt deexcitation γ-ray lines. Many γ-ray lines predicted theoretically[3] were observed below 8 MeV in solar flares by the gamma-ray experiments aboard SMM[1] and Hinotori[2]. However, the γ-ray line emission from the 15.11 MeV excited state of ^{12}C, which is also predicted to be produced in solar flares[4], was not detected. However, the ratio of the 15.11 MeV and of the 4.44 MeV prompt deexcitation lines of ^{12}C is of crucial interest for the determination of the accelerated particle spectrum in the 10-100 MeV range[4].

A key result of gamma-ray experiments[1,2] since 1980 is that ion

© 1988 American Institute of Physics

and electron acceleration or more precisely their interaction with matter is quasi simultaneous and can take place on a time scale of seconds or less. This is evidenced by the close simultaneity of the increase of the electron bremsstrahlung component at ~ 300 keV and of the count rate in the 4.1 - 6.4 MeV range which is predominantly (~ 90 %) due to resolved and unresolved nuclear γ-ray lines, thus characteristic of accelerated ions. In some events, the peak intensity of the higher photon energy can be reached later than at low energy[1,2], which is interpreted either by the concept of a second step acceleration of ions[5] or by different life times or propagation times of energetic electrons and ions from the acceleration site to the emitting region. However, the sensitivities of the detectors did not allow a detailed study of the time profiles of individual γ-ray lines. This is, however crucial for the understanding of particle acceleration in solar flares.

The SMM-GRS observations have also revealed a correlation between the 4-8 MeV excess fluence (time integrated flux) from resolved and unresolved gamma ray lines produced by energetic ions and the fluence above 270 keV resulting from electron bremsstrahlung[1]. Such a correlation suggests that nuclear line contributions could be present in all flares where emission above 270 keV is observed and that they could be detected with a more sensitive spectrometer.

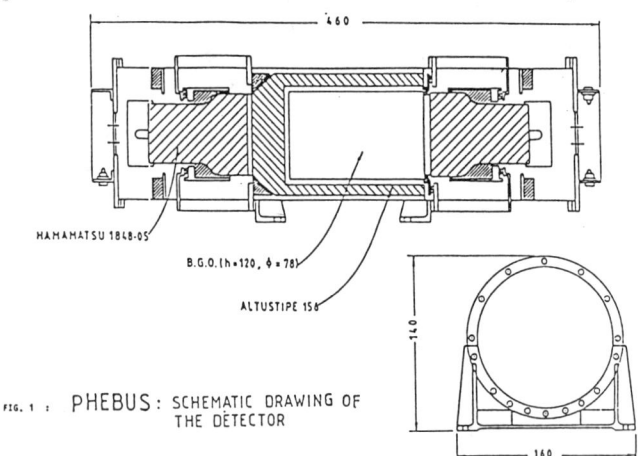

FIG. 1 : PHEBUS: SCHEMATIC DRAWING OF THE DETECTOR

In view of these problems, the GASOL experiment has been designed for providing solar γ-ray data with a better sensitivity than previous detectors. GASOL is part of both the SIGMA and PHEBUS experiments which will be flown on the Soviet GRANAT satellite. The launch is planned for November 1988 and the duration of the mission is 1.5 years. The SIGMA experiment (C.E.A. Saclay, France ; C.E.S.R. Toulouse, France) is devoted to X- and γ-ray images in the 30 keV - 2 MeV range. Its large CsI anticoincidence shield (4 cm thickness) will be used to record cosmic and solar bursts in the 0.1 - 8 MeV range (area for solar observations ≈ 3000 cm²). The PHEBUS experiment (C.E.S.R. Toulouse) will study cosmic and solar bursts between 75 keV and 124 MeV. We focus here on a preliminary presentation of the latter experiment. A complete description will be published later.

THE PHEBUS EXPERIMENT

1. The detector :
The PHEBUS experiment (Payload for High Energy Burst Spectroscopy) is designed to detect and record the time profiles and spectral characteristics of solar and cosmic gamma-ray bursts with a fine time resolution in the 75 keV – 124 MeV range. It consists of 6 BGO detectors and their associated electronics. Each detector (figure 1) includes a large BGO scintillator (78 mm in diameter, 120 mm in height) surrounded by a plastic (altustipe) anticoincidence jacket. On the GRANAT satellite, the detectors are directed along the 6 axes of a cartesian coordinate system in order to obtain a 4π sr field of view. Two of these detectors always have the Sun in their field of view with an angle of 90° between their axis and the solar direction, thus yielding a minimum geometrical area of 190 cm³. Moreover, depending on the pointing of the telescope of the SIGMA experiment, a third detector of the PHEBUS experiment can be directed towards the Sun (angle between the axis and the solar direction of 0°), thus increasing the geometrical area up to 235 cm². Each scintillator (BGO and plastic) is optically coupled to a 3" photomultiplier tube (PMT).

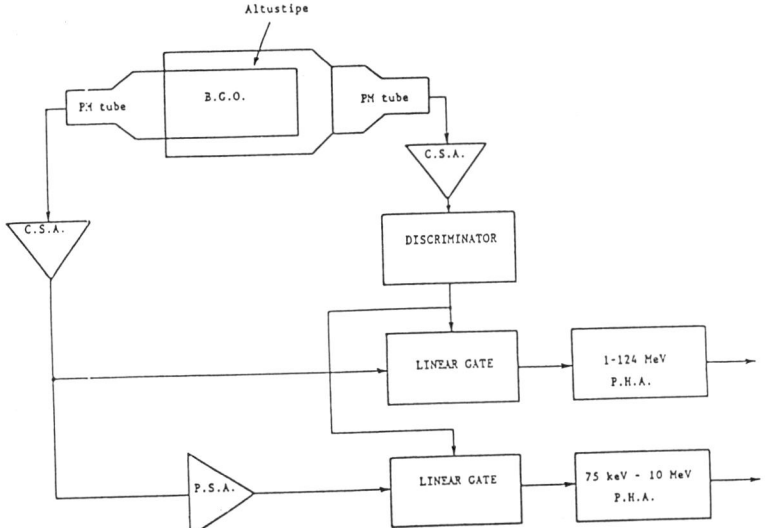

Fig. 2 : Diagram of the analog electronics

2. The analog electronics :
As shown on figure 2, the anode signal from the BGO PMT is amplified by a charge sensitive amplifier (C.S.A.). The output is fed directly to a 256 channel pulse height analyser (P.H.A.) for the 1-124 MeV range and to a 128 channel P.H.A. for the 75 keV – 10 MeV range.

The electronic chain associated with the plastic scintillator comprises a charge sensitive amplifier and a low level discriminator with a threshold energy of about 400 keV. The discriminator output provides a veto pulse for the linear gates of the pulse height analysers in order to prevent the analysis of signals due to charged particles.

3. The modes of operation :
   In the absence of a burst, the PHEBUS experiment operates in the survey mode which provides for each detector :
   - continuous records of the integral count rate in the 75-1600 keV range with a 64 sec time resolution.
   - two background spectra are obtained every 4 hours successively with and without the plastic anticoincidence in the total energy range (75 keV - 124 MeV) which is covered with 225 energy channels. These background spectra provide an in-flight detector calibration from the position of the naturally produced 511 keV line (interaction of background charged particles with the satellite) and of the broad peak at ~ 70 MeV corresponding to energy losses of ~ 300 MeV background protons in the BGO scintillator. The observation of this last peak requires that the plastic anticoincidence be disabled during this measurement.
   The PHEBUS instrument works in the burst mode when a count rate excess (about 8 σ above the background level) obtained with a 8 ms counter is detected in a 1/16, 1/4, 1 second interval (burst trigger).
   For each detector, the burst mode provides simultaneously :
1) a burst time history for 86.6 sec in the 75-1600 keV range. The time resolution is 1/128 sec for the first 7.9 sec, then 1/64 sec for the following 15.7 sec, then 1/32 sec for the remaining 63 sec.
2) time intervals for an accumulation of 24 counts in the 75-1600 keV range (time to spill mode) given by a 32768 Hz clock.
3) 176 spectra with 116 energy channels in the 75 keV - 124 MeV range. Each spectrum is recorded for a fixed number of counts between 75 and 1600 keV chosen to achieve a fine time resolution with adequate statistics. This step duration obviously depends on the burst intensity, but is expected to last less than 1 minute.
   The acquisition of these spectra is followed by the recording of 640 spectra with 40 energy channels in the 75 keV - 124 MeV range. The time resolution is constant and equal to 1 sec for the 26 channels recorded between 100 keV and 10 MeV and to 4 sec for the 14 channels between 10 MeV and 124 MeV. The total duration of this measurement is 640 sec and this record is mainly devoted to solar analysis.
   The various modes of operation are monitored by a program stored in the "PROM" of a 16 bit microprocessor (Intel 80C86). This program is repeated with a cycle of 4 hours : the survey and burst data are stored in a RAM. The memory is read out at the end of each 4 hour period for the survey mode and at the end of a burst for the burst mode. Information is then transmitted to the 128 Mbit bubble memory of the GRANAT satellite. Three bursts can be recorded during a 24 hour period.

4. Energy resolution and full energy peak efficiency :
   Preliminary calibrations of the PHEBUS detector have been carried out using radioactive sources for the low energy range ($E \leq 6.13$ MeV) and in-flight positron annihilation from the linear accelerator of Orme les Merisiers (France) for $E \geq 15$ MeV.
   The Full Width at Half Maximum (FWHM) energy resolution and the full energy peak efficiencies for 0 degree and 90 degrees (with respect to the detector axis) are presented in Figure 3.
   The low absolute efficiency observed at low energy at 0° is due to

absorption by materials (see Figure 1).

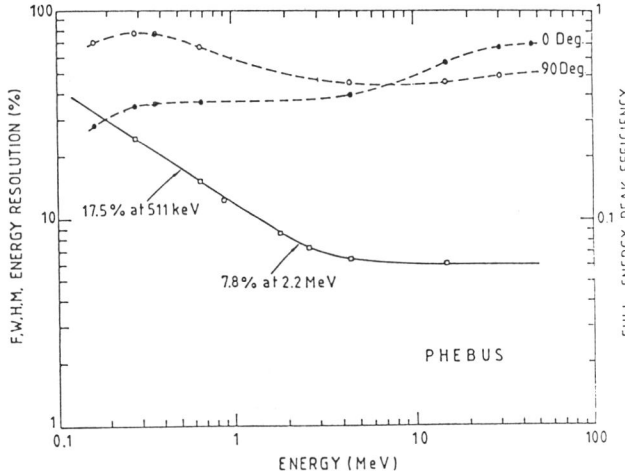

Fig. 3 : F.W.H.M. energy resolution and full energy peak efficiency as a function of the energy.

As an example, Figure 4 presents the detector response at 30 MeV for a beam (~ 15 mm in diameter) impinging on the BGO at 0° (near the center). At this energy, photons interact through pair production. Since the pair energy is high, the broad peak observed may be explained by the bremsstrahlung emission of the pairs. The background observed at low energy can be removed using a spectrum obtained with a beam not impinging on the BGO. In this case, the valley to peak ratio inferred is about 1-2 % near the BGO center and 20 % near the edge (measured only at 0°, in the detector axis direction).

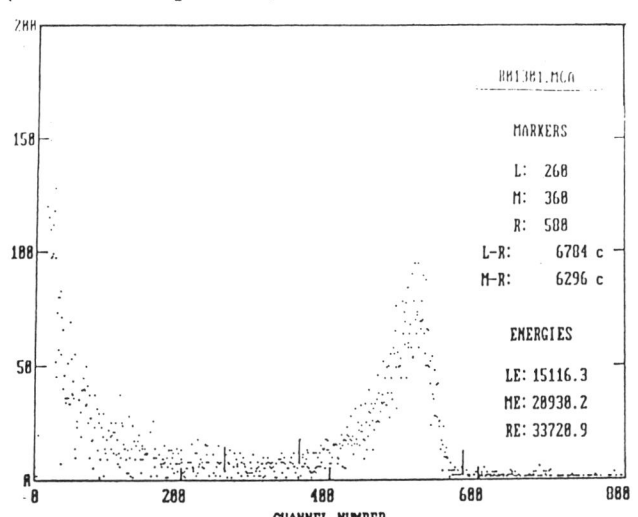

Fig. 4 : Detector response at 30 MeV, 0° (near the center).

5. Sensitivity :

Given the expected background spectrum and the energy resolution, the line sensitivity of the Phebus instrument is calculated using a detector area of ~ 200 cm² (lower limit of the geometrical area for solar observations), the full energy peak efficiency at 90° and a confidence level of 3 σ. The sensitivity for the high energy emission above 10 MeV has also been estimated with the same area and confidence level but using the total efficiency at 90° inferred from the linear absorption coefficient τ. The expected sensitivities are given in Table 2 for some solar γ-ray lines and for the continuum above 10 MeV. These sensitivity levels are on the order of some of the expected fluxes for time intervals of 1 sec estimated in solar flares. As an example, the 2.22 MeV and 4.44 MeV line strengths of the July 11, 1978 flare observed by HEAO 1[6] were estimated to be respectively 1 ± 0.3 photons cm⁻² s⁻¹ and 0.18 ± 0.07 photons cm⁻² s⁻¹ based on observations with a 4 minute integration time.

Table 2

Energy (MeV)	0.511	2.22	4.44	6.13	15.1	>10 MeV
Sensitivity ph cm⁻² s⁻¹	0.15	0.10	0.08	0.07	0.06	0.25

SCIENTIFIC PERSPECTIVE

The GASOL experiment will provide data in the 100 keV - 124 MeV range with better sensitivity than previous experiments, thus allowing data analysis with a good temporal resolution (≤ 1 sec up to 10 MeV and ≤ 4 sec above). This will lead to the detection of weaker events, thus increasing the number of observed events for any statistical study (such as center-to-limb effects) and telling us whether or not ions are accelerated in all flares. The experiment is expected to provide for the first time an accurate timing of the individual strong prompt γ-ray deexcitation lines as compared to the electron bremsstrahlung component and more generally to the various radiations observed from ground-based and space-borne instruments. This should yield strong constraints on the ion acceleration mechanism. Another goal is the detection of the ^{12}C deexcitation line at 15.11 MeV.

REFERENCES

1. E.L. Chupp : Ann. Rev. Astron. Astrophys., 22, 359 (1984).
2. M. Yoshimori, K. Okudaira, Y. Hirasima, et I. Kondo : Solar Phys., 86, 375 (1983).
3. R. Ramaty, B. Kozlovsky and R.E. Lingenfelter : Space Sci. Rev., 18, 341 (1975).
4. C.J. Crannell, H. Crannell and R. Ramaty : Astrophys. J., 229, 762 (1979).
5. T. Bai : in R.E. Lingenfelter, H.S. Hudson, D.M. Worall (eds), Gamma Ray Transients and Related Astrophysics, AIP, 315 (1982).
6. H.S. Hudson, T. Bai, D.E. Gruber, J.L. Matteson, P.L. Nolan and L.E. Peterson : Astrophys. J., 236, L91 (1980).

SPECTROSCOPIC CAPABILITIES OF THE SOLAR-A SATELLITE

M. Yoshimori
Department of Physics, Rikkyo University,
Toshima-ku, Tokyo 171, Japan

ABSTRACT

The Wide Band Spectrometer (WBS) will be flown on the Solar-A satellite to observe a flare photon spectrum from soft X-rays to gamma-rays. The WBS consists of three spectromers: a soft X-ray spectrometer of a gas proportional counter covering the 2 - 30 keV band, hard X-ray spectrometer of a NaI scintillation counter covering the 20 - 400 keV band and gamma-ray spectrometer of two BGO scintillation counters covering the 0.2 - 100 MeV band. The details of the spectral capabilities of the Wide Band Spectrometer are described.

INTRODUCTION

In the last epoch of solar maximum activity in the early 1980s, the Solar Maximum Mission (SMM) and the Hinotori satellite were launched to study high energy solar flare phenomena, and the understanding of plasma physics of solar flares, including energy storage and release, particle acceleration and solar terrestrial effects has been significantly advanced. These two satellite observations revealed that the most important direction for future flare observations lies in high resolution imaging in both soft and hard X-ray bands and in precise wide band spectrometry from soft X-rays to gamma-rays. To continue the systematic study of the solar flares, the Solar-A mission sponsored by the Institute of Space and Astronautical Science of Japan is scheduled for launch in August/September 1991, the next epoch of the solar maximum activity. In particular, the wide band spectral observation provides an essential key to solve the plasma heating, electron and ion acceleration and high energy photon emission at the flare site.

The primary scientific objectives of the wide band spectral observation are as follows. (1) The soft X-ray spectral observation in the 2 - 30 keV band determines temporal dependences of both the electron and ion temperature and the emission measure of hot plasma with high temporal resolution, and the lower cutoff energy of nonthermal electrons. (2) The hard X- and gamma-ray spectral observation in the 20 keV - 100 MeV band, together with radio and neutron observations, characterizes in great detail the electron and ion acceleration process up to GeV energies.

INSTRUMENT DESCRIPTION

The Solar-A Wide Band Spectrometer (WBS) consists of the following three subsystems: soft X-ray spectrometer (SXS), hard X-ray spectrometer (HXS) and gamma-ray spectrometer (GRS).

© 1988 American Institute of Physics

Further, a gamma-ray burst detector (GBD) is added to monitor non-solar X- and gamma-ray transient phenomena. The HXS also has the capability of detecting the non-solar transient phenomena.

(1) Soft X-ray spectrometer (SXS)

The SXS consists of a gas proportional counter filled with a gas mixture of Xe and CO_2 and detects soft X-rays in the 2 - 30 keV band. The SXS has two windows of different sizes SXS-1 and SXS-2 to cover flares of significant different soft X-ray fluxes. The SXS-1 has a large effective area suitable for small flares, while the SXS-2 has a small effective area suitable for large flares. The effective area curves of the SXS-1 and SXS-2 are shown in Fig. 1. The energy resolution of the SXS is about 1.5 keV (FWHM) at 5.9 keV. The in-flight energy calibration is achieved by the detection of 5.9 keV X-rays emitted by a ^{55}Fe radioactive source.

Each output pulse of the SXS-1 and SXS-2 anode wires, after amplification and shaping, is sent to a 128-channel pulse height analyzer (PHA). The 128-channel energy spectral data in the 2 - 30 keV band are recorded every 2 s. In addition, 2-channel (2 - 8 keV and 8 - 30 keV) pulse count data are recorded every 0.5 s to study the time history of soft X-ray emission. The pulse count data are used to monitor the occurence of solar flares.

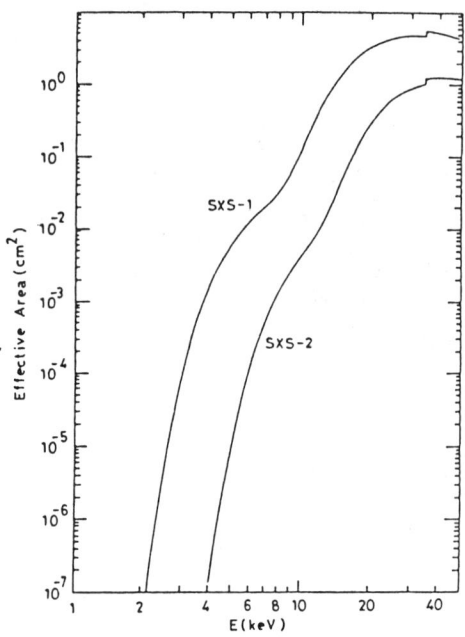

Fig. 1. The effective area curve of the SXS-1 and SXS-2.

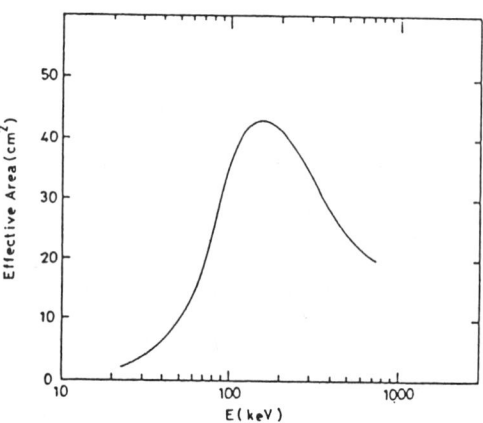

Fig. 2. The effective area curve of the HXS.

(2) Hard X-ray spectrometer (HXS)

The HXS consists of a NaI scintillator, which is 7.6 cm in diameter and 2.5 cm in thickness, optically coupled by a 3 in. diameter photomultiplier tube. It detects hard X-rays in the 20 - 400 keV band. The NaI scintillator is covered, by an aluminum absorber to suppress lower energy X-ray events. The effective area curve of the HXS is shown in Fig. 2. The energy resolution is about 10 keV (FWHM) at 60 keV. The in-flight energy calibration is achieved by the detection of 60 keV X-rays emitted by the Am-241 radioactive source. The Am-241 radioactive source decays by the simultaneous emission of 60 keV X-rays and 5.48 MeV alpha-particles. The alpha-particles are detected with two Si detectors producing calibration event tags. The 60 keV X-rays which interact in the HXS are detected in coincidence with these calibration tag signals.

The output pulse of the photomultiplier tube, after amplification and shaping, is sent to the 32-channel PHA. The 32-channel energy spectral data are recorded every 1 s. In addition, 2-channel (20 - 60 keV and 60 - 400 keV) pulse count data are recorded every 0.125 s to study the time history of hard X-ray emission. The hard X-ray pulse count data, together with the soft X-ray pulse count data, are used to monitor the occurrence of solar flares. Further, the HXS has the capability of detecting non-solar X- and gamma-ray transient phenomena such as cosmic gamma-ray bursts during the solar quiet time.

(3) Gamma-ray spectrometer (GRS)

The GRS consists of two identical BGO (bismuth germanate) scintillators and detect gamma-rays in the 0.2 - 100 MeV band. Each BGO scintillator, which is 7.6 cm in diameter and 5.1 cm in thickness, is optically coupled by the 3 in. diameter photomultiplier tube. The BGO scintillator has high density (7.13 g cm^{-3}) and high effective atomic number (74). These properties contributes towards improving the detection sensitivity to high energy gamma-rays. The scintillation efficiency (light output) of BGO scintillator, however, is about 20 % of that of NaI scintillator. The disadvantage gives rise to the reduction of energy resolution. The energy resolution (FWHM) of BGO scintillator is shown in Fig. 3. It is about 82 keV (12.4 %) at 662 keV. It is about twice that of the NaI scintillator. The response of the 7.6 cm diameter and 5.1 cm thick BGO scintillator was simulated from the Monte Carlo calculations[1]. The calculated effective area curves of (photopeak efficiency) x (detector area), (probability of energy losses > 10 MeV) x (detector area) and (probabilities of energy losses > 25 MeV) x (detector area) are shown in Fig. 4. The effective area curves of the SMM gamma-ray spectrometer[2,3] are shown in Fig. 4 for comparison.

Since the GRS covers the wide energy band of 0.2 - 100 MeV, the GRS output pulses are produced from both anode and dynode of the photomultiplier tube. The 0.2 - 10 MeV (GRS-L) and 8 - 100 MeV (GRS-H) gamma-ray pulses are produced from the anode and dynode, respectively. The GRS-L pulses, after amplification and shaping, are passed to the 128-channel quadratic space PHA.

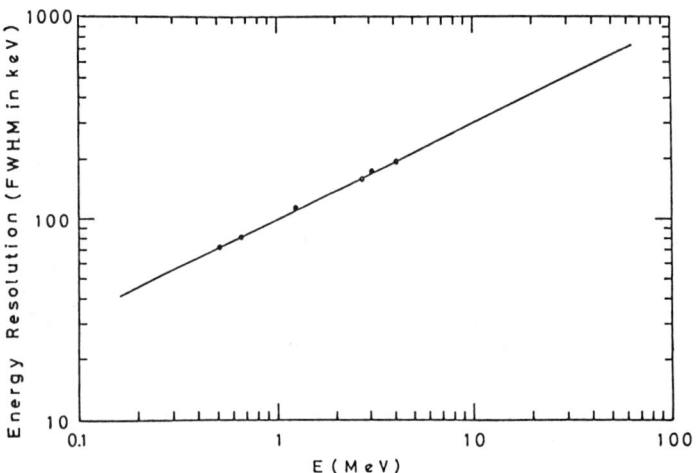

Fig. 3. The energy resolution (FWHM) of the GRS

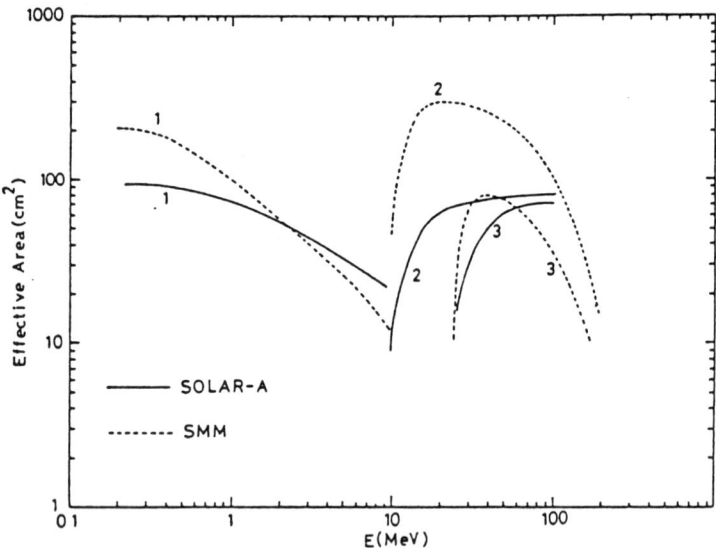

Fig. 4. The effective area curves of the Solar-A and SMM gamma-ray spectrometers. Numbers of 1, 2 and 3 attached to the curves correspond to (photopeak efficiency) x (detector area), (probability of energy losses > 10 MeV) x (detector area) and (probability of energy losses > 25 MeV) x (detector area), respectively.

The 128-channel energy spectral data in the 0.2 - 10 MeV band are recorded every 4 s. The GRS-H pulse, after amplification and shaping, are sent to the 16-channel PHA. The 16-channel energy spectral data in the 8 - 100 MeV band are recorded every 4 s. In addition, 2-channel pulse count data (0.2 - 0.7 MeV and 0.7 - 4 MeV) and 4-channel pulse count data (4 - 7 MeV, 7 - 10 MeV, 10 - 30 MeV and 30 - 100 MeV) are recorded every 0.25 s and 0.5 s, respectively, to study the time history of gamma-ray emission. The in-flight energy calibration in the 0.2 - 10 MeV band is achieved by the detection of 1.17 and 1.33 MeV gamma-rays emitted by the Co-60 radioactive source. The Co-60 radioactive source decays by the simultaneous emission of 1.17 and 1.33 MeV gamma-rays and beta-rays (the maximum energy is 300 keV). The beta-rays are detected with two Si detectors producing calibration event tags. The 1.17 and 1.33 MeV gamma-rays which interact in the GRS are detected in coincidence with these calibration tag signals. The in-flight energy calibration in the 8 - 100 MeV band, however, can not be carried out because of no radioactive source emitting gamma-rays of energies greater than 8 MeV.

The GRS also has the detection sensitivity to solar neutrons. The BGO scintillator was exposed to neutrons of energies of 15, 25, 35 and 45 MeV and the responses to the neutrons were experimentally studied. A gamma-ray peak at 14 MeV resulting from the neutron-induced reaction in the BGO scintillator was apparently detected. Although it is very difficult to determine the neutron energy from the BGO response, the neutron energy will be deduced from the sun-earth transit time for impulsive flares.

(4) Gamma-ray burst detector (GBD)

In addition to the SXS, HXS and GRS mentioned above, another NaI scintillation counter, which is 5.1 cm in diameter and 1 cm in thickness, is on board to monitor non-solar transient phenomena such as cosmic X- and gamma-ray bursts. This burst detector is oriented in the direction perpendicular to the sun. Therefore, it is insensitive to solar flares. Further, it serves as alarm for the south atlantic anomaly passage of the Solar-A. The 32-channel pulse height data in the 20 - 400 keV band and the 2-channel pulse count data (20 - 60 keV and 60 - 400 keV) are recorded every 1 s and 31 ms, respectively. The in-flight energy calibration is not carried out for the GBD. The WBS primary data are summarized in Table I. In addition to the primary data, the following house keeping data are recorded: SXS-1 and SXS-2 pulse counts of energies greater than 30 keV, HXS pulse counts in energies greater than 400 keV and GRS pulse count of energies greater than 100 MeV. Further, monitor data of high voltage biased to the photomultiplier tubes and the gas proportionalcounter and of the instrument temperature are recorded.

EXPECTED RESULTS

The Solar-A WBS is expected to produce the following results in the next epoch of solar maximum activity. The

Table I Summary of the WBS data

Data	Energy		Pulse height data	Pulse count data
SXS	2-30 keV	No. of channels	128	2
		Time resolution	2 s	0.5 s
HXS	20-400 keV	No. of Channels	32	1
		Time resolution	1 s	0.125 s
GRS-L	0.2-10 MeV	No. of channels	128	4
		Time resolution	4 s	0.25/0.5 s
GRS-H	8-100 MeV	No. of channels	16	4
		Time resolution	4 s	0.5 s
GBD	20-400 keV	No. of channels	32	2
		Time resolution	1 s	31 ms

SXS observation in the 2 - 30 keV band clarifies the temporal behavior of the electron temperature and emission measure of heated plasma and determines the cutoff energy of nonthermal electrons. The SXS observation will much contribute to solve the fundamental problem of plasma heating during solar flares. The HXS and GRS observations in the 20 keV - 100 MeV band characterize in great detail the electron and ion acceleration process and determine physical parameters such as the time scale and efficiency for the particle acceleration up to GeV energies. The wide band spectral observation from the SXS, HXS and GRS will provide essential keys to understand the plasma heating process and particle acceleration mechanism. The GRS will be able to detect more gamma-ray and neutron events in three years of the epoch of solar maximum activity, if the solar activity level for the coming solar cycle is assumed to be nearly the same as that of the last solar cyle in the early 1980s.

Further, the solar particle observation from the GEOTAIL satellite, the solar neutron observation from the ground-based neutron monitor and the radioheliograph obserbation at microwaves are planned in the early 1990s. Collaborative experiments of the Solar-A satellite with these facilities will provide very exciting results and solve questions of high energy solar flare phenomena.

REFERENCES

1. M. Yoshimori et al., Nucl. Instr. and Meth. (1988) in press.
2. D.J. Forrest et al., Solar Phys. 65 (1980) 15.
3. D.J. Forrest et al., Adv. Space Res. 6 (1986) 115.

THE HIGH ENERGY TRANSIENT EXPERIMENT
– HETE –
A MULTI-WAVELENGTH SURVEY MISSION FOR THE 1990's

G. Ricker, J. Doty, and S. Rappaport
MIT Physics Department and Center for Space Research
Cambridge, MA 02139

K. Hurley
University of California, Space Sciences Laboratory
Berkeley, CA 94708
and
Centre d'Etude Spatiale des Rayonnements
Toulouse, France

E. Fenimore and D. Roussel-Dupre
Los Alamos National Laboratory
Los Alamos, NM 87545

M. Niel and G. Vedrenne
Centre d'Etude Spatiale des Rayonnements
Toulouse, France

D. Lamb
Department of Astronomy and Astrophysics
University of Chicago
Chicago, IL 60637

S. Woosley
Board of Studies in Astronomy & Astrophysics
University of California at Santa Cruz
Santa Cruz, CA 95064

ABSTRACT

The High Energy Transient Experiment (HETE), an international mission currently planned for launch in 1992, is presented. Its prime objective is the multiwavelength study of gamma-ray bursts with UV, X-ray, and gamma-ray instruments. A unique feature of the mission is its capability to localize bursts with several arcsecond accuracy, in *near real-time* aboard the spacecraft, and to transmit these positions to the ground, thereby enabling rapid, sensitive follow-up studies. HETE will be launched from a Get Away Special (GAS) cannister on the Shuttle, utilizing a new spacecraft concept, the "Cheapsat".

INTRODUCTION

It has become increasingly apparent that the origin of cosmic gamma ray bursts (GRBs) can only be determined by identifying burster counterparts in other energy ranges. To date, slightly over half a dozen GRBs have positions determined to 10 square arcminutes or better [1]. Infrared [2], optical [3,4,5], and soft X-ray [6,7] searches have not revealed *any* convincing candidates for these GRBs. However, there now exist reports of six relatively bright optical transients which may be related to bursters [8,9,10,11,12]. Even apart from the important information which such observations convey about the physics of GRB sources, they can in principle serve the important function of bridging the gap between the relatively large (arcminute size) gamma ray error boxes, and the very small (arcsecond size) boxes needed to conduct deep optical searches (>25th magnitude) for the exceedingly faint quiescent burster counterparts. As all but one of these discoveries were made by examining archived photographic plates (obtained up to 78 years *before* the GRB), questions concerning possible proper motion, the association between the optical and gamma ray emission, and the time structure of the optical transient event, remain unanswered. The HETE (High Energy Transient Experiment) mission was conceived as an integrated approach to resolving these and other fundamental questions about bursters. It has been recommended, studied and/or redefined over the past 7 years by various groups[13,14,15,16,17]. In the following sections, we describe the recommended flight version of HETE, which addresses the need for small, rapidly disseminated error boxes, and spectral studies over a wide range of wavelengths; in this version, HETE utilizes some of the most up-to-date technology currently available for space applications.

INSTRUMENTATION

The HETE spacecraft contains three scientific instruments, whose properties are summarized in Table 1: an *omnidirectional gamma ray spectrometer*, a *wide-field X-ray monitor*, and an *ultraviolet transient camera array*. The gamma ray spectrometer is practically identical to an experiment which was developed in France for launch in mid-1988 aboard the Soviet Phobos probes. Each unit in the array utilizes cleaved NaI(Tl) scintillators to achieve excellent energy resolution (for a scintillator) down to about 6 keV. The energy range encompasses both regions in which line-like features have been reported (around 50 and 400 keV). Individually, the elements of the array have practically no spatial resolution, although an approximate measurement of burst arrival directions can be obtained by comparing responses of the different elements. The X-ray monitor is based upon a position sensitive Xe proportional counter with a Uniformly-Redundant Array (URA) mask. Its energy range (2-25 keV) partially overlaps that of the gamma ray instrument, but it achieves superior energy resolution. One objective of this instrument is to search for previously undetected line features in this portion of GRB energy spectra. The imaging capability of the X-ray monitor will provide burst positions accurate to about 6 arcmin. The UV camera utilizes a CCD chip that has been developed for low noise ground- and space-based missions, including the ASTRO-D and AXAF CCD sensors, while the overall experiment design concept has been proven in the ground-based Explosive Transient Camera [18]. It will localize bright 2 transient events occurring within its 2.5 sr field of view to an accuracy of <6 arcseconds. *Note that this localization is carried out onboard the spacecraft.*

Table I: Scientific Instruments for the HETE Mission

Omnidirectional γ-ray Spectrometer

Instrument Type	NaI(Tl); cleaved
Energy Range	6 keV to >1 MeV
Timing Resolution	4 ms
Spectral Resolution	~40% @ 6 keV
	~7% @ 662 keV
Detector Quantum Efficiency	80% @ 10 keV
	40% @ 1 MeV
Effective Area	120 cm^2 (total for 8 units)
Sensitivity ("10σ")*	3×10^{-8} erg cm^{-2} s^{-1} over the 8 keV-1 MeV energy range
Field of View	~2π sr
Angular Resolution	~π sr

Wide-Field X-ray Monitor

Instrument Type	Coded Mask With Position-Sensitive Proportional Counter
Energy Range	2 to 25 keV
Timing Resolution	1 ms
Spectral Resolution	~15% @ 6 keV
Detector Quantum Efficiency	90% @ 5 keV
Effective Area	~100 cm^2 (each of 4 units)
Sensitivity ("10σ")*	~8×10^{-9} erg cm^{-2} s^{-1} over the 2-10 keV range
Field of View	~2.5 sr (total for 4 units)
Angular Resolution ("1σ")	±6 arcmin

Ultraviolet Transient Camera Array

Instrument Type	UV CCD Detector with 25 mm f/1.5 optics
Energy Range	4 eV to 7 eV (1800 to 3100 Å)
Timing Resolution	~1 s
Spectral Resolution	N/A (Broadband Imaging Photometer)
Detector Quantum Efficiency	15% to 35%
Effective Area	3 cm^2 (each of 4 units)
Sensitivity ("10σ")	5×10^{-10} erg cm^{-2} s^{-1}
Field of View	~2.5 sr (total for 4 units)
Angular Resolution ("1σ")	± 3 arcsec

* Assuming a 250 keV thermal Bremsstrahlung spectra

Figure 1 compares the instrument sensitivities with measurements of transients in the optical, X-ray, and gamma-ray regions, and with predictions for the UV region. In the optical range, the transient observed by Pedersen et al.[10] from the March 5, 1979 GRB source is shown, as well as one of the archival transients discovered by Schaefer[8]. The predictions of Melia[19] and Melia et al.[21] are shown for the optical and UV ranges. Note that all fluxes are considerably greater than the UV camera sensitivity, shown for integration times of 1 to 1000 s. The X-ray and gamma-ray instrument sensitivities are compared to typical X-ray burst fluxes, the so-called "Jan 7" soft repeater (SGR 1806-20)[22,23], and the 1972 April 27 GRB which was observed over an exceptionally wide energy range [24]. The HETE sensitivity is sufficient to increase the number of well-located transients by about an order of magnitude over the mission, while decreasing the area of their error boxes by 1 - 2 orders of magnitude with respect to the areas which would be obtainable by interplanetary "triangulation" using the gamma ray emission alone. Excellent sensitivity and localization capability are not the only unique features of HETE, however. Four equally important aspects of the mission are:

1. Detection of transients in the UV region inaccessible to ground-based instruments. In this unexplored wavelength range, copious emission is predicted by several current models [19,20,21].

2. Rejection of many of the spurious sources of transient events such as satellite glints (whose UV spectrum should be a factor of ~20 below their visible spectrum) and meteors (since HETE operates above the atmosphere).

3. Detection of UV, X-, and gamma ray emission *simultaneously* aboard a single single spacecraft. This will resolve the important issue of the relative timing and time profiles of the emissions at these energies.

4. Transmission of exceptionally accurate burst position information to the ground in near real time (~seconds to ~minutes) to optical and other observatories, thereby allowing searches for short-lived emission to be carried out. This aspect is explained in more detail below.

THE HETE MISSION

HETE has been designed for launch from a Get Away Special (GAS) cannister on the Shuttle; it is based on a derivative of the "Cheapsat" concept of Defense Systems, Incorporated (DSI). Some of the mission specifications are presented in Table 2. GLOMR, a similar spacecraft to the one planned for HETE, was succcessfully launched from a GAS can in 1985; numerous other missions using this concept are now either launch ready or undergoing construction, and will be launched when Shuttle flights resume. Low cost, reliability, and simplicity are the key features of these systems. Simplicity of operation will also be an important feature of the ground-based portion of the mission. Following initial in-orbit checkout from a DSI ground station in McLean, VA, uplink commanding and data downlinking, as well as tracking and ephemeris maintenance, will be provided by a single MIT-operated ground station based on a PC-XT class controller.

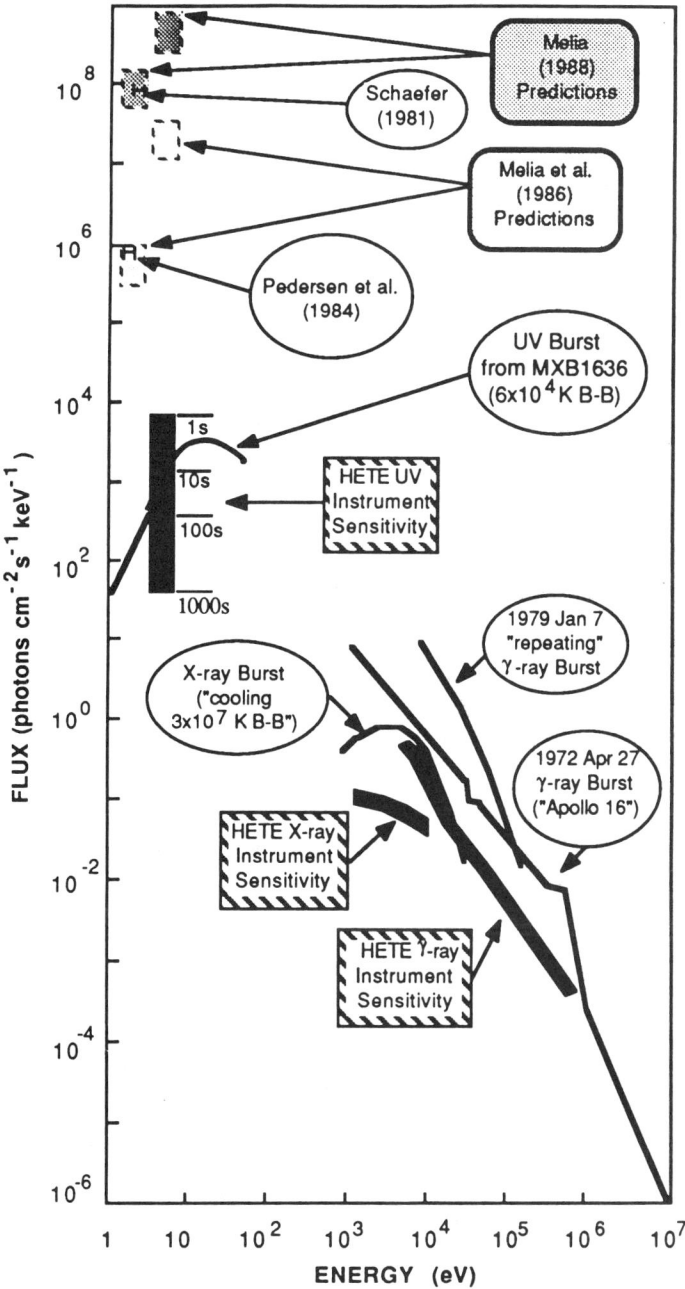

Figure 1. HETE instrument sensitivities compared to observed and predicted transient intensities.

Table II: The HETE Spacecraft and Mission

HETE Spacecraft

Dimensions(diameter x height)*	45 cm diameter x 34 cm (18 in diameter x 13.5 in)
Design Source	DSI "Cheapsat" - derivative
Stabilization	Single Axis, < 1 deg/hr Drift Rate
Orientation	
- Pitch and Yaw	Sun Point
- Roll	Earth's Magnetic Field
Attitude Control	Single Momentum Wheel With Magnetic Torquers
Configuration	Science Instruments on Anti-Sun Side (CCDs radiatively cooled to -40 deg C)
Solar Panels	Fixed, 1 m^2 area
Data Rate***	2000 bits/s
Antenna	Omnidirectional
Power Available to HETE instruments	30 W (orbit night) 100 W (orbit day)
Weight	
Spacecraft (including inert propellant)	145 lbs
Available for HETE instruments	105 lbs

HETE Mission

Orbital Specification:	
Shuttle Deployment	160 nm (nominal)
Operational Altitude	300 nm (circular)
Lifetime	> 5 years
Ground Station Location:	
During ejection and transfer orbit:	McLean, VA
During mission operations (2 options):**	
For 57° inclination orbit	Haystack Observatory (Westford, MA)
For 28.5° inclination orbit	Kitt Peak Observatory (Kitt Peak, AZ)
Telemetry	UHF to DSI-provided Automated Ground Station

---

*   Overall dimensions including spacecraft and instruments are 45 cm diameter x 89 cm (18 in diameter x 35 in high).

**   Inclinations between these values are also acceptable; these 2 cases are considered as explicit examples.

***   Orbit average, including spacecraft and instruments. During orbit night, the average data rate will be increased to 4000 bits/sec to accommodate the UV camera.

A NEW CONCEPT FOR DATA DISSEMINATION AND SHARING

A unique feature of HETE is the capability to downlink data along the line of sight to simple, inexpensive PC-based receive-only ground stations, using the Cheapsat "message forwarding" capability. This means that an observatory at practically any latitude can be equipped to receive burst information in real-time, automatically, during several passes extending over an interval of several hours per day (Figure 2). The arrangement for the standardized observatory station is illustrated schematically in Figure 3. As a minimum, it is intended to transmit burst times, positions (to several arcseconds), and UV, X-ray and gamma-ray fluxes to these ground stations. Thus sensitive follow-up searches for short-lived emissions can take place ~ minutes to ~ hours after the detection of a burst.

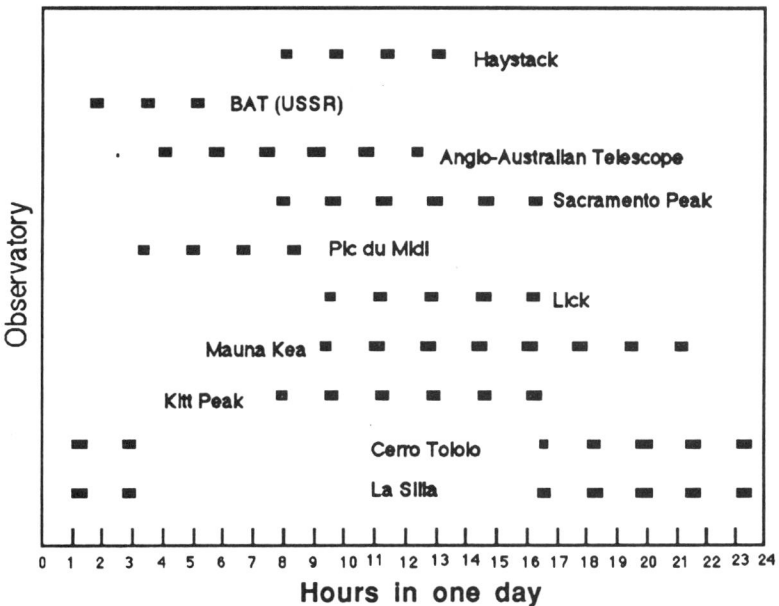

Figure 2 A plot of orbital passes for HETE over an entire day, for an orbit inclination of + 28°. The pass durations (length of short dashes) are shown for ten representative ground-based observatories, covering a wide range of longitudes. The number of successive "usable" passes (duration > 100 s) ranges from ~4 from high latitude observatories to ~8 for low latitude observatories. Each observatory could be equipped with an inexpensive automatic data receiver, comparable in complexity to the "WWV-B type" time service receivers which are standard equipment at observatories (Figure 3).

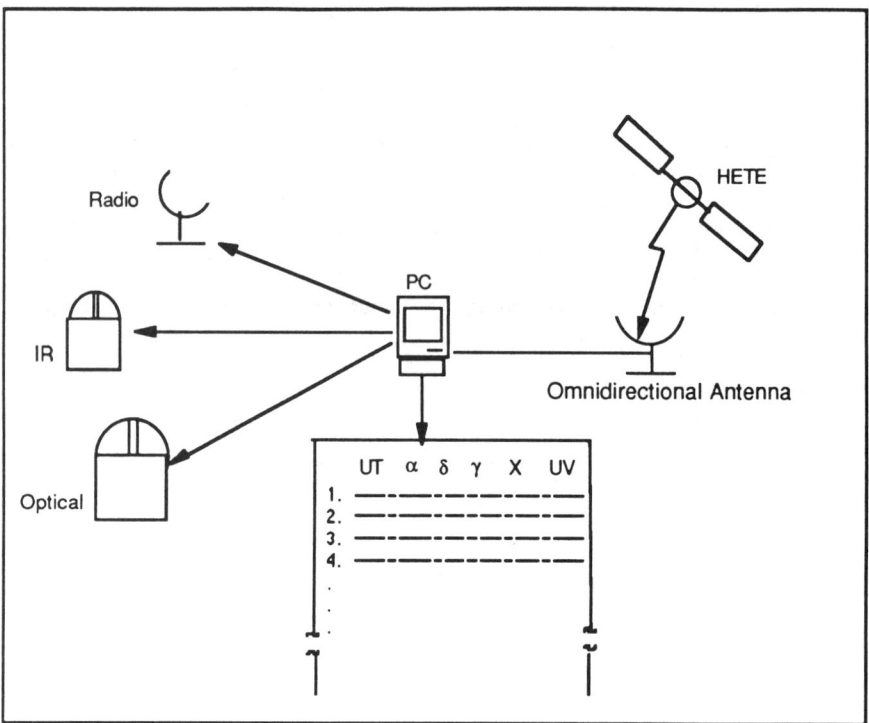

Figure 3. A concept for data dissemination and sharing in near real-time based upon HETE's message-forwarding capability. Simple, inexpensive PC-based receive-only ground stations can be acquired by observatories, allowing burst times, positions, and fluxes to be received automatically whenever the spacecraft is within the line of sight. For typical latitudes, this may occur for several passes spaced over an interval of several hours per day (Figure 2).

CONCLUSIONS

HETE is a conceptually simple, yet state-of-the-art mission which will revolutionize GRB studies by increasing the number of small error boxes, reducing their size, and making them available to the astronomical community in near real-time. Its multiwavelength approach will for the first time shed light upon important issues related to the GRB emission mechanism and source geometry. While this is the prime objective of HETE, the mission will also achieve several other objectives, such as monitoring the sky for flare stars and X-ray bursters, as well as the detection of entirely new classes of astronomical transients. The technology required for HETE has only recently come of age, but is fortunately readily accessible. The construction of the spacecraft and experiments would require only about 3 years, making it possible to have HETE operational in the early 1990's.

REFERENCES

1. J.-L. Atteia, C. Barat, K. Hurley, M. Niel, G. Vedrenne, W. D. Evans, E. E. Fenimore, R. W. Klebesadel, J. G. Laros, T. Cline, U. Desai, B. Teegarden, I. V. Estulin, V. M. Zenchenko, A. V. Kuznetsov, and V. G. Kurt, *Ap. J. Supp.* **64**, 305 (1987).

2. B. E. Schaefer, T. L. Cline, U. Desai, B. J. Teegarden, J.-L. Atteia, C. Barat, K. Hurley, M. Niel, W. D. Evans, E. E. Fenimore, R. W. Klebesadel, J. G. Laros, I. V. Estulin, and A. V. Kuznetsov, *Ap. J.* **313**, 226 (1987).

3. H. Pedersen, C. Motch, M. Tarenghi, J. Danziger, G. Pizzichini, and W. Lewin, *Ap. J. Lett.* **270**, L43 (1983).

4. B. Schaefer, P. Seitzer, and H. Bradt, *Ap. J. Lett.* **270**, L49 (1983).

5. C. Motch, H. Pedersen, S. A. Ilovaisky, C. Chevalier, K. Hurley, and G. Pizzichini, *Astron. Astrophys.* **145**, 201 (1985).

6. G. Pizzichini, M. Gottardi, J.-L. Atteia, C. Barat, K. Hurley, M. Niel, G. Vedrenne, J. G. Laros, W. D. Evans, E. E. Fenimore, R. W. Klebesadel, T. L. Cline, U. D. Desai, V. G. Kurt, A. V. Kuznetsov, and V. M. Zenchenko, *Ap. J.* **301**, 641 (1986).

7. M. Boer, J.-L. Atteia, M. Gottardi, K. Hurley, M. Niel, C. Barat, G. Pizzichini, K. Mason, G. Branduardi-Raymont, F. Cordova, J. G. Laros, W. D. Evans, E. E. Fenimore, R. W. Klebesadel, M. R. Sims, and C. Martin, *Astron. Astrophys.* (accepted), (1988).

8. B. Schaefer, *Nature* **294**, 722 (1981).

9. B. E. Schaefer, H. V. Bradt, C. Barat, K. Hurley, M. Niel, G. Vedrenne, T. L. Cline, U. Desai, B. J. Teegarden, W. D. Evans, E. E. Fenimore, R. W. Klebesadel, J. G. Laros, I. V. Estulin, and A. V. Kuznetsov, *Ap. J. Lett.* **286**, L1 (1984).

10. H. Pedersen, J. Danziger, K. Hurley, G. Pizzichini, C. Motch, S. Ilovaisky, N. Gradmann, W. Brinkmann, G. Kanbach, E. Rieger, C. Reppin, W. Trumper, and N. Lund, *Nature* **312**, 46 (1984).

11. R. Hudec, J. Borovicka, S. Danis, V. Franc, R. Peresty, and B. Valnicek, paper presented at the COSPAR/IAU Symposium on "The Physics of Compact Objects: Theory vs. Observations", Sofia, Bulgaria, July 1987, to be published in *Advances in Space Research*, 1988.

12. E. I. Moskalenko, G. V. Popravko, E. N. Kramer, I. S. Shestaka, A. N. Karnashov, V. V. Nazarenko, L. Ja. Skoblikova, V. F. Lemeschenko, S. V. Nazarenko, and Ju. M. Gorbanev, paper presented at the COSPAR/IAU Symposium on "The Physics of Compact Objects: Theory vs. Observations",

Sofia, Bulgaria, July 1987, to be published in *Advances in Space Research*, 1988.

13. Panel Discussion, in Gamma Ray Transients and Related Astrophysical Phenomena, Eds. R. Lingenfelter, H. Hudson, and D. Worrall, AIP Conference Proceedings No. 77, AIP Press, N.Y., 497 (1982).

14. G. Field, Chairman, in Astronomy and Astrophysics for the 1980's, Volume 1, Report of the National Research Council Astronomy Survey Committee, National Academy Press, Washington, DC, **142** (1982).

15. Evans ad hoc Committee Discussions, W. D. Evans and S. Woosley, 1982.

16. S. Woosley (Committee Chairman), in High Energy Transients in Astrophysics, Ed. S. Woosley, AIP Conference Proceedings No. 115, AIP Press, N.Y., **709** (1984).

17. E. P. Liang and V. Petrosian (Eds.) in Gamma Ray Bursts, AIP Conference Proceedings No. 141, AIP Press, N.Y., **63** (1986).

18. G. R. Ricker, J. P Doty, J. V. Vallerga, and R. K. Vanderspek, in High Energy Transients in Astrophysics, Ed. S. Woosley, AIP Conference Proceedings No. 115, AIP Press, N.Y., **669** (1984).

19. F. Melia, *Ap. J. Lett.* **324**, L1 (1988).

20. D. Hartmann, S. Woosley, and J. Arons, *Ap. J.* (accepted), (1988).

21. F. Melia, S. Rappaport, and P. Joss, *Ap. J. Lett.* **305**, L51 (1986).

22. J.-L. Atteia, M. Boer, K. Hurley, M. Niel, G. Vedrenne, E. E. Fenimore, R. W. Klebesadel, J. G. Laros, A. V. Kuznetsov, R. A. Sunyaev, O. V. Terekhov, C. Kouveliotou, T. Cline, B. Dennis, U. Desai, and L. Orwig, *Ap. J. Lett.* **320**, L105 (1987).

23. J. G. Laros, E. E. Fenimore, R. W. Klebesadel, J.-L. Atteia, M. Boer, K. Hurley, M. Niel, G. Vedrenne, S. R. Kane, C. Kouveliotou, T. L. Cline, B. R. Dennis, U. D. Desai, L. E. Orwig, A. V. Kuznetsov, R. A. Sunyaev, and O. V. Terekhov, *Ap. J. Lett.* **320**, L111 (1987).

24. A. Metzger, R. Parker, D. Gilman, L. Peterson, and J. Trombka, *Ap. J. Lett.* **194**, L19 (1974).

THE NUCLEAR ASTROPHYSICS EXPLORER, A PROPOSED EXPLORER MISSION FOR THE 1990s

J. L. Matteson
Center for Astrophysics and Space Sciences
University of California, San Diego, La Jolla, CA 92093

B. J. Teegarden
Laboratory for High Energy Astrophysics
Goddard Space Flight Center, Greenbelt, MD 20771

W. A. Mahoney
Earth and Space Sciences Division
Jet Propulsion Laboratory, Pasadena, CA 91109

ABSTRACT

The Nuclear Astrophysics Explorer was proposed for NASA's Explorer Concept Study Program by an international collaboration of 25 scientists from 9 institutions. The proposal was accepted and a one year study began in June 1988. The Nuclear Astrophysics Explorer would obtain high resolution observations of gamma-ray lines, $E/\Delta E \sim 1000$, at a sensitivity of $\sim 3 \times 10^{-6}$ ph/cm^2-s, in order to study fundamental problems in astrophysics such as nucleosynthesis, supernova dynamics, and neutron star and black hole physics. The instrument would operate from 15 keV to 10 MeV and use an array of cooled Ge spectrometers in a very low background configuration. A 10° FWHM field of view would contain a versatile coded mask system which would provide 4° imaging and the ability to efficiently measure diffuse emission and study it with 2° resolution.

INTRODUCTION

Gamma-ray lines are the most direct probe of cosmic nuclear processes. High resolution spectroscopy of these lines can provide basic information on fundamental astrophysical problems such as nucleosynthesis, supernova dynamics, and neutron star and black hole physics. Gamma-ray line observations have, in fact, already made major contributions to our understanding of a variety of astrophysical objects and phenomena. Table I contains a summary of these observations which were made with instruments that had relatively low sensitivity and, in many cases, low energy resolution. The observations' statistical significance is $> 10\sigma$ for the solar flare lines and 3-10σ for most of the others. Recent advances in technology now permit us to build an Explorer-class high resolution gamma-ray spectrometer that is a factor of 100 more sensitive than those previously flown. Such a major advance in observational capability will

TABLE I ASTROPHYSICAL GAMMA RAY LINE OBSERVATIONS

Process	Observed Energy	Source	Flux ph/cm²-s	Ref
$e^{\pm}$ Annihilation	511	Galactic Center	$0.6\text{-}1.8 \times 10^{-3}$	10-12
Radiation	511	Interstellar Gas	$\sim 2 \times 10^{-3}$/rad	13
	511	Solar Flares	up to ~ 0.1	14-16
	400-460	Gamma Ray Bursters	up to 70	17-19
(Redshifted)	~ 400	CrabPulsar Transient	$2\text{-}7 \times 10^{-3}$	20,21
	~ 413	10June74 Transient	7×10^{-3}	22,23
	500-2000	Cygnus X-1	up to 2×10^{-2}	24-25
Radioactive Decay				
^{56}Co(Ecg)^{56}Fe	847	Supernova 1987A	$\sim 10^{-3}$	27-32
	1238	" "	$\sim 10^{-3}$	27,29-32
	2598	" "	$\sim 10^{-3}$	31
^{26}Al($\beta^+\gamma$)^{26}Mg	1809	Interstellar Gas	4.8×10^{-4}/rad	33,34
Nuclear Excitation				
^{56}Fe (p,p'γ)	847	Solar Flares	up to ~ 0.05	14-16
^{24}Mg (p,p'γ)	1369	" "	up to ~ 0.08	14-16
^{20}Ne (p,p'γ)	1634	" "	up to ~ 0.1	14-16
^{28}Si (p,p'γ)	1779	" "	up to ~ 0.08	14-16
^{12}C (p,p'γ)	4438	" "	up to ~ 0.1	14-16,35
^{16}O (p,p'γ)	6129	" "	up to ~ 0.1	14-16
Neutron Capture				
^{1}H (n,γ)^{2}H	2223	Solar Flares	up to ~ 1	14-16,35,36
^{1}H (n,γ)^{2}H	2223	10June74 Transient	1.5×10^{-2}	22,23
(Redshifted)	1790	" "	3×10^{-2}	22,23
^{56}Fe (n,γ)^{57}Fe	5947	" "	1.5×10^{-2}	22,23
(Redshifted)				
Cyclotron Emission	30-70	Gamma Ray Bursters	up to 3	18,37-39
& Absorption in	20-58	X-Ray Pulsators	$1\text{-}3 \times 10^{-3}$	40-44
$\sim 10^{12}$ gauss fields	73-79	Crab Pulsar Transient	4×10^{-3}	21,45,46

not only lead to much deeper insight into the problems already studied, but will also open many other astrophysical questions to study, and doubtless lead to unexpected discoveries as well. These prospects motivated us to propose a high resolution gamma-ray spectrometer for NASA's Explorer Concept Study Program. Our proposal, along with 3 others, has recently been accepted for a 1 year study beginning in June 1988. We call our investigation the Nuclear Astrophysics Explorer (NAE). The NAE would operate over the 15 keV to 10 MeV energy range, with a sensitivity to narrow lines of $\sim 3 \times 10^{-6}$ ph/cm²-s and an energy resolution, $E/\Delta E$, of ~ 1000.

SCIENTIFIC MOTIVATION

Gamma-ray lines are produced in astrophysical objects by nuclear transitions, electron-positron annihilation, neutron capture and electron transitions between quantized levels in strong magnetic fields. Unique astrophysical information is encoded in line energies, shapes, and intensities; not only do lines indicate the presence of specific nuclei or electron-positron pairs, but the line parameters, i.e. intensities, centroid shifts, widths and profiles, contain information on abundances, bulk velocities, gravitational potentials, densities, temperatures, and the spectra of the exciting particles. Furthermore, the high transparency of matter to gamma-rays allows the lines to be used as tracers of processes anywhere in the Galaxy, except for stellar interiors. Gamma-ray line astrophysics has been reviewed by many authors[1-9].

As can be seen from Table I, gamma-ray lines have been observed from many physical processes. Also they have been observed from a wide variety of astrophysical objects and phenomena: supernova 1987a, nucleosynthesis, the galactic plane, the galactic center, solar flares, x-ray pulsators, the vicinity of a black hole and gamma-ray bursts. The relatively large fluxes, $\sim 10^{-4}$ to 10^{-3} ph/cm^2-s, assure that the NAE would use these lines and others that it would discover to attack a number of major problems in astrophysics: nucleosynthesis in supernovae and novae, including the discovery of young, unknown supernova remnants in the Galaxy and the sites and rates of galactic nucleosynthesis; compact objects including the enigmatic, time-variable source of electron-positron annihilation radiation in the galactic center region, Cyg X-1 and the ubiquitous gamma-ray bursters; the mixing of the interstellar gas and the elemental abundances of the interstellar dust and solar particles, the acceleration and interactions of solar flare accelerated particles and low energy cosmic rays; and the physics of the intense, $\sim 10^{12}$ gauss, magnetic fields and hot plasmas of accreting neutron stars in x-ray pulsators.

All of the results to date have been obtained with instruments of limited sensitivity, greater than a few x 10^{-4} ph/cm^2-s, and, in many cases, limited energy resolution, $E/\Delta E \sim 20$. As a result, generally only the intensities have been determined to a useful accuracy and the information carried by the other line parameters has not yet been exploited. However, observations with the NAE would obtain detailed measurements of the parameters of known lines and many weaker lines as well. It would simultaneously obtain 1) a hundred fold improvement in sensitivity to point sources of narrow lines, i.e. $\sim 3 \times 10^{-6}$ ph/cm^2-s, and a diffuse source sensitivity of $\sim 2 \times 10^{-5}$ ph/cm^2-s-rad, 2) an energy resolution that is less than most lines' predicted widths in order to reveal line profiles and centroid shifts, i.e. $E/\Delta E \sim 1000$ and 3) an angular resolution of a few degrees in order to map diffuse emission, resolve source complexes and locate sources.

INSTRUMENT CONCEPT

The NAE instrument concept is shown in Figure 1 and its parameters are given in Table II. It contains 9 large, ~ 300 cm^3 each, Ge detectors in a heavily

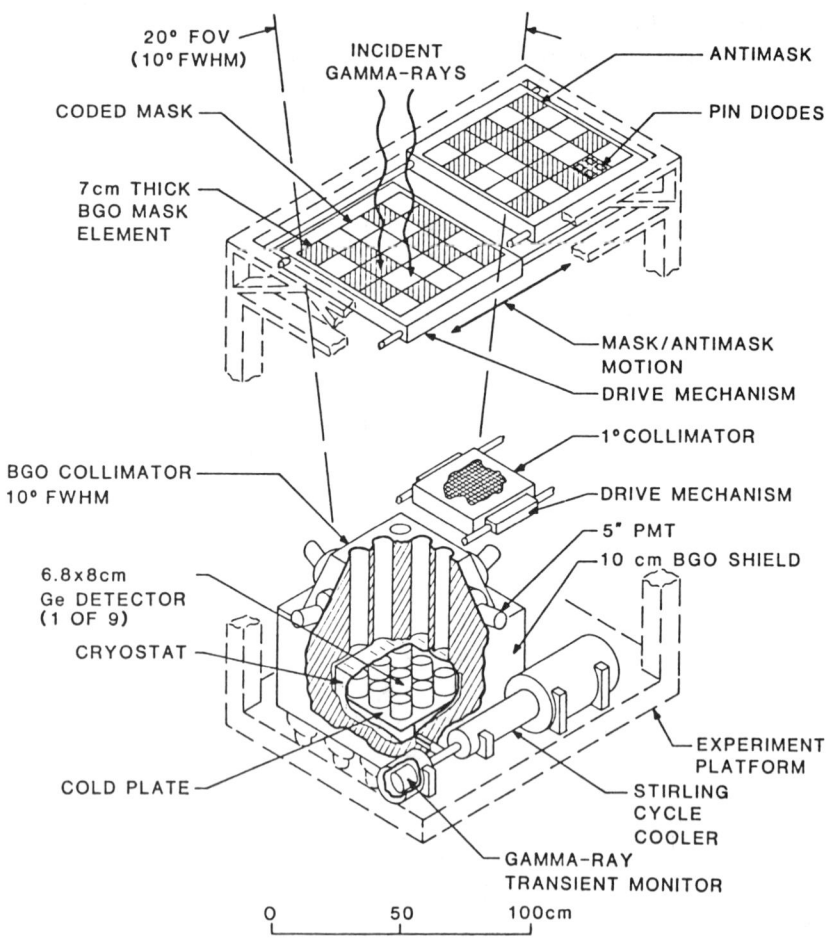

Figure 1. The Nuclear Astrophysics Explorer instrument concept. The mask/antimask is located 120 cm above the germanium detectors.

TABLE II NAE INSTRUMENT PARAMETERS

Energy Range:	15 keV to 10 MeV
Energy Resolution:	2 keV at 1 MeV
Sensitivity ($T=10^6$ s):	2 to 3 x 10^{-6} ph/cm^2-s
Field of View:	10° FWHM (Image Mode)
	20° FWOM
	1° FWHM (Fine Collimator Mode)
Angular Resolution:	4° FWHM (Image Mode)
	2° FWHM (Knife-Edge Mode)
Instrument Size:	1m(W) x 1.3m(L) x 1.8m(H)
Cryostat Thermal Load:	5W
Power:	500 W (50% electronics, 50% refrigerator)
Mass:	1075 kg
Bit Rate:	~ 15 kbps
On-Board Memory:	~ 1 Mbyte

shielded 3x3 array, and a single Ge detector with a nearly 4 π steradian field of view, which serves as a gamma-ray transient monitor. These are cooled by a Stirling Cycle refrigerator. The 3x3 array has very low background, shown in Figure 2, because of 3 essential features. 1) The detectors use position sensitivity, obtained through axial segmentation and pulse shape discrimination, to discriminate against induced radioactivity in the detectors themselves, the dominant background component in heavily shielded instruments at ~ 1 MeV. 2) A very thick, 10 cm anticoincidence shield made of bismuth germanate (BGO) attenuates the 1 MeV ambient gamma-ray background by a factor of 100. 3) A 10° FWHM field of view, defined by apertures in the BGO shield, reduces the 1 MeV background due to aperture gamma-rays to the level of the residual, non-rejected detector radioactivity. Imaging and aperture chopping for suppression of background systematics are simultaneously obtained through the use of a 5 x 5 element coded mask/antimask system. The mask code, produced by BGO elements that are opaque to gamma rays, produces a shadowgram on the detector array which is deconvolved by a matrix multiplication to produce a sidelobe-free map with 4° angular resolution. The mask and antimask are alternately placed in the aperture during imaging observations. The mask/antimask combination can also be moved together in and out of the aperture to obtain either 1) a totally blocked or totally open aperture, in order to modulate and detect diffuse flux with high efficiency, or 2) a "knife-edge" which is smoothly swept over the aperture to modulate and detect structure in diffuse sources, as well as locate point sources, with 2° angular resolution. A 1° FWHM collimator that is effective below 150 keV can be placed in the aperture to improve the sensitivity and angular resolution at low energies.

In order that the instrument concept could be accomodated by the "Explorer envelope", it was necessary to consider the tradeoff between sensitivity and imaging, since each of these affect the weight, size and cost. The conclusion was to strive for the best possible sensitivity, with adequate angular resolution to resolve the stronger sources and complement the results expected from the GRO. Our goal is to be able to measure the profile of the predicted 847 keV line from explosive nucleosynthesis of ^{56}Co in Type I supernovae in the Virgo cluster, where they are estimated to occur at least 3 times per year[47]. A narrow line sensitivity of $\sim 3\times10^{-6}$ ph/cm^2-s is necessary for this.

As a consequence of its very low background, the NAE would become background limited, at 1 MeV, at very low flux levels, $< 10^{-5}$ ph/cm^2-s. It

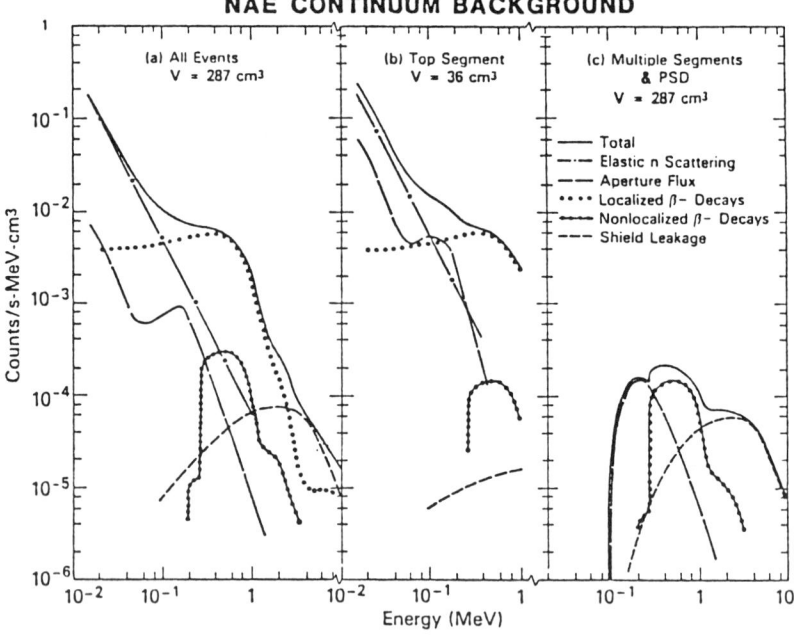

Figure 2. The predicted NAE background for 3 operating modes. In (a) the detectors operate without position sensitivity. In (b) events are required to occur only in the top 1 cm thick detector segment, reducing the background by a factor of ~ 8 while providing good efficiency below ~ 200 keV. In (c) events are required to have a multiple site signature by depositing energy in more than one of the eight 1 cm thick segments or having the pulse shape of a radially dispersed energy loss. These methods elimate the background produced by localized β^--decays due to induced radioactivity, allowing a factor of ~ 20 reduction in the total background and good efficiency above ~ 200 keV.

would obtain sensitivity to larger fluxes very rapidly. Only 1/2 hour would be required to reach 2×10^{-4} ph/cm^2-s, the limiting flux for previous instruments, and 1 day to reach 10^{-5} ph/cm^2-s. With 10^6 s of good data, the sensitivity limit of $\sim 3 \times 10^{-6}$ ph/cm^2-s would be reached. This sensitivity is shown in Figure 3 along with those of the HEAO-3 and GRO/OSSE experiments. The NAE will have a sensitivity and energy resolution that are 10 and 30 times better at 1 MeV, than the OSSE. Its sensitivity would be 100 times better than the HEAO-3. This results from the careful consideration of the many factors which affect sensitivity. These and their values are listed in Table III. The NAE's

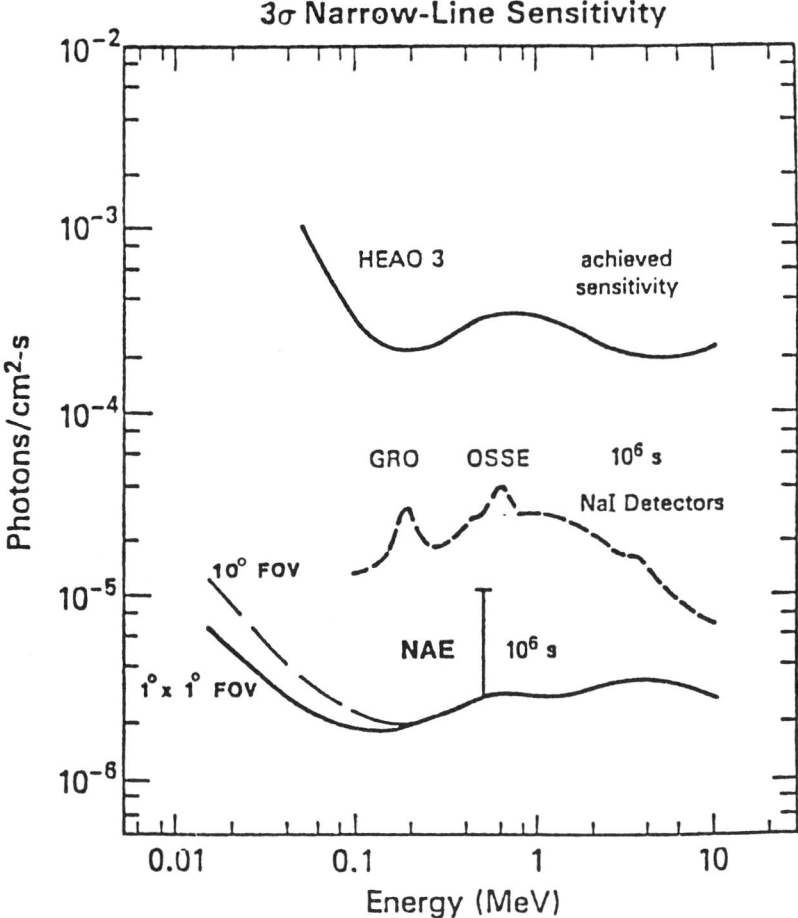

Figure 3. The predicted sensitivity of the NAE to narrow gamma-ray lines in a 10^6 sec observation. For comparison the predicted sensitivity of the GRO/OSSE and the achieved sensitivity of the HEAO-3, which typically obtained $\sim 2 \times 10^5$ seconds of data on a source.

TABLE III SENSITIVITY FACTORS SCALING HEAO 3 TO NAE AT 1 MeV

Item		Sensitivity Factor
Larger Detector Volume (6 x)		2.5
More Time on Individual Targets due to Pointed S/C (6 x)		2.5
More Livetime in Lower Background Orbit (2 x)		1.4
Higher Resolution Detectors		1.1
Larger Signal Modulation		1.1
Lower Instrumental Background		
Lower inclination orbit (28° vs 45°)	1.3	
Better Compton suppression*	1.3	
Less passive mat., lower shield thresh.	1.2	
Less shield leakage*	1.7	
Segmentation and PSD	2.8	
Smaller field of view	1.1	
Product	10.6	
TOTAL INSTRUMENT SENSITIVITY FACTOR (Product of Factors)		112.

* Due to improved BGO shield. HEAO C1 shield was 6.6 cm CsI.

sensitivity and versatility will allow it to be used to pursue many astrophysical problems with relatively brief observations, e.g. a few hours to a few days. Therefore, the observing program would extend beyond the instrument team and involve a large number of scientists who would use the NAE as a facility.

The NAE instrument concept incorporates several new design features, e.g. very large volume Ge detectors, a large array of detectors, a massive BGO anticoincidence shield, the moving mask/antimask and a long-life mechanical refrigerator. All of these are presently being proven in the laboratory or balloon flight instruments.

FEASIBILITY STUDY PHASE PLANS

The NAE concept is based on the work of research groups at 9 institutions in the US and France which have been pursuing these objectives for up to 20 years through studies in theoretical astrophysics, laboratory developments and a series of increasingly advanced instruments flown in space and on high altitude balloons. This group of approximately 25 scientists, plus supporting technical staffs is now beginning a 1 year feasibility study of the NAE mission, to be completed in May 1989. The study will fulfill several objectives. All mission requirements will be developed. The instrument, spacecraft, integration, launch, and ground system, will be defined. A management plan, risk assessment and risk offset plan will be developed. An estimated cost for the mission will be developed. NASA will review the study phase final reports and decide which programs will go into development for spaceflight. The announcement of this selection is expected in late 1989 and the earliest spaceflight opportunity is expected to occur in 1995.

ACKNOWLEDGEMENTS

This paper is based on the proposal for the NAE study, which resulted from the work of many people. The contributions of Neil Gehrels, Reuven Ramaty, Jim Higdon, Bill Wheaton, Michael Pelling, Richard Lingenfelter and Marvin Leventhal were particularly important.

REFERENCES

1. D.D. Clayton, in Gamma-Ray Astrophysics, eds. F. W. Stecker and J.I. Trombka (NASA SP-339, Washington, DC, 1973) p. 263.
2. R. Ramaty and R. E. Lingenfelter, Ann. Rev. Nucl. Part. Sci. 32, 235 (1982).
3. J. L. Matteson, Adv. Space Res. 3, No. 4, 135 (1983).
4. R. E. Lingenfelter and R. Ramaty, in Conference Papers of 19th International Cosmic Ray Conference, Vol. 5, 1985, p. 19.
5. L. E. Peterson, in these proceedings.
6. R. E. Lingenfelter, in these proceedings.
7. J.C. Higdon, in these proceedings.
8. R. Ramaty, in these proceedings.
9. E. L. Chupp, in these proceedings.
10. R. C. Haymes et al., Ap. J. 201, 593 (1975).
11. M. Leventhal et al., Ap. J. (Letters) 225, L11 (1978).
12. G. R. Riegler et al., Ap. J. (Letters) 248, L13 (1981).
13. G. H. Share et al., Ap. J. 326, 717 (1988).
14. E. L. Chupp et al., Nature 241, 333 (1973).
15. M. Yoshimori et al., Solar Physics 86, 375 (1983).
16. E. L. Chupp et al., Ann. Rev. Astr. Ap. 22, 359 (1984).
17. E. P. Mazets et al., Nature 282, 587 (1979).
18. E. P. Mazets et al., Nature 290, 378 (1981).
19. B. J. Teegarden and T. L. Cline, Ap. J. (Letters) 236, L67 (1980).
20. M. Leventhal et al., Ap. J. 216, 491 (1977).
21. C. A. Ayre et al., Mon. Not. R. Astr. Soc. 205, 285 (1983).
22. A. S. Jacobson et al., in Gamma-Ray Spectroscopy in Astrophysics, eds. T. L. Cline and R. Ramaty (NASA TM 79619, Greenbelt, MD, 1978) p. 228.
23. J. C. Ling in Gamma-Ray Transients and Related Astrophysical Phenomena, eds. R. E. Lingenfelter, H. S. Hudson and D. M. Worrall (AIP, New York,

1982) p. 143.
24. P. L. Nolan and J. L. Matteson, Ap. J. 265, 389 (1983).
25. J. C. Ling et al., Ap. J. (Letters) 321, L117 (1987).
26. J. C. Ling, in these proceedings.
27. M. Matz et al., IAU Circular No. 4568 (1988).
28. W. Sandie et al., IAU Circular No. 4526 (1988).
29. W. R. Cook et al., IAU Circular No. 4527 (1988).
30. W. A. Mahoney et al., IAU Circular No. 4584 (1988).
31. A. C. Rester et al., IAU Circular No. 4535 (1988).
32. S. Barthelmy et al., IAU Circular No. 4593 (1988).
33. W. A. Mahoney et al., Ap. J. 286, 578 (1984).
34. G. H. Share et al., Ap. J. (Letters) 292, L61 (1985).
35. H. S. Hudson et al., Ap. J. (Letters) 236, L91 (1980).
36. T. Prince et al., Ap. J. (Letters) 255, L81 (1982).
37. B. R. Dennis et al., in Gamma-Ray Transients and Related Astrophysical Phenomena, eds. R. E. Lingenfelter, H. S. Hudson and D. M. Worrall (AIP, New York, 1982) p. 153.
38. G. J. Hueter in High Energy Transients in Astrophysics, ed. S. E. Woosley (AIP, New York 1984) p. 373.
39. T. Murakami et al., Nature, in press (1988).
40. J. Trümper et al., Ap. J. (Letters) 219, L105 (1978).
41. W. A. Wheaton et al., Nature 282, 240 (1979).
42. D. E. Gruber et al., Ap. J. (Letters) 240, L127 (1980).
43. J. Tueller et al., Ap. J. 279, 177 (1984).
44. G. S. Maurer et al., Ap. J. 254, 271 (1982).
45. J.C. Ling et al., Ap. J. 231, 896 (1979).
46. M.S. Strickman et al., Ap. J. (Letters) 253, L23 (1982).
47. S. E. Woosley et al., Comments Nucl. Part. Phys. 9, 185 (1981).

ABOUT THE ABILITY OF GRASP TO MEASURE DIFFUSE GAMMA-RAY LINE SOURCES

Ph.Durouchoux[1]; G.Bignami[2]; A.Dean[3]; N.Lund[4]; B.McBreen[5]; V.Schonfelder[6]; B.Swanenburg[7]; G.Vedrenne[8]; C.Winkler[9].

-[1] CEN Saclay BP N°2 F 91,Gif sur Yvette France - [2] IFCTR/CNR Via Bassini 15,20133 Milano,Italy. -[3] Physics Dept.univ.of Southampton, UK. -[4] Danish Space Research Inst.Copenhagen,Denmark - [5] Physics Dept.University College,Dublin 4,Ireland - [6] Max Planck Institute for Physics and Astrophysics D-8046 Garching,FRG. - [7] ROL,Leiden,The Netherlands. -[8] CESR,9 av du colonel Roche 31029 Toulouse,France -[9] ESA/ESTEC Postbus 299,22400 AG Noordwijk,The Netherlands.

ABSTRACT

The GRASP mission (*Gamma-ray Astronomy with Spectroscopy and Positioning*) is currently under study as an ESA space astronomy mission to be launched in the mid 90's.GRASP is designed as a high quality spectral imager ($E/\Delta E \approx 1000$ at 1 Mev) with positioning to the arc minute level within a large field of view ($\approx 7°$) which operates over a wide spectral range (30 Kev-100 Mev) with a 3σ sensitivity of typically 10 mcrab or better over the entire operational range within an observational period of $\approx 10^5$ seconds. In this paper,we will mainly discuss the capability of the instrument with respect to the study of both point source and diffuse source measurements.

© 1988 American Institute of Physics

I INTRODUCTION

The GRASP telescope, which may be the first high resolution spectral imager to operate in the gamma-ray region in space will have the following features:

- a wide operational bandwidth (30 Kev to 100 Mev) which for the first time, links X-ray and γ ray astronomy.
- high resolution spectrometry over the range 30 Kev to 10 Mev ($E/\Delta E \approx 1000$ at 1 Mev)
- accurate source positioning (typically $\approx$ 1 arc min) within a field of view of $\approx$ 50 square degrees.
- high sensitivity for both extended and point sources (typically 10 mcrab at 3 σ in 10^5 s)

These wide ranging goals are achieved by the use of combination of a coded aperture mask and a position sensitive detector plane. The gamma-ray detector plane consists of an assembly of CsI(Tl) scintallators and germanium solid state detectors arranged into an overall position sensitive array.

II DESCRIPTION OF THE GRASP TELESCOPE

The principal characteristics of the GRASP instrument (fig 1) are:

- a mosaic of 24 planar type hyperpure Ge detectors for low energy photon detection and photon localization. Each planar has orthogonal strips etched on both sides of the detector, so that a gamma-ray which interacts, liberates electron-hole pairs, which are separately collected at the upper and lower electrodes, thereby localizing the interaction position with a spatial accuracy of 9 mm in the XY plane. The total geometric area of the planars is 510 cm^2.
- a stack of 5 planars is located under each position sensitive planar, which works as a large volume detector for high energy photons (> few hundreds of kev to 10 Mev).

The overall spectral resolution is of the order of 1.7-2 Kev at 1 Mev and the geometric area 620 cm^2. The configuration, advantageous in terms of the rejection of β decay background, has a low total background level and is resistant to radiation damage.

- a Stirling cycle cooler system associated with a passive radiator for the solid state spectrometer make a long mission possible (>3 years)

- an active shield for the solid state spectrometer consisting of an array of CsI scintillators (mean thickness: 12 cm)

Another part of the γ ray detector plane consists of an array of 3 D position sensitive CsI(Tl) scintillators capable of both locating and measuring the energy deposited by particle interactions throughout the detector volume, corresponding to a sensitive area of ≈ 2500 cm^2.

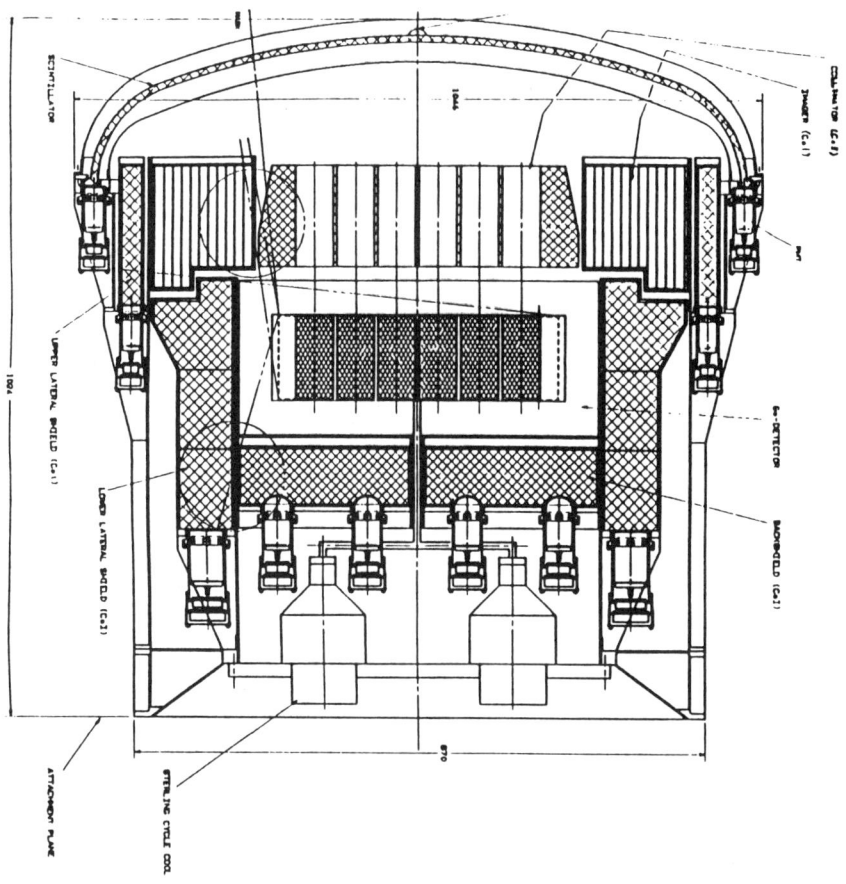

Fig 1 Schematic diagram of the GRASP telescope

- an hexagonal URA mask located about 4 meters "above" the detector plane gives a point source location capability of the order of the one arc minute.

The narrow line and broad line sensitivities of the germanium and CsI detectors for 10^5 s. and 10^6 s. observation periods are shown in fig 2 and 3.Typically these 3σ sensitivities are of the order of a few times 10^{-6} photons. $cm^{-2}.s^{-1}$ at 1 Mev for narrow lines and a few times 10^{-7} photons.$cm^{-2}.s^{-1}$ at 10 Mev for broad lines.

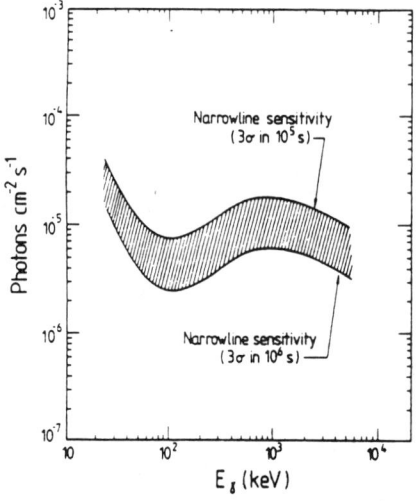

Fig 2: Narrow line sensitivity of the GRASP telescope (Ge detectors)

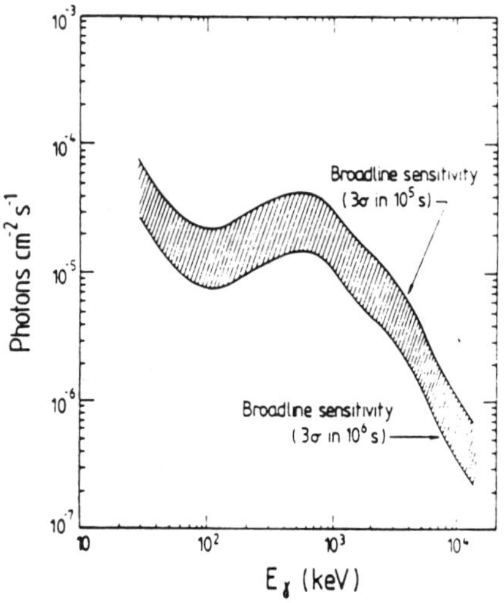

Fig 3: Broadline sensitivity of the GRASP telescope (CsI(Tl)) detectors

The continuum sensitivity for GRASP is presented in fig 4 (where a 10 mcrab spectrum is shown as reference)

III ASTROPHYSICAL OBJECTIVES

1- Extragalactic objects

The study of active galaxies is to be one of the major features of the mission.The unambiguous identification of a large number of active galaxies will lead directly to the compilation of a gamma-ray luminosity function for these objects in the region of the spectrum where their liminosity is at a maximum.Futhermore the detailed study of red-shifted electron-positron annihilation lines from these distant sources has fundamental cosmological implications.

Precise measurements of both the line and continuum spectra will provide a revealing probe of the physics in the vicinity of the compact objects associated with extragalactic nuclei.

Furthermore,a study of explosive nucleosynthesis in local

supernovae (<10 Mpc) will be possible.

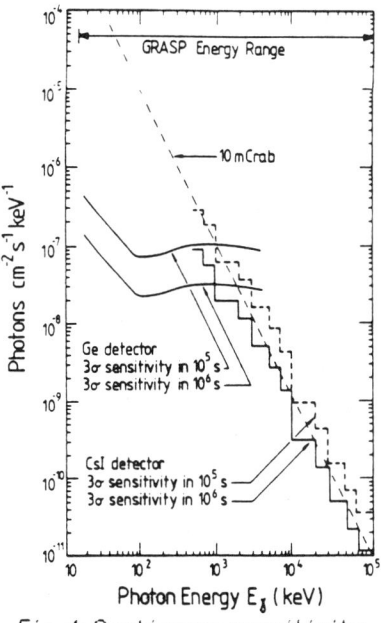

Fig 4: Continuum sensitivity of the GRASP telescope

2- Galactic objects

In the context of our galaxy, GRASP will discover new gamma-ray sources, map extended objects, locate point sources precisely (≈ 1 arc min) analyse their emission spectra with high resolution and study the variability of a wide variety of spectral objects, with special emphasis on the galactic center.

A picture of the distribution of the recent products nucleosynthesis in the galaxy will be derived by mapping key emission lines Al^{26} and $e^- + e^+$ annihilation.

This has direct bearing on the study of stellar nucleosynthesis e.g. in the Red Giant and Wolf-Rayet phases.

Also galactic novae are exciting targets for high resolution spectroscopy studies, as they are potential explosive nucleosynthesis sites. Finally, γ rays coming from the interaction of cosmic rays with the interstellar medium are interesting goals for observations.

IV MISSION AND SCENARIO

This mission is under phase A study at the European Space Agency for a final selection which will take place in November 1988. This phase A is accompanied by industrial studies with the aim to accomodate GRASP on an existing space platform (ROBUS, ISOBUS,..) which might allow about 10^3 pointings in the sky, lasting 10^5 to 10^6 s. This would maximise the scientific output of the mission, and offer a significant opportunity of a beneficial fallout on the wider astronomical community through an Associate Observer Program which will be the subject of preliminary discussions during the GRASP Workshop scheduled at the Observatoire de PARIS-MEUDON on May 31st-June 1st & 2nd 1988.

ANGAS: A NEW SPACEBORNE HIGH RESOLUTION GAMMA-RAY SPECTROMETER

G. H. Nakano, J. R. Kilner, M. J. Murphy, M. H. Vartanian
Lockheed Palo Alto Research Laboratory, Palo Alto, CA 94304

G. P. Lasché
DARPA, Arlington, VA 22209

ABSTRACT

The Advanced Nuclear Gamma-ray Analysis System (ANGAS) is a spaceborne, high resolution gamma-ray spectrometer which incorporates a rotational aperture synthesis imaging system. The primary sensor array comprises 19 large, cryogenically cooled, high purity n-type Ge detectors sensitive to gamma-rays from about 20 keV to 10 MeV. These sensors are surrounded by a massive NaI(Tl) shield that provides Compton and external background suppression and serves independently as a large omni-directional spectrometer with medium resolution. Holes drilled through the top of the shield define a 15° FWHM field-of-view for the Ge sensors. An angular resolution of about 1.2° FWHM is obtained by the imaging system. The instrument will be flown in low earth orbit on the P86-2 USAF satellite in early 1991 and is designed to operate in space for 3.4 years. A brief description of the instrument and its sensitivity to cosmic gamma-ray sources is presented.

INTRODUCTION

The detection and precise identification of gamma-ray line spectra provide direct evidence of the high energy processes and resulting nuclear interactions taking place at various astrophysical sites. The recent discoveries of line emission from SN1987a in the LMC is a cogent case in point.[1,2] These efforts have included balloon-borne observations using high resolution Ge spectrometers[1] as part of two campaigns from Alice Springs, Australia, in May and October of this year (1987). Numerous theoretical studies and light curve predictions[3] have been made for the emissions expected from the decay of ^{56}Co. Doppler-broadened line shapes have also been discussed.[4] Sensitivity to such features can only be accomplished with the use of high resolution spectrometers which provide the best means of identifying gamma-ray line emissions unambiguously.

A number of balloon-borne instruments to perform high resolution spectroscopy have been flown in the past several years but only a few have been built and flown as spacecraft payloads.[5,6] These initial spaceborne efforts all used Ge(Li) or p-type Ge sensors, both types of which suffered various degrees of radiation damage. The mission lifetimes were also relatively short due to cooler limitations. Although several important observational results were obtained with such spaceborne detectors,[5,7] it is clear that a Ge spectrometer system with greater sensitivity and directionality is needed for continued progress in gamma-ray astrophysics.

The ANGAS instrument represents the next generation of high resolution gamma-ray spectrometers to be flown as part of the US Space Program. The instrument is being developed at the Lockheed Palo Alto Research Laboratory and is scheduled for launch in early 1991 on the P86-2 (Starscan) satellite being built by the Ball Aerospace Systems Division in Boulder, Colorado. The purpose of this paper is to present a brief description of the ANGAS instrument and a preliminary estimate of its detection sensitivity.

INSTRUMENTATION

The ANGAS instrument depicted in Figure 1 is designed to provide fine resolution spectral data in the energy range between $\simeq 20$ keV and 10 MeV with significantly greater sensitivity than that of previous Ge spectrometers flown in space. The 19 n-type, radiation-resistant detectors provide a total geometrical area of 373 cm^2 in the non-imaging mode (i.e. with the collimators open), and a sensitive volume of 2050 cm^3. The use of one or more segmented-electrode detectors is planned, depending on their successful development and availability. In that event, the outer electrode will be divided into two segments with the front section 15 mm deep and the remaining 41 mm of the crystal making up the back section. The energy signal is collected from the common central contact and threshold discriminators connected to the outer channels locate the site(s) of the interactions. Enhanced sensitivity to spectra below about 400 keV can be gained in this configuration because these photons are predominantly detected in the front segment whereas the background is uniformly distributed throughout the volume. In addition, the sensitivity to gamma-rays up to about 2 MeV can be increased through the use of the so-called "duode" or coincidence mode whereby spectra are accumulated for only those events undergoing Compton scatters between both sections of the sensors. The Ge detectors are mounted in a common vacuum cryostat to minimize radial dimensions and are arranged to accommodate a retractable rotational modulation collimator mechanism for the imaging system.

The NaI(Tl) shield weighs 250 kg and is divided into five major sections with a quasi-two layer design which optimizes shielding efficiency, Compton suppression, and the identification of escape events. Two stacked, disk-shaped sections (with collimator

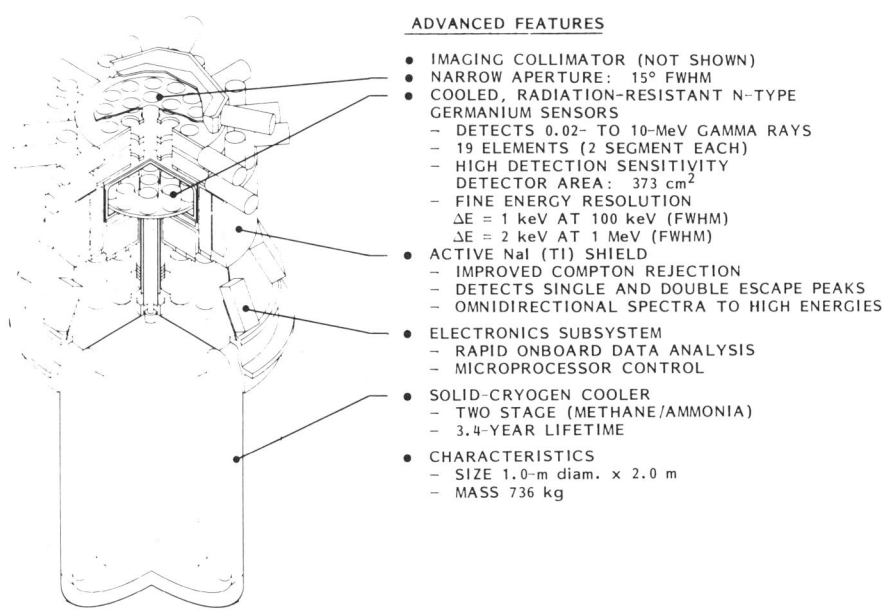

Fig. 1. ANGAS Instrument Isometric View.

boreholes) form the upper shield, two concentric annular sections comprise the side shield, with a fifth section covering the bottom. A thin plastic scintillator is used to veto in-aperture charged particle events. All sections are optically independent with selectable threshold settings to control counting rates for optimum performance. A total of 52 two-inch photomultiplier tubes (Hamamatsu Model 1847) are coupled to the scintillator elements.

The two annular NaI sections are further divided into four 90° quadrants. This feature allows determination of the source direction for transient events, such as gamma-ray bursts, by analyzing the counting rates in each sector of the outer annulus. An angular resolution of 2-3° can be achieved by this method. The onboard data analysis system includes a burst detection and memory function described below. Additional advantage in making spectral measurements can be obtained by the use of coincidence logic between different sections of the scintillator. For each of the five major NaI sections, two separate 256 channel spectra are recorded for the energy ranges from 100 keV to 10 MeV and 100 keV to 100 MeV and stored in dual memory banks. These data are read out once every 2 seconds and 10 seconds, respectively.

The ANGAS electronics system processes input signals from the Ge sensors, NaI shield, and four background monitors (not discussed in this paper). The 19 Ge detectors have separate amplifiers and pulse height analyzers to minimize noise and provide redundancy. The five NaI shield sections are viewed by either 8 or 12 photomultiplier tubes each, with individual sections separately summed and pulse height analyzed. A light pulser calibration system is used to balance the PM tube gains and optimize the resolution. Two microprocessors are used; one serves to collect and format the data while the other performs data analysis. The sensor data for each Ge event are reported as a multi-byte word which includes the energy deposited in each NaI detector involved with the event. Multiple Ge sensor data may either be summed to obtain the total energy or used to determine the polarization of the gamma-rays from the source. Counting rates from every sensor are sampled each frame (every 0.032 seconds) and spectra from the NaI shields are collected every two seconds. Associated with the data analysis microprocessor is a one megabyte memory used to store NaI spectra and counter data at four times the above rates. When an on-board algorithm identifies a gamma burst, it and its precursor are retained in this memory and later included in the output data stream. About five percent of the 52 kbit/sec ANGAS output data is allocated to the on-board analyzer output.

The 19 Ge detectors will be cooled to a temperature of about 70K by the two-stage solid methane/ammonia cooler shown in Figure 1. To achieve optimum resolution, the spectral preamplifier FET's will be coupled to the solid ammonia reservoir and maintained at 150K. The cooler weighs approximately 270 kg which includes 100 kg of ammonia and 50 kg of methane. The design incorporates advanced thermal isolation techniques and is expected to provide a 3.4 year lifetime.

IMAGING SYSTEM

The ANGAS experiment will employ the technique of rotational aperture synthesis[8] to provide a direct imaging capability at gamma-ray energies. In contrast to other coded-aperture systems, this method is naturally suited to the use of large volume, non-positional Ge detectors, allowing the acquisition of high spectral resolution data simultaneously with the image. As shown in Figure 2a, each collimation hole is covered top and bottom by a series of equally wide and spaced tungsten bars approximately 0.75 cm thick. One pair of detectors has grids with one bar across the aperture, another pair has two bars, etc., up to a total of nine bars (with the nineteenth aperture fully open). Heuristically, each detector looking out at the sky through the grids sees a

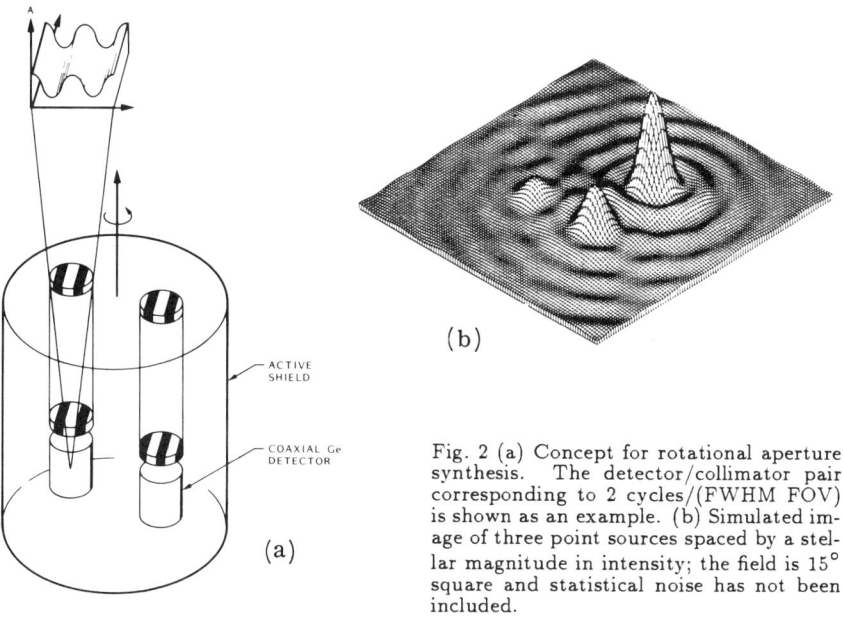

Fig. 2 (a) Concept for rotational aperture synthesis. The detector/collimator pair corresponding to 2 cycles/(FWHM FOV) is shown as an example. (b) Simulated image of three point sources spaced by a stellar magnitude in intensity; the field is 15° square and statistical noise has not been included.

quasi-sinusoidal fringe pattern whose spatial frequency is proportional to the number of bars across the aperture. Two detectors for each spatial frequency are required for an accurate subtraction of the large DC background component, and the grid "phasing" is that which maximizes the reconstructed signal.[8,9] In the ANGAS imaging mode, the P86-2 satellite will spin about the optical axis of the instrument, making the situation analogous to Earth rotational aperture synthesis in radio astronomy. By taking a suitable linear combination of count rates from each detector pair, one measures essentially the Fourier components of the sky brightness distribution with an optimum sampling coverage of the two-dimensional spatial frequency domain. The image is reconstructed digitally by applying the inverse transform, producing a point source response with 1.2° FWHM resolution and good sidelobe characteristics (Figure 2b). The raw image may be effectively enhanced using standard methods such as CLEAN or maximum entropy. Extensive computer simulations of the imaging performance have been carried out and verified by laboratory measurements for a variety of image scenes and signal-to-background ratios.

DETECTION SENSITIVITY

Figure 3 displays the ANGAS narrow line sensitivity a for a point source on-axis in both the imaging and non-imaging modes. Also shown for comparison are the published sensitivities for the HEAO-3[6] and GRO/OSSE[10] soft gamma-ray experiments. The ANGAS 3σ flux sensitivity for either mode can be expressed:[9]

$$F_{3\sigma} \simeq \frac{5.5}{\varepsilon_{NI,I} \cdot A} \left(R_B \cdot \Delta E_{FWHM} / T_{obs} \right)^{1/2} \quad \text{photons}/(\text{cm}^2 \cdot \text{s})$$

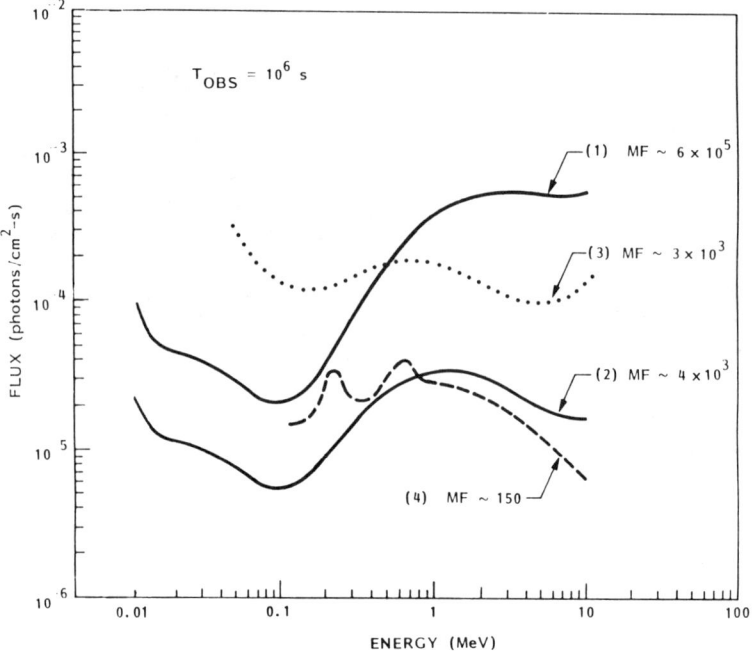

Fig. 3. 3σ Statistical Flux Sensitivity (1) ANGAS imaging mode. (2) ANGAS non-imaging mode. (3) HEAO-3 10^5 sec survey sensitivity. (4) GRO/OSSE. The "multiplex factors" (MF) are the number of independent spatial×spectral bins for the instrument or mode (see text).

where, for the non-imaging case, ε_{NI} is the detector photopeak efficiency, A the total geometrical area, T_{obs} the sum of source and background observation times, R_B the background rate, and ΔE_{FWHM} the detector resolution. The numerical factor is derived from reference 11, and background-limited statistics are assumed. The curve includes the sensitivity enhancement ($\sim 25\%$ below 2 MeV) assuming that all nineteen detectors are segmented. For the imaging mode, the signal is understood as the peak height of the point response, which is affected by off-axis vignetting effects and loss of contrast at high energies due to grid transmission. The noise is the statistical fluctuation in this peak height from one observation to the next. This noise is constant over the field and equal to $\simeq \sqrt{2 \cdot B_T}$, where B_T is the total background count in ΔE_{FWHM}.[9] The sensitivity formula given above then follows, where the various signal attenuation factors are contained in the imaging photopeak efficiency, ε_I. Below a few hundred keV, for which the tungsten is effectively opaque, $\varepsilon_I \simeq 1/4 \cdot \varepsilon_{NI}$ on-axis due to the geometric blockage of the aperture by the grids. At higher energies and for field points off-axis (but within the field-of-view), ε_I has been determined using Monte Carlo simulations, and is implicit in the ANGAS imaging mode curve of Figure 3. The response to high energy sources outside the field-of-view will be treated elsewhere.

Finally, to provide a fair comparison between the different capabilities shown in Figure 3, we have included for each curve a "multiplex factor," defined as the number of spatial pixels multiplied by the total number of spectral resolution elements pertaining to that instrument or mode of operation. In the spectral domain, it represents the importance of high resolution in providing unambiguous identifications of complicated, closely-spaced line structure; in the spatial domain, it is the advantage gained by simultaneously imaging a large number of pixels compared to a scanning mode of operation or to a model-dependent deconvolution of source structure. This factor accounts for the large advantage that ANGAS has in this respect compared to the other experiments, which would not otherwise be reflected in the standardized display of instrumental sensitivity which has been adhered to in the Figure. Finally, we briefly note that ANGAS in its imaging mode should be relatively immune to systematic errors caused by orbital variations in the background counting rates since the rotational modulation method is basically equivalent to on source/off source "chopping" at the satellite spin frequency (~ 1 min^{-1}).

CONCLUSIONS AND ACKNOWLEDGEMENTS

Because of its unique combination of spectral, spatial, and polarimetric measuring capabilities, the ANGAS flight experiment will fulfill an important need in gamma-ray astrophysics for the coming decade. The data obtained should provide new insight and understanding of nucleosynthesis and other high-energy processes occuring at a wide variety of astrophysical sites.

We wish to acknowledge the dedication and hard work of the engineering staff at the Lockheed Space Sciences Laboratory in the design and assembly of this experiment. The ANGAS program, sponsored by DARPA through the Office of Naval Research, would not be possible without the continuing support of Dr. R. Alewine at DARPA and Mr. R. Gracen Joiner at ONR. The research presented here was supported in part by the Lockheed Independent Research Program.

REFERENCES

1. W.Sandie, G.Nakano, L.Chase, G.Fishman, C.Meagan, R.Wilson, W. Paciesas, and G.Lasché, this Workshop (1987) and George Mason Workshop (1987).

2. S.M. Matz, G.H. Share, M.D. Leising, R.L.Kinzer, E.L.Chupp, and C. Reppin, this Workshop (1987); W.R. Cook, D.M. Palmer, and T.A. Prince, ibid.

3. S.E.Woosley, P.A.Pinto, P.A.Martin, and T.A.Weaver, Ap.J. $\underline{318}$, 644 (1987), and subsequent publications; N.Gehrels, C.J.MacCallum, and M.Leventhal, Ap.J., in press; K. Nomoto, T.Shigeyama, and K.Hashimoto, Proc. ESO Workshop on SN1987a, ed. J.Danziger, in press.

4. K.W.Chan and R.E.Lingenfelter, Ap.J. $\underline{318}$, L51 (1987).

5. G.H.Nakano, W.L.Imhof, J.B.Reagan, and R.G.Johnson, in Gamma-ray Astrophysics, ed. P.W.Stecker and J.I.Trombka, p.71 (1973); G.H. Nakano and W.L.Imhof, in Gamma-ray Spectroscopy in Astrophysics, ed. T.L.Cline and R.Ramaty, p.529 (1978); B. Teegarden, G.Porreca, D.Stilwell, U.D.Desai, T.L.Cline, and D.Hovestadt, ibid, p. 516.

6. W.A.Mahoney, J.C.Ling, A.S.Jacobson, and R.M.Tapphorn, NIM, $\underline{178}$, 363 (1980).

7. G.R.Riegler, J.C.Ling, W.A.Mahoney, W.A.Weaton, J.B.Willett, and A.S.Jacobson, Ap.J. $\underline{248}$, L13 (1981); W.A.Mahoney, J.C.Ling, W.A.Weaton and A.S.Jacobson, Ap.J. $\underline{286}$, 585 (1984).

8. L.N.Mertz, G.H.Nakano, and J.R.Kilner, JOSA A, $\underline{3}$, 2167 (1986).

9. M.H.Vartanian, G.H.Nakano, J.R.Kilner, and G.P.Lasché, (1988), in preparation.
10. J.D. Kurfess et al., Adv. Space Research, Vol. 3, No.4, 109 (1983).
11. J.B.Willett, J.C.Ling, W.A.Mahoney, and A.S.Jacobson, in Gamma-ray Spectroscopy in Astrophysics, ed. T.L.Cline and R.Ramaty, p.450 (1978).

Late Note: Recent budgetary cutbacks have forced the cancellation of the Starscan satellite, requiring a modification of the ANGAS instrument towards a lighter weight design with fewer detectors and therefore less sensitivity. However, as suggested by the editors, it has been agreed to proceed with the publication of this paper since the lightweight version would be similar in most respects to the design described above.

THE GAMMA-RAY IMAGING SPECTROMETER (GRIS) INSTRUMENT AND PLANS FOR OBSERVING SN 1987A

J. Tueller, S. Barthelmy, N. Gehrels, and B. J. Teegarden
Goddard Space Flight Center, Greenbelt, MD 20771

M. Leventhal
AT&T Bell Laboratories, Murray Hill, NJ 07974

C. J. MacCallum
Sandia National Laboratories, Albuquerque, NM 87185

ABSTRACT

The Gamma-Ray Imaging Spectrometer (GRIS) is a powerful second-generation high-resolution gamma-ray spectrometer. It consists of an array of seven large (typically >200 cm^3) n-type Germanium detectors surrounded by a thick (15 cm) NaI active shield. Its energy range is 0.02 to 10 MeV. A new detector segmentation technique will be employed to reduce the detector background. The β-decay background component, which is expected to be dominant in the 0.2-2 MeV range, will be suppressed by roughly a factor of 20. The 3σ GRIS sensitivity to a narrow Fe line at 847 keV (expected to be the most intense from a supernova) will be ~ 2x10^{-4} photons/cm^2-s for an 8 hr observation of the LMC over Alice Springs, Australia with unsegmented detectors. The instrument in simplified form will be ready to observe SN 1987A in early 1988.

INTRODUCTION

The GRIS project was created in 1982 as the result of a scientific review panel recommendation that NASA support a development program of high-sensitivity, high-resolution gamma-ray detectors for future flight opportunities. GRIS was one of two instruments competitively selected for this balloon SRT program. It is a second-generation instrument that incorporates new advances in detector technology and new methods in gamma-ray background suppression. The GRIS sensitivity will be 5-10 times better than the best previous high-resolution spectrometers.

INSTRUMENT DESCRIPTION

Figure 1 shows the overall layout of the GRIS experiment and Table 1 summarizes its characteristics. The basic instrument is an array of seven coaxial Germanium detectors surrounded by a thick NaI shield[1]. An active NaI mask/antimask system is used for imaging and precise source location[2]. The seven individual Germanium detectors

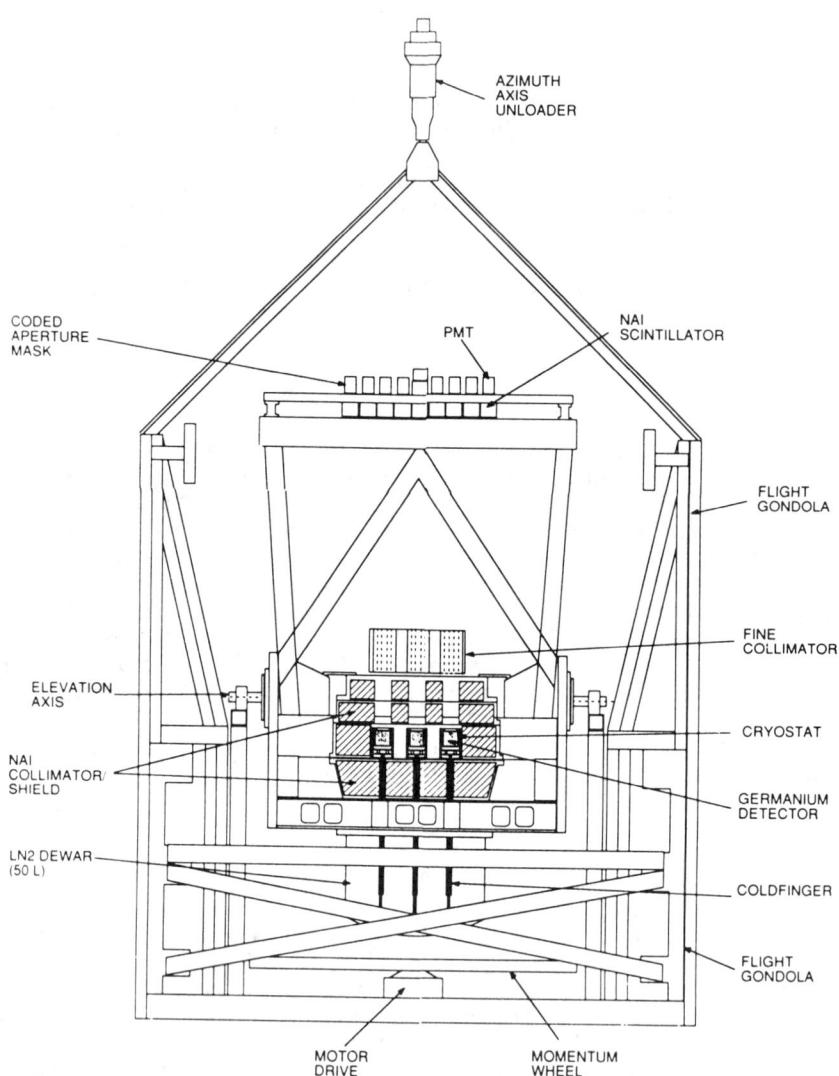

Figure 1 - Cross section of GRIS

TABLE 1
GRIS PARAMETERS

Energy Range	0.02 to 10 MeV
Energy Resolution	2.0 to 2.6 keV @ 1.1 MeV
No. of Germanium Detectors	7
Detector Type	High-purity n-type coaxial
Detector Size	Typ. 6.8 cm diam. x 7.0 cm lngth
Total Detector Area	230 cm^2
Total Detector Volume	1540 cm^3
Collimation	20° or 30°, NaI collimator 3°x3°, removable fine collimator
3σ Narrow-Line Sensitivity* (E=847 keV, T= 8 hr)	2.0×10^{-4} ph/cm^2-s (w/o seg) 8.0×10^{-5} ph/cm^2-s (w/ seg)
3σ Continuum Sensitivity* (E=200 keV, ΔE/E=0.1, T=8 hr)	1.2×10^{-5} ph/cm^2-s-keV
Experiment Weight	1750 kg (no-mask configuration)
Telemetry Rate	56 kbps
On-board Data Storage	Redundant tape recorders 20 hour capability

*Assuming a single-transit SN 1987A observation from Alice Springs, Australia.

are each enclosed in their own low-mass cryostat. This design (rather than a single large multi-detector cryostat) was chosen for two principle reasons: 1) minimization of internally produced background and 2) ease of replacement of individual detectors for repairs, upgrades and modifications. A new technique to suppress the β-decay background component is being implemented in the GRIS experiment[3]. It takes advantage of the fact that the β-decays deposit their energy at a single localized site in the Germanium crystal, whereas the gamma rays in general lose their energy through multiple interactions.

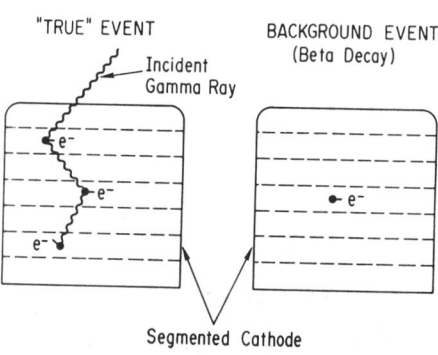

Fig 2 - Suppression of β-decays using multiple-segment mode.

A cathode segmentation technique is used to effectively slice up the detector into a stack of ~ 1 cm thick layers. Individual signals detected from each segment will then indicate whether we had a single- or multiple-point interaction (see Figure 2). A sensitivity improvement of a factor of two is expected at the 847 keV ^{56}Co line. This technique will not be implemented on the first GRIS flight due to the compressed schedule necessary for observing SN 1987A. Figure 3 shows the calculated 3σ GRIS sensitivity to narrow lines as a function of energy.

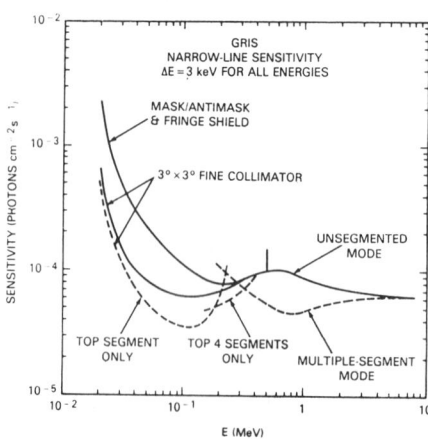

Fig 3 - GRIS sensitivity for 8 hour observation at 3 g/cm² depth over Australia.

The array of Germanium detectors is surrounded by a thick (15 cm) NaI active collimator/shield. Detailed background calculations[4] have shown that thick active shielding is required to minimize the high-energy (> 1 MeV) and 511-keV backgrounds. The active NaI collimator can be configured to have either a 20° or 30° field-of-view, the choice of which depends on the observing program for the flight in question. For observations of SN 1987A the 20° FOV configuration will be used.

The GRIS experiment includes an active, uniformly-redundant mask/antimask system to generate sky maps over a 9° x 15° field-of-view (see Figure 4). The location of SN 1987A is already precisely known and there are no other known gamma-ray sources within 20° of the LMC. There is therefore no scientific need for the GRIS mask for the SN 1987A observation. In order to lighten the payload and achieve a higher altitude, the mask will not be included in this flight.

(a)

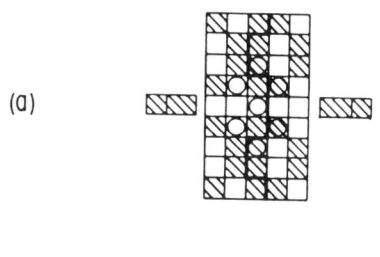

The flight gondola (also shown in Figure 1) provides a pointed platform for the instrument. An azimuth-over-elevation digitally-controlled pointing system orients the instrument using a magnetic reference to an absolute accuracy of 0.3°. A CID camera will be used to image star fields during the night and the sun during the day in order to accurately determine the pointing direction. This information will be used to correct the instrument pointing by command sent from the ground. On-board data storage and stored observing program capabilities have been included to allow GRIS to continue operating outside of line-of-sight telemetry range.

(b)

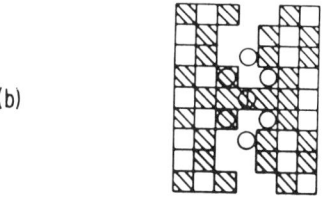

Fig 4 - GRIS mask (a) and antimask (b).

SN 1987A OBSERVATION PLANS

GRIS has been selected to participate in NASA's ongoing SN 1987A program and will be flown in the April 1988 and November 1988 wind turn-around campaigns over Alice Springs, Australia.

REFERENCES

1. B. J. Teegarden et al., 19th ICRC 1, 213 (1985)
2. N. Gehrels et al., ibid 3, 303 (1985).
3. N. Gehrels et al., IEEE Trans. on Nucl. Sci. NS-31, 307 (1984).
4. N. Gehrels, Nucl. Inst. Meth. A239, 324 (1985).

CHARACTERISTICS OF EXITE AND PLANS FOR SN1987A

Corbin E. Covault, Joao Braga[1] and Jonathan E. Grindlay

Center for Astrophysics, 60 Garden Street, Cambridge, MA 02138

ABSTRACT

The development, integration, and testing of the Energetic X-ray Imaging Telescope Experiment (EXITE) are described. EXITE is a balloon-borne hard x-ray imaging payload with high sensitivity in the 20-300 keV band, angular resolution of 22 arcmin (with source locations to $\sim$ 2 arcmin) in a 3.4° FWHM field-of-view and energy resolution of 9% at 122 keV. Details of the detector system, a 34 cm diameter NaI scintillation crystal with image intensification and position-sensitive readout, are described as well as the integration and testing of EXITE into a new balloon gondola with stable ($\sim$1 arcmin) pointing. EXITE is able to carry out high-sensitivity ($\sim$ 10 mCrab) observations of SN1987A to measure the Compton-degraded x-ray and gamma ray continuum, to search for the 122 keV cobalt-57 line, and (at 0.1 msec time resolution) to search for the new pulsar thought to have been produced in the SN1987A event.

INTRODUCTION

EXITE is a new balloon-borne hard x-ray telescope[1,2,3] for high energy astrophysics and is now fully assembled and ready for its first flight to observe SN1987A in April or May, 1988, as part of the NASA balloon campaign from Alice Springs, Australia. Herein we describe the instrument, its performance characteristics and results of initial testing and calibration. We also describe briefly the gondola, data recording and command systems and their testing. We then outline plans for observing 1987A and describe several of EXITE's advantages for making these observations.

AN OVERVIEW OF THE EXITE INSTRUMENT

The EXITE detector consists of a 34.5 cm diameter, 6.4 mm thick NaI(Tl) crystal bonded to the curved entrance window of a large two-stage image intensifier tube manufactured by Thomson-CSF. The curved crystal is hermetically sealed behind a thin (10 mil) aluminum entrance window. The image intensifier first stage accelerates and de-magnifies primary photoelectrons onto a small phosphor screen which is optically coupled to the second stage through a fiber optic plug. The second stage terminates in a position-sensitive silicon PIN diode. Charge from the diode is collected from 4 readouts allowing for accurate determination of both the energy and position of incident photons on the crystal.

The detector system is contained within a cylindrical graded passive shield, and rear cap, to shield against both primary and secondary fluorescence photons from the atmospheric and cosmic background in the 20-300 keV band of interest. The shield is

---

[1] 1. on leave "Instituto de Pesquisas Espaciais", Brazil

made of layered lead (2 mm), tin (1 mm), and copper (1 mm). The front, or entrance aperture, of the detector is collimated by two criss-crossed 1-D linear slat arrays. The collimator blades are made of lead (0.4 mm) laminated in a copper (0.3 mm) casing, and are spaced 13 mm apart for a field-of-view of 3.4° FWHM, which includes one full cycle of the coded aperture mask. The entire detector is surrounded on all sides by an active shield of plastic scintillator (1.3 cm NE-102 on the four sides and rear; 2 mm NE-102 on the front entrance aperture) to anti-coincidence charged particle background. This assembly is housed within a thin stainless steel pressure vessel to maintain the detector system in a one atmosphere room-temperature nitrogen gas environment. The front entrance to the vessel is covered with a 15 mil mylar window which is pressure reinforced with an external netting of kevlar cord.

EXITE imaging is accomplished by a rectangular URA mask[4] located 2 meters in front of the detector. The primary mask pattern is an 11×13 pattern with a 13 mm element size, identical to the collimator blade spacing. This pattern is permuted to approximately 4 cycles within the mask to cover the area within a 60 cm diameter circular support frame. Elements of the mask are graded and were made by epoxying together layers of lead, tin, and copper (with same thicknesses as the graded shields) and then epoxying the completed mask elements into the array between two thin (1 mm) layers of transparent lexan.

The detector assembly is supported on an elevation mount within the EXITE gondola frame. Hydraulic shock absorbers support the shielded detector to cushion the image intensifier from the sudden decelerations of both parachute deployment and landing impact. Additionally, a sophisticated hydraulic and pneumatic shock absorber system has been installed within the gondola frame and landing skid to reduce initial landing and toppling shocks (See Figure 1). The alt-az pointing system on the gondola is stabilized to within 1 arcmin by two high-speed gyros (azimuth and elevation) for inertial reference. Absolute azimuth pointing is accomplished with two magnetometers in coordination with ground calculations to compensate for local magnetic field fluctuations. Inertial drift corrections can be made during the flight by pointing checks using two on-board television cameras, (one equipped with a neutral density filter for planet or solar observations during a day-time flight.)

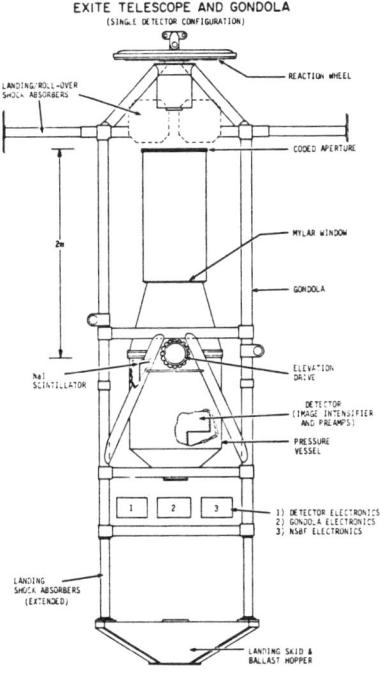

Fig. 1: Overall layout of EXITE

Every aspect of flight time experiment control will be managed through the ground support MicroVAX computer, which will provide real-time display of system housekeeping and primary science data. Experiment command sequences will also be composed and relayed by the MicroVAX. This system will allow for real-time data analysis, including rapid image generation and energy histograms. Also, the MicroVAX will spool raw telemetry data to cartridge tape, providing convenient back-up to NSBF data storage systems. Additionally, we have developed a remote down-range ground system based on either (or both) LSI-11 and PC computers for reading and displaying housekeeping data and sending commands. This extra station will allow EXITE to extend the duration of its flights.

EXITE has been designed for good sensitivity in the 20-300 keV energy range. The overall EXITE characteristics for observations near the zenith are described in Table I. X-rays entering the detector are occulted by mask elements and attenuated (at the lowest energies) by intervening layers of material including 2 mm thick lexan support for the mask, 2 mm plastic scintillator, and the 10 mil aluminum window. These effects, together with the collimator transmission and raw detector efficiency, are combined to generate the plot of effective area as a function of energy shown in Figure 2. The plot includes the atmospheric attenuation for a float altitude of ~ 3 grams/cm^2 and an observing elevation of $45°$ for the supernova, as expected for observations from Alice Springs.

Table I: EXITE Parameters at 122 keV

Effective area (through mask, etc.)	~ 350 cm^2
Sensitivity (5 σ, 10^4 sec)	$\sim 10^{-5}$
photons/cm^2-sec-keV Spatial Resolution	~ 6 mm (FWHM)
Energy Resolution	$\sim 9\%$ (FWHM)
Angular Resolution	22 arcmin
Source Locations	~ 2 arcmin
Time Resolution	≤ 0.1 ms
Telemetry Data Rate	80.0 kbs
Field-of-View	$3.4°$ (FWHM)

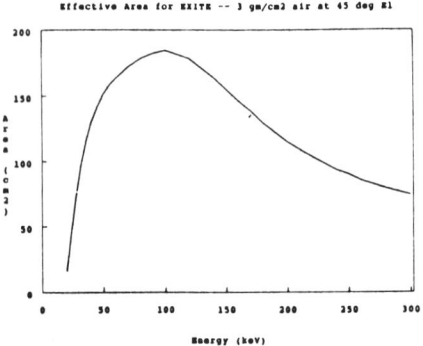

Fig. 2: Effective area for EXITE on SN1987A.

RECENT DEVELOPMENT OF THE EXITE SYSTEM

Initial testing and development of the detector was completed[2,3] by 1986, and component fabrication and subsystem assembly were accomplished by Spring 1987. During Summer, 1987, the EXITE detector and gondola systems were integrated for the first time. The gondola feedback loops have been balanced and adjusted for flight configuration. Gondola aquisition and inertial modes were tested using a laser to determine pointing accuracy and stabilization. The measured pointing stability was ~1 arcmin. Also, gondola housekeeping telemetry, including digital and multiplexed experiment diagnostics, have been verified and can be displayed in real time on either or both the MicroVAX and PC computers.

Futher extensive tests have been accomplished on the detector and associated telemetry systems, including tests of adjustable discriminator thresholds, time-tagging, slewed events, and detector rates. The active shields have also been tested, including operations of 4 HV power supplies and 12 phototubes with commandable onboard discriminator circuits for anti-coincidence vetoeing. Finally the real-time flight control software, including data backup, housekeeping and primary science data display, and experiment commands, have been thoroughly exercised.

Pre-flight energy calibration of the EXITE detector was accomplished with the use of several x-ray emitting radio-isotopes, and pre-amp gains and offsets were directly measured using both test pulses and radio-isotopes. A typical cobalt-57 spectrum is shown in Figure 3. During flight, an absolute spectral calibration will be accomplished by observations of the Crab Nebula x-ray source. Additionally, we plan to place a cadmium-109 calibration source on-board for in-flight energy calibration.

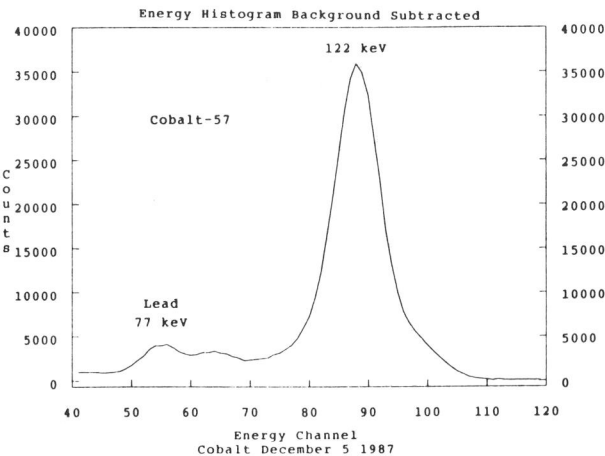

Fig. 3: Cobalt-57 (122 keV) spectrum.

The detector system distortion was mapped as a function of position by illuminating the detector at various grid locations within the collimator. This mapping was used in conjunction with fully illuminated detector images to measure the radial gain variation across the detector. Corrections for pre-amp offsets and flat-fielding for gain variations will be applied to raw data before image generation and spectral analysis. A complete description of these test and analysis techniques will be given in a forthcoming paper.

Tests of the imaging systems were also accomplished. Figure 4 shows a perspective plot of a reconstructed image of a cobalt-57 point source illuminating one mask cycle area on the detector from a distance of 25 meters. Newly installed software will allow for use of nearly the entire detector area for imaging.

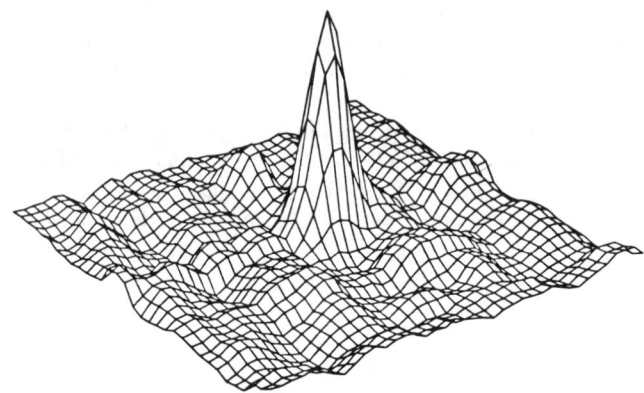

Fig. 4: Cobalt-57 (122 keV) image.

This is done by binning the detector events into 23 overlapping arrays, each of which corresponds to one full 13×11 URA mask pattern, in such a way that 92% of the round detector area is used. The final image that is is produced from summing the 23 correlated images improves signal-to-noise by a factor of ∼1.8 over the single mask pattern.

THE EXITE CAMPAIGN TO OBSERVE SN1987A

Current models[5,6] of SN1987A predict that the present energetics of the expanding shell are dominated by gamma emissions from radioactive cobalt-56 from 0.07 $M_\odot$ of nickel-56 produced in the supernova. This scenario is confirmed by the optical and UV observations of the bolometric light curve which shows a near-perfect exponential decay with a half-life time scale of 79 days, identical to the decay rate of cobalt-56. The recently observed decrease of the optical decay from the exponential suggests the shell is now becoming transparent to the primary gamma ray (cobalt-56) heat source.

During the initial stages of the expansion, the cobalt-56 gamma ray line emissions do not penetrate the optically thick layers of the outer shell. However, as the shell expands and thins, we expect to be able to detect a hard x-ray continuum spectrum resulting from Compton down-scattering of these gamma ray lines. In August 1987 the Japanese Ginga satellite[7] and the Kvant module aboard the Soviet space station Mir[8] first detected evidence of such x-rays from the vicinity of SN1987A. As reported at this Workshop, recent results from the GRS experiment aboard NASA's Solar Maximum Mission and other balloon flights flown during the Fall 1987 NASA balloon campaign have detected preliminary evidence for direct cobalt-56 line emissions at 847 keV and 1238 keV, as well as the Compton continuum at lower energies.

EXITE has high sensitivity in the 20-300 keV energy range to measure the expected Compton continuum spectrum. Figure 5 shows the expected EXITE sensitivity (5σ) for a $\sim$3 hour exposure on SN1987A together with the measured spectrum and models as reported by the Mir investigators[8]. EXITE has several significant advantages over other hard x-ray or gamma ray instruments for the SN1987A campaign. Perhaps most important is the higher resolution imaging capability, which should allow EXITE to completely isolate the spectrum of SN1987A from LMC X-1 at $0.6°$ separation. It is possible that the original spectra of SN1987A as observed by the nonimaging Ginga and Mir experiments may be contaminated by flux from LMC X-1, which as a black hole candidate has both a soft and hard x-ray spectrum[9].

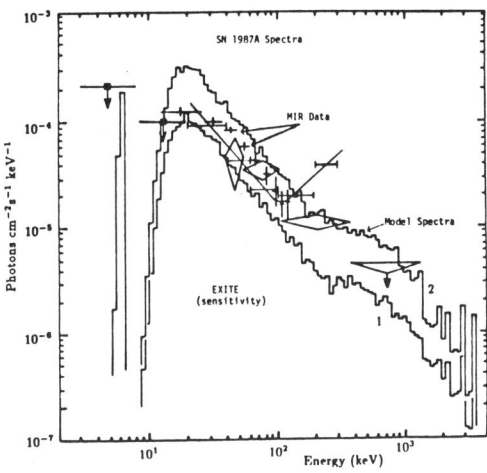

Fig.5: Expected sensitivity (5σ, for a 10^4 sec observation) on SN 1987A vs. spectrum and models reported by Sunyaev et al.[8]

EXITE can therefore better constrain the physical properties, such as mass and compositional mixing, of the expanding shell. EXITE will also search for 122 keV line emission from cobalt-57 produced in the supernova. Present models[6] predict cobalt-57 line strengths during April 1988 of about 5×10^{-5} photons/cm^2-sec, or below EXITE's estimated sensitivity threshold to this line ($\sim 2\times 10^{-4}$ photon/cm^2-sec). However, the production of the parent isotope nickel-57 in the supernova is sufficiently uncertain as to make the search worthwhile.

Finally, the EXITE event sequencing system will time-tag each incoming photon to better than 100 microseconds (relative to a temperature-controlled quartz oscillator, with relative stability of $\sim 10^{-8}$), allowing the search for a rapid x-ray pulsar as expected from a neutron star produced in the supernova. While no such periodic variations have yet been detected, the reports that the optical light curve has recently deviated downward from the exponential suggest that the shell is becoming transparent, and therefore the optical path to the center of the supernova may soon be thin enough for direct observations of the core. EXITE's energy range of 20-300 keV is well-suited for searching for the initial appearance of the pulsar, which is expected[10] to be first detectable at energies ≥ 20 keV.

CONCLUSIONS

The EXITE system is completely assembled and ready to fly. The initial objective for its first flight is to measure the performance of the detector and gondola systems. The primary science objective is of course SN1987A, for which EXITE can set contraints not available from previous or other planned observations. These include:

- High resolution imaging, allowing us to measure the spectrum of SN1987A down to 20-30 keV free of contamination from LMC X-1.
- The ability to search for the pulsar, again free from LMC X-1, at energies where it may first appear.
- A simultaneous measurement of the hard x-ray spectrum from the black hole candidate LMC X-1.

Finally, if we obtain a long flight ($\geq$ 12-16 hours), we hope to observe the galactic center region. This would provide new constraints on the spectra and locations of galactic center/galactic bulge sources. We might also observe the active galaxy Cen A between the SN1987A and galactic center observations.

Acknowledgements

We thank J. Gomes, G.Nystrom, V. Kuosmanen, F. Licata, L. Coyle and J. D'Angelo for technical support. We also thank N. Kronenfeld, M. Chansky, and C. Liu for their efforts in assembly and software support. C.E. Covault gratefully acknowleges the National Science Foundation Graduate Student Fellowship Program for support. J. Braga gratefully acknowledges the Smithsonian Pre-doctoral Fellowship Program and the CNP Fellowship Program (Brazil) for support. This work is partially supported by NASA grant NAGW-624.

REFERENCES

1. Grindlay, J.E., M.R. Garcia, R.I. Burg, and S.S. Murray, *IEEE Transactions in Nuclear Science*, **33**, 750 (1986).
2. Garcia, M.R., J.E. Grindlay, R.I. Burg, S.S. Murray, and J.W. Flanagan, *IEEE Transactions in Nuclear Science*, **33**, 735 (1986).
3. Garcia, M.R., *PhD Thesis*, Harvard University (1987).
4. Grindlay, J.E. and Murray, S., in *NASA TM 83848*, (S. Holt, ed.), p. 349 (1981)
5. Woosley, S.E., these proceedings.
6. Woosley, S.E., Ensman, and Pinto P., *Nature*, in press (1988).
7. Dotani, T. et al, *Nature*, **330**, 230 (1987).
8. Sunyaev, R., et al, *Nature*, **330**, 227 (1987).
9. Marshall, F. and White, N., *Ap. J.*, **281**, 354 (1984).
10. Sutherland, P., Y. Xu, and R. McCray, these proceedings.

PERFORMANCE OF LAPEX AND ITS SPECTROSCOPIC CAPABILITIES IN THE 20-300 KEV ENERGY BAND TO OBSERVE SN1987a

F. Frontera, A. Basili, D. Dal Fiume, T. Franceschini, G. Landini,
E. Morelli, M. Pamini, J.M. Poulsen, S. Silvestri
Istituto TE.S.R.E. del C.N.R. − Bologna, Italy

E. Costa, D. Cardini, A. Emanuele, A. Rubini
Istituto di Astrofisica Spaziale del C.N.R. − Frascati, Italy

INTRODUCTION

Recent observations of SN1987a both in the 1-10 keV[1] and in the 10-350 keV energy range[2] detected X-ray emission from the source with a very hard spectrum, a power law with $\alpha \sim 1.4$, and a flux of ~10 mCrab at 30 keV. We describe the performances of the LAPEX experiment[3] for observation of SN1987a. In the 20-300 keV operative energy band of LAPEX, the following goals can be achieved: detection of emission lines due to Co^{57} (122 keV) and Ti^{44} (67.9 and 78.4 keV), elements that could be produced in the supernova explosion; measurement of the comptonized spectrum from the expanding ejecta; investigation on possible coherent pulsations due to a newly born pulsar down to timescales of ~0.1 ms.

In the following, a thorough description of the payload and of its performances will be given.

DESCRIPTION OF THE PAYLOAD

We describe the main features of the detector. A complete description of the project has already been reported[3,4].

The detector consists of an array of 16 square phoswich units, each with a front surface of 145x145 mm^2, grouped 4 by 4. Each unit is made of a sandwich of 6 mm NaI(Tl) and 50 mm CsI(Na), viewed by a 5" PMT through a tapered lead-glass light pipe. The design of the single unit is the result of in-flight tests of different configurations[5] and of Monte-Carlo calculations of the propagation of scintillation UV photons generated by X-ray absorption in the crystals. Above each group of 4 units is mounted a graded collimator (Pb-Sn-Cu) with an hexagonal field of view of 3° FWHM. Each of the 4 collimators, one for each group, can be independently rocked by 8°, to give simultaneous measure of source and background. The total geometric collecting area through the collimators is 3100 cm^2. The 4 units of each group and the rocking collimator above them are contained in a box (5 sides) of plastic scintillator viewed by five 3.5" PMTs, to reject background due to charged particles. A passive graded shield is inserted between the phoswich units and the plastic scintillator. Figure 1 shows an overall view of the payload.

The energy resolution of the detector is about 17% at 60 keV. A continuous gain control and equalization is achieved using tagged sources of Am^{241} embedded in a plastic scintillator, one source for each group of 4 units. The 60 keV photons from Americium are tagged with a coincidence signal from the plastic scintillator, given by the α-particles emitted in coincidence with the X-ray photons. The signals of the 60 keV photons detected in the phoswich units are properly flagged before being passed to the data handling system.

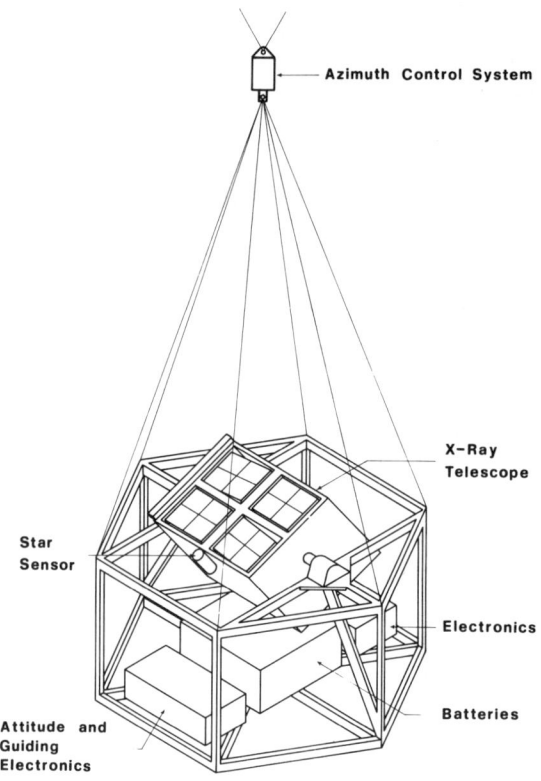

Fig. 1. An overall view of the payload.

The absolute calibration of the entire detector is obtained with four linear sources of Ce^{139} that, on command, scan the field of view of the instrument. Background rejection is achieved using both the phoswich technique, and an anticoincidence system which rejects signals simultaneous with signals coming from the plastic scintillators and/or from the other three units in the same group.

Signals from phoswich units are pulse-height analysed in 256 energy channels, giving a channel width of ~ 1 keV in the 20-300 keV energy band.

Data are acquired by a PCM encoder at a frequency determined by the telemetry rate. A FIFO buffer will derandomize the data and will allow an event-by-event transmission without losses. For each event, occurrence time (9 bits expanded on ground with reference to telemetry format), pulse height (8 bits), pulse shape (7 bits) and detection unit identification (4 bits) will be transmitted. In parallel with this transmission mode, time integrated spectra of each detection unit will be transmitted. The integration time of these spectra is about 25 s for a 81 kbit/s transmission rate. Housekeepings data and various counters will also be transmitted.

An on-board computer based on a MC68010 μP executes the observation plan, with time-tagged commands for azimuth and zenith pointings and for the rocking of the collimators. It also sends commands to the various subsystems at the appropriate times. Measurement of the actual pointing of the telescope is obtained using 14 bit absolute shaft encoders.

The azimuth control system, based on magnetometer sensor, permits a stabilization of the gondola within about 10'. A star sensor will be employed to measure the absolute pointing direction and to correct the pointing if required.

The ground support equipment is a Motorola VME-10 based on a MC68010 μP. It monitors in real time the payload status, performing a quick-look analysis of the technical and scientific data. It also controls the proper functioning of the on-board computer and prepares commands and the observation plan for uploading.

EXPECTED PERFORMANCES

The strategy of observation is based on rocking the collimators in order to control the background variability with repetitive ON-source OFF-source sequence. A conservative estimate of systematic errors due to the variation of exposed area, atmospheric depth and viewing solid angle is about 0.3% of the measured background count rate. The 3 sigma sensitivity of the instrument is shown in Fig. 2. This estimate is based on measured background count rates from an in-flight test[5] performed from Trapani (Italy), scaled for the different cut-off rigidities of Trapani and of Alice Springs. We assume an integration time of 10000 s and a float altitude of 3 mbar, corresponding to a residual atmospheric depth of 4 g/cm^2 in the line of sight of SN1987a. This sensitivity corresponds to a count rate of ~ 1% of the background count rate, well above the quoted estimated systematic errors. In Fig. 2 we also report a simulation of a photon spectrum from SN1987a that can be measured by LAPEX, assuming the same intensity and spectral shape as that measured by Sunyaev et al.[2].

The spectroscopic capabilities of LAPEX are shown in Fig. 3, where the sensitivity to emission lines from 60 to 200 keV is plotted as a function of the continuum flux in the same energy band for an observing time of 10000 s.

The sensitivity of LAPEX to coherent pulsations in the 20-60 keV energy band (an optimal choice for signal-to-noise ratio) was estimated using as a function representative of coherent pulsations the function

$$r(t) = r_o * (1 + A * \sin(\omega t + \phi)) \quad (1)$$

as in Leahy et al.[6]. Assuming an X-ray flux as that measured by

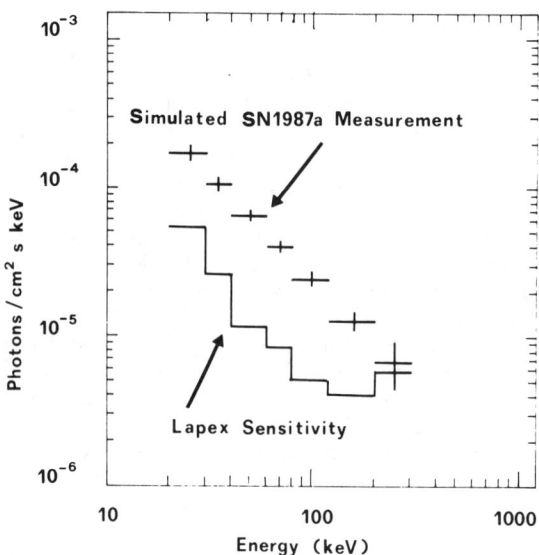

Fig. 2. The 3 sigma sensitivity of LAPEX (10000 s of observing time - 4 g/cm² atmospheric depth - Alice Springs) as a function of energy (lower curve) and the hard X-ray spectrum of SN1987a as can be measured by LAPEX assuming the source flux measured by Sunyaev et al. (1987).

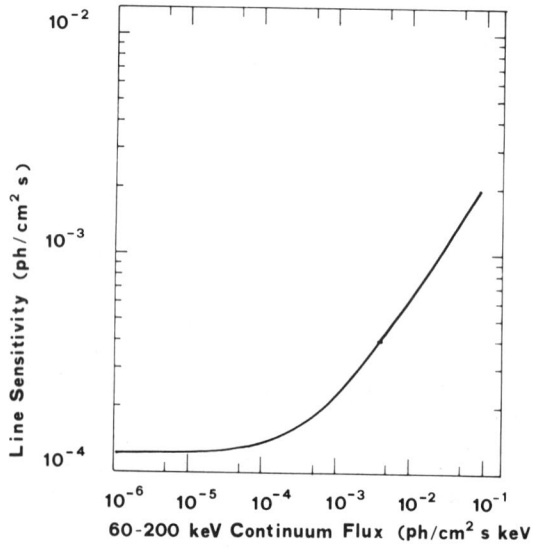

Fig. 3. The 5 sigma sensitivity of LAPEX (10000 s of observing time - 4 g/cm² atmospheric depth - Alice Springs) to emission lines as a function of continuum source emission in 60-200 keV.

Sunyaev et al.[2], and an integration time of 10000 s, at a confidence level of 99.87% the amplitude of modulation A that can be detected is about 0.025 (see Fig. 4).

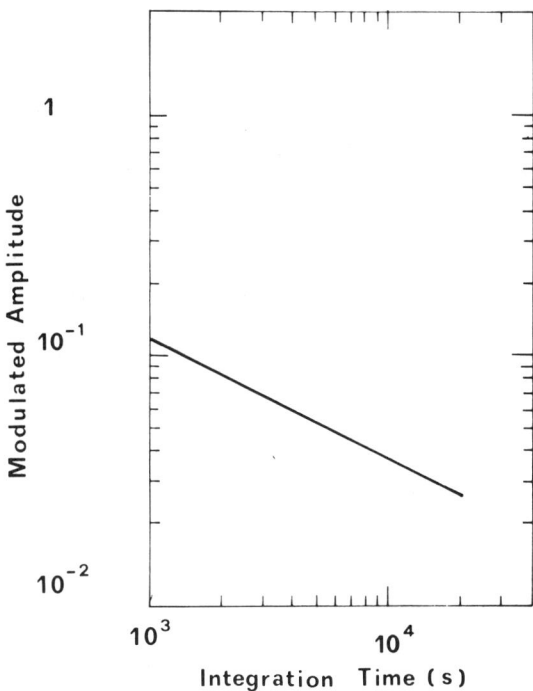

Fig. 4. The sensitivity of LAPEX (10000 s of observing time - 4 g/cm^2 atmospheric depth - Alice Springs - 99.87% confidence level) to coherent pulsations from SN1987a in 20-60 keV.

REFERENCES

1. Dotani T. et al., Nature, 330, 230 (1987).
2. Sunyaev R. et al., Nature, 330, 227 (1987).
3. Frontera F. et al., Nuovo Cimento, 7C, 656 (1984).
4. Frontera F. et al., Proc. 19th ICRC, OG 9.3-1, 355 (1985a).
5. Frontera F. et al., Nucl. Instr. and Meth., A235, 573 (1985b).
6. Leahy, D.A. et al., Ap. J., 266, 160 (1983).

A HIGH-RESOLUTION GAMMA-RAY, HARD X-RAY, AND NEUTRON SPECTROMETER FOR SOLAR FLARE OBSERVATIONS

R. P. Lin
Space Sciences Laboratory, University of California, Berkeley, CA 94720

ABSTRACT

We describe a high resolution gamma-ray, hard X-ray, and neutron instrument (HIGRANS) designed to study the acceleration of >10 MeV ions and >15 keV electrons in solar flares through high resolution spectroscopy of the gamma-ray lines and hard X-ray and gamma-ray continuum. HIGRANS will cover the energy range 15 keV to 20 MeV with keV spectral resolution sufficient for accurate measurement of all parameters of the expected gamma-ray lines with the exception of the neutron capture deuterium line. The instrument consists of an array of high-purity, n-type coaxial germanium detectors (HPGe) cooled to $<90°$K and surrounded by a bismuth germanate (BGO) anticoincidence shield. Electrical segmentation of the HPGe detector into a thin front segment and a thick rear segment, together with pulse-shape discrimination, provides optimal dynamic range and signal-to-background characteristics for flare measurements. Neutrons and gamma-rays up to $\sim 0.1-1$ GeV can be detected and identified with the combination of the HPGe detectors and rear BGO shield. HIGRANS can also be flown on long duration balloon flights for solar flare studies.

INTRODUCTION

High resolution spectroscopy of flare gamma-ray line and hard X-ray and gamma-ray continuum emission can provide a qualitatively new window on flare particle acceleration processes. This is because essentially all the nuclear gamma-ray lines produced by the energetic ions, and many of the important features of the hard X-ray continuum produced by the energetic electrons, are unresolved by present spacecraft detectors (Figure 1). High resolution measurements from ~ 10 keV to $\gtrsim 20$ MeV are required for:

1) Nuclear line spectroscopy, including determination of line shapes and asymmetries, to provide detailed information on the shape of the energy spectrum and angular distribution of the accelerated ions.[1]
2) Flare hard X-ray continuum spectroscopy to provide the detailed shape of the energy spectrum of the accelerated electrons which very likely carry a large fraction of the total flare energy.[2] High resolution is required to resolve the steep spectrum of the emission from the high temperature "superhot" plasmas[3] in the flare and to identify and resolve the sharp breaks[4] in the accelerated electron spectrum. These features appear to be critical clues to the acceleration mechanism.
3) Very high sensitivity measurements of the narrow neutron capture deuterium line at 2.223 MeV to determine whether even small flares can accelerate ions to >10 MeV energy.[5]
4) Resolution of the many expected nuclear gamma-ray lines to provide a powerful new method of obtaining solar elemental abundances.[6]

© 1988 American Institute of Physics

5) Determination of the shape and temporal evolution of the positron annihilation line at 511 keV to give information on the temperature and density of the annihilation region.[7]

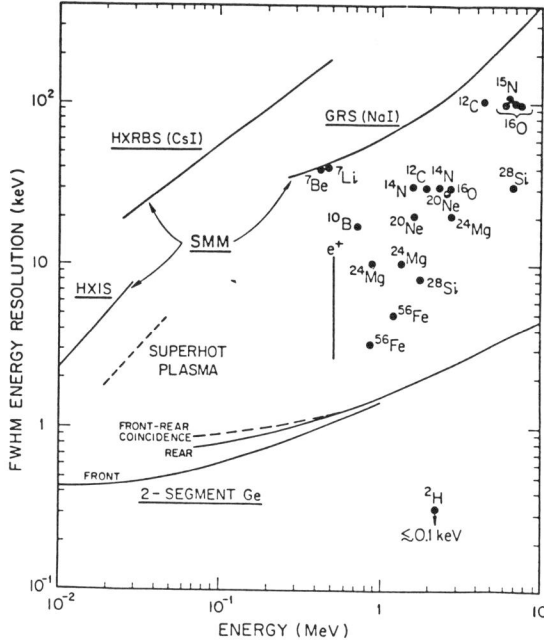

Fig. 1. The spectral resolution of the HIGRANS 2-segment HPGe detectors is compared with that of the hard X-ray and gamma-ray instruments on SMM. The typical line widths expected for gamma-rays in solar flares are shown, as well as the resolution required to resolve the steep super-hot thermal component (dashed line).

With high sensitivity measurements the temporal evolution of the accelerated ions and electrons can be closely followed, and the role of electron acceleration in very small transient releases of energy, i.e., microflares,[8] and their relationship to coronal heating, can be explored.

Below we describe the design for a High Resolution Gamma-Ray, Hard X-ray, and Neutron Spectrometer (HIGRANS). HIGRANS cleanly separates the $\lesssim 200$ keV hard X-ray region from the gamma-ray line region (0.4–20 MeV) so the high spectral resolution needed for line measurements can be maintained even in the presence of the very intense [$\gtrsim 10^4$ (cm^2 sec)$^{-1}$ above 10 keV] flare hard X-ray fluxes. HIGRANS will also be able to separate high energy gamma-rays from neutrons and provide moderate spectral resolution measurements up to ~0.1–1 GeV to study high energy pion and neutron production.[9]

HIGRANS is part of the proposed Solar High-Energy Astrophysical Plasmas Explorer (SHAPE), whose primary scientific goal is to understand the impulsive release of energy, efficient acceleration of particles to high energies, and rapid transport of energy—fundamental processes that occur in solar flares, gamma-ray bursters, supernova remnants, the galactic center region, and active galaxies. Solar flare studies are the centerpiece of SHAPE, because in flares these high energy processes not only occur routinely but also can be observed in unmatched detail at most energies. The powerful new instruments of SHAPE can locate the regions of particle acceleration and energy release, determine the structure and conditions in those regions, characterize in great detail the accelerated particle distributions, follow the subsequent transport of energy through the plasma, and thereby identify the operative physical mechanisms. Besides HIGRANS, the SHAPE payload includes GRID (Gamma-Ray Imaging Device), a Fourier-transform X-ray and gamma-ray imager having arcsecond spatial resolution and covering the energy range from a few keV to 1 MeV; LEIS (Low-Energy Imaging Spectrometer), an imaging spectrometer with arcsecond spatial resolution operating at UV and EUV wavelengths; and SIPS (Soft X-ray Impulsive Phase Spectrometer), a Bragg crystal spectrometer with high spectral resolution and great sensitivity at wavelengths between 1 and 9 Å.

INSTRUMENT DESCRIPTION

HIGRANS consists of an array of 12 dual-segment, high-purity germanium (HPGe) detectors in a BGO scintillator annulus and back shield/detector assembly (Figure 2). The detector system is enclosed in a plastic-scintillator charged-particle shield. The HPGe detectors are contained in a cryostat and cooled to an operating temperature of 90°K with a two-stage solid-cryogen (methane/ammonia) refrigerator via a cold finger. Table I gives the instrument parameters.

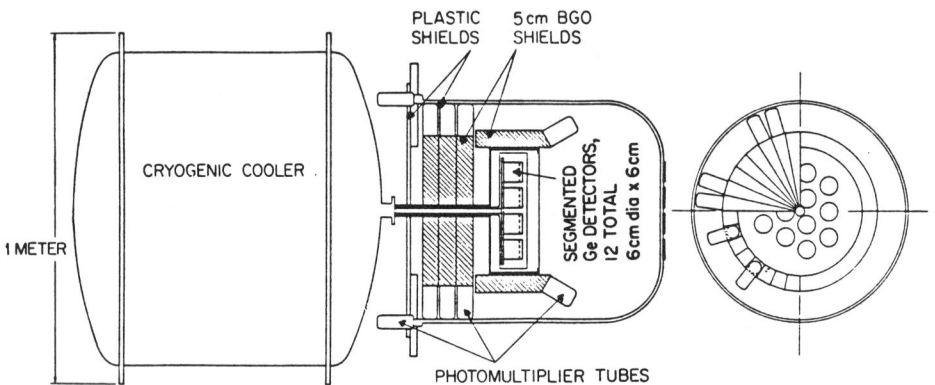

Fig. 2. Schematic of HIGRANS instrument.

Segmented HPGe Detectors. The closed-end HPGe detectors are fabricated from n-type material, 6 cm in diameter and 6 cm thick, and are operated in the reverse bias mode, i.e., the holes are collected by the outside electrode to minimize radiation damage from long term exposure to energetic particles in space. Based on accelerator tests, no detectable degradation is expected for this type of reverse bias detector in a several-year space mission in low earth orbit.[10]

Each detector has multiple collecting electrodes which divide it into two distinct volumes, or segments, according to the electric field pattern (Figure 3). In the central 1.0 cm diameter hole which extends to within ~6 mm of the top, two separate contacts are provided.[11] The top contact collects charge from the upper 1 cm segment of the detector, and the long lower contact collects charge from the

Table I. HIGRANS Instrument Parameters

Energy range	
Gamma-rays and hard X-rays	10 keV to 20 MeV (high-resolution HPGe)
	20 MeV to >100 MeV (medium resolution BGO)
Neutrons	20 MeV to 1 GeV
Energy resolution	0.6 to 5 keV FWHM (high-resolution HPGe)
Total detector area	340 cm^2 HPGe, 1400 cm^2 BGO
Field of view	120° FWHM
Shields	5-cm-thick BGO on sides, three 5-cm thick GTO at rear, 1-cm-thick plastic scintillator over 4π steradians
Germanium detector cooling	90°K for 3 years with solid-cryogen cooler

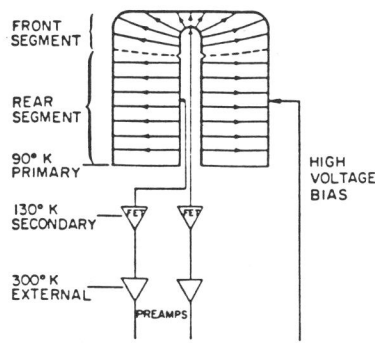

Fig. 3. Schematic of a dual-segment HPGe detector.

bottom ~5 cm coaxial segment. The curved outer surface and top surfaces are implanted with boron to make a very thin (~0.3 micron) window for X-rays.

The top segment alone is used at low energies ($\lesssim$150 keV) where photoelectric absorption dominates (Figure 4). Photons are absorbed in the top ~1 cm segment, while Compton scattered photons and detector background are rejected by anticoincidence with the adjacent bottom segment of the detector. Therefore, this mode has the excellent background rejection properties of a phoswich type scintillation counter.[12]

Higher energy ($\gtrsim$150 keV) photons are detected primarily in the thick bottom segment alone mode, with a smaller fraction also detected via top-bottom coincidences (Figure 4). The bottom segment is shielded by the top segment from low energy, $\lesssim$10^2 keV protons. Even in the largest flares the peak count rates of the bottom segments is $\lesssim$2×10^4 c/s per detector. Each detector has independent signal paths for the front and rear segments as indicated in Figure 3. Each signal path includes a cooled FET wide bandwidth charge sensitive preamplifier, followed by dual shaping amplifiers.[13] A slow shaper-amplifier (~10 μs time constant) is used for pulse height analysis, and a fast shaper-amplifier-discriminator (~400 ns) supplies

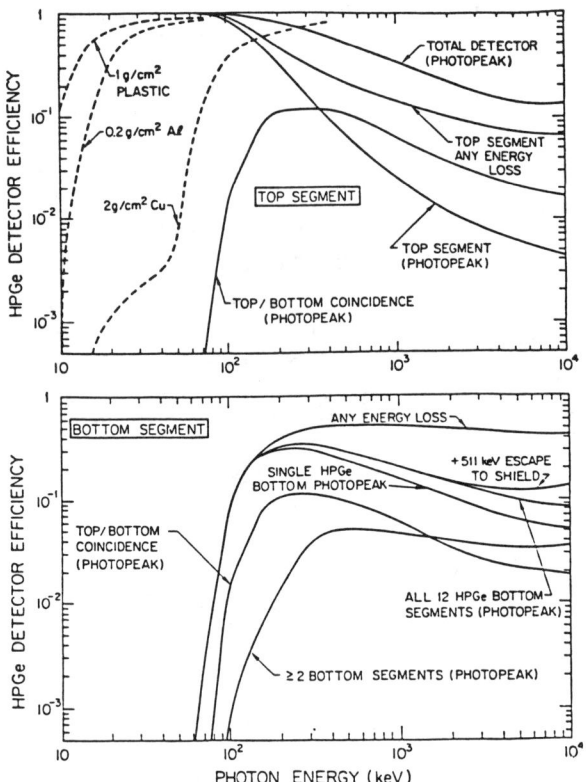

Fig. 4. The photopeak efficiency for the top segment (upper plot) and bottom segment (lower p[lot] [of] the HPGe detector is shown here for different modes. In the upper plot the total (top and bott[om seg]ments) 12-detector array photopeak efficiency and the effect of various absorbers (dashed lines) [are] shown. The lower plot shows the increase in efficiency at high energies from the addition of a [Be] shield window.

fast pulses for coincidence, pileup rejection, timing, and rate accumulations[.] [The] pile-up rejection system allows operation at rates up to 5×10^4 s^{-1} per d[etector] without resolution degradation.

To accommodate the hard X-ray fluxes of the largest flares, various ab[sorbers] are placed in front of the plastic shield, ~30 cm from the HPGe detecto[r, for] attenuation in the solar direction (see Figures 2 and 4). These absorbers blo[ck less] than 20% of the full 120° FWHM field of view, so cosmic gamma-ray bursts w[ill gen]erally be unaffected.

Table II shows that in a flare the size of the 27 April 1981 gamma-ray [flare,] several hundred to a couple thousand counts would be obtained by HIGRANS [in each] of a dozen major lines (including four lines of Fe). Flare bremsstrahlung con[tinuum] will usually dominate the detector background, leading to line-to-continuu[m ratios] of $l/c \approx 1$ to 10, except for the broad Li-Be combined lines where $l/c \approx 0.2$. [The] lines will be detected at ~10–60 σ's significance, permitting line shapes[, asym]metries, etc. to be accurately determined. Even for a flare with line fluenc[es] than 30 times smaller, most of these lines could be detected at the 3σ level if the l/c remained the same. For detection of narrow lines HIGRANS is ~5–10 times as sensitive as the SMM GRS instrument.

Table II. Solar Flare Gamma-Ray Lines

Line Energy (MeV)	Excited Nucleus	FWHM (keV)	Fluences at Earth in FWHM		Counts in HIGRANS (1000 s)			HIGRANS 3σ line sensitivity fluence ph/cm^2 ($10^3/10^2$ s)
			Lines	Flare Continuum (ph/cm^2)	Line	Flare Continuum	Detector Background	
Prompt Lines								
~0.45	^{7}Li,^{7}Be	62	20	105	3.7×10^3	1.1×10^4	1.7×10^3	0.67/0.2
0.847	^{56}Fe	4.5	5	2	630	250	40	0.15/0.05
0.931	^{55}Fe	5	2	2	230	230	35	0.15/0.05
1.238	^{56}Fe	9	2.5	2	250	200	35	0.2/0.06
1.317	^{55}Fe	13	2.5	2	240	190	45	0.2/0.06
1.369	^{24}Mg	18	7.5	3	690	280	55	0.25/0.08
1.634	^{20}Ne	23	13	2.5	1.1×10^3	200	45	0.25/0.08
1.779	^{28}Si	25	8	2	630	180	45	0.25/0.08
2.313	^{14}N	54	6	3	410	205	50	0.3/0.1
4.438	^{12}C	115	10	1.5	510	80	20	0.25/0.08
6.129	^{16}O	120	10	1	440	44	10	0.2/0.06
Delayed Lines								
0.511	e^+	2–10	25	—	4.3×10^3	—	44–220	~0.25/~0.12
2.223	^{2}H	3(0.1)*	60(4.3)*	—	4×10^3	—	3	0.1/0.05

* This line has an intrinsic width of ~0.1 keV, so the HIGRANS instrument FWHM resolution (3 keV) is substituted. The 2.223 MeV fluence observed for the 27 April 1981 flare is highly attenuated because the flare is located near the limb; the value of 60 is for a comparable flare within ~70° of disk center.

The most sensitive indicator of ion acceleration is the narrow nuclear line of neutron capture deuterium at 2.223 MeV, which is delayed from the impulsive phase by the thermalization time for the neutrons. HIGRANS can detect at 3σ significance a 2.223 MeV line fluence $\gtrsim 10^3$ times smaller than the 4 August 72 event. If all flares accelerate ions to tens of MeV energy and the gamma-ray emission scales as the microwave burst intensity, then ~10–15 flares should be detected per month in the 2.223 MeV line near solar maximum.

Good background rejection is important for the 511 keV positron annihilation and the 2.223 MeV neutron capture deuterium line, which usually lasts well past the impulsive phase. Also SMM observations show that following the impulsive phase there is a delayed phase in some (perhaps many) flares which is dominated by nuclear emissions.[15] The segmented HPGe detector provides the basic configuration for powerful background rejection techniques in the gamma-ray line region from a few hundred keV to several MeV (see [16,17] for details).

High Energy Gamma-ray and Neutron Detector and Shield. The 5 cm thick BGO shield annulus and the three rear 5 cm thick sections are composed of a matrix of BGO units (Figure 2), each with its own photomultiplier tube, preamplifier, discriminator, high-voltage power supply, LED light pulser, and servo-gain-control electronics. At energies below 20 MeV, the BGO assembly provides shielding for the HPGe array against gamma-ray background. It also provides limited collimation for analysis of radiation from cosmic sources using techniques analogous to those for the SMM GRS. The annulus and first rear section also reject gamma-rays which are

Compton scattered in the HPGe and deposit energy in the BGO above the shield threshold level. The forward layer of the rear shield is also pulse-height analyzed over the energy range from 100 keV to 20 MeV, thus complementing the HPGe data with high photofraction measurements of broad lines and the continuum.

At energies above 20 MeV, the three rear sections together with the HPGe array and the charged particle shield form a multilayer high-energy gamma-ray and neutron spectrometer. The relative energy loss in these layers will be used to differentiate between events produced by gamma-rays and neutrons. The feasibility of this technique has been proven on a smaller scale in the SMM GRS. The improvement in the neutron sensitivity of HIGRANS, compared with that of GRS, is a factor of 10 or more. The sensitivity to gamma-rays above 20 MeV shows a similar improvement of a factor of 5 or more. This improvement is a result of a much larger area and energy containment of the HIGRANS shield and the added benefits of higher operating energy, better energy resolution, and more sophisticated shield logic to deal with self-gating effects.

HIGRANS ON LONG DURATION BALLOON FLIGHTS

Because of the Challenger disaster and the current bottleneck in the Explorer program, it appears unlikely that any NASA spacecraft for solar flare studies can be launched for the next solar maximum. Long duration balloon flights (LDBF) offer an attractive way to obtain high energy measurements during the next solar maximum, since energetic photons and neutrons are able to penetrate the upper layers of the earth's atmosphere. Large, powerful instruments, up to ~3000 lbs. total payload weight, can be carried by the present standard 28.4 million cu. ft. balloons to altitudes of ~130,000 ft. (40 km). At that altitude there is less than 3 g/cm^2 of overlying atmosphere, so high quality hard X-ray and gamma-ray measurements down to ~15 keV are possible.

For standard zero-pressure balloons the temperature of the gas, and therefore the balloon altitude, is controlled by the radiation received from the Sun and the Earth. Thus the balloon is at high altitude during sunlight hours but drops during nighttime. If the balloon initially reaches a high daytime float altitude it will remain above the tropopause in its day-night excursions under normal conditions without ballast drops. Then, in this simple RAdiation COntrolled balloON (RACOON) mode,[18] flight durations are limited only by balloon lifetime and gas losses (which can be offset by ballast drops). During the three-month summer season at mid-latitudes, strong stable zonal winds flow with high velocity approximately along latitudinal lines, so circumglobal LDBFs of 12–20 days are feasible.

In 1983 a 15 million cubic ft. balloon with an ~1200 lb. payload, designed to search for solar flare neutrons, made a circumnavigation of the globe in the southern hemisphere in ~18 days.[19] In 1987, two exploratory RACOON balloon flights were launched from Alice Springs, Australia. One payload carried a complement of hard X-ray and gamma-ray detectors, including both liquid-nitrogen cooled germanium detectors for high spectral resolution and large-area phoswich scintillation detectors for high sensitivity, for observations of microflares and flares from the sun.[20] The standard, 28.4 million cubic ft., 0.8 mil polyethylene, zero-pressure balloon developed for normal short-duration flights was used. The payload was designed for automated continuous 24 hours/day operation with a solar cell power system, a pointing and navigation system, data and telemetry systems, etc.

The balloon started westward at an average rate of ~30° longitude per day (~130 km/hour), but slowed down as the balloon altitude decreased. The payload was commanded to cut down over Brazil after 12 days, and was recovered with relatively minor damage. All the detectors, the experiment data system, and electronics appeared to have functioned perfectly. The data transmission to the GOES spacecraft appears to have been intermittent. Only ~10% of those data have been recovered. All of the data, however, were stored in the on-board VCR tape. Analysis of those recorded data is underway.

Table III gives the expected frequency for detecting gamma-ray lines from flares under the assumption of ~12 hours solar observation of gamma-rays per day with HIGRANS on a high altitude balloon in the summer season. Nuclear gamma-ray line emission from flares is assumed to scale as the hard X-ray peak flux. Since a typical LDBF around the world is 15–20 days, there is a good chance in a single flight of obtaining a flare energetic enough to provide good spectroscopy of several nuclear lines, and gamma-ray lines should be detectable in several flare events. If every flare accelerates ions to $\geq$10 MeV then the very narrow 2.223 MeV neutron capture deuterium line should be detected every couple of days.

Table III. HIGRANS Event Probabilities*

	events/day
High resolution spectroscopy flare event ($>10\sigma$ in several lines)	~1/20 days
Detection of several gamma-ray lines in a flare	~1/5 days
Detection of narrow 2.223 MeV neutron-capture line, if every flare accelerates >10 MeV ions	~1/2 days
Hard X-ray continuum fast spectroscopy flare (~1 sec detailed spectra)	~2/day
Hard X-ray microflares	$\geq 10^2$/day

* Assumes scaling with hard X-ray peak flux distribution measured by SMM.

ACKNOWLEDGMENTS

This research was funded in part by NASA grants NAGW-516 and NAGW-449. I wish to acknowledge the efforts of the SHAPE proposal team (Dr. Brian Dennis, P.I.), and, in particular, the HIGRANS experiment team.

REFERENCES

1. R. Ramaty and R. E. Lingenfelter, Ann. Rev. Nucl. Sci. **32**, 235 (1982).
2. R. P. Lin and H. S. Hudson, Solar Phys. **50**, 153 (1976).
3. R. P. Lin, R. A. Schwartz, R. M. Pelling, and K. C. Hurley, Astrophys. J. Lett. **251**, L109 (1981).
4. R. P. Lin and R. A. Schwartz, Astrophys. J. **312**, 462 (1987).
5. E. L. Chupp, Ann. Rev. Astron. Astrophys. **22**, 359 (1984).
6. R. J. Murphy, R. Ramaty, D. J. Forrest, and B. Kozlovsky, Proc. 19th Intern. Cosmic Ray Conf, La Jolla, **4**, 240 and 253 (1985).

7. C. J. Crannell, G. Joyce, R. Ramaty, and C. Werntz, Astrophys. J. **210**, 582 (1976).
8. R. P. Lin, R. A. Schwartz, S. R. Kane, R. M. Pelling, and K. C. Hurley, Astrophys. J. **283**, 4211 (1984).
9. D. J. Forrest, W. T. Vestrand, E. L. Chupp, E. Rieger, J. Cooper, and G. Share, *19th Intern. Cosmic Ray Conf. Papers*, NASA Conf. Publ. 2376, Vol. 4 (NASA, Washington, D.C., 1985), p. 146.
10. R. H. Pehl, NASA Tech. Memorandum **79619**, 473 (1978).
11. P. N. Luke, IEEE Trans. Nucl. Sci. **NS-31**, 312 (1984).
12. J. L. Matteson, P. L. Nolan, W. D. Paciesas, and R. M. Pelling, Space Sci. Instr. **3**, 491 (1977).
13. D. A. Landis, F. S. Goulding, and R. M. Pehl, IEEE Trans Nucl. Sci. **NS-18**, 115 (1970).
14. D. J. Forrest, *Positron-Electron Pairs in Astophysics*, M. L. Burns, A. K. Harding, and R. Ramaty, eds. (AIP, New York, 1983), p. 3.
15. D. J. Forrest, W. T. Vestrand, E. Rieger, and G. H. Share, Bulletin American Astr. So. **18**, 697 (1986).
16. J. Roth, J. H. Primbsch, and R. P. Lin, IEEE Trans. Nucl. Sci. **NS-31**, 367, 1984.
17. D. M. Smith, M. Shapshak, R. Campbell, J. H. Primbsch, and R. P. Lin, this volume (1988).
18. V. Lally, Proc. XXIV COSPAR Conf., Ottawa, **I**, 1.4 (1982).
19. R. Koga, F. M. Frye, Jr., A. Owen, B. V. Denehy, D. Mace, and J. Thomas, *19th Intern. Cosmic Ray Conf. Papers*, NASA Conf. Publ. 2376, Vol. 4 (NASA, Washington, D.C., 1985), p. 142.
20. R. P. Lin, D. W. Curtis, J. H. Primbsch, P. Harvey, W. K. Levedahl, D. M. Smith, R. M. Pelling, F. Duttweiler, and K. C. Hurley, Solar Physics, accepted for publication (1988).

PROMETHEUS I: RICE UNIVERSITY'S NEW GAMMA RAY TELESCOPE

R. C. Haymes, J. E. Fitch and B. Sen

Rice University, Houston, TX USA

and

S. Averin, Moscow Physical Engineering Institute, Moscow, USSR

ABSTRACT

Now nearing completion, Prometheus I is a (0.1 - 5) MeV balloon-borne actively collimated mosaic of 11 × 11 independent scintillation counters. It improves sensitivity by rejecting background events due to instrument activation. Each counter is an optically isolated 1.2 cm × 1.2 cm × 5.0 cm (thick) NaI crystal and separate photomultiplier. A beta ray emitted by an activated nucleus in the mosaic causes only one counter to pulse, but a gamma ray causes several counters to pulse simultaneously. Otherwise unvetoed events that involve only one counter are rejected in data analysis. The 30 cm thick active collimator constructed of plastic scintillator vetoes events that are due to charged particles and to gamma radiation from outside its 28° FWHM aperture. Plastic scintillator has a small cross section for activation and its spallation products all emit charged particles simultaneously with the emission of any gamma ray. Surrounding the active collimator in all directions, as well as lining it, is a 2.54 cm thick LiF passive slow-neutron shield. At midlatitudes, Monte Carlo simulations have shown that the 6-hour 3-sigma 1 MeV gamma ray sensitivity is approximately 5×10^{-5} photons $cm^{-2} s^{-1}$, assuming that all the background due to activation and leakage is rejected or accounted for.

With some sensitivity loss, 1° imaging will be feasible when a thick coded mask is placed in the aperture. Monte Carlo simulations have shown that image blurring by Compton scattering is tolerable.

INTRODUCTION

Gamma ray astronomy in the (0.1 - 5) MeV energy interval has great potential for revealing qualitatively different information about the physical nature of cosmic sources. Rice University hopes to undertake systematic gamma ray spectroscopy of Type One supernovae, as well as other sources. A 3σ sensitivity of better than 10^{-4} photons $cm^{-2} sec^{-1}$ will be needed (Arnett, 1982; Gehrels et al., 1987). Given six hours observing time (i.e., the time obtainable in a balloon flight), detection of a gamma ray line this weak would be feasible with an aperture of modest dimensions, say 200 cm^2, but only if an observing instrument with the sensitivity of an actively collimated telescope, but free of the systematic error due to instrument activation by the particle radiation environment, is used.

PROMETHEUS

In Figure 1 Prometheus I is shown in solid lines. Prometheus I is presently an actively collimated mosaic of 11 × 11 = 121 independent scintillation counters. Each counter consists of a 1.2 cm × 1.2 cm × 5.0 cm (thick) thallium activated NaI crystal, viewed by a separate photomultiplier (PMT). The output pulses from each PMT are routed to 121 separate CAMAC analog-to-digital converters. Inflight energy calibration will use the atmospheric-scintillator 0.5 MeV feature and the 1.46 MeV line from the PMT glass.

Only events which deposit all of their energy in the mosaic are eligible for acceptance as photons. Because of Compton scattering, a gamma ray in the 0.1-5 MeV energy interval that is totally absorbed in the mosaic will cause several of the mosaic

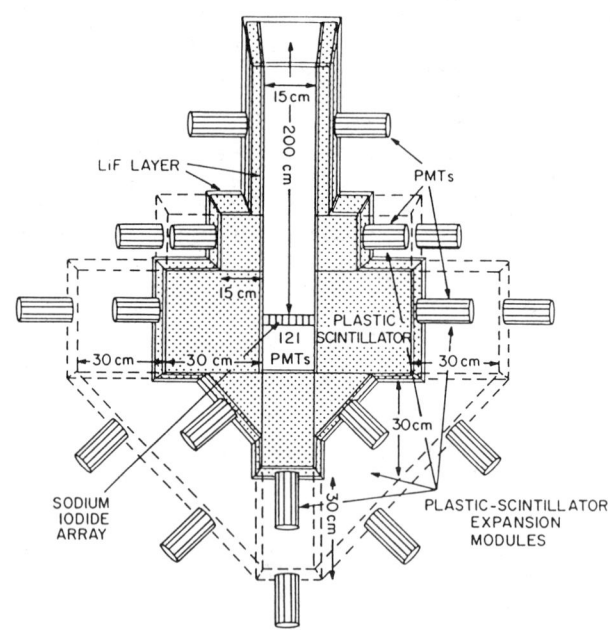

Figure 1. Schematic sketch of the new gamma ray telescope. The coded aperture's mask is not shown. The dashed lines show a future enhanced version that would have improved sensitivity because of thicker shielding.

elements to simultaneously pulse. Monte Carlo simulations have shown that a typical number N is three interactions at an energy of 1.0 MeV; N increases with increasing energy.

ACTIVATION AND ITS REJECTION

The proton-spallation and neutron-capture induced activity is primarily beta ray emission. The betas may be emitted at any time, including times long after any veto pulse from the active collimator has terminated. The betas have a variety of end-point energies, up to 2 MeV or so. The mean energy of the mixture of all the emitted betas is 1.18 MeV. This component is responsible for the "beta bulge," which is a continuum component of the background that dominates by two orders of magnitude the transatmospheric spectrum near 1 MeV in the count-rate energy spectra from existing narrow beam actively collimated detectors. The beta bulge, which extends from about 0.4 MeV to about 2.5 MeV, peaks at 1 MeV and covers just the energies of greatest interest for the cobalt 56 problem. A beta ray will cause only one array element to become illuminated. Otherwise unvetoed events in which only one mosaic element is illuminated are rejected, as being due to beta rays. Internal-origin photons with energies ≤ 0.20 MeV are likely to be totally absorbed in only one element of the mosaic. If a beta ray emission is simultaneously accompanied by such a photon, the event would be rejected.

Neglecting leakage, activation γ-rays from the mosaic that are totally absorbed limit the sensitivity of Prometheus at all energies. This is also true for beta rays accompanied by simultaneous ≥ 0.20 MeV gamma rays. To estimate this remaining background we scale the data obtained with the previous Rice instrument (Haymes et al., 1975), assuming that all the γ's that were measured by the previous instrument were instrumental in origin, because no zenith angle variation was detected and the brightest cosmic sources (i.e., the Crab Nebula and the Galactic center region) contributed only ~5% to the total counting rate, at energies around 1 MeV. We assume no β ray contribution outside the beta bulge and no γ-ray contribution within the bulge.

Activation background measured by the detector depends linearly on the (mass x solid angle) product of the NaI that is the source. The mass of the previous instrument's NaI collimator was about 60 times that of its NaI detector. We assume the detection efficiency of the previous detector was the same, at a given energy, for photons originating in the collimator as it was for photons originating in the previous detector itself. Gamma-ray-only collimator activation events such as electron capture (EC) decays contributed to the previous detector's count rate. Those photons originating in the detector were already inside, and thus "saw" 4π steradians of detector, but the solid angle subtended by the detector at the collimator was approximately one steradian. To the flux of EC γs from the collimator should be added the not-well-known gamma rays from the decay of metastables formed in the collimator that have half lives longer than the several microsecond duration of the veto pulse. Its neglect gives us a lower limit on the fraction of the background gamma rays that originated in the collimator. This lower limit is $(60/1) \times (1/4\pi) = 5$; a lower limit is that activation gamma rays from the collimator's mass were about five times more intense than those from the detector. If the metastables added an equal amount, a more realistic value would be ~ 1/10. The mass of the actively collimated NaI mosaic in Prometheus is approximately the same as the mass of the previously-used NaI detector; the activation background due to the mosaic is taken equal to that due to the previous detector. The instrumental background is the sum of the activation and leakage backgrounds. Since the absorption thicknesses of the two active collimators are comparable (see below), their leakage backgrounds at a given energy will be approximately equal. We conclude that the mosaic's activation background, at energies outside those of the beta bulge, will be roughly 0.1 of the previous detector's background counting rate.

ACTIVE COLLIMATOR

Thick blocks of plastic scintillator (minimum thickness = 30 cm; Bicron Type BC - 416) are used for the active collimator. The minimum plastic scintillator thickness for externally-incident photons is 2 gamma ray mean free paths at 1 MeV.

Plastic has a relatively small cross section for activation, and a charged particle is simultaneously emitted with the emission of a gamma ray by activated plastic. Plastic scintillator activation pulses are therefore vetoed. The threshold level for detection by the plastic scintillator has to be set as low as feasible, because Compton scattering with its continuum of pulse heights dominates the photon-matter interaction. Laboratory testing of the minimum feasible threshold is now starting, with an initial value of 0.04 MeV.

NEUTRONS

Completely surrounding Prometheus' plastic-scintillator collimator on all sides (including the aperture), is a one-inch thick layer of lithium fluoride powder, with a measured 1.4 gm cm^{-3} bulk density. The layer acts as a slow neutron passive shield, through the nuclear reaction ^{6}Li(n,α)T. LiF spallation cross sections were computed from the formulae given by Silberberg and Tsao (1973) and cause a 3σ sensitivity contribution of 2×10^{-5} photons cm^{-2} sec^{-1} MeV^{-1}, except at 0.51 MeV, where it is 7×10^{-5} cm^{-2} sec^{-1}, given six hours.

Lithium with the natural isotopic enrichment is used. The omnidirectional thermal neutron flux at 5 gm cm^{-2} atmospheric depth and the geomagnetic latitude of Texas is 0.05 neutrons cm^{-2} sec^{-1} (e.g., Armstrong et al. (1973)). When this is folded with the capture cross sections of Li and I, the result is that the 1 MeV, 3σ gamma ray sensitivity contribution resulting from formation of iodine 128 in the mosaic is in the interval 7×10^{-6} to 4×10^{-5} photons cm^{-2} sec^{-1}. The quasi-continuum extending in energy up to 6.8 MeV that results from prompt ^{128}I deexcitation gamma rays, will be vetoed by the interaction of one or more of the photons with the collimator and/or pulses from (n,p) scatters.

Deposition of energy in the collimator, as by recoil protons from scattering of fast neutrons, generates a veto pulse. If we are able to set the detection threshold of the collimator sufficiently low, these events will be detected, vetoing the fast neutron background. The plastic scintillator is 5 scattering mean free paths thick for 10 MeV neutrons. Inelastic scatters in the collimator give rise to carbon 12 gamma rays at 4.43 MeV. If the ^{12}C gamma ray does not interact with the collimator, the recoiling lowered-energy neutron is likely to scatter at least once before escaping the system. Neutron-capture gamma ray photons at 2.23 MeV will be either prevented by the LiF shield or vetoed by the pulses arising from (n,p) scatters of fast neutrons.

SENSITIVITY

Monte Carlo simulations have been run, assuming that all activation background was successfully rejected and that leakage of ambient gamma radiation was the only residual source of background. The atmospheric gamma ray spectra deduced by Gehrels (1985) for Palestine, Texas, were assumed.

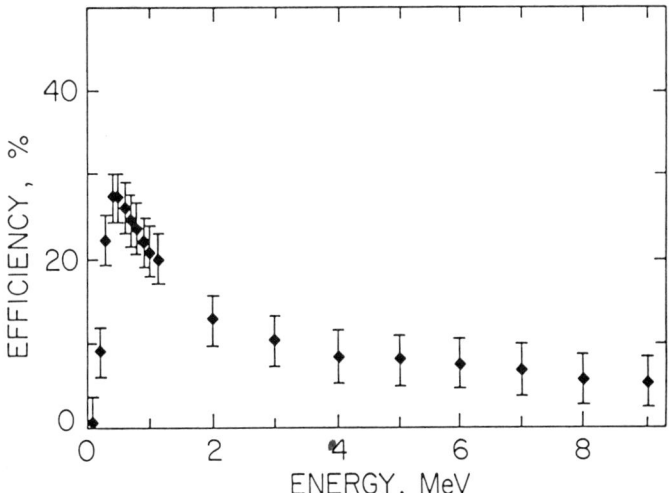

Figure 2. Monte Carlo calculations of detection efficiency vs. energy, when the beta ray rejection technique is applied. At the four energies available to us, measurements with laboratory radioactive sources agreed with the calculations to within the ± 20% uncertainty in absolute source strength.

Inside the beta bulge energies, the 3σ sensitivity of Prometheus I is limited by leakage gamma rays to ~5 x 10^{-5} photons cm^{-2} sec^{-1}, given 3 hours on source, 3 hours off source. This sensitivity must be regarded as optimistic, because it assumed complete elimination of activation background. Earlier, we showed that the residual detector activation at energies outside the beta bulge is expected to be ~0.1 of the previous instrument's background. That six-hour sensitivity is therefore expected to be 1 x 10^{-4} photons cm^{-2} sec^{-1}. Prometheus II may achieve 3σ sensitivities of 2 x 10^{-5} and 5 x 10^{-5} photons cm^{-2} sec^{-1}, respectively, given six hours.

Figure 2 depicts the efficiency vs. energy curve for on-axis gamma rays. The sharp drop at E ≤ 0.4 MeV is because of the photoelectric, single-crystal events. Figure 3 shows that, at E ≈ 1 MeV, the expected sensitivity should be adequate to detect the cobalt 56 energy spectra from extragalactic Type One supernovae.

Because the 0.122 MeV spectral line from cobalt 57 is also of astrophysical interest, the N ≥ 2 acceptance condition will not be automatically applied at this energy. The 3σ sensitivity of Prometheus to cobalt 57 γ-rays will be the same as similar-size actively collimated detectors, or about 1 x 10^{-3} photons cm^{-2} sec^{-1}, given six hours observing time.

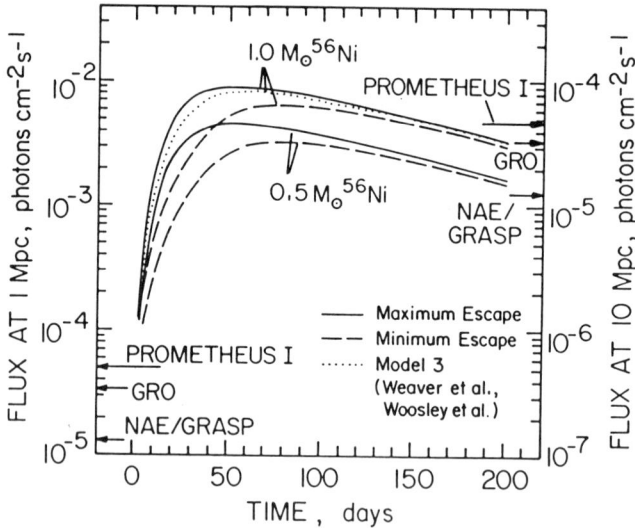

Figure 3. Expected 0.847 MeV iron-line light curves for two hypothetical Type One supernovae, located at 1 Mpc and 10 Mpc respectively. The 3σ sensitivities projected for future gamma ray astronomy space missions are shown (Gehrels et al., 1987). We have added to their curves the 3σ sensitivities expected for Prometheus.

IMAGING CAPABILITY

An imaging capability would be useful in minimizing confusion of sources. In collaboration with Prof. T. P. Kohman of the Carnegie-Mellon University, we are performing computer investigations of the results expected if a thick coded mask were placed in the aperture of Prometheus. Using a cell size for the mask equal to the size of the mosaic elements and placed just in front of the collimator, a geometrical angular resolution of 3° results, implying a bright-source spatial resolution of ~1° when pointing errors are considered.

Monte Carlo simulations have shown that the centroid of the light emitted by the mosaic is within 0.6 crystals of the site of the first Compton interaction, at all energies. This suggests MeV coded aperture imaging will not be seriously degraded by Compton blurring of the image.

REFERENCES

Armstrong, T. W., Chandler, K. C. and Barish, J. 1973, J.G.R., 78, 2715.
Arnett, W. D. 1982, in Supernovae: A Survey of Current Research (J. J. Rees and R. J. Stoneham, eds.) D. Reidel Publ. Co., Dordrecht, p. 231.
Gehrels, N. 1985, Nucl. Instr. Meth., A239, 324.
Gehrels, N., Leventhal, M. and MacCallum, C. J. 1987, Ap. J., 322, 215.
Haymes, R. C., Hall, R. D., Walraven, G. D., Meegan, C. A., Shelton, D. H. and Djuth, F. T. 1975, Ap. J., 201, 593.
Silberberg, R. and Tsao, C. H. 1973, Ap. J. Suppl., 25, 315.

DEVELOPMENT OF A LIQUID XENON TIME PROJECTION CHAMBER FOR GAMMA-RAY ASTRONOMY

Elena Aprile and Masayo Suzuki
Columbia Astrophysics Laboratory
Departments of Physics and Astronomy
Columbia University, New York, New York 10027

ABSTRACT

Work is in progress at the Columbia Astrophysics Laboratory on the development of a liquid xenon imaging chamber for gamma-ray astronomy. Such an instrument will combine high detection efficiency, good energy and position resolution with a large effective area. These properties, together with the excellent background rejection capability provided by 3-dimensional imaging, will allow orders of magnitude better sensitivity than currently used instruments, for the detection of weak gamma-ray sources in space.

We have started with studies of the maximum energy resolution achievable with noble liquid ionization chambers. A value of 26 keV FWHM for 1MeV electrons has been measured in liquid argon, dominated mostly by recombination straggling on highly ionizing delta electrons produced in large number along the primary ionizing track.

We are presently working on liquid xenon to establish the most efficient gas purification scheme and the best spectroscopic performance. The initial result of about 38 keV FWHM for the energy resolution of the dominant 570 keV gamma rays from ^{207}Bi, measured with a gridded ionization chamber, is very encouraging.

Our final goal is the construction of a large area (~1000 cm^2) liquid xenon imaging chamber with 20 keV FWHM energy resolution at 1 MeV and a few millimeters spatial resolution, which will be capable of exploring a wide range of astrophysics phenomena expected to produce gamma-rays in the energy range of 0.1 to 10 MeV. As a first application of this instrument we are planning on a spring 1990 balloon flight out of Australia to observe the 122 keV gamma-ray line of ^{57}Co produced by the supernova 1987A with a 3σ sensitivity of 2×10^{-5} ph cm^{-2}sec^{-1}.

MOTIVATION AND OBJECTIVES

The ultimate limitations to the sensitivity practically achievable with the technologies currently employed in gamma-ray astronomy have led us to concentrate on the use of noble liquids for the next generation of imaging spectrometers. Liquid xenon is particularly promising for the detection of gamma-rays, because of its high density (3.1g/cm^3), high atomic number (54) and its high electron mobility. The small Fano factor (0.04), small electron diffusion constant (~50 cm^2/s) and the good scintillation properties also promise excellent energy and position sensitivity. Furthermore a detector can be built in any size or shape and is not damaged by high radiation levels.

© 1988 American Institute of Physics

Of the different types of liquid xenon filled detectors, a time projection chamber (TPC) in the ionization mode, triggered by the primary scintillation light, appears to be the most advantageous. The major technical challenges are the high liquid purity and the low noise electronics requirements.

Among the large number of high energy processes which result in emission of gamma ray lines in the energy range of 0.1 to 10 MeV, explosive nucleosynthesis from SN1987A remains one of the most tempting and will be the objective of our initial scientific program. After the first observations of the 847 keV line from the decay of ^{56}Co reported at this conference (S. M. Matz et al., W. Sandie et al.), instruments with sensitivities in the 10^{-5} ph cm^{-2} sec^{-1} range will be needed in the future to detect the lower energy gamma-ray lines at 14 keV, 122 keV and 136 keV from the ^{57}Co -> ^{57}Fe decay process. The 122 keV line is of particular interest for the light it can shed on isotopic production in Type II Supernovae and for the clues it will yield in interpreting the infrared spectrum of SN1987A. The measured 847 keV line strength and the optical light curve both imply an explosive yield of 0.07 M_o of ^{56}Ni in SN1987A. The infrared spectrum shows lines of both Co II and Ni II about 250 days after the explosion. The short half-life of ^{56}Ni requires that the infrared nickel line arise from the isotope ^{57}Ni, while the cobalt line is assumed to be predominantly ^{56}Co, by far the most abundant isotope produced. However, ionization balance and radiative transfer effects make it difficult to use these two lines to imply directly the 57/56 isotope ratio of iron-peak elements produced in the explosion.

A direct detection of the atomic mass 57 species through observation of the 122 keV ^{57}Co -> ^{57}Fe gamma-ray line, then, will provide a direct measure of this ratio while producing an important additional constraint on the interpretation of the IR spectrum.

We can calculate the expected strength of this line by using the quoted overproduction rate of isotopes compared to solar abundances expected for explosive nucleosynthesis in a Type II event (Woosley and Weaver 1982). The ^{57}Fe/^{56}Fe overproduction ratio is 2.3 and the solar abundance ratio is 0.024. Thus, we expect that 2.3 x 0.024 x 0.07 $M_o \sim 4 \times 10^{-3}$ M_o or N = 8×10^{52} nuclei of ^{57}Fe were produced by SN1987A.

For the spring of 1990, at a time t = 3.1 years after the event, we calculate an unattenuated 122 keV line flux of 3.6×10^{-4} ph cm^{-2} sec^{-1} from the relation:

$$F = 0.26 \left[\frac{N}{10^{51}}\right]\left[\frac{\eta}{1.0}\right]\left[\frac{1\,yr}{\tau}\right]\left[\frac{1\,kpc}{D}\right]^2 e^{-t/\tau} \text{ ph cm}^{-2}\text{s}^{-1}$$

where the branching ratio $\eta = 0.88$, the average lifetime $\tau = 1.1$ yrs and the distance to the LMC is taken as D = 50 kpc.

At the time of our observation, more than three years after the event, the attenuation in turbulent core and expanding envelope will be much reduced. It is clear, however, that the value for the 122 keV line flux calculated above represents an upper limit to the expected flux to be observed at Earth, and that sensitivities substantially below 10^{-4} ph cm^{-2} sec^{-1} will be required to make astrophysically important measurements of this line.

For an on-source integration time of 3×10^4 s, an effective area of 1000 cm^2, an energy resolution of 7 keV and a conservative background rate of 10^{-4} ph cm^{-2} sec^{-1} keV^{-1}, our instrument will have a 3 σ sensitivity $S = 2 \times 10^{-5}$ ph cm^{-2} sec^{-1} at 122 keV.

With the sensititivity of existing NaI (Tl) and Ge gamma-ray detectors the observation of this line at the 3-5 σ level will be problematic. However a single balloon flight of the proposed liquid xenon instrument could detect the 122 keV emission from ^{57}Co at the 50 σ level, providing crucially important new constraints on isotope production ratios, turbulence and mixing in the expanding ejecta, and the interpretation of the IR line spectra.

Other interesting phenomena which require instruments with sensitivities in the 10^{-6} ph cm^{-2} sec^{-1} range include galactic center emission of positron annihilation radiation and cosmic line emission from ^{22}Na production in Novae.

Although less energetic than supernovae by several orders of magnitude, galactic novae are thought to be the site of significant production of ^{22}Na, which decays to an excited state of ^{22}Ne at 1.275 MeV. No evidence of any 1.275 MeV emission of celestial origin has been reported so far. The best upper limit to the mass of ^{22}Na produced by a recent nova is presently 7×10^{-7} M$_o$ (Leising et al. 1988). The upper limit to the flux of the 1.275 MeV line from the galactic center is 1.2×10^{-4} ph cm^{-2} sec^{-1}, corresponding to a mass limit of 3×10^{-6} M$_o$ for the accumulated debris of ~100 novae which would have occurred during the mean lifetime of ^{22}Na.

By taking into account the much lower diffuse background rate in the 1 MeV energy region and the excellent rejection capability provided by the liquid xenon detector internal imaging, our 3 σ line sensitivity of 7×10^{-6} ph cm^{-2} sec^{-1} would enable us to detect in a single balloon flight 1.7×10^{-7} M$_o$ of ^{22}Na from the galactic center region, or 1.6×10^{-8} M$_o$ from a single nova at a distance of 3 kpc. These sensitivities limits are well matched to the expected production of ~ 10^{-7} M$_o$ of ^{22}Na per nova.

Future detailed studies of the 511 keV gamma-ray line which results from the annihilation of positrons from the galactic center also require instruments with a 3 σ sensitivity of the order of 10^{-6} ph cm^{-2} sec^{-1} and a spatial resolution of few degrees. Both of these requirements are easily met by the proposed liquid xenon imaging detector, which could well resolve the ambiguity concerning the diffuse and variable point source components of the galactic annihilation line.

THE LIQUID XENON TIME PROJECTION CHAMBER

A liquid xenon ionization chamber operated in the time projection mode (TPC) is a high density, high resolution detector, continuously sensitive to any ionizing event. Gamma-rays will interact with xenon producing secondary electrons through photoelectric, Compton or pair production processes. These electrons will ionize and excite the liquid creating a large number of electron-ion pairs and scintillation photons. Under a uniform electric field, the ionization electrons are drifted from their point of creation to a collection electrode structure. To minimize losses in the ionization signal, the liquid has to be free from electronegative impurities which would reduce the signal by attachment processes, the electric field has to be strong enough to reduce the electron-ion recombination process, and the transparency of the shielding grid has to be maximized.

To obtain the spatial development of the energy release, a segmented anode or orthogonal wire planes will be used. The X and Y information will be extracted from the signals induced on these sensing electrodes, while the third coordinate, parallel to the electric field direction, will be obtained from the drift time. If a reference time t_o is provided, one knows the absolute position of the ionizing event and a complete 3-dimensional image of the particle trajectory can be reconstructed. Since liquid xenon is an excellent scintillator, the detection of the primary scintillation light with its fast time constant, can be used to obtain the starting time (t_o) of the event. The total energy deposited by the event is obtained by summing the signals from the individual sensing X-Y electrodes. A totally sensitive detector of very large area can therefore be built with excellent energy and spatial resolution.

RESULTS TO DATE AND FUTURE PLANS

Central to the operation of an ionization chamber is the problem of achieving and maintaining the high purity of the liquid which is required to avoid a significant loss of signal and therefore of spectral resolution. To address this and other technical problems which most severely challenge the successful performance of this type of detector and to understand the basic mechanisms which are mostly responsible for the discrepancy between measured and expected energy resolution, we have built a gridded ionization chamber and associated gas purification and handling system. The chamber has been filled with high purity liquid argon and liquid xenon and tested with different radioactive sources.

The electric field dependence of the charge yield and of the energy resolution of electrons, gamma-rays and alpha particles has been extensively measured. Typical energy spectra obtained in both liquids with an internal ^{207}Bi source are shown in the following figures. In liquid argon an intrinsic energy resolution of 26 keV FWHM is observed for the dominant 976 keV conversion electron line of ^{207}Bi. This value is the best achieved so far, but is still a factor of seven larger than the theoretical prediction of 4 keV based on the Fano factor(T. Doke 1982).

This discrepancy can be explained if we take into account the production of delta-electrons along the primary ionizing track. The rate of recombination on these low energy delta-tracks is much stronger than on the primary track.

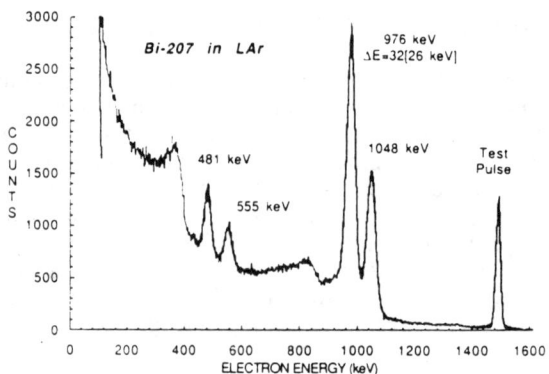

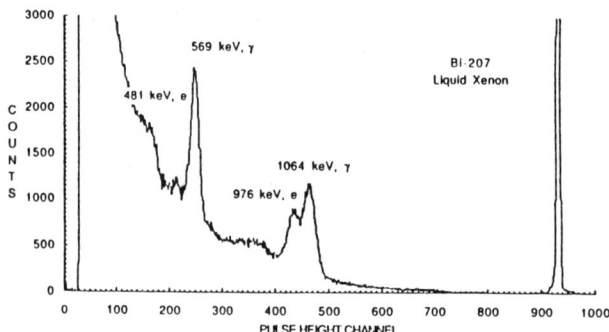

Recombination fluctuations in the number of collected electrons from delta-tracks dominate the observed energy resolution, producing much larger fluctuations than those expected from the Fano factor. This hypothesis is consistent with our data and for the first time provides a good explanation of the limited energy resolution results obtained in high density rare gas chambers under typical drift fields. Details of this work may be found in two recent publications (Aprile et al. 1987; Aprile et al. 1988).

Although we can exclude attachment effects on the argon results, we are still improving the purification system for xenon and measuring the influence of different detector construction materials on the purified gas. The higher boiling point and the higher oxygen attachment cross section (Bakale et al. 1976) make liquid xenon much more difficult to purify and keep 'clean' with respect to liquid argon. While these studies continue with small chambers, we are constructing a larger scale prototype detector with an active area of about 300 cm^2 and a segmented anode to study the position resolution and demonstrate three-dimensional track imaging in liquid xenon. The chamber will be equipped with a CaF_2 window to transmit the primary scintillation light which is centered around 170 nm. A UV sensitive photomultiplier tube will detect this light which will be used for triggering the ionizing event. The initial active volume of this prototype will be about two liters.

CONCLUSIONS

To improve the detection sensitivity to weak gamma-ray line emitting cosmic sources one needs an instrument with high stopping power, high energy and position resolution and an efficient background reduction capability. In this respect, a large area liquid xenon imaging chamber is sufficiently promising to warrant a detailed investigation.

We have made substantial progress in understanding the basic processes which characterize noble liquid filled detectors and with the additional research program currently underway at Columbia we will be able to prove the feasibility of a large area liquid xenon imaging spectrometer as a fundamental new capability for gamma-ray astronomy.

This work is supported by DARPA/ONR and partially by NASA.

REFERENCES

E. Aprile, W. Ku, J. Park and H. Schwartz, Nucl. Instr. and Meth. A261, 519(1987).

E. Aprile, W. Ku and J. Park, to appear in IEEE Trans. Nucl. Sci. (February 1988).

G. Bakale, U. Sowada and W.F. Schimdt, J. Phys. Chem. 80, 255 (1976).

T. Doke, A. Hitachi, S. Kubota, A. Nakamoto and T. Takahashi, Nucl. Instr. and Meth. 134, 353 (1976).

T. Doke, Nucl. Instr. and Meth. 196, 87 (1982).

M.D. Leising, G.H. Share, E.L. Chupp and G. Kanbach, to appear in Ap. J. 328, (May 1988).

S.M. Matz, G.H. Share, M.D. Leising, E.L. Chupp, W.T. Vestrand, W.R. Purcell, M.S. Strickman and C. Reppin, submitted to Nature (January 1988). See also these Proceedings.

W. Sandie, G. Nakano, L. Chase, G. Fishman, C. Meegan, R. Wilson, W. Paciesas and G. Lasche, see these Proceedings.

S.E. Woosley and T.A. Weaver, Essays in Nuclear Astrophysics, eds. C.A. Barnes, D.D. Clayton and D.N. Schramm (1982).

THE GAS SCINTILLATION DRIFT CHAMBER AS A HARD X-RAY DETECTOR

A. Parsons, B. Sadoulet, S. Weiss, D. Smith, K. Hurley and R.P. Lin
Physics Department and Space Sciences Laboratory
University of California, Berkeley

G. Smith
Lawrence Berkeley Laboratory

ABSTRACT

Gas scintillation chambers have excellent energy resolution, but so far have been limited to small areas and low pressures because of the window used to collect the scintillation light.

We are developing a new read-out method based on wave shifter light fibers which overcomes these limitations, while preserving the good energy resolution of gas scintillation. In addition, this scheme should provide excellent spatial resolution and good imaging capabilities.

We are presently testing the scheme and will present our latest results, including the first experimental demonstration of the wave shifter fiber read-out. As an example of the potential of the technique, we will describe our current design for a detector of 40 cm diameter filled with 4 g/cm^2 of xenon. The energy resolution is expected to be 2.2% FWHM at 122 keV and through the identification of the K shell photon, the background will be an order of magnitude lower than current detectors.

INTRODUCTION

The xenon Gas Scintillation Drift Chamber (GSDC) with wave shifter fiber readout offers the unique combination of good spectral resolution (only ~ 2 to 3 times worse than cooled germanium), large area (~1 m^2), large stopping power (~10 g/cm^2), extremely low background and excellent spatial resolution (down to ~ 150 μm at low energies). GSDC's should thus prove to be a very effective tool for the study of hard X-ray and soft gamma-ray emission in the energy range from 20 keV to a few MeV. The capabilities of the instrument are excellent in the low energy region (20 keV to 150 keV), but the power of the instrument also extends to significantly higher energies. At energies above a few hundred keV, the GSDC becomes a powerful Compton telescope. In this region the energy resolution is still excellent and the Compton interaction position can be constructed accurately (to 300 μm rms). The direction of the Compton scattered electron and the energy and position of absorption of the scattered photon can be measured. This limits the position of the source to an arc of an annulus.

As a first use of the instrument, we are beginning to design and prepare a prototype GSDC for a spring 1990 balloon flight. The primary objective of the flight will be to detect and measure the intensity of the low energy gamma-ray lines of ^{57}Co produced by the supernova SN1987A. In the future, the GSDC should also prove to be an effective tool for the study of many high energy astrophysics phenomena such as the prominent diffuse extragalactic background, Active Galactic Nuclei, cosmic gamma ray bursts, and high energy processes occurring on compact objects. Recently, we have proposed[1] a new detector concept based on a GSDC and a hemispherical

© 1988 American Institute of Physics

coded aperture mask. Such a detector would be able to monitor half of the sky all of the time without having to share time between various sources as in a small field of view telescope. Such continuous all sky monitoring over the energy range 20 keV to several MeV, together with the large gains in sensitivity, spatial resolution and spectral resolution may lead to the discovery of many new and exciting transient phenomena.

At present, a technical prototype of the gas scintillation drift chamber has been constructed and is yielding encouraging results.

THE DETECTOR

The gas scintillation drift chamber depends upon two processes whereby UV light is emitted. When an incident X-ray interacts with the xenon gas by photoionization, Compton scattering or possibly pair production, a small amount of UV scintillation light, the so-called primary scintillation, is produced. With the detection of this primary scintillation, we should be able to locate the event in the drift direction. The electrons extracted in the initial interaction are then drifted towards a readout plane, where they generate scintillation light in a high field region (reduced field of 3 kV/cm/atm). Detection of this secondary scintillation light allows two dimensions of spatial imaging (perpendicular to the drift direction) and its intensity is proportional to the energy initially deposited within the chamber. We prefer this detection method over the more conventional method based on an electron avalanche onto detection wires because it maintains the intrinsic energy resolution of a gas detector.

The technical prototype GSDC is designed to handle 40 atm of xenon gas. To date the chamber has been operated at the lower pressure of 5 atm with the chamber exposed to an ^{241}Am source.

LIGHT COLLECTION SCHEMES

We have been testing two different detection schemes for observing the secondary scintillation light. To normalize ourselves to the current literature on GSDC's, we have used the conventional configuration based on an Anger camera geometry with sixteen one-half inch phototubes set behind a quartz window. However, to increase the stopping power of the instrument, high pressures and deep conversion regions are needed. The high pressure requirements along with the need for large area detectors makes this Anger camera readout scheme impractical for astrophysical applications. This problem should be bypassed with an alternate technique being developed which involves a crossed array of wave shifting light fibers to detect the scintillation light [2].

The fibers are doped with dimethyl POPOP and are covered with an external coating of p-Terphenyl (200 µg/cm^2). Thus the UV light is shifted into the blue by the two wave shifters and is picked up at the ends of the fibers by the phototubes. The typical optical efficiency of such an apparatus should approach the efficiency of the Anger camera geometry. In a study done using a UV monochromator, we used a beam of 1735Å UV photons to measure the collection efficiency of the fibers. Our result estimated the efficiency of the coated light fibers to be $(3.3\pm1)\times10^{-3}$ photoelectrons per UV photon [3]. The fibers can sit very close to the scintillation gap and so they can see almost half the scintillation light. Thus the position reconstruction should be simpler since the light transfer function is much more sharply peaked.

POTENTIAL CAPABILITIES OF THE GSDC

The potential capabilities of the GSDC are very attractive. Because the generous light output is linearly proportional to the initial number of electron-ion pairs produced, the energy resolution is completely dominated by the initial ionization process (~22.6 eV/ion-electron pairs for xenon). Furthermore, above 29 keV when the K-shell escape photon is detected, there is significant energy resolution improvement because the escape photon's energy is well known. Figure 1 is a plot of the computed ultimate energy resolution for a GSDC.[4] We see that an energy resolution of only 2 to 3 times worse than cooled germanium detectors can be obtained.[5]

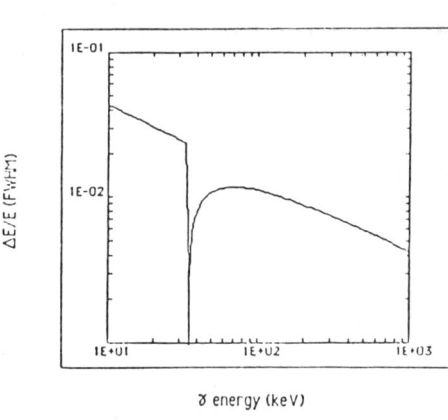

Figure 1. A plot of the computed energy resolution of the GSDC.

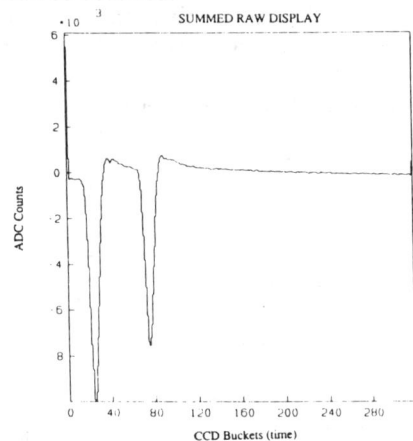

Figure 2. A sample raw display of a typical double pulse event.

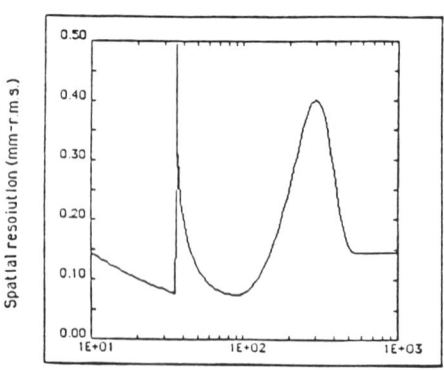

Figure 3. A plot of the calculated position resolution of the GSDC.

The K shell escape photon signature provides an excellent background rejection method. The imaging capability of the detector will be used here to identify the presence of the K shell X-ray emitted in ~80% of the photoelectric absorption events. This K shell signature discriminates against all charged particle and neutron interactions, including β^- decays. We expect background rejection in the 35-150 keV range to be far better than any existing detector system. Figure 2 is a sample raw event recorded by the chamber in the Anger camera geometry. The two pulses represent both the initial interaction and the reabsorption of a K shell escape X-ray. The specific event shown in the figure is fairly typical of the two pulse events we see in our chamber operating at 5 atm and viewing the 60 keV X-ray from an ^{241}Am source. The clean separation of the two pulses demonstrates that the identification of this K shell signature will be relatively easy.

The GSDC offers the possibility of excellent spatial resolution. If the primary scintillation is detected, we can use it as time zero and operate the GSDC as a time projection chamber. This in addition to the light fiber readout in the xy-plane will provide three dimensional imaging capability. Using the light fiber array at low energies, spatial resolution of 150 µm is possible. Figure 3 is a plot of the estimated spatial resolution of the GSDC as a function of the X-ray energy. [4]

CURRENT STATE OF DEVELOPMENT

We have made significant progress recently and have demonstrated for the first time the operation of a wave shifter fiber readout in our technical prototype chamber [6]. Our main findings so far are:

- The light collection efficiency of the fibers in a real chamber environment is observed with X-ray pulses to be $(2\pm0.5)\times10^{-3}$ photoelectrons/initial UV photon. This agrees with our measurement of $(3.3\pm1)\times10^{-3}$ made with the UV monochromator and is much higher than our conservative design estimate[2] of 0.6×10^{-3} photoelectrons/initial UV photon.

- The pattern of the pulses observed on the fibers which are currently grouped in strips of 8 mm clearly demonstrates the possibility of xy-imaging.

- The time profile of X-ray pulses readily displays the presence of 2 pulses in the case of the emission of a K shell escape photon in both the Anger camera and wave shifter fiber readout. Figure 2 is an example of such a display.

- Using the wave shifter fibers, there appears to be some degradation of the gas without recirculation leading to a decrease in pulse height of 20% per day. However, we have shown that passing the gas through a molecular sieve restores the light yield, even after 50 days of operation; thus a proper recirculation system should solve this problem. This system is being implemented at present.

There are, however, a few features that we do not understand as yet:

- The overall light yield obtained with our chamber in the

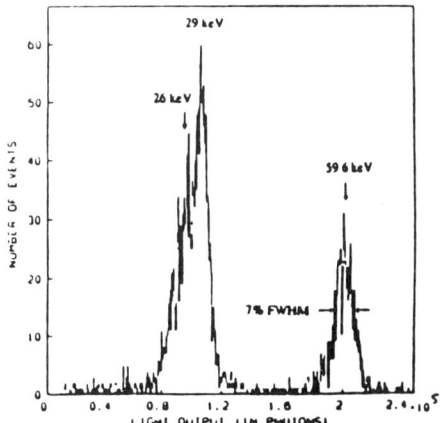

Figure 4. A position corrected spectrum recorded using the conventional Anger camera geometry.

conventional window geometry is about 5 times smaller than quoted in the literature [5, 7]. This result may be due to a problem of gas purity. This light yield is responsible for the 7% FWHM resolution at 60 keV that we obtain instead of the 3.2% observed by Nguyen Ngoc et al.[5] under similar conditions. Figure 4 is a sample position corrected spectrum recorded for the Anger camera geometry with 5 atm Xe and 3 kV/cm/atm scintillation field.

- So far we have not been able to detect primary scintillation with our ^{241}Am 60 keV source in the Anger camera geometry. This is not surprising because the technical prototype was not designed for this purpose.

We are currently working to solve these problems. In the coming months we plan to:
- Operate the chamber at higher pressure and develop a surface treatment of the scintillation mesh spacers to prevent electrical breakdown along the surfaces at the higher electric fields that will be necessary at the higher pressures.
- Work on changing the gas composition to limit diffusion and increase the drift velocity of the electrons without decreasing the light yield significantly. For astrophysical applications, it is important to increase the drift velocity to decrease the dead time that will result from the detection of cosmic rays.

FUTURE PLANS

The results from our technical prototype are encouraging enough to begin with the design of the scientific instrument.

We hope to be able to use the GSDC to detect low energy gamma ray lines produced by explosive nucleosynthesis within the supernova SN1987A. The primary objective of a balloon flight to take place in 1990 will be to measure the flux of the 122 keV line produced from the decay of ^{57}Co to ^{57}Fe. ^{57}Co is a daughter of ^{57}Ni which is expected to be produced in the explosion along with the much more abundant isotope ^{56}Ni which decays to ^{56}Co. Other experiments are set to search for the 847 keV line which is produced in the decay of ^{56}Co to ^{56}Fe. Thus the measurement of both the 847 keV and the 122 keV line intensities will provide a good estimate of the amount of ^{56}Ni and ^{57}Ni produced in the supernova, and will thus help determine whether this supernova is typical of the events that cause galactic nucleosynthesis.

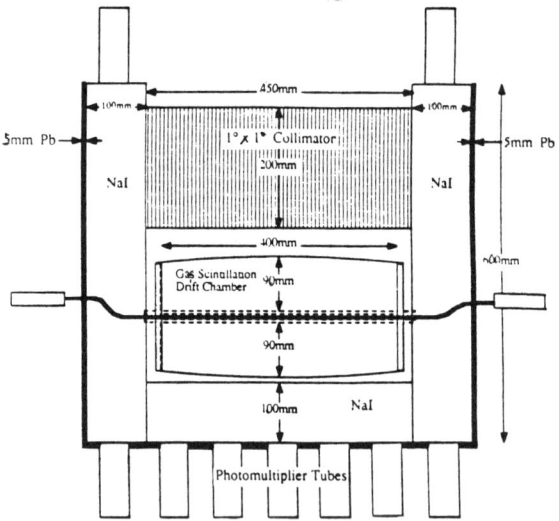

Figure 5 A schematic of the scientific prototype that we plan to use to look at the lines from SN1987A.

Calculations based on a standard spherically symmetric model for a 15 $M_\odot$ type II supernova [8,9] with 0.07 $M_\odot$ of ^{56}Ni and the solar abundance ratio of ^{57}Fe/^{56}Fe predict a 122 keV line flux that is barely detectable (~3 to 5σ) for the best balloon-borne germanium detector arrays but could be detected at ~15-20σ by our GSDC instrument. At the 122 keV line of ^{57}Co, the predicted energy resolution would be 2.2% FWHM. The total detection background would be 1×10^{-3} (keVs)$^{-1}$ which would give a 3σ detector threshold of 5×10^{-6} (cm^{2}s)$^{-1}$.

Figure 5 is a schematic diagram of the proposed detector. This GSDC will be surrounded by ~10 cm thick NaI anti-coincidence shields covered on the outside with a

0.5 cm thick Pb sheath. The field of view is limited to 1°x1° FWHM with a tantalum/tin/copper graded Z slat collimator. Monte Carlo simulations of this shield and collimator including the passive material in the detector, together with the K shell photon identification in the GSDC, show that the background in the 35 to 150 keV range is less than 2×10^{-6} γ-$(cm^2 \, sec \, keV)^{-1}$, about ten times less than any other detector. Note, however, that these Monte Carlo simulations do not include the background resulting from secondary production due to cosmic ray activation of the instrument.

As previously mentioned, an Explorer type mission, the High Energy All Sky Imager (HEASI) Explorer [1] has been proposed using the GSDC with wave shifter fiber readout in conjunction with a hemispherical coded aperture mask. The mask would be made in the form of approximately half a geodesic dome so that the detector would be constantly monitoring nearly the entire half sky away from the earth. In the low energy region (20-150 keV), the reconstruction accuracy of the photon conversion point (~150 μm) is smaller than the size of the holes of the mask which will determine the angular resolution (see Appendix I of Reference 1). The size of individual sky pixels in the coded mask pattern is 150 arcsec^2. The localization performance will be somewhat degraded in the higher energy ranges, yet the energy resolution will still be very good.

Thus it is clear that the Gas Scintillation Drift Chamber is an extremely promising tool for the understanding of many high energy astrophysics phenomena. There is much work to be done, but the potential of such an instrument is very great.

ACKNOWLEDGEMENTS

It is a pleasure to acknowledge support from the NASA Innovative Research Grant and NASA Grant No. NAGW516. We have additional support from the NASA Graduate Student Researchers Program (SW) and a Graduate Opportunity Fellowship from the University of California at Berkeley (AP).

REFERENCES

1. B. Sadoulet, R.P. Lin, T. Prince, K. Hurley and W. Pietsch, "High Energy All Sky Imager (HEASI) Explorer," NASA proposal, University of California preprint, 1986.
2. B. Sadoulet, "The Case for Gas Scintillation Drift Chambers,"University of California preprint, March 15, 1985.
3. S. Weiss, "Report of Test of Light Fiber Collection Efficiency Using UV Monochromator at SSL," unpublished, March 24, 1987.
4. B. Sadoulet, R.P. Lin and S. Weiss, IEEE Trans. in Nucl. Sc., NS-34, no. 1, p. 52 (1987).
5. H. Nguyen Ngoc, J. Jeanjean, H. Itoh and G. Charpak, Nucl. Instr. and Meth., 172, pp. 603-608 (1980).
6. B. Sadoulet, S. Weiss, A. Parsons, R.P. Lin,and G. Smith, IEEE Trans. in Nucl. Sc., NS-35, no. 1, in press.
7. D. F. Anderson *et al.,* Nucl. Instr. and Meth., 163, p. 125 (1979).
8. T. A. Weaver and S. E. Woosley, Ann. N.Y. Acad. Sci., 336, p.335 (1980).
9. T. A. Weaver and S. E. Woosley, in Supernova Spectra, eds. R. Meyerott and G. H. Gillespie, (New York, Am. Inst. Phys., 1980) p. 15.

PULSE SHAPE DISCRIMINATION FOR BACKGROUND REDUCTION IN GERMANIUM DETECTORS

D. M. Smith,* M. Shapshak,* R. Campbell, J. H. Primbsch, R. P. Lin
Space Sciences Laboratory, University of California, Berkeley, CA 94720

P. N. Luke, N. W. Madden, R. H. Pehl
Lawrence Berkeley Laboratory, University of California, Berkeley, CA 94720

ABSTRACT

We report the results of laboratory tests of a pulse shape discrimination technique for reducing β^--decay background in germanium detectors. The technique distinguishes single-site from multiple-site energy deposits in the detector through differences in the shape of the collected current pulse. At 816 keV, >83% of single-site events can be rejected, while retaining ~70% of typical photon events.

INTRODUCTION

Improved sensitivity is the primary goal in the design of high resolution gamma-ray spectrometers for astrophysical sources, and reduction of background is an important means of reaching that goal. In modern germanium detector systems, collimation and anticoincidence shielding are used to reduce the background from diffuse cosmic and atmospheric flux entering the aperture and leaking in from other directions. A major source of background, particularly in the energy range 200 keV to 2 MeV, is the decay of radioactive nuclei produced by primary and secondary cosmic rays interacting in the detector. Many of these decays result in one or more prompt photons which will trigger the anticoincidence shield, but β^--decays will generally be completely contained in the detector. β^--decay electrons have a short mean free path in germanium and thus deposit all their energy in one location, while photons of the same energy generally Compton scatter several times before final photoabsorption (Figure 1), resulting in energy deposit at multiple sites in the detector.

Dividing one electrode of a coaxial detector into multiple segments provides one method for identifying single- and multiple-site events.[1] Here we show that analysis of the shape of the current pulse can also provide an effective means of determining whether an interaction occurred at a single site or at multiple sites. An analysis of the two techniques and the results of Monte Carlo simulations of both were given in a previous paper.[2]

The 5.5 cm diameter by 5.5 cm long closed-end reverse-electrode coaxial germanium detectors used here are electrically divided at the central electrode into an ~1.0 cm thick top segment and an ~4.5 cm thick rear segment. The top segment alone is used for detection of low energy (<150 keV) photons where photoelectric absorption dominates. A small fraction of multiple-site photon events will be identified simply by front/rear segment coincidence, but most of the multiple-site interactions will be contained entirely in the rear coaxial segment. An energy

---

* Also Physics Department, University of California, Berkeley.

© 1988 American Institute of Physics

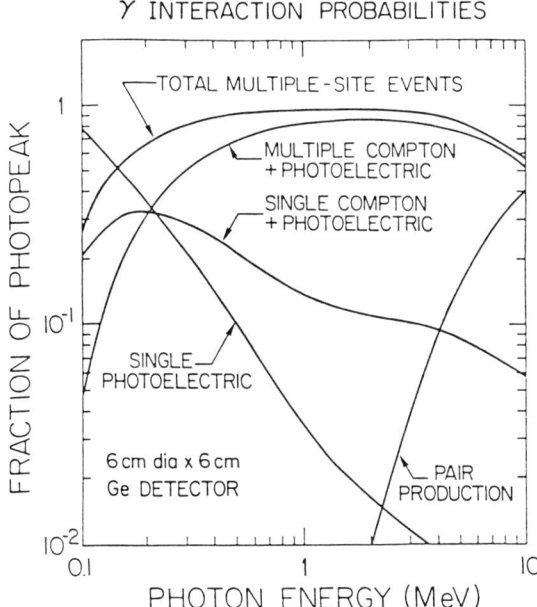

Fig. 1. Fraction of the photopeak contributed by different energy loss mechanisms for photons normally incident upon the face of a 6 cm diameter, 6 cm length coaxial detector.

deposit in the detector will free charge carriers, electrons and holes, which will drift along electric field lines until they are collected at the electrodes. The motion of these charge carriers induces a current at the electrodes, which reaches a maximum as they enter the region of highest field; for a detector free of any impurities and with a cylindrical symmetry, this region would be near the center contact. Thus the shape of the current pulse as a function of time depends on the radial position of the energy deposit. For a given total energy deposit, a multiple-site interaction spread over different radial locations will result in a different current pulse shape than a single-site interaction. Thus, in principle, pulse shape discrimination can be used to separate single-site (β^--decay) from multiple-site (photon) interactions.

EXPERIMENTAL PROCEDURE

We have generated single and multiple interactions in the laboratory. The experimental setup is shown in Figure 2. To create single-site events, monochromatic photons from a gamma source are Compton scattered in the primary germanium detector and then collected in an auxiliary germanium detector. Single Compton scatter events in the primary detector are identified by three criteria: energy deposit must occur in coincidence in the two detectors; the photon involved must pass through two sets of collimators to ensure that a scatter of $90° \pm 3°$ has occurred; and the energy deposit in each detector must fall within 14 keV of the energy expected from the Compton formula for a 90° scatter of a photon of that energy. The allowed errors in angle and energy are equivalent for an 1173 keV incident photon. Monte Carlo simulations indicate that ~95% of the events at this energy satisfying these criteria will be single Compton scatters leaving 816±14 keV.

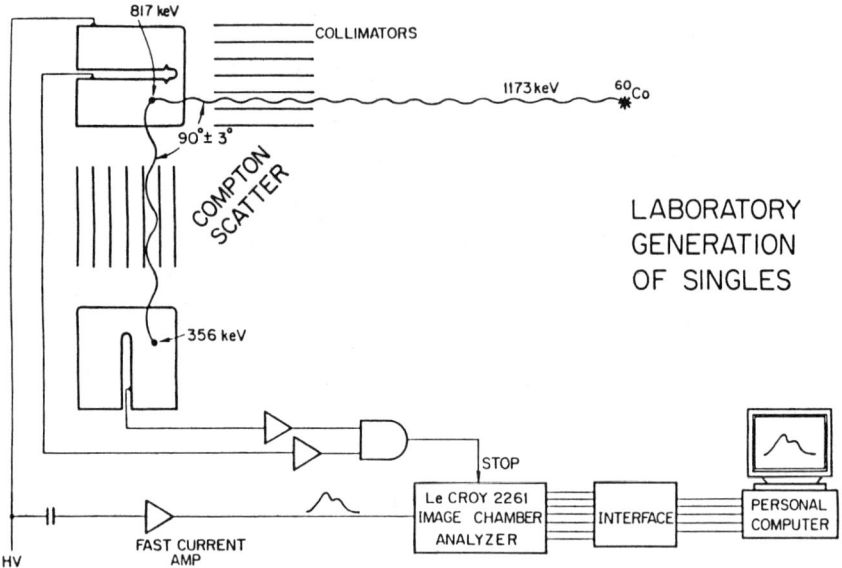

Fig. 2. The experimental set-up to generate single-site interactions. The 1173-keV gamma-ray from the ^{60}Co source is Compton scattered through a single 90° scatter to obtain a single 816 keV energy deposit in the rear coaxial segment.

For comparison a gamma-ray source of approximately the same total photon energy (for example 835 keV) as the single energy deposit is used. The 835 keV photons are incident on the top surface of the detector, and only photopeak events are accepted. Monte Carlo simulations (Figure 1) show that ~95% of these events are expected to be multiple-site events. Table I shows the gamma-ray source lines which we have used to obtain a range of energies.

The shape of the current pulse is obtained by a high speed current preamplifier capacitatively coupled to the high voltage electrode of the detector. The current pulse is digitized by a 10-bit LeCroy 2261 image chamber analyzer which samples every 10 ns. An accurate measurement of the total energy of the event is obtained from a separate charge preamplifier coupled to the central electrode. Figure 3 shows typical pulse shapes from the laboratory measurements and from computer simulations. Typical current pulse durations are ~200–300 ns.

Single Compton scatter events collected in the laboratory are used to define the current pulse shape characteristic of single-site interactions. One pulse shape will not, however, suffice for all events. The radius at which a single deposit occurs determines the drift time of the carriers to the central electrode, and thus the time-to-peak of the current pulse shape (Figure 4a). In addition, different pulse shapes trigger the digitizing process at different times. We have therefore divided the total set of singles into 19 subsets by time-to-peak, each of which is subdivided by three or four different values of start time. Since ~5 percent of the events which meet the

Table I. Pulse Shape Discrimination Efficiency

	Gamma-ray source	Line energy (keV)	Single Compton scatter energy (keV)	Number of events	% Identified as single	% Identified as multiples
Singles	^{54}Mn	835	518	4375	88%	—
Multiples	^{22}Na	511	—	4400	—	80%
Singles	^{60}Co	1173	816	16650	83.3%	—
Multiples	^{54}Mn	835	—	18850	—	68.1%
Singles	^{226}Ra	1765	1368	5470	90%	—
Multiples	^{226}Ra	1378	—	4400	—	72%
Singles	^{226}Ra	2204	1789	1017	100%	—
Multiples	^{226}Ra	1765	—	2200	—	80%

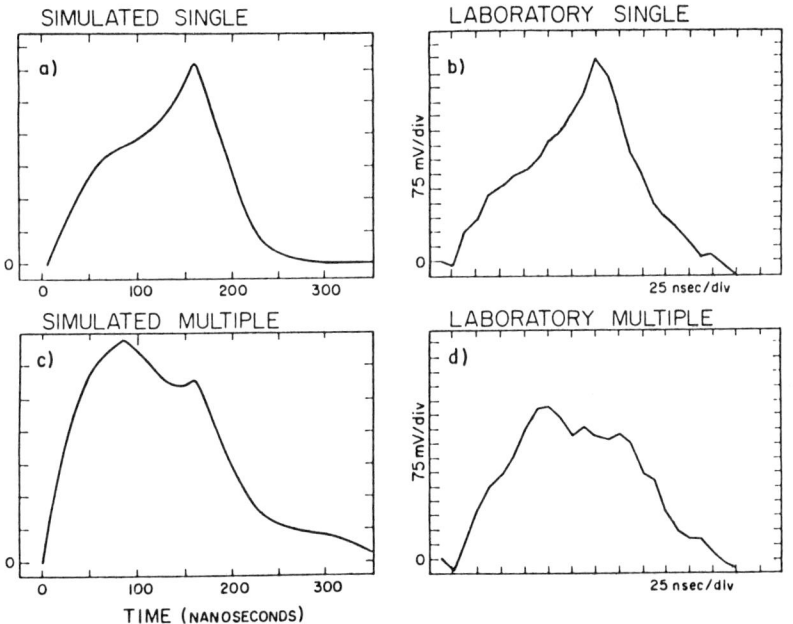

Fig. 3. Pulse shapes for typical multiple- and single-site interactions. *(a)* Simulated single at a radius of 2.7 cm. *(b)* Laboratory single. *(c)* Simulated multiple. *(d)* Laboratory multiple.

criteria for a single scatter will, in fact, be multiples, pulses with obviously anomalous shapes are removed. A pulse shape envelope which brackets the remaining pulses both above and below is then defined. Pulses which fall outside this envelope at any point are then considered to be multiple-site events. The laboratory multiples, generated by photons which are totally absorbed in the detector, are used to test the performance of this envelope, after first being normalized so that the total energy (area under the pulse shape) is equal to that of the singles which generated the envelope. Since each pulse is normalized by its energy before being compared to a pulse shape envelope, in principle envelopes generated at one energy should serve for all energies. However, at low energies electronic noise becomes relatively more important, so that wider envelopes are necessary. We have therefore generated sets of pulse shape envelopes at the various energies in Table I.

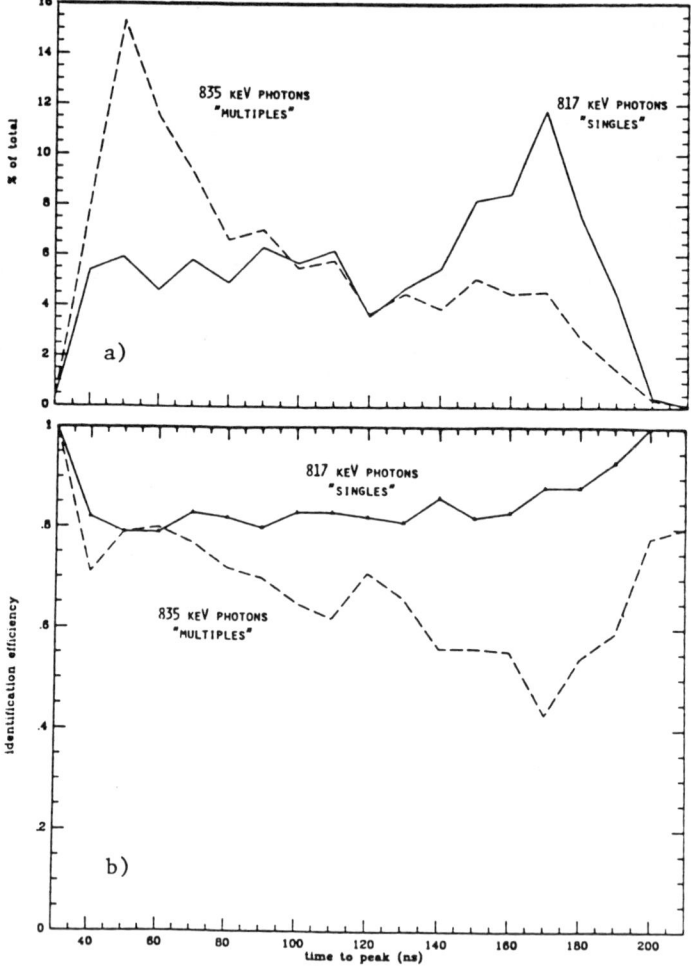

Fig. 4. (a) Distribution of single and multiple events as a function of time-to-peak. (b) Efficiency for identifying single and multiple events as a function of time-to-peak.

RESULTS

The percentage of the supposed single Compton scatter events believed to be truly singles and thus used to create the envelopes, and the percentage of multiple-site photopeak events correctly identified by being outside the singles' pulse envelopes are shown in Table I, averaged over all start times and times-to-peak for each energy. The variation of discrimination ability with time-to-peak of the pulses (i.e., the radius at which the event occurred) is shown in Figure 4b. For the single

Compton scatter events the percentage given is a lower limit to the actual efficiency for identifying true single-site events since a few percent of the scatter events produced in the laboratory will be multiples. Similarly a small fraction of the photon photopeak absorption events will be single-site interactions.

FIRST APPLICATION

This technique will be applied to a single germanium detector in an upcoming long duration balloon flight described elsewhere in this volume.[3] The pulse envelopes obtained at 816 keV and 1368 keV will be used for analysis of the 847 and 1238 keV lines expected from the decay of ^{56}Co in SN 1987A, the 518 keV pulse envelope for 511 keV positron annihilation radiation from the supernova and the galactic center region, and the 1789 keV envelopes in the region of the 1809 keV line from galactic ^{26}Al. An extensive study of the probable sources of background at float altitude[4] estimates that β^--decays contribute ~35–40% of the continuum background at these energies in an instrument with similar shielding. By observing the number of single-site events at float altitude we will determine the β^--decay contribution to background. This will be the first such experimental separation of background components.

Future detector systems will be designed with more effective anticoincidence shielding and collimation so that β^--decay will dominate their background counting in this energy range. For such an instrument, the data in Table I imply that improvements in sensitivity of at least 2.3, 1.7, and 2.3 at 518, 816, and 1368 keV, respectively, could be achieved with pulse shape discrimination. (The sample at 1789 keV is too small to give an accurate measure of the discrimination which can actually be achieved.) The gain in sensitivity could be even greater, since the number of singles which will be falsely accepted as photon events in actual operation is very small; the events rejected from the samples of singles in Table I were mostly true multiples, an artifact of using photon interactions to mimic β^--decays. Pulse shape analysis might also eventually be used in conjunction with multiple segmentation, providing two dimensions of spatial resolution within a coaxial detector.

ACKNOWLEDGMENTS

This research was funded in part by NASA grant NAGW-449.

REFERENCES

1. L. Varnell and R. H. Pehl, this volume.
2. J. Roth, J. H. Primbsch, and R. P. Lin, IEEE Trans. Nucl. Sci. **NS-31**, 367 (1984).
3. M. Pelling, J. Matteson, L. Peterson, R. Lin, K. Hurley, R. Pehl, G. Vedrenne, and M. Niel, this volume.
4. N. Gehrels, Nuc. Inst. Meth. **A239**, 324 (1985).

PERFORMANCE OF A FIVE-SEGMENT COAXIAL N-TYPE GERMANIUM DETECTOR

L.S. Varnell, J.C. Ling, W.A. Mahoney
Jet Propulsion Laboratory, Pasadena, CA 91109

R.H. Pehl, C.P. Cork, D.A. Landis, P.N. Luke,
N.W. Madden, D.F. Malone
Lawrence Berkeley Laboratory, Berkeley, CA 94720

ABSTRACT

A five-segment n-type coaxial detector with associated test cryostat and electronics has been fabricated and tested. Measurements were made of detector efficiency as a function of the number of segments in which interactions occur and the energy threshold of the segment discriminators. The performance of the detector agrees well with predictions. A second detector is now being fabricated along with a rugged mount, flight cryostat, and electronics for use in a future balloon flight.

INTRODUCTION

Since 1981 the High Energy Astrophysics Group at JPL has been working with the Semiconductor Detector Group at LBL to develop a segmented coaxial Ge detector for gamma-ray astronomy. The purpose of segmentation is to locate radiation interaction sites in the detector so that internal background produced in space by particle reactions in the detector can be identified and removed from the gamma-ray signal of astronomical sources. A drawing of the segmented coaxial detector with associated electronics is shown in Figure 1. Gamma rays in the energy range from 150 to 8000 keV interact primarily by Compton scattering followed by photoelectric absorption. In the process of Compton scattering, the energy of the gamma rays is deposited in the detector in several locations separated by distances of the order of one centimeter. By contrast, electrons from the beta decay of radioisotopes produced by particle reactions with Ge, the principal internal background, have a range of the order of one millimeter. If the segments are of the order of one centimeter thick, the beta particles nearly always interact in only one segment, while the gamma rays usually interact in more than one segment. By requiring that an event interact in two or more segments, gamma rays can be selected and internal radioactivity rejected. The segmented detector can be used in another mode for photons with energy less than

© 1988 American Institute of Physics

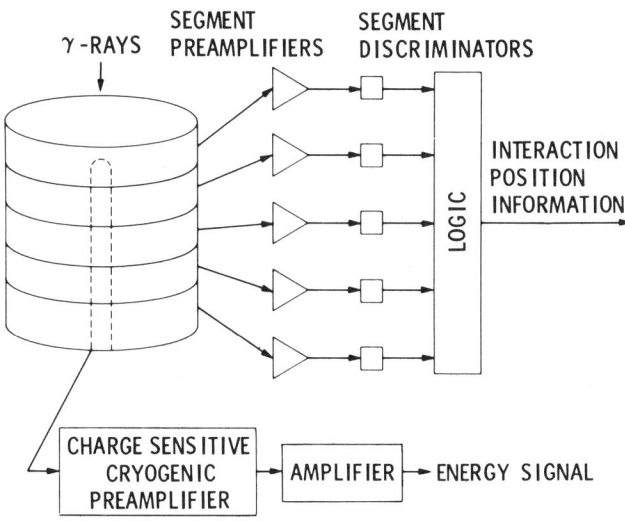

Fig. 1. Diagram of the five-segment coaxial Ge gamma-ray detector. The signal for all energy deposited in the detector is taken from the inner electrode. Energy deposited in a given segment produces a signal in the segment signal chain, which can then be analyzed or used to trigger a discriminator, so that segments in which interactions occur can be identified.

100 keV. At these low energies, where the photon is completely stopped in the first segment, events which interact in the first segment and no other are accepted. The detector then has the same efficiency as an unsegmented detector but only 1/5 the background (for a five-segment detector). If the detector is fabricated with a thin boron-implanted outer p^+ electrode (approximately 0.3 micrometers thick), the efficiency can be high for photons down to 5 keV. Moreover, considerations of radiation damage require using reverse-electrode (p^+ outer electrode) coaxial detectors in space. In a space environment, where high-energy protons are the principal cause of resolution degradation through radiation damage, the conventional-electrode (n^+ outer electrode) coaxial detector begins to degrade at a fluence 30 or more times lower than the reverse-electrode detector. This means that a reverse-electrode detector can remain in space for several years without degradation, while the resolution of a conventional-

electrode detector deteriorates markedly in a few months. A previous paper[1] described the problem of detector background in a space environment and gave the results of measurements made with a planar Ge detector divided into two segments. This paper describes measurements made with a five-segment coaxial detector of a size typical of the detectors to be used in arrays in future space instruments. A detailed study of the background in balloon-borne germanium detectors has been given by Gehrels[2].

DETECTOR PERFORMANCE

The detector was fabricated from a cylindrical crystal approximately 55 mm in diameter and 55 mm long, with a net electrically active impurity concentration that varied from 1.0 to 1.5 x 10^{10} donors per cm^3. To evaluate the crystal, it was first made into an ordinary closed-end coaxial detector which depleted at 2300 V and had energy resolution of 1.8 keV FWHM at 1332 keV at an operating bias of 4000 V. The peak shape was symmetrical, indicating that charge trapping was not significant. After this evaluation, the crystal was etched to remove the outer electrode and a segmented electrode was produced using a tape mask during the ion implantation process. The outer p^+ electrode was produced by boron ion implantation, the inner n^+ electrode by Li diffusion. Each segment electrode is about 1 cm long separated from the neighboring segment electrodes by a gap about 1 mm wide.

The detector was mounted in a cradle with its axis horizontal and contact made to the outer segment electrodes with bellows springs. The five segments were DC coupled to cooled FET charge-sensitive preamplifiers. The inner electrode has a cooled FET preamplifier but was AC coupled so that the detector bias of 3000 V could be applied. The energy resolution using the center electrode was 2.6 keV FWHM at 1332 keV, 1.3 keV FWHM at 60 keV, and the efficiency was 26% of a 7.6 cm x 7.6 cm NaI detector at 1332 keV with a source distance of 25 cm. The energy resolution measured for the five segments through the segment signal chains varied from 2.1 to 2.5 keV FWHM at 1332 keV and from 1.4 to 1.7 keV FWHM at 60 keV. The detector holder and electronics were not optimized for best resolution but were designed for ease of mounting and removal of the detector during testing. Care was taken to shield the signals from the separate segments to avoid crosstalk, and no problem which could be attributed to crosstalk was observed. A detailed description of some of the techniques used in the cryostat and electronics is given in the article by Pehl *et al* [3].

To verify the operation of the segmented detector, measurements were made of the full-energy-peak (FEP) efficiency versus energy for various requirements on the number of segments in which interactions occur and on the energy threshold of the segment discriminators (LLDs). The laboratory spectra were taken with radioactive sources using the energy signal from the inner electrode gated by a coincident logic

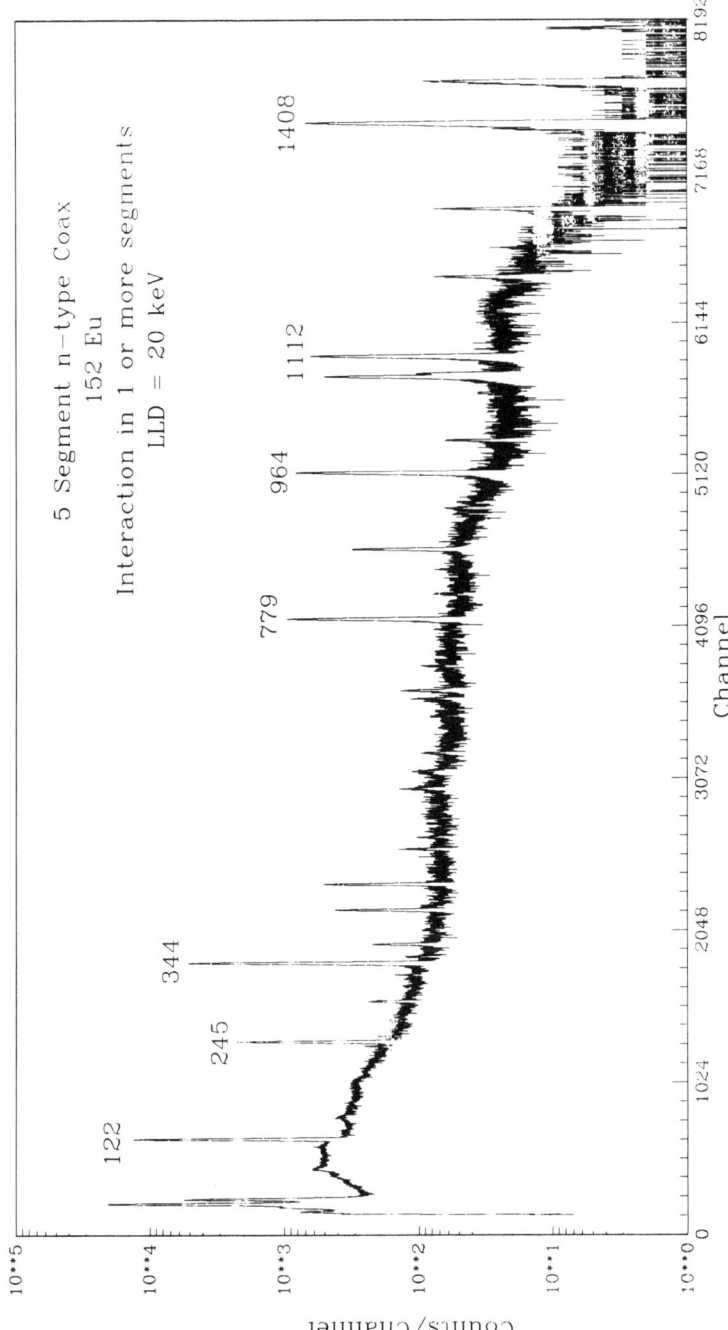

Fig. 2. Spectrum for ^{152}Eu for the segmented coaxial detector. All energy signals are accepted for accumulation, since all events will interact in one or more segments.

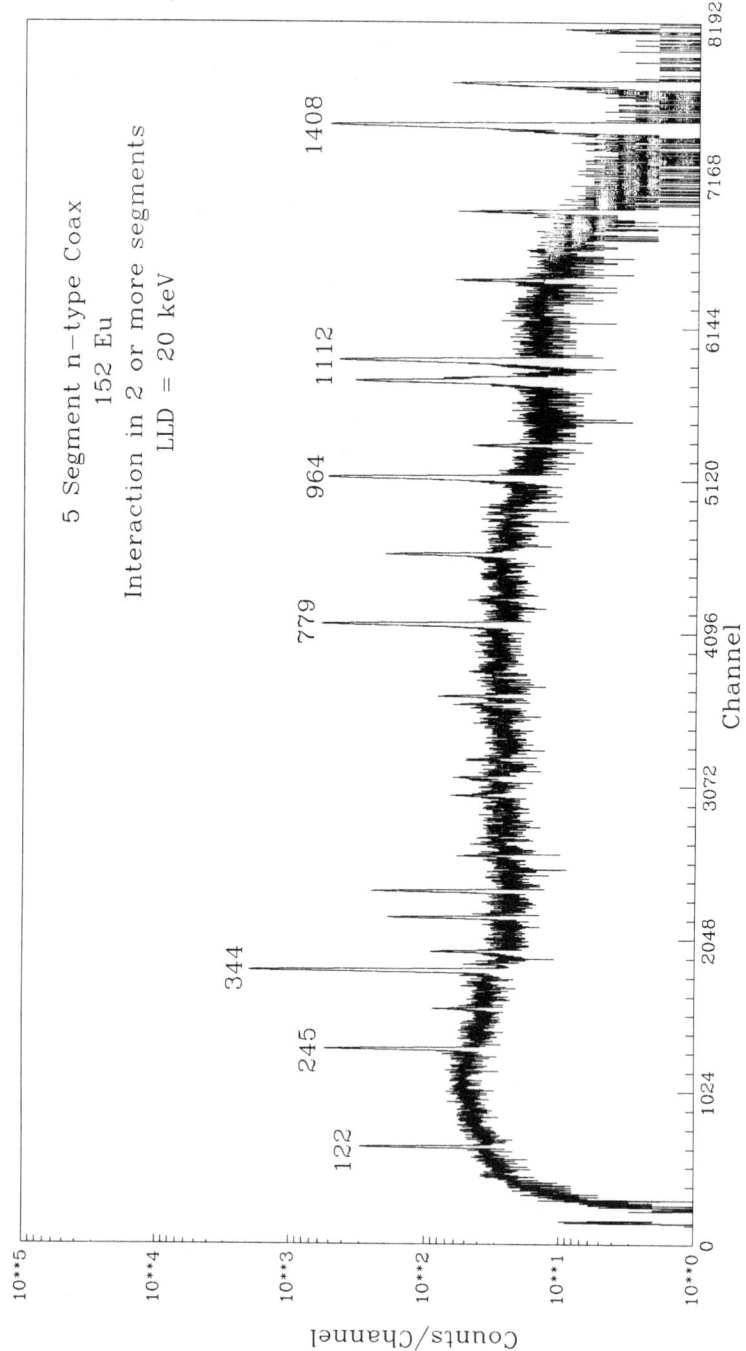

Fig. 3. Spectrum for ^{152}Eu accumulated with the requirement that an interaction occur in two or more segments. Segment discriminators were set at 20 keV.

signal from the segment discriminators. To obtain the logic signal, the outputs of the five discriminators were sent to a universal coincidence module, where the minimum number of input signals required for an output signal could be selected from one to five, corresponding to interactions in any one or more segments, any two or more, up to interactions in all five segments. The output of the universal coincidence was used to gate the pulse-height analyzer for accumulation of spectra from energy signals from the inner electrode. The energy level of the segment discriminators could be adjusted from 20 keV upwards. Figures 2 and 3 show spectra of ^{152}Eu, which has strong lines from 40 to 1408 keV. The spectrum of Figure 2 was accumulated with a requirement that an interaction occur in one or more segments, equivalent to a spectrum with no logic requirements, as in an ordinary detector. The spectrum of Figure 3 was taken with the requirement that interactions occur in two or more segments with a discriminator level of 20 keV. Such a requirement would be used in flight to eliminate

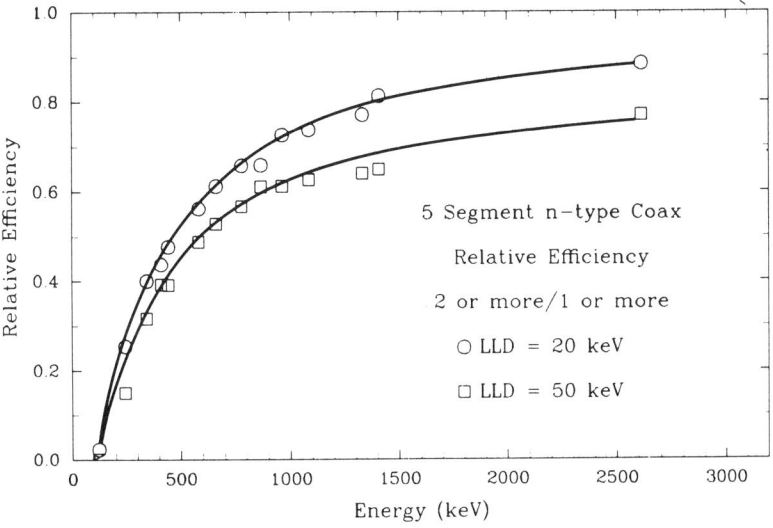

Fig. 4. Relative FEP efficiency versus energy for the segmented coaxial detector with the requirement that interactions occur in two or more segments. The relative efficiency is obtained by dividing the FEP area for the spectrum requiring interactions in two or more segments by the FEP area for the free spectrum (one or more segments with interactions). Segment discriminators were set at 20 and 50 keV. The smooth curves are drawn to guide the eye.

particle-induced background which deposits energy predominantly in one segment. The efficiency of the two-or-more-segments mode is lower especially at low energies because some gamma rays will interact and deposit all their energy in one segment. Figure 4 shows the ratio as a function of gamma-ray energy of the efficiency of the segmented detector operating in the two-or-more-segments mode to the efficiency of an ordinary detector of the same size. The experimental values are obtained by dividing the area of a given line in Figure 3 by the area of the line corresponding in Figure 2. For an LLD setting of 20 keV the relative efficiency varies from almost zero at 120 keV to 0.9 at 2614 keV. Gamma rays of energy below 100 keV are stopped almost completely in the first segment. These measurements were repeated for higher settings of the segment discriminators to observe the effect of discriminator setting on efficiency. As can be seen from Figure 4, higher discriminator settings result in a significant reduction of the FEP efficiency, even at energies above 1 MeV. This was not expected on the basis of our original Monte Carlo calculations, which showed that the efficiency did not decrease significantly for segment discriminator threshold settings up to 50 keV. We are preparing further calculations which follow the scattered radiation to lower energies more realistically.

The work reported here represents a significant advance in Ge detector technology. New techniques were developed for detector fabrication, detector mounting, cryostat electrical feedthroughs, and front-end electronics. The excellent performance of the detector in the laboratory has been verified. The true test of the detector's effectiveness for gamma-ray astronomy must come from balloon or space flight. A side-by-side comparison of the segmented detector with a standard detector or a detector using other means of background suppression is necessary for future progress in detector technology. Work is near completion on a second five-segment detector and flight mount, cryostat, and electronics for a future balloon flight.

ACKNOWLEDGEMENTS

The research described in this paper was performed at the Jet Propulsion Laboratory, California Institute of Technology, under contract with the National Aeronautics and Space Administration, and at the Lawrence Berkeley Laboratory, University of California, under contract with the U.S. Department of Energy No. DE-AC0376SF00098.

Funding for this research was provided by grants from the JPL Director's Discretionary Fund and the NASA Planetary Instrument Definition and Development Program.

REFERENCES

1. L. S. Varnell, J. C. Ling, W. A. Mahoney, A. S. Jacobson, R. H. Pehl, F. S. Goulding, D. A. Landis, P. N. Luke, N. W. Madden, IEEE Trans. Nucl. Sci., 31, 300 (1984).
2. N. Gehrels, Nucl. Inst. Meth., A239, 324 (1985).
3. R. H. Pehl, N. W. Madden, D. A. Landis, D. F. Malone, C. P. Cork, IEEE Trans. Nucl. Sci., 32, 22 (1985).

BOLOMETERS AS HIGH-RESOLUTION GAMMA SPECTROMETERS

George Simpson
Space Science Center, University of New Hampshire, Durham, NH 03824

ABSTRACT

A significant advance in nuclear gamma-ray spectroscopy could be made if detectors capable of measuring Doppler shifts at MeV energies were available. With this goal in mind, we have investigated the prospects for constructing gamma-ray bolometers. We discuss the advantages and disadvantages of this approach, drawing on recent progress in their application to X-ray astronomy and neutrino detection.

INTRODUCTION

Bolometers, which measure radiant energy by converting photons into heat, are interesting to study because they raise the possibility of achieving very high resolution. This has been demonstrated at X-ray energies by the GSFC group who, using a silicon boule at 304 mK, have achieved a resolution of 17 eV at 6 keV[1,2,4]. This is more than a factor of 10 improvement over the best resolution previously achieved with solid state detectors. Their detector is unsuitable for gamma-ray work, having an active volume of only 2.5×10^{-6} cm^3. Blas Cabrera[3] also took up the bolometer concept, proposing that a neutrino detector be built using many 1 kg blocks of silicon at 1 mK. He suggested that the required thermometry be performed using either superconducting thin films or SQUIDs.

This paper discusses the potential of bolometers as spectrometers for gamma-ray astronomy. The scope of the discussion is restricted to defining one feasible design for a single bolometer unit, which corresponds to the individual detectors of conventional gamma-ray spectrometers (eg: NaI or Ge crystals). A brief discussion of the background conditions which will be encountered in orbit, and some possible background suppression strategies are given. We are trying to answer the question: "Is it possible, in principle, to use a bolometer to perform high-resolution astronomical spectroscopy at gamma-ray energies?"

SCIENTIFIC PROMISE

Recent solid-state detector gamma-ray results have shown that the widths of the galactic 511[5,6,7] and 1809[8] keV lines are less than the resolution of these devices (~1 keV). It is reasonable therefore to expect that narrow widths will be a characteristic of many other astrophysical lines, as yet undiscovered. Some of the questions which could be addressed if a spectrometer with sufficient sensitivity and resolution to measure Doppler shifts at gamma-ray energies became available include the following:

SOLAR: <u>Sites of nuclear activity on the sun</u>: If the line-of-sight velocities of the excited nuclei emitting gamma-rays could be determined, then by correlating the velocities with those found at other wavelengths, the interaction regions could be uniquely identified. Different lines, such as those from positron annihilation and neutron capture, may well be produced at different places in the solar atmosphere.
<u>Electron temperature at the annihilation site</u>: If we could measure the width of the

positron annihilation line, we could determine the electron temperature, and thus open the door to gamma-ray and radio correlation studies

BURSTS: Neutron star mass: It would be possible to make precise measurements of gravitational red-shifts in gamma-ray burst spectra, if we had high enough resolution. These data would pin down the ratio of object mass to emission radius. Assuming that the emission comes from the surface, and assuming that we know the density from the equation of state, we could then determine the radius of the star.

PULSARS: With a very high resolution spectrometer, we might find Doppler shifts in pulsar emission which are a function of the phase. This would give a very strong clue as to the emission geometry.

PHYSICAL PRINCIPLES

In this section, we discuss the physical principles which govern bolometry at MeV energies. One of the key property of bolometers which makes them attractive for gamma-ray astronomy is the fact that they use the bulk properties of matter, rather than surface effects. This is important because the photon cannot be confined to a small volume. A second property is the very low specific heat which some substances show near 0K.

1. Specific Heats: The Debye-Somerfeld Equation shows that the specific heat of the lattice approaches zero as the third power of the temperature, while that due to electrons is linear in the temperature. For very pure Germanium and Silicon, the electronic specific heat is negligible. At .0001K (0.1 mK), the specific heats of these elements actually fall as low as 1 electron volt per degree kelvin per gram. The figure below shows the thermal capacity as a function of temperature in this range, for the Germanium detector described below.

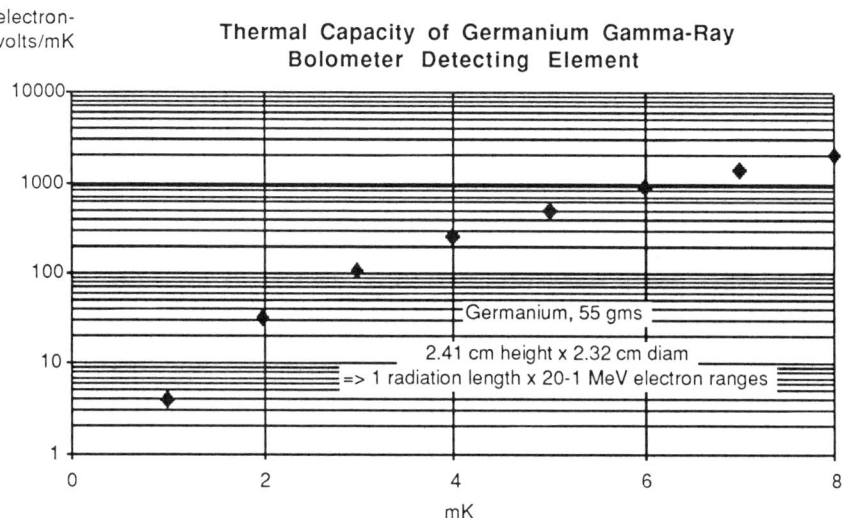

2. Gamma-Ray Absorption: We require that gamma-rays of the energies of interest have a probability (>0.5) of being totally absorbed in the detector. To achieve this, we

need a thickness of at least one interaction length, and we also need the linear dimensions to be large compared to the range of recoil electrons. A cylindrical element of Germanium, 2.3 cm in diameter and 2.4 cm in height, while only one interaction length long, meets these requirements.

3. Recovery: Because the thermal resistance from an object to a Helium bath is a very strong function of surface quality, one usually provides thermal links between the element and the cooling surface. This strategy not only promotes a uniform response over the volume of the bolometer element, but also allows the recovery rate to be controlled. The parameters of the thermal link must be chosen to provide heat flow adequate for the bolometer element to recover quickly. They should have the highest possible thermal conductivity, combined with the lowest possible specific heat. The next figure shows a configuration which could be used in a gamma-ray telescope.

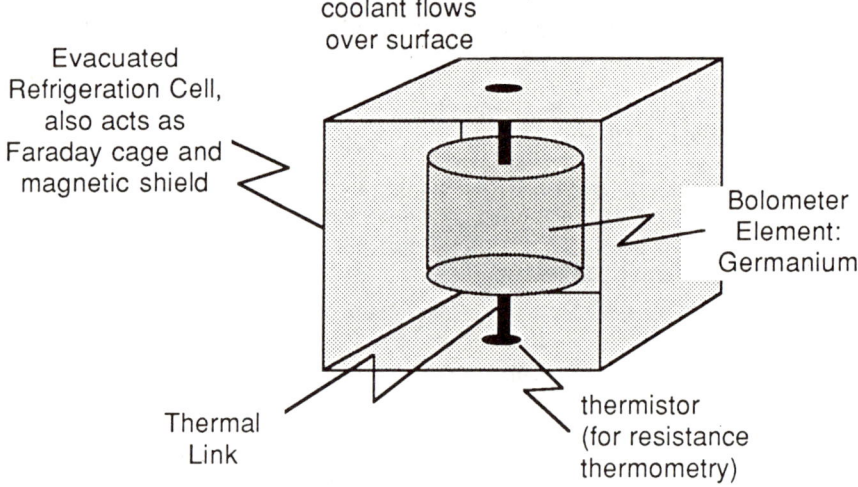

4. Thermometry: The thermometry requirements for this instrument pose a challenge. We must measure the total heat flow of a pulse which lasts only microseconds. We will assume, for the present discussion, that equipment is available which can measure the temperature of the bolometer element with an accuracy of 1mK across the range from 1 mK to 1K, within the instrument response time. One strategy to achieve this requirement is to use noise thermometry to establish the absolute temperature baseline, and resistance thermometry to follow the thermal pulse. Since it is common to use doped Germanium as the resistor element, the possibility exists to combine the resistor with the bolometer itself. This would resolve the often troublesome issue of thermal contact between the sample and the thermometer.

A PRELIMINARY DESIGN

The preliminary design given in this section shows that reasonable parameters can be found allowing such a device to function as a gamma-ray spectrometer. The next step, a detailed optimization making use of engineering data for existing devices, will be published in a follow-on paper. The parameters of this design are as follows:

1. Material: Germanium is chosen, because of its low specific heat in the millikelvin regime, and its high gamma-ray absorption coefficient. Silicon, for example, requires more than 5 times as much mass as Germanium, to achieve the same stopping power for gamma-rays.
2. Mass: 55 gms provides one radiation length.
3. Shape: A right cylinder, of diameter approximately equal to the height, is chosen for this design exercise.
4. Operating Temperature: 1-5 mK
5. Link Properties: Pure copper links are chosen to give the highest possible thermal conductivity combined with the lowest thermal capacity. These quantities are a function of temperature, so the net instrument response will be non-linear.

DESIGN MODELS

1. Response model: The response model describes what happens to the detector when it is hit by a gamma-ray or other particle. It models the transfer of energy from the infalling photon into heat, and it predicts the response of the instrument as a function of time after the pulse. We have used STELLA, an interactive graphical modelling tool, for this purpose. The STELLA model of our system is shown below. Each of the graphical elements contains a relation specifying its response to its inputs. The boxes integrate the flows into and out of them, which are shown as outlined arrows with valves. The circles contain relations governing the local properties, and the arrows show the functional dependencies. For example, heat enters the bolometer element (U_bolo_element) from gamma-ray energy deposits (power_input), and leaves it due to cooling via the thermal links (cooling_power). The specific heats and conductivities of the links and the bolometer elements are represented graphically. The most important of the equations are shown below the figure. We used this simple model to predict the response of our system to gamma-rays of given energies, adjusting parameters such as link area and bolometer mass until we achieved acceptable performance.
2. Gamma-Ray transport model: We have used a gamma-ray transport program to determine the photofraction and distribution of energy deposits in our bolometer element.
3. Resolution and Background Models: For purposes of the present discussion, we assume that the achievable resolution limit is given by the value already achieved with X-ray detectors. An investigation of the sources of the various background contributions, and what can be done to suppress each of them, will be completed in the near future.

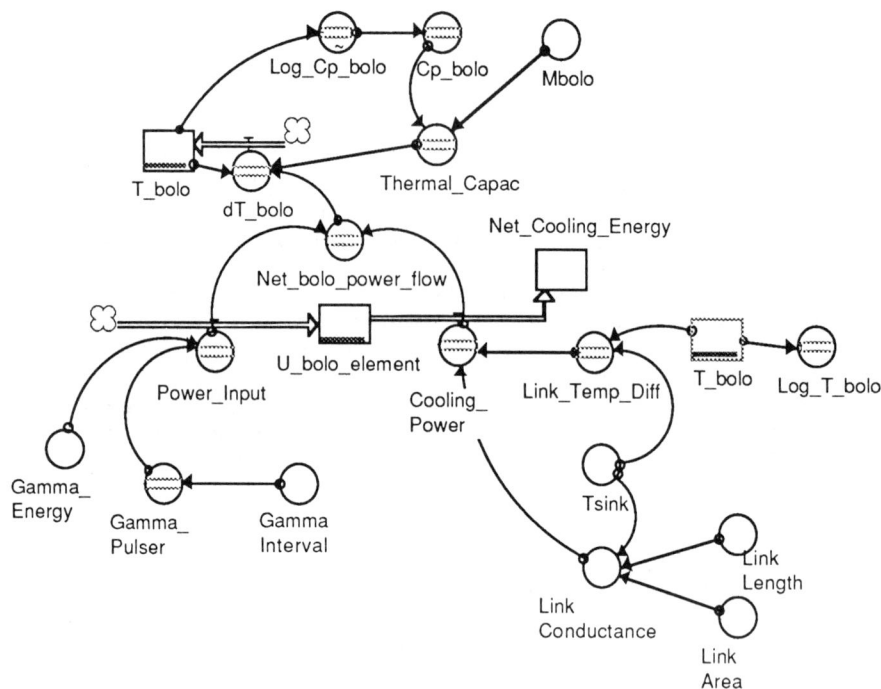

Key Model Equations
Net_Cooling_Energy = Net_Cooling_Energy + dt * (Cooling_Power)
    INIT(Net_Cooling_Energy) = 0
T_bolo = T_bolo + dt * (dT_bolo)
    INIT(T_bolo) = 25 {mK}
U_bolo_element = U_bolo_element + dt * (-Cooling_Power + Power_Input)
    INIT(U_bolo_element) = 0 {eV: arbitrary reference level}
Cooling_Power = IF(Link_Temp_Diff>0) THEN
    Link_Temp_Diff*Link_Conductance ELSE 0 {eV per microsecond}
dT_bolo = Net_bolo_power_flow/Thermal_Capac {mK}
Gamma_Pulser = PULSE(1,3,Gamma_Interval)
Link_Temp_Diff = T_bolo-Tsink {mK}
Net_bolo_power_flow = Power_Input-Cooling_Power {ev/microsecond}

MODEL PREDICTIONS

1. <u>Response model</u>: The thermal pulse time constants, as a function of time, were optimized by adjusting the properties of the thermal link in the model. Pulse profiles such as that seen below were typical. The important point is that reasonably fast pulses can be achieved, with the conductivities and specific heats which exist at these temperatures.

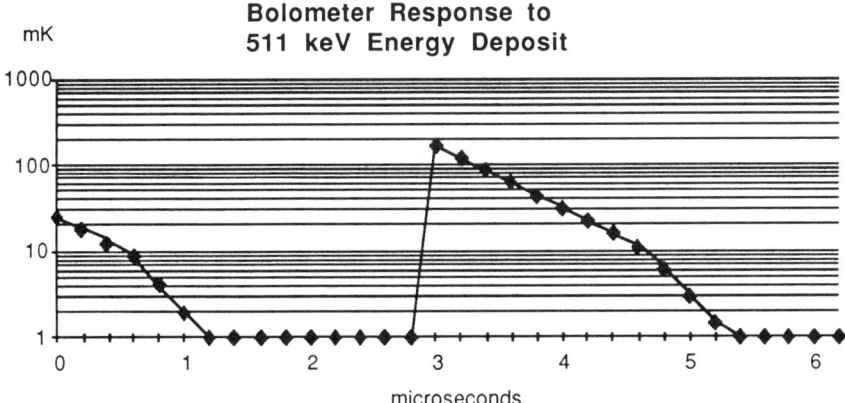

2. <u>Gamma-Ray Transport Model</u>: The figure below shows the efficiencies for the various interaction types, as a function of photon energy, for a single bolometer element. It is clear from the figure that we would like to have a thicker detector, or a material with a higher photopeak efficiency.

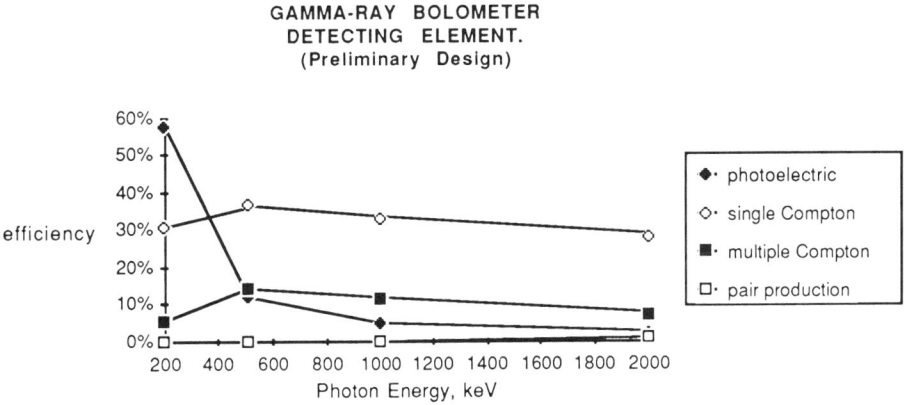

3. <u>Background</u>: The small bolometer element size defends it against excessive dead-time due to charged particle events. Our detector has a geometrical factor of ~82 cm^{-2} sr. The expected rate of charged particle events through the element is therefore ~10/second. Recovery times faster than 1 millisecond are required, to avoid having gamma-ray events ride on the tail of charged particle events. Activation background is the most serious problem for this system; our models do not yet address this issue.

4. Resolution Model: The resolution model is not yet complete. But considering that with higher temperatures, McCammon et al. achieved 17 eV at 6 keV, it should be possible to achieve similar values in the gamma ray regime.

SUMMARY

The advantages of the bolometer for gamma-ray spectroscopy are its very high ultimate resolution and its simplicity of concept. Some disadvantages are the relatively modest sensitivity which each individual element may have (too avoid pileup due to charged particle events), the necessity of cooling to millikelvin temperatures, and the need for extremely good electrical, acoustical, and magnetic isolation, to achieve the ultimate resolution.

REFERENCES

1. S.H. Moseley, J.C. McCammon, and D. McCammon, J.Appl.Phys. 56(5),1257 (1984)
2. D. McCammon, S.H. Moseley, J.C. Mather, and R.F. Mushotzky, J.Appl.Phys. 56(5), 1263 (1984)
3. B. Cabrera, L.M. Krauss, and F. Wilczek, Phys. Rev. Letters, 55 (1), 25 (1985)
4. S.H. Moseley, R.L. Kelley, R.J. Schoelkopf, A.E. Szymkowiak, D. McCammon, and J. Zhang (in press)
5. M. Leventhal, C.J. MacCallum, and P.D. Stang, Ap.J. 225, L11 (1978)
6. M. Leventhal, C.J. MacCallum, A.F. Huters, and P.D. Stang, Ap.J. 240, 338 (1980)
7. G.R. Riegler, J.C. Ling, W.A. Mahoney, W.A. Wheaton, J.B. Willett, A.S. Jacobson, and T.A. Prince, Ap.J. 248, L13 (1981)
8. W.A. Mahoney, J.C. Ling, W.A. Wheaton, and A.S. Jacobson, Ap.J. 286, 578 (1984)

MODEL INDEPENDENT SPECTRAL DECONVOLUTION

Bradley E. Schaefer
NASA, Goddard Space Flight Center, Code 661, Greenbelt, Md. 20771

ABSTRACT

This paper proposes a new method for deconvolving gamma-ray spectra in a model independent manner. The advantages of this algorithm are that the required computation is easy, quick, intuitive, and accurate. The disadvantage is that the spectral resolution is typically equal to the FWHM of the detector's photopeak. Essentially, the method sacrifices high frequency information about the source spectrum and gains certainty in the deconvolution. Such a method should not be used for line studies, but is ideal for continuum studies. In brief, the algorithm bins the observed count spectrum to some optimal resolution (whereupon the detector response matrix can be inverted with certainty) and the source spectrum is determined by a multiplication with the inverse of the response matrix.

INTRODUCTION

When faced with the matrix equation for gamma-ray detectors, the obvious first try to solve for the source spectrum is to multiply the count spectrum by the inverse of the detector response matrix. Unfortunately, as is common knowledge, this approach leads to poorly conditioned results. This spectral deconvolution problem is just one of a large class of such problems, called 'inverse problems', for which a solution can not be well determined. When expressed in matrix notation, these problems are characterized by significant off diagonal terms. Another way of looking at the problem is that the determinant of the instrumental matrix is not significantly different from zero so that a meaningful inversion is not possible. Fundamentally, the problem arises because only a finite set of observations is being tried to constrain an infinite class of possible solutions. In practical situations, the problem is exacerbated by the presence of errors. The final result is that a large number of possible solutions can result in observations essentially identical to those obtained.

Since the problem is so widespread throughout astronomy[1] and science, many possible methods have been developed to handle the inverse problem. The general experience has been that no one technique is best for all problems and that many techniques are best in various individual cases. All of the proposed techniques either in some way assume properties for possible solutions or loose information.

The traditional method of deconvolving gamma-ray spectra involves the assumption of a specific mathematical form for the acceptable solutions. The parameters of the chosen model are then optimized in the chi-square sense so that the count spectrum predicted for the model closely matches the observed count spectrum. This is a reasonable procedure if the true form of the spectrum is already known. However, if the real source spectral shape is not known, then the fit has only dubious merit. This is illustrated quite sharply in my own field of gamma-ray bursters, where many spectral shapes can be successfully fit to any one burst spectrum and where about ten different models have been successfully applied to at least some spectra. Fenimore, Klebesadel, and Laros[2] have pointed out how 'obliging' the resulting spectrum is for the traditional technique. That is, if a soft (or hard) spectral shape is assumed, then the deduced photon spectrum will also be soft (or hard). Another disadvantage of the traditional method is that the derived photon spectrum is occasionally used by outside investigators to fit physical models - an entirely incorrect procedure.

Recently, Loredo and Epstein[3] have developed the Backus-Gilbert algorithm with application to gamma-ray spectroscopy. They find that this is a fully reliable method for inverting spectra. One interesting fundamental result from the Backus-Gilbert formalism is that there is a trade off between spectral resolution and accuracy. The formalism developed by Laredo and Lamb allows the experimenter to choose the desired trade off. So, if high resolution is used, then the derived source spectrum will be very uncertain with large error bars. If the resolution is degraded, then the source spectrum can be well determined.

ALGORITHM

The idea behind my algorithm is that it may be possible to obtain a well determined source spectrum by some means other than the Backus-Gilbert formalism. The above mentioned trade off suggests that this can be accomplished by degrading the resolution (i.e., binning the data and the detector response matrix). When a real matrix has very fine bins, any two successive columns have no significant differences and hence the determinant must be effectively zero. However, when the real matrix is binned sufficiently, any two successive columns will be significantly different. Hence the determinant will be greatly different from zero and the matrix invertable. With an invertable matrix, the source spectrum can be found directly by multiplying the observed count spectrum by the inverse matrix.

The advantages and disadvantages of my method can be easily deduced from the above description. This method has the advantage that it is easy to understand and is intuitive as to why it works. The method is computationally fast since a relatively small matrix needs to be inverted only once per detector and only one matrix multiplication is needed per spectrum. The primary advantage is that the resulting spectrum should be well determined, that is, there are no oscillations or 'obliging' shifts as found when full resolution is retained. The primary disadvantage comes from the requirement of degrading the resolution. We can estimate the degree of degradation required by realizing that the binning must make adjacent columns of the response matrix different. Since most of the response for gamma-ray detectors is in the photopeak, a bin size equal to the FWHM of the photopeak should serve to distinguish each column. So my algorithm is not optimal for line studies, but works well continuum studies.

When line features are used in simulation tests, a narrow line has some of its flux redistributed into adjacent bins. This is an illustration of why the algorithm should not be used for line studies. But it also shows that line flux should not be included in a spectrum deconvolved for purposes of studying the continuum. A reasonable way to handle line flux is to subtract it out of the count spectrum before deconvolution. The line energy and flux should be determined by some other method (such as the traditional method) which can accurately determine the line parameters.

My algorithm (and every other algorithm) always runs into trouble for the lowest (first) energy bin. This is easy to see because photons in the 'zeroth' bin will contribute to the first bin since the edge of the photopeak still contains significant response. This trouble is worst for very soft spectra, as the relative contribution to the first bin from the zeroth bin is large. The result is that the flux in the first bin will be overestimated and the error will propagate to one or two more bins. One method to remove this source of error involves the inclusion of an estimate for the the number of counts that would have been observed in the zeroth bin had it been measured. Such an estimate can be made by extrapolating the lowest actually observed bins in count space. Also needed is a response matrix that includes the zeroth bin. Then when the source spectrum is calculated, the effect of the error in extrapolating the zeroth bin will be such that the error in the first bin will be a second order effect. Naturally, the computed source spectrum for the zeroth bin has little information content.

The error bars for the source spectrum can be easily calculated from simple propagation of errors. If $\underline{D}$ is the response matrix, and $\underline{S}$ is the source matrix, then the count matrix, $\underline{C}$, will be $\underline{D}x\underline{S}$. The deduced source spectrum for an observed count spectrum is

$$\underline{S}_i = \sum \underline{D}_{ij}^{-1} x \underline{C}_j \qquad (1)$$

with a variance of

$$\sigma_s^2 = \sum (\underline{D}_{ij}^{-1})^2 x (\underline{C}_j + \underline{B}_j) \qquad (2)$$

where $\underline{B}$ is the observed background count matrix.

TESTS

Since my primary interest is in gamma-ray spectroscopy with the Burst and Transient Source Experiment (BATSE) on Gamma Ray Observatory, I will address all of my algorithm tests to this experiment. Specifically, I will use the detector response matrix calculated by Geoff Hueter for the spectroscopy detector (3" thick and 5" in diameter NaI(Tl) crystals).

The test procedure consisted of detailed modelling of the various physical processes that go into the production of an observed count spectrum and the subsequent comparison of the deduced source spectrum

with the input source spectrum. These tests incorporate the effects
of photon noise in the count spectrum as well as the effects of errors
in the response matrix measurement by means of a Monte Carlo
technique. That is, a given test is performed a large number of times
(typically 100) where in each run a different set of random errors is
included. The actual values of the errors are randomly selected from
a distribution which is appropriate for the process being modelled.
The photon noise of the count spectrum is taken from a Poisson
distribution. The variance of the measurement of each element in the
detector responce matrix is taken to be the Poisson variance from the
number of 'photons' accumulated by Hueter's Monte Carlo program (that
calculated the matrix in the first place) divided by ten. The
division by ten is a crude attempt to account for the fact that the
final matrix will be a smoothed version, where perhaps ten bins will
contribute to the smoothing process for each element.

So the detailed testing procedure for the program consisted of
the following steps: (1) A source spectrum is assumed and the real $\underline{S}$
is calculated by appropriate integrals over the full resolution
bins. (2) The background flux $\underline{B}$ is similarily calculated for a power
law of index -2.5 with a normalization factor appropriate for the
BATSE spectroscopy detectors. (3) Read in the full resolution $\underline{D}$ as
calculated by Hueter. (4) Calculate a full resolution $\underline{C}$ by
multiplying $\underline{D}$ and $\underline{S}$. (5) Bin $\underline{S}$, $\underline{B}$, $\underline{D}$, and $\underline{C}$ to the optimal
resolution (see the next section). (6) Now start the Monte Carlo
runs over which the statistics of the errors will be determined.
(7) Add noise to the true $\underline{D}$ to get the matrix, $\underline{D}'$, as used by the
observer. (8) Invert $\underline{D}'$ to get $\underline{D}'^{-1}$. (9) Add noise to $\underline{C}$ to get the
actual observed count spectrum $\underline{C}'$. (10) Extrapolate the lowest
observed bins of $\underline{C}'$ in log space to deduce the zeroth bin. (11)
Calculate the deduced photon spectrum as $\underline{S}'=\underline{D}'^{-1}x\underline{C}'$. (12) Calculate
the error bars as given by equation 2. (13) Calculate the difference
between the deduced and real source photon spectrum, $\underline{S}'-\underline{S}$. (14)
Return to step 6. (15) Compile the statistics on $\underline{S}'-\underline{S}$.

This test procedure has been applied to a large number of
spectral shapes with a wide range of fluences. The results from these
tests show the algorithm to be reliable. The distibution of $\underline{S}'-\underline{S}$ has
a variance closely equal to the variance from equation 2 in all
cases. The mean value of $\underline{S}'-\underline{S}$ is in general small compared to the
calculated scatter.

The algorithm will sucessfully reproduce line features.
However, the flux in the line cannot be reliably determined since some
of the flux is redistributed in neighboring bins. Typically, this
redistribution will occur at the level of one calculated error bar.

The trouble with the lowest energy bin is found to be greatest
for the steepest spectrum, say those with power law indices of -3.
The steepest spectra may have an error of up to a factor of two in the
first bin, and noticeable systematic deviations for up to four more
bins. Fortunately, many spectral shapes soften at lower energies as
is observed for gamma-ray bursters. The typical error for the first
bin for an average burst spectrum may be 10%, while the second and
third bin will have a smaller error. Hard spectra have no noticeable

error. In summary, the algorithm has difficulty in the lowest two or so bins for very soft spectra.

If the resolution has not been degraded sufficiently, then this algorithm will not work. So for example, when the resolution is set to half the FWHM of the photopeak, large oscillations and ridiculous solutions will be found. One symptom of this is that the variation of the determinant of D' is roughly equal to the determinant itself. (In other words, the determinant is not significantly different from zero and the matrix cannot be usefully inverted.) When the binned resolution is degraded to where the determinant is much greater than its uncertainty, the process becomes stable and accurate. This happens roughly when the resolution equals the FWHM of the photopeak. This resolution is sufficient for the algorithm to work, but it may not be optimal for defining the spectrum.

OPTIMAL RESOLUTION

Two effects can affect the optimal resolution: First, the slope of the spectrum will affect the optimal resolution. Suppose a spectrum is quite flat, then clearly it is better to trade off energy resolution for small vertical error bars. Similarily, for steep spectra, it is best to use a high resolution, despite the large vertical error bars. Second, the flux in a spectrum must affect the optimal resolution. For example, an intense spectrum can suffer a resolution somewhat better than the photopeak FWHM because the trend to increase the vertical error bar is offset by the large number of photons and by the greater number of spectral determinations. Similarily, it is better to have larger bins if the counts are few. I will now derive the optimal resolution.

For model fitting where measured values have errors in only one dimension (the vertical direction), the standard least squares fitting technique is best. I have not been able to find in the literature a procedure which is applicable to cases where errors occur in both dimensions. However, the following is a reasonable procedure which generalizes the least squares method.

In the normal mode, χ^2 is defined as the sum of the square of the vertical distances, d, in units of the uncertainty, σ_d. My generalization to two dimensions replaces the vertical distance in this definition with the perpendicular distance. Let me be more specific. Suppose that we are trying to fit a model to a set of data points over some spectral range W. The resolution across this range for all bins will be R. Let us work in log-log space, since realistic spectra will be more nearly linear. Let W be sufficiently small that locally the model will have a uniform slope. The number of counts in each bin, N, will be ARE^{α}, where E is the photon energy in 1keV and A is a normalization constant. We can convert to log-log space with the definitions $Y=\log(N)$, $X=\log(E)$, and $\beta=\log(AR)$, so that $Y=\beta+\alpha X$. Let a specific observed point be (X_0, Y_0). The distance and uncertainty between this point and the model line is

$$d = |(Y_0-\beta-\alpha X_0)/(1+\alpha^2)^{0.5}|, \qquad (3)$$

$$\sigma_d^2 = (\alpha^2\sigma_X^2 + \sigma_Y^2)/(1+\alpha^2). \tag{4}$$

So I can define a generalized chi-square as

$$\chi^2 = \sum (d/\sigma_d)^2 = W<(Y_0 - \beta - \alpha X_0)^2>/R(\alpha^2\sigma_X^2 + \sigma_Y^2). \tag{5}$$

The local best fit line can be evaluated by minimizing chi-square with respect to α and β. Equation 5 has an obvious generalization for any fitting function which may be best for future gamma ray spectral fits to use. The optimal binning can also be established by setting the partial differential of χ^2 with respect to R equal to zero. The term in the numerator is not a function of R, so we have

$$0 = \partial [R(\sigma_Y^2 + \alpha^2\sigma_X^2)]/\partial R. \tag{6}$$

Now the variances in this last equation must be expressed as a function of R,

$$\sigma_Y = 0.434\sigma_N/N, \quad \sigma_X = 0.434R/2E. \tag{7}$$

From my Monte Carlo simulations for the BATSE spectroscopy detectors,

$$\sigma_N^2/N = 4\Delta/R \tag{8}$$

where Δ is the FWHM of the photopeak. Equation 8 is roughly accurate for all energies and fluences for Δ/R greater than a half. For the BATSE crystals, the value of Δ rises with energy roughly as $E^{0.66}$ 1.33keV. So the optimal value of R is

$$R/\Delta = 1.2A^{-0.25}E^{-\alpha/4}|\alpha|^{-0.5}(1\text{keV})^{-0.25}. \tag{9}$$

This relation quantifies the relationships with fluence and spectral slope as discussed at the start of this section.

CONCLUSIONS

I have developed a new spectral deconvolution algorithm for use with gamma-ray spectroscopy. The advantages of the method are that it is easy to understand, easy to implement, computationally quick, accurate, and model independent. The cost to obtain these advantages is that the detector resolution must be degraded to roughly that of the FWHM of the photopeak.

I hope that in the future, investigators will publish photon spectra produced with this algorithm. This will allow outside researchers (with no knowledge of the response matrix) to meaningfully fit physical models to the continuum.

1. I. Craig and J. Brown, Inverse Problems in Astrophysics (Hilger, Bristol, 1986).
2. E. Fenimore, R. Klebesadel, and J. Laros, proceedings of the XXIV COSPAR Meeting, Ottawa Canada, May 1982.
3. T. J. Loredo and R. I. Epstein, Astrophys. J., in press (1988).

WHEN AND WHY BACKGROUND SUBTRACTION MUST BE DONE BEFORE DATA ACCUMULATION IN GAMMA-RAY SPECTROSCOPY

Wm. A. Wheaton, A. S. Jacobson, J. C. Ling, and W. A. Mahoney
Jet Propulsion Laboratory, California Institute of Technology
4800 Oak Grove Drive, Pasadena, CA 91109

ABSTRACT

Two operations common to spectroscopy data analysis are the accumulation of data and subtraction of background. These operations do not necessarily commute, and in some important situations accumulation followed by subtraction introduces serious systematic errors. This occurs when the background is strongly variable but must be estimated from data taken before or after the source observation. Then long data accumulations may cease to converge to a unique count rate, since none exists due to the variable background. Instead, the value obtained reflects the particular way the background was sampled. Because of dead time variations and data selection cuts, the source and background sampling may differ, so that the expectation value of the difference of the source and background accumulations ceases to equal the true net source count rate. However, if the background is subtracted for each source scan and only the net spectra accumulated, the effect is suppressed. We present simulations comparing the two methods, and briefly discuss practical means of subtracting background before the main accumulation is performed.

INTRODUCTION

At energies above about 10 keV in high-energy astronomy the background becomes highly variable, so that, with the decreasing source to background ratios often encountered, the danger of systematic error of background subtraction becomes acute. To overcome this problem in data from the Jet Propulsion Laboratory High Resolution Gamma-Ray Spectrometer[1] on the Third High Energy Astronomy Observatory (HEAO 3), we have developed an analysis method which reverses the familiar "accumulate-then-subtract" order of operations. We call this technique the "scan-by-scan" method[2-5]. In this paper we illustrate the nature of the problem, and the power and limitations of our approach, with simulations of a simple counting experiment in the presence of a variable background, analyzed by both methods.

SYSTEMATIC ERROR OF BACKGROUND SUBTRACTION

Figure 1 shows the HEAO 3 background spectrum. Also shown is a spectrum of the Crab, one of the strongest hard x-ray sources in the sky. The source rate is nowhere more than 35% of the background, dropping to less than a percent over 1 MeV. For any systematic error introduced by background subtraction to be well below the 1 σ Poisson noise, also shown for a typical observation[6], the background must be removed to less than one part in 4000 at 1 MeV. Therefore the magnitude and character of its variability is critical. In general the background is a strong function of the geomagnetic coordinates, orientation, and activation history of the spacecraft. The orbital variables cause it to change on a characteristic time scale of about 10 min for low earth orbits such as that of HEAO 3. The orientation dependence introduces frequencies related to the spacecraft spin period. As a result, the behavior

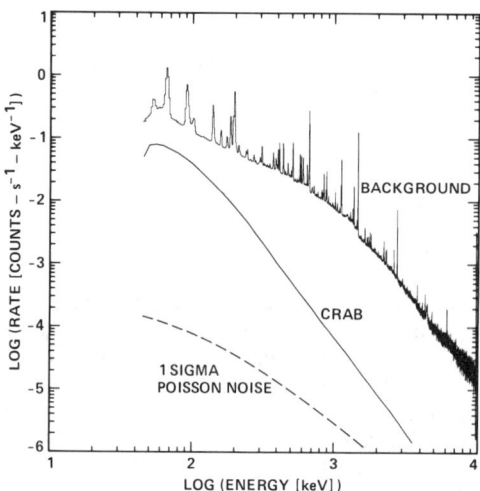

Figure 1. Strong source (Crab) and background spectra for HEAO 3. The dashed curve indicates 1 σ Poisson errors obtained for a representative observation of a continuum source.

of the background is sufficiently complex that it is typically impractical to model it reliably to better than a few per cent accuracy.

Therefore, rather than attempt to construct very accurate global models of the background, experimenters usually simply measure it. If the exposure to the background is much larger than to the source, as it often is, the statistical inaccuracy of the background estimate is relatively small, and the danger of systematic error due to a bad global model is overcome.

However, because of the background's large and complex variability, there still remains the danger that the data which are collected to estimate it will not in fact be representative of conditions of the source observation, to the high precision required to achieve results limited by Poisson statistics alone. Furthermore, unlike Poisson errors, the distribution of any resulting systematic errors of background subtraction is unknown, so that quantitative confidence bounds cannot easily be placed on their magnitude.

SUBTRACT-ACCUMULATE APPROACH

Data from scanning x-ray experiments are frequently analyzed by accumulating counts in azimuthal bins until sufficient data have been collected to give good statistics. In the simplest method, the flux from a source at some given position is derived by designating an azimuthal region around the source position as "source", and adjacent regions as "background". The count rates in the source and background regions are then computed from the counts and exposure times in each, and subtracted to estimate the net source rate over background. A slightly more sophisticated technique, introducing no essential change, is to fit the data, as by a least-squares algorithm, to the response expected from a point source at the given position, taking the instrument aperture function into account. Either method may be summarized by saying that the procedure is first to accumulate, then subtract background; we will sometimes refer to them collectively as the "superposition" method.

While this method has worked well at x-ray energies where the orbital variation in background is not too extreme, it may encounter difficulties above 10 keV. If the source and background accumulations include data in which the detector count rate varied, the "average" count rates obtained in the two regions are not physically well-defined, but may be dominated by the particular mix of background conditions which happened to be included.

The first line of defense of the superposition analysis method against systematic background subtraction error has been to carefully select the data so as to reject intervals in which the background variation is a problem. Unfortunately at high energy this variation is pervasive, so that when one attempts to formulate selection criteria only a small fraction of the data survives. Thus there is a trade-off between systematic errors and counting statistics errors. Eventually the data set can be so severely restricted that the Poisson error grows larger than the systematic error. This approach is clearly not very satisfactory because the information in much of the data is discarded. Also, the problem can actually be aggravated by the selection, because it may contribute to non-uniformity in sampling the source and background data.

This is because, as shown below, an essential element in the effect is the fact that scans across the source are generally not complete, but are affected by variable dead time and by the operation of the many data selection and rejection tests, such as on earth blockage, magnetic latitude, South Atlantic Anomaly, or data transmission errors, which are always required to analyze a real experiment. But when the estimated rates obtained for the source and background regions are affected by the particular mix of data included, then the error in their difference, the source count rate estimate, may not be derived from the square root of the numbers of counts, but rather from the haphazard statistics of the selection process.

The alternative we have developed for HEAO 3 reverses this usual accumulate-subtract sequence of analysis. By this method, the source flux and its error are estimated in a separate fit for each scan. Thus we subtract background for each scan individually, before accumulating scans to obtain significant answers. Although the statistical significance of the data from each scan is negligible, good statistics are recovered by the accumulation of many scans. By this means the association of the source and background data for each scan is preserved until the background has been subtracted, and its variability removed. Therefore the wide variation of the background from one scan to another does no harm. Furthermore, residual systematic error can be directly checked by histogramming the flux estimates obtained for single scans and comparing the width of the distribution with that expected due to counting statistics. One may view this approach as simply an example of the statistical method, standard in research outside the physical sciences, of pairing data samples (here read source data for each scan) with controls (here background data for the same scan) carefully matched to be similar in parameters not of primary interest which might somehow affect the results.

SIMULATIONS

We can illustrate and compare the two alternative methods by means of simple simulated experiments. We model an experiment consisting of 1000 scans across a source, generating Monte Carlo count data and exposure times for each. The simulations have been idealized for clarity, with only the minimum complexity needed to capture the effects under consideration. The idealized model is just the classic counter experiment, in which a source rate is measured by accumulating background ("B") and source plus background ("S+B") runs. There are therefore only two bins, B and S+B, in each scan. No fitting is involved, only subtraction. The essential modifications to an ideal experiment are that, first, the background is here taken as randomly variable so that it is not the same from one scan to another but is constant within any single scan, and second the effective exposure time for both the B and S+B data also varies randomly.

A single simulated experiment of 1000 scans constitutes a "trial". We analyze the same count data for each trial in the two different ways, generating an estimate of the source count rate and its error for each. We repeat this for 100 trials, and study the distribution of the estimates so obtained. If the analysis method were biased we would find that the histogram of estimated rates is significantly shifted from the true value input to the simulation. While bias is a danger by either method if certain analysis errors are committed, we shall not consider these further here, and neither of the methods studied gives biased results. However, if the histogram of estimates is significantly wider than theoretically expected from Poisson statistics, the experimenter is likely to be misled, because in real life, she has access to just the single result of her one experiment, corresponding to one trial in our simulation.

Figure 2 shows histograms of 100 such experiments, each consisting of 1000 scans across a constant 0.02 c s^{-1} source, plus a variable (uniformly distributed, $\delta b/\bar{b}$ = 30% RMS) background count rate with mean $\bar{b}$ = 1.0 c s^{-1}. The top panel shows the result when the data were analyzed by accumulating counts and live time for each trial, then subtracting B from S+B to estimate the source rate. The bottom panel

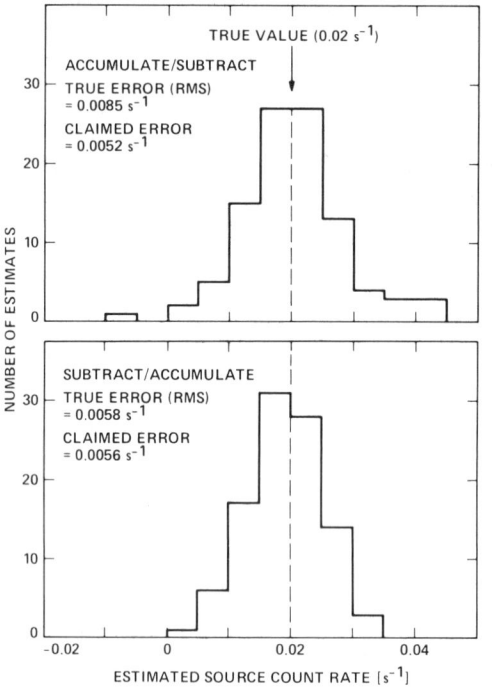

Figure 2. Histogram of the count rate estimates obtained for 100 Monte Carlo trials of a simulated observation consisting of 1000 scans past a constant source, in the presence of a strong background with 30% RMS variability, when analyzed by accumulating before subtraction (top) and by subtraction before accumulation (bottom). Systematic error broadens the former to 1.4 times the Poisson noise, whereas the latter is unaffected.

shows the result analyzed by subtracting B for each scan, then accumulating weighted averages of these estimates for 1000 scans. The 40% excess error in the superposition analysis in this case means, of course, that about half the data have been in effect thrown away. Much more serious however is the fact that the superpositon method claims to have a smaller error (essentially just the Poisson error) than it really does. Thus, (for this particular run) the superposition histogram has three estimates apparently more than 3 σ above the true mean while the subtract - accumulate order gives none. (On Gaussian statistics, a good approximation here, 0.14 would be expected.) Similarly, an apparent 4 σ effect observed via superposition would most likely be reduced to less than 3 σ by subtracting before accumulating.

Figure 3 summarizes the dependence of the systematic error on background and exposure variability. For clarity we have simply plotted the ratio of the RMS error of the estimates obtained via the superposition and subtract-accumulate approach. However, the absolute error was in fact always consistent with the claimed Poisson error for the subtract, then accumulate, trials. The scatter of the plotted points gives an indication of the sampling error in the simulation, each point representing 1000 scans of 100 trials each. In Figure 3a, the RMS background variability, $\delta b/\overline{b}$, ranged from 0 to 55%. The effect is eliminated if the source and background exposure are always the same for each scan. Figure 3b shows how the effect depends on exposure variability. The latter is expressed as a percent of the nominal maximum accumulation times for each scan. (These were 300 sec for B and 100 sec for S+B. Then, for example, 50% variability in Figure 3b means that the live time was chosen randomly between 150 and 300 sec for B, and independently between 50 and 100 sec for S+B, as if dead time variations or data selection conditions were operating.)

Of course subtraction before accumulation cannot guarantee against systematic error due to background variation which occurs within a scan, although histogramming the flux estimates at least allows one to detect it. We have also carried out a simulation series similar to those of Figure 3, but allowing variable background

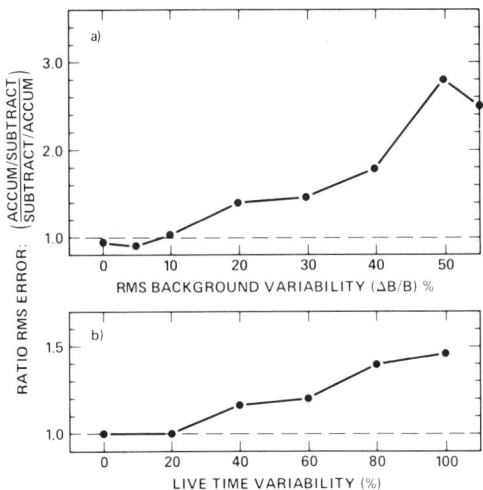

Figure 3. Comparison of the subtract-accumulate and accumulate-subtract analysis methods as a function of a) RMS background variability; b) exposure variability.

within each scan. Although it was somewhat lengthy to describe in detail, the results were in complete accord with expectations.

CONCLUSION

Subtraction of background prior to data accumulation offers one useful way to combat systematic error of background subtraction. It is applicable in situations where the background is strongly variable, is not sampled in the same way as the source data, and the background variation is not dominated by time scales short compared to the sampling intervals. Depending on the details of experiment design, some instruments are strongly affected by these considerations, and some are largely immune. Some of these details are subject to the control of the designer, and some are not; thus one may expect that as the problem of systematic error in background subtraction is more widely appreciated, future experiments will be less susceptible. However instruments which achieve high sensitivity by the use of large exposure -- notably space experiments -- are ikely to be troubled by poor S/B ratio, and it seems very doubtful that large exposure will ever be given up in the effort to observe weaker sources, no matter how successful attempts to reduce the background may be. Thus, the idea of subtraction before accumulation should continue to be of value in the future.

The wide range of susceptibility to systematic error of background subtraction among the various types of experiments in high-energy astronomy implies a corresponding range of suitability to our approach. Focusing instruments, such as Einstein, EXOSAT, and AXAF, are likely to be almost immune, and of course low-energy instruments in general seem to be less affected. Modulation collimator and coded aperture experiments are strongly protected by their rapid chopping of the source flux. Similarly, fast-spinning spacecraft (eg, OSO) may be affected less than slow spinners (eg, HEAO 1 and HEAO 3). Slower scanning and chopped observations, experiments using earth or lunar occultation, and sky-map analyses, are all especially liable to encounter problems with background variability, and for these especially subtraction of the background before accumulation should give improved results.

The principal technical obstacles in implementing this method are, first, the requirement to fit to Poisson data very efficiently and with very small numbers of counts, both of which are essential for high-resolution spectroscopy, and second, the determination of the weights for averaging over scans. The former has been satisfactorily solved for HEAO 3 by recasting the standard least-square fit formalism into one which is linear in the observed counts, and the second, by taking care to use weights which are not correlated with the data in the averages. A description of the essential core of the fitting method we use has been given by Wheaton et al.[7]; a fuller treatment of this and the weighting is in preparation[8].

ACKNOWLEDGMENTS

W.A.W. thanks Frank Primini and Duane Gruber for many enlightening discussions. We also wish to thank Alfred Dunklee and Robert Radocinski for their essential contributions to solving the many practical problems encountered in implementing the scan-by-scan analysis system for HEAO 3. This work was performed by the Jet Propulsion Laboratory, California Institute of Technology, under contract with the National Aeronautics and Space Administration.

REFERENCES

1. W. A. Mahoney, J. C. Ling, A. S. Jacobson, and R. M. Tapphorn, Nuc. Instr. and Meth. 178, 363, (1980).

2. G. R. Riegler, J. C. Ling, W. A. Mahoney, W. A. Wheaton, J. B. Willett, A.S. Jacobson, and T. A. Prince, Ap. J. (Letters) 248, L13, (1981).

3. J. C. Ling, W. A. Mahoney, W. A. Wheaton, A. S. Jacobson, and L. Kaluzienski, Ap. J. 275, 307, (1983).

4. A. P. Marscher, K. Brecher, W. A. Wheaton, J. C. Ling, W. A. Mahoney, and A. S. Jacobson, Ap. J. 281, 566, (1984).

5. W. A. Mahoney, J. C. Ling, W. A. Wheaton, and A. S. Jacobson, Ap. J. 286, 578, (1984).

6. J. C. Ling, W. A. Mahoney, W. A. Wheaton, and A. S. Jacobson, Ap. J. (Letters) 321, L117, (1987).

7. W. A. Wheaton, J. C. Ling, W. A. Mahoney, and A. S. Jacobson, BAAS 15, 940, (1983).

8. W. A. Wheaton, A. L. Dunklee, J. C. Ling, W. A. Mahoney, and A. S. Jacobson, in preparation (1988).

AIP Conference Proceedings

		L.C. Number	ISBN
No. 1	Feedback and Dynamic Control of Plasmas – 1970	70-141596	0-88318-100-2
No. 2	Particles and Fields – 1971 (Rochester)	71-184662	0-88318-101-0
No. 3	Thermal Expansion – 1971 (Corning)	72-76970	0-88318-102-9
No. 4	Superconductivity in d- and f-Band Metals (Rochester, 1971)	74-18879	0-88318-103-7
No. 5	Magnetism and Magnetic Materials – 1971 (2 parts) (Chicago)	59-2468	0-88318-104-5
No. 6	Particle Physics (Irvine, 1971)	72-81239	0-88318-105-3
No. 7	Exploring the History of Nuclear Physics – 1972	72-81883	0-88318-106-1
No. 8	Experimental Meson Spectroscopy –1972	72-88226	0-88318-107-X
No. 9	Cyclotrons – 1972 (Vancouver)	72-92798	0-88318-108-8
No. 10	Magnetism and Magnetic Materials – 1972	72-623469	0-88318-109-6
No. 11	Transport Phenomena – 1973 (Brown University Conference)	73-80682	0-88318-110-X
No. 12	Experiments on High Energy Particle Collisions – 1973 (Vanderbilt Conference)	73-81705	0-88318-111–8
No. 13	$\pi\text{-}\pi$ Scattering – 1973 (Tallahassee Conference)	73-81704	0-88318-112-6
No. 14	Particles and Fields – 1973 (APS/DPF Berkeley)	73-91923	0-88318-113-4
No. 15	High Energy Collisions – 1973 (Stony Brook)	73-92324	0-88318-114-2
No. 16	Causality and Physical Theories (Wayne State University, 1973)	73-93420	0-88318-115-0
No. 17	Thermal Expansion – 1973 (Lake of the Ozarks)	73-94415	0-88318-116-9
No. 18	Magnetism and Magnetic Materials – 1973 (2 parts) (Boston)	59-2468	0-88318-117-7
No. 19	Physics and the Energy Problem – 1974 (APS Chicago)	73-94416	0-88318-118-5
No. 20	Tetrahedrally Bonded Amorphous Semiconductors (Yorktown Heights, 1974)	74-80145	0-88318-119-3
No. 21	Experimental Meson Spectroscopy – 1974 (Boston)	74-82628	0-88318-120-7
No. 22	Neutrinos – 1974 (Philadelphia)	74-82413	0-88318-121-5
No. 23	Particles and Fields – 1974 (APS/DPF Williamsburg)	74-27575	0-88318-122-3
No. 24	Magnetism and Magnetic Materials – 1974 (20th Annual Conference, San Francisco)	75-2647	0-88318-123-1
No. 25	Efficient Use of Energy (The APS Studies on the Technical Aspects of the More Efficient Use of Energy)	75-18227	0-88318-124-X

No. 26	High-Energy Physics and Nuclear Structure – 1975 (Santa Fe and Los Alamos)	75-26411	0-88318-125-8
No. 27	Topics in Statistical Mechanics and Biophysics: A Memorial to Julius L. Jackson (Wayne State University, 1975)	75-36309	0-88318-126-6
No. 28	Physics and Our World: A Symposium in Honor of Victor F. Weisskopf (M.I.T., 1974)	76-7207	0-88318-127-4
No. 29	Magnetism and Magnetic Materials – 1975 (21st Annual Conference, Philadelphia)	76-10931	0-88318-128-2
No. 30	Particle Searches and Discoveries – 1976 (Vanderbilt Conference)	76-19949	0-88318-129-0
No. 31	Structure and Excitations of Amorphous Solids (Williamsburg, VA, 1976)	76-22279	0-88318-130-4
No. 32	Materials Technology – 1976 (APS New York Meeting)	76-27967	0-88318-131-2
No. 33	Meson-Nuclear Physics – 1976 (Carnegie-Mellon Conference)	76-26811	0-88318-132-0
No. 34	Magnetism and Magnetic Materials – 1976 (Joint MMM-Intermag Conference, Pittsburgh)	76-47106	0-88318-133-9
No. 35	High Energy Physics with Polarized Beams and Targets (Argonne, 1976)	76-50181	0-88318-134-7
No. 36	Momentum Wave Functions – 1976 (Indiana University)	77-82145	0-88318-135-5
No. 37	Weak Interaction Physics – 1977 (Indiana University)	77-83344	0-88318-136-3
No. 38	Workshop on New Directions in Mossbauer Spectroscopy (Argonne, 1977)	77-90635	0-88318-137-1
No. 39	Physics Careers, Employment and Education (Penn State, 1977)	77-94053	0-88318-138-X
No. 40	Electrical Transport and Optical Properties of Inhomogeneous Media (Ohio State University, 1977)	78-54319	0-88318-139-8
No. 41	Nucleon-Nucleon Interactions – 1977 (Vancouver)	78-54249	0-88318-140-1
No. 42	Higher Energy Polarized Proton Beams (Ann Arbor, 1977)	78-55682	0-88318-141-X
No. 43	Particles and Fields – 1977 (APS/DPF, Argonne)	78-55683	0-88318-142-8
No. 44	Future Trends in Superconductive Electronics (Charlottesville, 1978)	77-9240	0-88318-143-6
No. 45	New Results in High Energy Physics – 1978 (Vanderbilt Conference)	78-67196	0-88318-144-4
No. 46	Topics in Nonlinear Dynamics (La Jolla Institute)	78-57870	0-88318-145-2
No. 47	Clustering Aspects of Nuclear Structure and Nuclear Reactions (Winnepeg, 1978)	78-64942	0-88318-146-0
No. 48	Current Trends in the Theory of Fields (Tallahassee, 1978)	78-72948	0-88318-147-9

No. 49	Cosmic Rays and Particle Physics – 1978 (Bartol Conference)	79-50489	0-88318-148-7
No. 50	Laser-Solid Interactions and Laser Processing – 1978 (Boston)	79-51564	0-88318-149-5
No. 51	High Energy Physics with Polarized Beams and Polarized Targets (Argonne, 1978)	79-64565	0-88318-150-9
No. 52	Long-Distance Neutrino Detection – 1978 (C.L. Cowan Memorial Symposium)	79-52078	0-88318-151-7
No. 53	Modulated Structures – 1979 (Kailua Kona, Hawaii)	79-53846	0-88318-152-5
No. 54	Meson-Nuclear Physics – 1979 (Houston)	79-53978	0-88318-153-3
No. 55	Quantum Chromodynamics (La Jolla, 1978)	79-54969	0-88318-154-1
No. 56	Particle Acceleration Mechanisms in Astrophysics (La Jolla, 1979)	79-55844	0-88318-155-X
No. 57	Nonlinear Dynamics and the Beam-Beam Interaction (Brookhaven, 1979)	79-57341	0-88318-156-8
No. 58	Inhomogeneous Superconductors – 1979 (Berkeley Springs, W.V.)	79-57620	0-88318-157-6
No. 59	Particles and Fields – 1979 (APS/DPF Montreal)	80-66631	0-88318-158-4
No. 60	History of the ZGS (Argonne, 1979)	80-67694	0-88318-159-2
No. 61	Aspects of the Kinetics and Dynamics of Surface Reactions (La Jolla Institute, 1979)	80-68004	0-88318-160-6
No. 62	High Energy e^+e^- Interactions (Vanderbilt, 1980)	80-53377	0-88318-161-4
No. 63	Supernovae Spectra (La Jolla, 1980)	80-70019	0-88318-162-2
No. 64	Laboratory EXAFS Facilities – 1980 (Univ. of Washington)	80-70579	0-88318-163-0
No. 65	Optics in Four Dimensions – 1980 (ICO, Ensenada)	80-70771	0-88318-164-9
No. 66	Physics in the Automotive Industry – 1980 (APS/AAPT Topical Conference)	80-70987	0-88318-165-7
No. 67	Experimental Meson Spectroscopy – 1980 (Sixth International Conference, Brookhaven)	80-71123	0-88318-166-5
No. 68	High Energy Physics – 1980 (XX International Conference, Madison)	81-65032	0-88318-167-3
No. 69	Polarization Phenomena in Nuclear Physics – 1980 (Fifth International Symposium, Santa Fe)	81-65107	0-88318-168-1
No. 70	Chemistry and Physics of Coal Utilization – 1980 (APS, Morgantown)	81-65106	0-88318-169-X
No. 71	Group Theory and its Applications in Physics – 1980 (Latin American School of Physics, Mexico City)	81-66132	0-88318-170-3
No. 72	Weak Interactions as a Probe of Unification (Virginia Polytechnic Institute – 1980)	81-67184	0-88318-171-1
No. 73	Tetrahedrally Bonded Amorphous Semiconductors (Carefree, Arizona, 1981)	81-67419	0-88318-172-X

No. 74	Perturbative Quantum Chromodynamics (Tallahassee, 1981)	81-70372	0-88318-173-8
No. 75	Low Energy X-Ray Diagnostics – 1981 (Monterey)	81-69841	0-88318-174-6
No. 76	Nonlinear Properties of Internal Waves (La Jolla Institute, 1981)	81-71062	0-88318-175-4
No. 77	Gamma Ray Transients and Related Astrophysical Phenomena (La Jolla Institute, 1981)	81-71543	0-88318-176-2
No. 78	Shock Waves in Condensed Matter – 1981 (Menlo Park)	82-70014	0-88318-177-0
No. 79	Pion Production and Absorption in Nuclei – 1981 (Indiana University Cyclotron Facility)	82-70678	0-88318-178-9
No. 80	Polarized Proton Ion Sources (Ann Arbor, 1981)	82-71025	0-88318-179-7
No. 81	Particles and Fields –1981: Testing the Standard Model (APS/DPF, Santa Cruz)	82-71156	0-88318-180-0
No. 82	Interpretation of Climate and Photochemical Models, Ozone and Temperature Measurements (La Jolla Institute, 1981)	82-71345	0-88318-181-9
No. 83	The Galactic Center (Cal. Inst. of Tech., 1982)	82-71635	0-88318-182-7
No. 84	Physics in the Steel Industry (APS/AISI, Lehigh University, 1981)	82-72033	0-88318-183-5
No. 85	Proton-Antiproton Collider Physics –1981 (Madison, Wisconsin)	82-72141	0-88318-184-3
No. 86	Momentum Wave Functions – 1982 (Adelaide, Australia)	82-72375	0-88318-185-1
No. 87	Physics of High Energy Particle Accelerators (Fermilab Summer School, 1981)	82-72421	0-88318-186-X
No. 88	Mathematical Methods in Hydrodynamics and Integrability in Dynamical Systems (La Jolla Institute, 1981)	82-72462	0-88318-187-8
No. 89	Neutron Scattering – 1981 (Argonne National Laboratory)	82-73094	0-88318-188-6
No. 90	Laser Techniques for Extreme Ultraviolt Spectroscopy (Boulder, 1982)	82-73205	0-88318-189-4
No. 91	Laser Acceleration of Particles (Los Alamos, 1982)	82-73361	0-88318-190-8
No. 92	The State of Particle Accelerators and High Energy Physics (Fermilab, 1981)	82-73861	0-88318-191-6
No. 93	Novel Results in Particle Physics (Vanderbilt, 1982)	82-73954	0-88318-192-4
No. 94	X-Ray and Atomic Inner-Shell Physics – 1982 (International Conference, U. of Oregon)	82-74075	0-88318-193-2
No. 95	High Energy Spin Physics – 1982 (Brookhaven National Laboratory)	83-70154	0-88318-194-0
No. 96	Science Underground (Los Alamos, 1982)	83-70377	0-88318-195-9

No.	Title		
No. 97	The Interaction Between Medium Energy Nucleons in Nuclei – 1982 (Indiana University)	83-70649	0-88318-196-7
No. 98	Particles and Fields – 1982 (APS/DPF University of Maryland)	83-70807	0-88318-197-5
No. 99	Neutrino Mass and Gauge Structure of Weak Interactions (Telemark, 1982)	83-71072	0-88318-198-3
No. 100	Excimer Lasers – 1983 (OSA, Lake Tahoe, Nevada)	83-71437	0-88318-199-1
No. 101	Positron-Electron Pairs in Astrophysics (Goddard Space Flight Center, 1983)	83-71926	0-88318-200-9
No. 102	Intense Medium Energy Sources of Strangeness (UC-Sant Cruz, 1983)	83-72261	0-88318-201-7
No. 103	Quantum Fluids and Solids – 1983 (Sanibel Island, Florida)	83-72440	0-88318-202-5
No. 104	Physics, Technology and the Nuclear Arms Race (APS Baltimore –1983)	83-72533	0-88318-203-3
No. 105	Physics of High Energy Particle Accelerators (SLAC Summer School, 1982)	83-72986	0-88318-304-8
No. 106	Predictability of Fluid Motions (La Jolla Institute, 1983)	83-73641	0-88318-305-6
No. 107	Physics and Chemistry of Porous Media (Schlumberger-Doll Research, 1983)	83-73640	0-88318-306-4
No. 108	The Time Projection Chamber (TRIUMF, Vancouver, 1983)	83-83445	0-88318-307-2
No. 109	Random Walks and Their Applications in the Physical and Biological Sciences (NBS/La Jolla Institute, 1982)	84-70208	0-88318-308-0
No. 110	Hadron Substructure in Nuclear Physics (Indiana University, 1983)	84-70165	0-88318-309-9
No. 111	Production and Neutralization of Negative Ions and Beams (3rd Int'l Symposium, Brookhaven, 1983)	84-70379	0-88318-310-2
No. 112	Particles and Fields – 1983 (APS/DPF, Blacksburg, VA)	84-70378	0-88318-311-0
No. 113	Experimental Meson Spectroscopy – 1983 (Seventh International Conference, Brookhaven)	84-70910	0-88318-312-9
No. 114	Low Energy Tests of Conservation Laws in Particle Physics (Blacksburg, VA, 1983)	84-71157	0-88318-313-7
No. 115	High Energy Transients in Astrophysics (Santa Cruz, CA, 1983)	84-71205	0-88318-314-5
No. 116	Problems in Unification and Supergravity (La Jolla Institute, 1983)	84-71246	0-88318-315-3
No. 117	Polarized Proton Ion Sources (TRIUMF, Vancouver, 1983)	84-71235	0-88318-316-1

No. 118	Free Electron Generation of Extreme Ultraviolet Coherent Radiation (Brookhaven/OSA, 1983)	84-71539	0-88318-317-X
No. 119	Laser Techniques in the Extreme Ultraviolet (OSA, Boulder, Colorado, 1984)	84-72128	0-88318-318-8
No. 120	Optical Effects in Amorphous Semiconductors (Snowbird, Utah, 1984)	84-72419	0-88318-319-6
No. 121	High Energy e^+e^- Interactions (Vanderbilt, 1984)	84-72632	0-88318-320-X
No. 122	The Physics of VLSI (Xerox, Palo Alto, 1984)	84-72729	0-88318-321-8
No. 123	Intersections Between Particle and Nuclear Physics (Steamboat Springs, 1984)	84-72790	0-88318-322-6
No. 124	Neutron-Nucleus Collisions – A Probe of Nuclear Structure (Burr Oak State Park - 1984)	84-73216	0-88318-323-4
No. 125	Capture Gamma-Ray Spectroscopy and Related Topics – 1984 (Internat. Symposium, Knoxville)	84-73303	0-88318-324-2
No. 126	Solar Neutrinos and Neutrino Astronomy (Homestake, 1984)	84-63143	0-88318-325-0
No. 127	Physics of High Energy Particle Accelerators (BNL/SUNY Summer School, 1983)	85-70057	0-88318-326-9
No. 128	Nuclear Physics with Stored, Cooled Beams (McCormick's Creek State Park, Indiana, 1984)	85-71167	0-88318-327-7
No. 129	Radiofrequency Plasma Heating (Sixth Topical Conference, Callaway Gardens, GA, 1985)	85-48027	0-88318-328-5
No. 130	Laser Acceleration of Particles (Malibu, California, 1985)	85-48028	0-88318-329-3
No. 131	Workshop on Polarized ^{3}He Beams and Targets (Princeton, New Jersey, 1984)	85-48026	0-88318-330-7
No. 132	Hadron Spectroscopy–1985 (International Conference, Univ. of Maryland)	85-72537	0-88318-331-5
No. 133	Hadronic Probes and Nuclear Interactions (Arizona State University, 1985)	85-72638	0-88318-332-3
No. 134	The State of High Energy Physics (BNL/SUNY Summer School, 1983)	85-73170	0-88318-333-1
No. 135	Energy Sources: Conservation and Renewables (APS, Washington, DC, 1985)	85-73019	0-88318-334-X
No. 136	Atomic Theory Workshop on Relativistic and QED Effects in Heavy Atoms	85-73790	0-88318-335-8
No. 137	Polymer-Flow Interaction (La Jolla Institute, 1985)	85-73915	0-88318-336-6
No. 138	Frontiers in Electronic Materials and Processing (Houston, TX, 1985)	86-70108	0-88318-337-4
No. 139	High-Current, High-Brightness, and High-Duty Factor Ion Injectors (La Jolla Institute, 1985)	86-70245	0-88318-338-2

No. 140	Boron-Rich Solids (Albuquerque, NM, 1985)	86-70246	0-88318-339-0
No. 141	Gamma-Ray Bursts (Stanford, CA, 1984)	86-70761	0-88318-340-4
No. 142	Nuclear Structure at High Spin, Excitation, and Momentum Transfer (Indiana University, 1985)	86-70837	0-88318-341-2
No. 143	Mexican School of Particles and Fields (Oaxtepec, México, 1984)	86-81187	0-88318-342-0
No. 144	Magnetospheric Phenomena in Astrophysics (Los Alamos, 1984)	86-71149	0-88318-343-9
No. 145	Polarized Beams at SSC & Polarized Antiprotons (Ann Arbor, MI & Bodega Bay, CA, 1985)	86-71343	0-88318-344-7
No. 146	Advances in Laser Science–I (Dallas, TX, 1985)	86-71536	0-88318-345-5
No. 147	Short Wavelength Coherent Radiation: Generation and Applications (Monterey, CA, 1986)	86-71674	0-88318-346-3
No. 148	Space Colonization: Technology and The Liberal Arts (Geneva, NY, 1985)	86-71675	0-88318-347-1
No. 149	Physics and Chemistry of Protective Coatings (Universal City, CA, 1985)	86-72019	0-88318-348-X
No. 150	Intersections Between Particle and Nuclear Physics (Lake Louise, Canada, 1986)	86-72018	0-88318-349-8
No. 151	Neural Networks for Computing (Snowbird, UT, 1986)	86-72481	0-88318-351-X
No. 152	Heavy Ion Inertial Fusion (Washington, DC, 1986)	86-73185	0-88318-352-8
No. 153	Physics of Particle Accelerators (SLAC Summer School, 1985) (Fermilab Summer School, 1984)	87-70103	0-88318-353-6
No. 154	Physics and Chemistry of Porous Media—II (Ridge Field, CT, 1986)	83-73640	0-88318-354-4
No. 155	The Galactic Center: Proceedings of the Symposium Honoring C. H. Townes (Berkeley, CA, 1986)	86-73186	0-88318-355-2
No. 156	Advanced Accelerator Concepts (Madison, WI, 1986)	87-70635	0-88318-358-0
No. 157	Stability of Amorphous Silicon Alloy Materials and Devices (Palo Alto, CA, 1987)	87-70990	0-88318-359-9
No. 158	Production and Neutralization of Negative Ions and Beams (Brookhaven, NY, 1986)	87-71695	0-88318-358-7

No. 159	Applications of Radio-Frequency Power to Plasma: Seventh Topical Conference (Kissimmee, FL, 1987)	87-71812	0-88318-359-5
No. 160	Advances in Laser Science–II (Seattle, WA, 1986)	87-71962	0-88318-360-9
No. 161	Electron Scattering in Nuclear and Particle Science: In Commemoration of the 35th Anniversary of the Lyman-Hanson-Scott Experiment (Urbana, IL, 1986)	87-72403	0-88318-361-7
No. 162	Few-Body Systems and Multiparticle Dynamics	87-72594	0-88318-362-5
No. 163	Pion–Nucleus Physics: Future Directions and New Facilities at LAMPF (Los Alamos, NM, 1987)	87-72961	0-88318-363-3
No. 164	Nuclei Far from Stability: Fifth International Conference (Rosseau Lake, ON, 1987)	87-73214	0-88318-364-1
No. 165	Thin Film Processing and Characterization of High-Temperature Superconductors	87-73420	0-88318-365-X
No. 166	Photovoltaic Safety (Denver, CO, 1988)	88-42854	0-88318-366-8
No. 167	Deposition and Growth: Limits for Microelectronics (Anaheim, CA, 1987)	88-71432	0-88318-367-6
No. 168	Atomic Processes in Plasmas (Santa Fe, NM, 1987)	88-71273	0-88318-368-4
No. 169	Modern Physics in America: A Michelson-Morley Centennial Symposium (Cleveland, OH, 1987)	88-71348	0-88318-369-2

.